TASI 2012

Searching for New Physics at Small and Large Scales

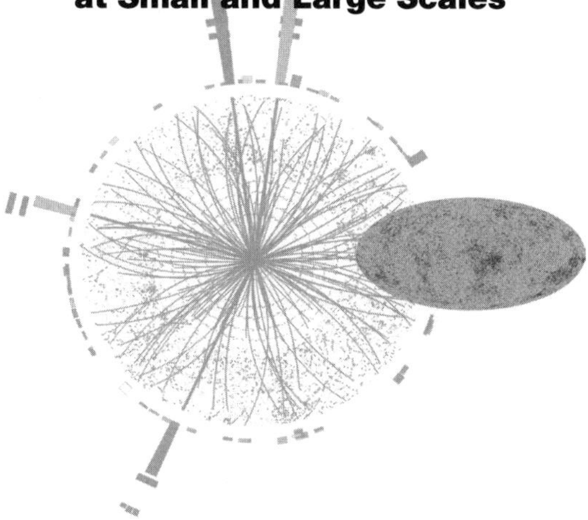

TASI 2012

Searching for New Physics at Small and Large Scales

Proceedings of the 2012 Theoretical Advanced Study Institute
in Elementary Particle Physics

Boulder, Colorado, 4 – 29 June 2012

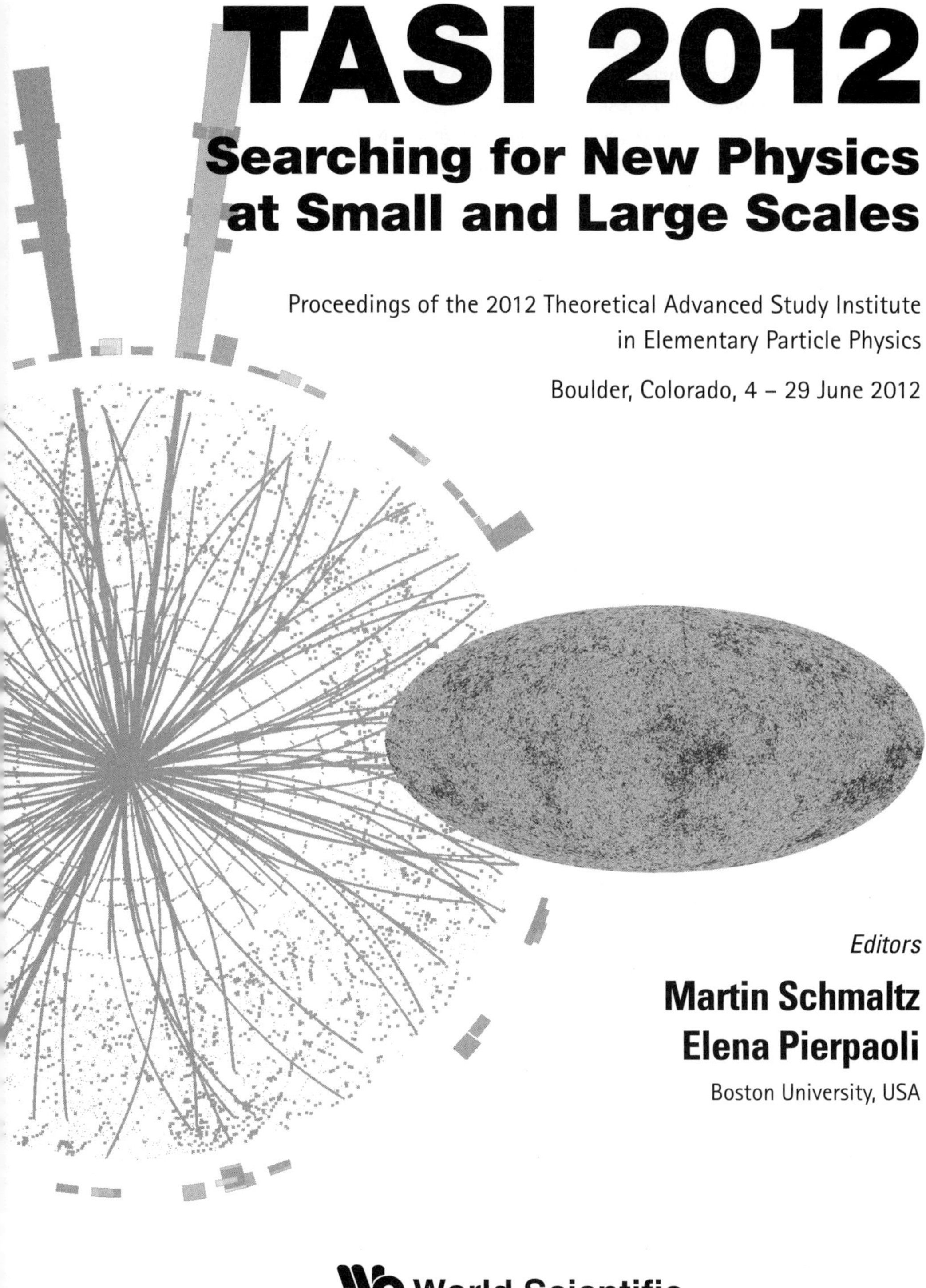

Editors

Martin Schmaltz
Elena Pierpaoli

Boston University, USA

World Scientific

NEW JERSEY • LONDON • SINGAPORE • BEIJING • SHANGHAI • HONG KONG • TAIPEI • CHENNAI

Published by

World Scientific Publishing Co. Pte. Ltd.

5 Toh Tuck Link, Singapore 596224

USA office: 27 Warren Street, Suite 401-402, Hackensack, NJ 07601

UK office: 57 Shelton Street, Covent Garden, London WC2H 9HE

Library of Congress Cataloging-in-Publication Data

TASI (Conference) (2012 : Boulder, Colo.)

Searching for new physics at small and large scales TASI 2012 : proceedings of the 2012 Theoretical Advanced Study Institute in Elementary Particle Physics, Boulder, Colorado, 4–29 June 2012 / editors, Martin Schmaltz & Elena Pierpaoli, Boston University, USA.

pages cm

ISBN 978-9814525213 (hardcover : alk. paper)

1. Nuclear physics--Congresses. I. Schmaltz, Martin, editor of compilation. II. Pierpaoli, Elena, editor of compilation. III. Title.

QC770.T35 2012

539.7'2--dc23

2013021879

British Library Cataloguing-in-Publication Data

A catalogue record for this book is available from the British Library.

Printed in Singapore by World Scientific Printers.

Preface

The Theoretical Advanced Study Institute (TASI) in elementary particle physics has been held each summer since 1984. TASI was first organized in Ann Arbor, and was held in subsequent years in New Haven, Santa Cruz, Santa Fe and Providence. Since 1989, TASI has been permanently hosted by the University of Colorado in Boulder. Each year TASI attracts more than fifty of the most promising advanced theory graduate students in the United States, and several graduate students from abroad. Occasionally, TASI also accepts experimental students and beginning postdocts. The emphasis of the topics at TASI has shifted from year to year, but typically there have been courses of lectures in phenomenology, field theory, string theory, mathematical physics, and particle-astrophysics, as well as seminars on related theoretical and experimental topics. TASI has been highly successful, especially in introducing the students to a much broader range of ideas than they normally experience at their home institutions while working on their (often narrow) dissertation topics. The title of TASI-2012 was *Searching for New Physics at Small and Large Scales.*

The program comprised two parallel lecture series on Particle Physics and on Cosmology. The Particle Physics series covered LHC related experimental techniques, phenomenology, as well as basics in Physics Beyond the Standard Model. It consisted of the following lecture courses:

Peter Skands (CERN): *Introduction to QCD*
Jesse Thaler (MIT): *Super-tricks for Superspace*
Michael Peskin (Stanford): *Weak Interactions and Higgs, Theory*
Yuri Gershtein (Rutgers): *LHC and Higgs, Experiment*
Veronica Sanz (York & CERN): *Collider physics*
Aaron Pierce (Michigan): *SUSY at the LHC*
Michele Papucci (Berkeley): *SUSY Model Building*
Jesse Shelton (Yale): *Jet substructure and new physics*
Rouven Essig (SUNY Stony Brook): *Dark Matter from Particle Physics*

The other lecture series focused on modern cosmology with special attention to the topics of Dark Matter, the microwave background and alternatives to the Standard Model of Cosmology:

Edmund Bertschinger (MIT): *Introduction to Cosmology*
Robert Caldwell (Dartmouth): *Dark Energy*
Fabian Schmidt (Caltech): *Modified gravity*
Julien Lesgourgues (EPFL Lausanne & CERN): *cosmological perturbations and dark matter*
Leonardo Senatore (Stanford): *Inflation*
Shirley Ho (CMU & Berkeley): *What can we learn from Large Scale Structure of the Universe?*
Stefano Profumo (UC Santa Cruz): *DM constraints from astrophysical data*

Included in the program were also two excellent public lectures

Robert Caldwell: *Einstein's Island and the Invisible Cities of Cosmology*
Michael Peskin: *The Higgs boson, is it real?*

The Scientific Program was organized by Elena Pierpaoli and Martin Schmaltz. We were provided with invaluable assistance by the local directors K.T. Mahanthappa and Tom DeGrand. Tom also continued his tradition of leading TASI participants on weekend excursions into the Rocky Mountains. We would especially like to acknowledge Susan Spika for efficient secretarial help; Gurjeet Singh for help in the day-to-day running of TASI; Daniel Wegman for organizing student seminars; and the National Science Foundation, U.S. Department of Energy and University of Colorado for providing the facilities and financial support.

Finally, we would like to thank the Scientific Advisory Board for inviting us as the program directors. It was a great pleasure to witness the stimulating lectures and interact with such talented and curious students. We hope that TASI 2012 contributed to their development as scientists who will make exciting discoveries in the decades to come.

E. Pierpaoli and M. Schmaltz

Contents

Tasi 2012 Group Photo: Identification List

1 Peter Winslow
2 Kevin Givens
3 Chris Rosen
4 Oscar Henriksson
5 Ennio Salvioni
6 Jan Heisig
7 Martin Schmaltz
8 Jesse Thaler
9 Tom DeGrand
10 Jack Kearney
11 Valerie Domcke
12 David Schaich
13 Zachary Thomas
14 Liang Dai
15 Daliang Li
16 Gustavo Marques Tavares
17 Laurent Canetti
18 David Yaylali
19 Ivan Rodriguez
20 Gauthier Durieux
21 Sohyun Park
22 Peisi Huang
23 Durmus Karabacak
24 Kristopher Healey
25 Jonathan Cornell
26 Aditya Yechan Gunja
27 Daniel Wegman
28 Matthew Walters
29 Andrea Thamm
30 Wei Xue
31 Erik Perkins
32 Daniele Bertolini
33 Ayon Patra
34 Melinda Andrews
35 Jiayin Gu
36 McCullen Sandora
37 Angelo Monteux
38 Michael Park
39 Marko Simonovic
40 Matthew Lewandowski
41 Elisa Mariani
42 Garrett Goon
43 Ahmed Ismail
44 Simon Knapen
45 Jing Ren
46 Daisy Mak
47 Ran Huo
48 Yiming Xu
49 Pawin Ittisamai
50 Jiajun Liao
51 Mikhail Solon
52 Chris Murphy
53 Austin Joyce
54 Rick Gupta
55 Joe Bramante
56 Sean Downes
57 Peter Skands
58 Kim Boddy
59 Grant Larsen
60 Sujeet Akula
61 Edmund Bertschinger
62 K.T. Mahanthappa
63 Catalina Gomez Sanchez
64 Jessica Cook
65 Elena Pierpaoli
66 Mathieu Cliche
67 David A. Eby
68 Robert Caldwell
69 Zachary Burell
70 Walter Tangarife-Garcia
71 Chris Brust
72 Masha Baryakhtar

Chapter 1

Dark Energy Models

Robert R. Caldwell

Department of Physics & Astronomy, Dartmouth College
6127 Wilder Laboratory, Hanover, New Hampshire 03755, USA
E-mail: robert.r.caldwell@dartmouth.edu

These lectures present an introduction to dark energy cosmology. We begin with a review of the Standard Cosmological Model, in which the accelerated cosmic expansion is due to a cosmological constant. We then turn to models of quintessence, a light cosmic scalar field, and explore the particle physics origins and cosmological behavior.

Contents

1. The Standard Cosmological Model

The mathematical description of our universe is encoded in the Standard Cosmological Model, consisting of the Hot Big Bang theory of the universe, with three significant additions: inflation, as the theory of the birth of the universe and origin of fluctuations; dark matter, as the major source of mass on galactic scales and larger; and dark energy, as the source of the accelerated cosmic expansion. These lectures will focus on the physics of dark energy.

The Standard Cosmological Model builds upon the Hot Big Bang, which in turn is based upon general relativity manifest in Einstein's equations, quantum theory as realized through the Standard Model of particle physics, and the Cosmological Principle. We hold these founding theories and principles to be true. Hence, we interpret observations and experiments through the lens of a Robertson-Walker spacetime. Based upon a vast catalog of data, we now know that we live in a spatially flat, low density, accelerating universe. That is, the spacetime metric is

$$ds^2 = -dt^2 + a^2(t)(dx^2 + dy^2 + dz^2)$$

whereby the geometry of the spatial sections is Euclidean or planar, and thereby referred to as "flat". The scale factor $a(t)$ is determined by the time-time Einstein equation, also known as the Friedmann equation,

$$3H^2 = 8\pi G\rho$$

where $H = d\ln a/dt$ and ρ represents the energy density in all forms of matter and radiation. By low density, we mean that ρ due to all Standard Model particles and fields plus dark matter adds up to just $\sim 30\%$ of the critical density:

$$\sum_{i=SM,DM} \rho_i \sim 0.3\rho_c = \frac{3H^2}{8\pi G}.$$

Absent any spatial curvature, the validity of the Friedmann equation implies $\sim 70\%$ of the energy density of the universe is otherwise unaccounted for or missing. And by accelerating, we mean that the second derivative of the scale factor is positive, $\ddot{a} > 0$. This kinematic property of the metric is usually characterized by the so-called deceleration parameter: $q \equiv -(\ddot{a}/a)/H^2$. Expressed in terms of the cosmological fluid with energy density ρ and pressure p, then $q = \frac{1}{2}(1 + 3p/\rho)$. That the deceleration parameter has been measured to be negative implies that the missing energy has strong, negative pressure. We call this stuff "dark energy" for lack of a better name.

The numbers k, Ω, and q have been regarded since the time of Hubble as the keys to unlocking the secrets of our world model. The spatial curvature is represented by k, which we now know to be zero or at least negligibly small. Geometry is no longer the "arbiter of fates" as believed by a previous generation of cosmologists. The fraction of critical density in the form of matter and radiation is Ω, which we now know to be $\simeq 0.3$. This is in stark contrast to the longtime prediction $\Omega = 1$ of the generation of particle physicists who shifted their attention from inner space to outer space. The rate of deceleration of the Big Bang is represented by q, which we now know to be ~ -0.5. We are not living among the dying embers of the Big Bang, but rather grasping at uncertainty as to where the cosmic acceleration will take us.

The mystery that these three numbers held for previous cosmologists has been taken up by a new enigma. *What is the equation of state of dark energy?* In terms of a parameter, we want to know the value of w, defined as the ratio of the homogeneous pressure to energy density of the dark energy. These lectures are about dark energy, and the predictions that various dark energy theories make for the parameter w. There are a number of very good review articles [1–3] and even textbooks [4] on this subject. We focus our attention on the cosmological constant, and quintessence or a cosmic scalar field. The material and referencing are admittedly idiosyncratic, but the depth should give students a solid background upon which to build. The approach taken in these lectures is to focus on a few basic calculations, with the goal of introducing students to the basic tools of dark energy models.

2. The Cosmological Constant

The default assumption is that dark energy is due to a cosmological constant. Hence,

$$\rho_\Lambda = \Omega_\Lambda \frac{3H_0^2}{8\pi G} = \frac{\Lambda}{8\pi G}.$$

The cosmological constant, Λ, has units of curvature or inverse-length squared. For a constant energy density, the equation of state in this case is $w = -1$. This is in excellent agreement with current data.

Every student should take a moment to figure out what are all the numbers and what they mean. The energy density fraction is $\Omega_\Lambda \simeq 0.7$, The Hubble distance is $H_0^{-1} \simeq 3000h^{-1}$ Mpc, the Hubble parameter is $h \simeq 0.7$, a megaparsec is 1 Mpc$= 3 \times 10^{22}$ m, and of course $\hbar c = 197$ MeV-fm. As an exercise, use these numbers to figure out the energy scales (in GeV or eV) of the Hubble distance, the cosmological constant, and the energy density. You should be able to show that $H_0 \sim 2 \times 10^{-42}$ GeV and $\rho_\Lambda \sim (0.002\ \text{eV})^4$. These numbers should be puzzling to a particle physicist. Why is the universe so old, in terms of Standard Model energy scales? Why is the energy scale of the cosmological constant so tiny, again in terms of Standard Model energy scales?

If dark energy is a cosmological constant, then we turn to Zeldovich [5] for the first explanation. The elementary particle physics vacuum carries an energy density

$$\rho = \sum_i g_i s_i \int \frac{d^3k}{(2\pi)^3} \sqrt{k^2 + m_i^2}$$

where the sum is over all i species of mass m_i, with degeneracy factors g_i and $s_i = \pm 1$ for bosons/fermions. The integral is divergent. Any cutoff or regularization, in the method of Pauli-Villars as carried out by Zeldovich, at an energy scale associated with the Standard Model yields an energy density far in excess of observation. The integral would have to be cut off at the milli-eV scale. This is a challenge to explain.

It seems there is little one can say about the cosmological constant, although that has not stopped the authors of thousands of articles. But perhaps it is useful to keep the following in mind. If dark energy is the cosmological constant, is there a symmetry or principle that enables it and endows it with this tiny energy? Or, if the dark energy is not a cosmological constant, then what makes Λ so small, or even zero, that we can neglect it? In either case, is there any other testable consequence of such a symmetry or principle? Students who are interested in thinking about the cosmological constant problem (or CCP) further are encouraged to start by reading two classic references, Steven Weinberg's Loeb Lectures at Harvard, which formed the basis of a 1989 Review of Modern Physics article [6], or Stefan Nobbenhuis' classification of proposed solutions to the CCP [7].

3. Dynamical Dark Energy

The smallness of the dark energy, as measured in terms of Standard Model energy scales, leads to the coincidence problem as illustrated in Fig. 1. Why are we living in an epoch during which the dark energy has just overtaken the matter energy density? A less anthropocentric way of posing this question is to ask: Why does the dark energy overtake the matter energy density so late in the history of the universe? If this question has an answer, then perhaps the answer is that the dark energy is not a constant but is instead dynamical. Hence, we consider the possibility that the cosmological constant is zero (or negligible) and look to a scalar field theory of dark energy.

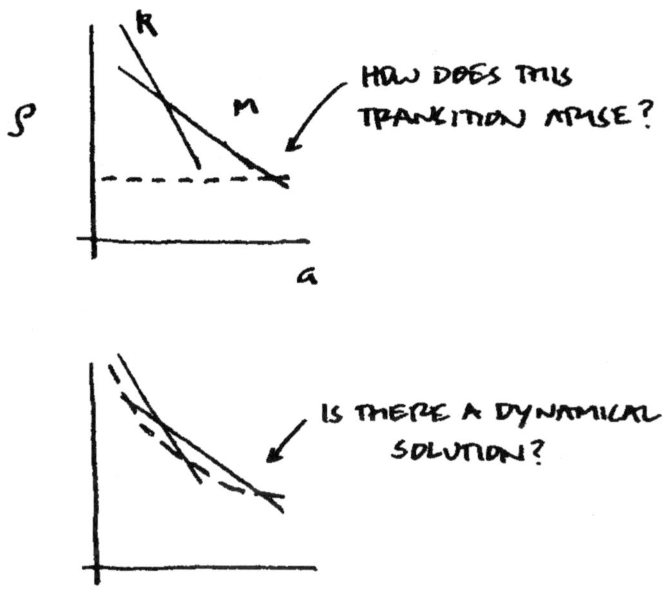

Fig. 1. The coincidence problem illustrated.

To start our study of dynamical scalar fields, we consider a theory proposed by Wetterich [8] in 1988. This consists of a theory with dimensionless parameters only, in which Newton's constant is made dynamical. The classical Standard Model is present but with all masses turned off or the Higgs self-coupling set to zero. This theory is dilatation invariant at the classical level. The dimensionful fundamental constants are induced by vev's of scalar fields or quantum effects. That is, we propose the action $S = \int d^4x \sqrt{-\tilde{g}} \tilde{\mathcal{L}}$ where

$$\tilde{\mathcal{L}} = \frac{1}{\ell^2}\tilde{R} - 4\frac{\omega}{\ell^4}\tilde{g}^{\mu\nu}\partial_\mu \ell \partial_\nu \ell + \mathcal{L}_{SM}^{m \to 0}.$$

We augment this with a scalar field ϕ that has a constant vacuum potential at its minimum, $\phi_{\min} : V'(\phi_{\min}) = 0$ so that $V(\phi_{\min}) \neq 0$. The full action for this theory is

$$S = \int d^4x \sqrt{-\tilde{g}} \left[\tilde{\mathcal{L}} - \frac{1}{2}\tilde{g}^{\mu\nu}\partial_\mu \phi \partial_\nu \phi - V(\phi) \right].$$

To study this theory, it is convenient to perform a conformal transformation to the Einstein frame. That is, rescale the metric so that the Einstein-Hilbert action is recovered for gravity.

Questions that often arise are "what is the right frame?" or "what is the physical frame?" when a conformal transformation is performed. For those who are puzzled, the answer is perhaps disappointing. There is no preferred or correct frame. The same phenomena may be described in either frame, though the way these phenomena are manifest may be subtle. Let me simply point out that the clocks and rods that are used by observers in the Einstein frame are not the same clocks and rods used by observers in the Jordan or non-minimally coupled frame. A simple system that one can study to appreciate my message is a scalar field plus electromagnetism, or even scalar electrodynamics, in a Robertston-Walker spacetime $ds^2 = a^2(\tau)(-d\tau^2 + d\vec{x}^2)$ which is conformal to Minkowski spacetime. With two metrics $\tilde{g}_{\mu\nu}$ and $g_{\mu\nu}$ related by a factor $a^4(\tau)$, then it is clear that distances and times have different meaning in the two frames. One needs to adopt a careful definition to be able to identify the physical electric or magnetic fields.

To proceed, we use the conformal transformation $\tilde{g}_{\mu\nu} = \Omega^2 g_{\mu\nu}$, $\tilde{R} = \Omega^{-2}R - 6\Omega^{-3}\Box\Omega$ and to be specific we set $\ell = \Omega\sqrt{16\pi G}$. Many of these expressions and more can be found in the appendix of Sean Carroll's text-book [9]. We also make an integration by parts, to convert $\Omega^{-1}\Box\Omega$ into $-\Omega^{-2}(\partial\Omega)^2$. We finally arrive at a new look at our theory:

$$S = \int d^4x\sqrt{-g}\left[\frac{R}{16\pi G} + \mathcal{L}' + \Omega^4\mathcal{L}_{SM}^{m\to 0}\right]$$

$$\mathcal{L}' = -\frac{1}{16\pi G}\left(\frac{6+4\omega}{\Omega^2}(\partial\Omega)^2\right) - \frac{1}{2}\Omega^2(\partial\phi)^2 - \Omega^4 V.$$

We are not done yet. Provided that $\omega \neq -3/2$, then we can reparameterize ℓ or Ω as $\Omega = A\exp(\sqrt{\frac{4\pi G}{3+2\omega}}\chi)$, and then eliminate A by shifting χ. We find

$$S = \int d^4x\sqrt{-g}\left[\frac{R}{16\pi G} - \frac{1}{2}(\partial\chi)^2 - \frac{1}{2}(\partial\phi)^2 W(\chi) + \left(\mathcal{L}_{SM}^{m\to 0} - V\right)W^2(\chi)\right],$$

where $W(\chi) = \exp(\alpha\chi/2M_P)$. This is a nice result, in that our original theory has become a fairly standard looking scalar field theory, with χ being the dilaton. And here is the main point: if our scalar field theory of ϕ contains a cosmological constant $V(\phi_{\min}) \neq 0$, or a cosmological constant is generated by $\mathcal{L}_{SM}$, then its effect on spacetime as seen in the Einstein frame is modulated by $W^2(\chi)$. Since V has a minimum in ϕ we can futher reduce the theory to

$$S = \int d^4x\sqrt{-g}\left[\frac{R}{16\pi G} - \frac{1}{2}(\partial\chi)^2 - V_0 e^{\alpha\chi/M_P}\right]$$

where $V_0 = \mathcal{L}_{SM}^{m\to 0} - V(\phi_{\rm min})$. This is Wetterich's proposal for a solution to the cosmological constant problem, to dynamically adjust it via the χ field. Although one can poke holes in this idea, it is educational to continue to examine the consequences.

The evolution of the above system for χ is to be evaluated in a Robertson-Walker spacetime with power-law expansion $a(t) \propto t^n$ and $n = 2/(3+3w_B)$ where w_B is the equation of state of the dominant, background fluid. With metric $ds^2 = -dt^2 + a^2(t)d\vec{x}^2$ the equation of motion is

$$\frac{d^2\chi}{dt^2} + 3H\frac{d\chi}{dt} + \frac{\alpha}{M_P}V_0 e^{\alpha\chi/M_P} = 0.$$

Making an ansatz $\chi = A\ln Bt$ then one can show that $A = -2M_P/\alpha$ and $B^2 = \alpha^2 V_0/(2M_P^2(3n-1))$. The energy density and pressure of the scalar field are $\rho = \frac{1}{2}\dot{\chi}^2 + V = 6n(M_P/\alpha t)^2$ and $p = (4-6n)(M_P/\alpha t)^2$. The cute result is that the equation of state is $w_\chi = p/\rho = w_B$, so the scalar field exactly mimics the dominant background. The χ energy density as a fraction of the critical density is $\Omega_\chi = \rho_\chi/(3H^2/(8\pi G)) = 24\pi(1+w_B)/\alpha^2$. This means for $\alpha^2 > 24\pi(1+w_B)$ we have a self-consistent solution, with the χ field in the background. Hence, what we have is a scaling solution for the χ field that maintains an equation of state that mimics the background. Provided α is sufficiently large, where $\alpha = \sqrt{64\pi/(3+2\omega)}$, meaning $-1/2 < \omega < 0$, then the cosmological constant, no matter how large, is dynamically adjusted to be small.

The real lesson from this model is the scaling behavior. An exponential potential can help adjust a cosmic scalar field so that it remains everpresent at a trace level. If, at some point late in its evolution, some secular evolution of α causes it to drop below the threshold for scaling, $\alpha^2 < 24\pi(1+w_B)$, then the scalar field becomes potential energy dominated, with $w \to -1$ and $\Omega_\chi \to 1$. Such behavior would provide for an interesting dark energy candidate.

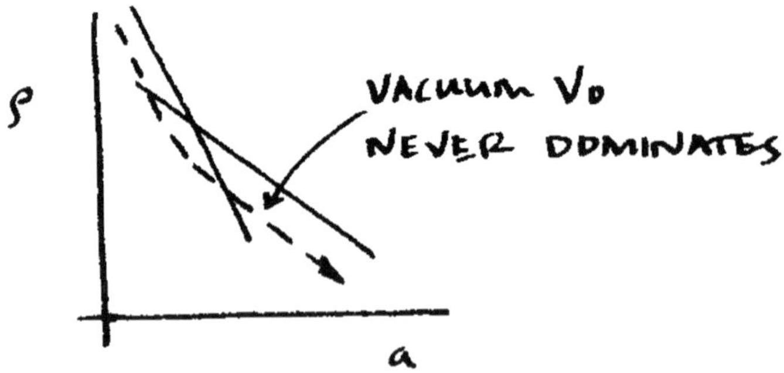

Fig. 2. Dynamical adjustment of the cosmological constant using a scaling field.

4. Quintessence

We consider a cosmic scalar field as a model for dynamical dark energy. Much in the same way that the inflaton drives primordial inflation, a quintessence [10] field can drive the present-day cosmic acceleration. Potentials that have been tried out include: $V = \frac{1}{2}m^2\phi^2$, $M^4(1+\cos(\phi/f))$, $M^{4+n}\phi^{-n}$, etc. Since $w = (\dot{\phi}^2 - 2V)/(\dot{\phi}^2 + 2V)$, and since we require $w \simeq -1$, then $\dot{\phi}^2 \ll V$. In order that it dominate the cosmic energy density at present, we need $V \sim M_P^2 H_0^2$. And in order that it does not cluster significantly within the Hubble volume, then it should have a mass $\sqrt{V''} \lesssim H$. If the potential is of the form $V = \frac{1}{2}m^2\phi^2$, then these two conditions imply $\phi \sim M_P$. Check for yourself, this is a ridiculously tiny mass and an obscenely large field strength. It is a challenge to build a particle physics model of such a light field, with Planckian amplitude, that nevertheless remains dark. [11–13] However, we will let nature be the final arbiter, since a scalar is the simplest field that can accelerate our universe.

We move on to a discussion of the general behavior of cosmic scalar fields. Given a potential, there is a one-to-one mapping of the scalar field attributes onto an equation of state history $w(t)$. The energy density for a given w trajectory is

$$\rho(a) = \rho(a_0)\exp\left[3\int_a^{a_0} d\ln a'(1+w(a'))\right].$$

To be simple, most studies of the effect of quintessence on cosmology assumes a slowly varying equation of state, so that setting $w =$ constant or else allowing a slow drift such as $w = w_0 + w_a(1 - a/a_0)$ seems like a reasonable first approximation. There is a good reason, since rapid variations lead to detectable features in the cosmic microwave background, large scale structure, or the supernova magnitude-redshift diagram that have not been seen. Moreover the data are consistent with a slowly varying w.

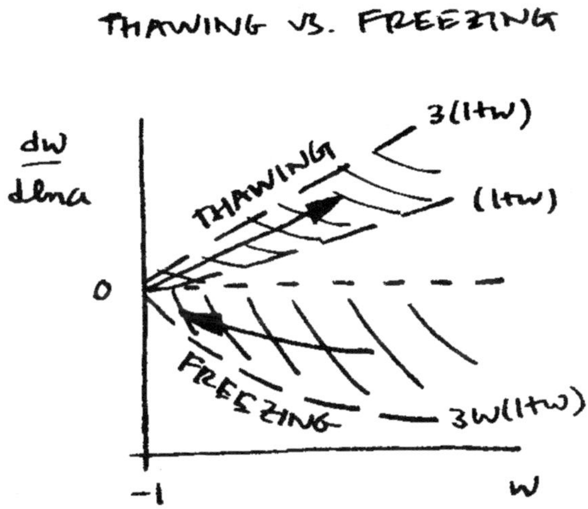

Fig. 3. A phase space portrait of thawing and freezing models of dark energy.

A typical scalar field does not give a constant equation of state. However, we can identify two typical equations of state based on the behavior, dubbed "thawing" and "freezing" [14] as illustrated in Fig. 3. In the former case, the scalar field has been frozen by Hubble friction for most of cosmic history until recently, when it began to roll down its potential. Consequently, the equation of state starts at $w = -1$ and grows more positive as time progresses. In the latter case, the field has been rolling down it potential but at an ever slowing rate so that w is approaching -1. The freezing models include the tracker field, which we will examine shortly. The thawing models, which are somewhat simpler and more familiar, as they include linear, quadratic, and quartic potentials, suggest a simple behavior for the equation of state. That is, a wide range of thawing fields behave with w satisfying $dw/d\ln a = N(1 + w)$ for $1 \leq N \leq 3$. The solution is $w = -1 + (1 + w_0)(a/a_0)^N$, where w_0 is the present-day equation of state.

The elephant in the room for most models of quintessence is that the cosmic acceleration requires fine tuning. For a simple quadratic potential, the mass and initial field strength must be chosen carefully. A class of freezing fields, however, solve the sensitive dependence on initial conditions thanks to an attractor solution [15]. Because the equation of state lurks slightly more negative than the background, these models have been dubbed "trackers". [16], an example of which is the potential $V = M^{4+n}\phi^{-n}$. To take a closer look, the equation of motion is $\ddot{\phi} + 3H\dot{\phi} + V' = 0$. In a background spacetime with a power-law scale factor $a \propto t^{2/(3+3w_B)}$, we take an ansatz $\phi = At^B$. It is easy to find A and B so that the kinetic and potential energy scale with the same power of t. In this case, the equation of state is $w = -1 + (1 + w_B)n/(2 + n)$, which means that for small n a slightly negative equation of state can be achieved. This solution persists as long as the field remains subdominant. However, because the equation of state is more negative than the background, the scalar field energy density will eventually catch up and overtake the background. When this occurs, $w \to -1$.

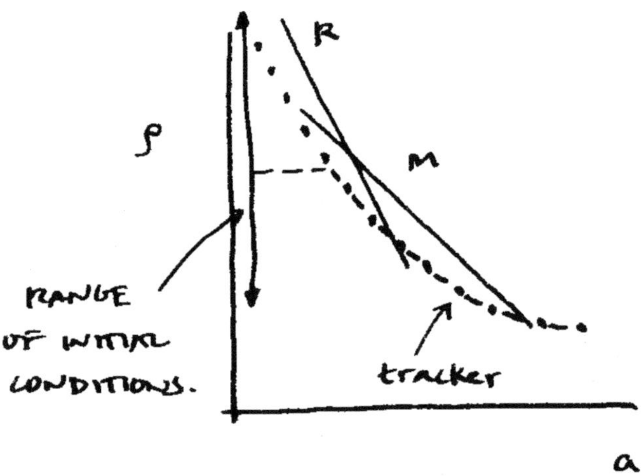

Fig. 4. The tracking model.

A more general analysis of this potential reveals several interesting features. [17] First, the tracking equation of state depends more simply on the curvature of the potential. Defining $\Gamma = V''V/V'^2$ then the equation of state is $w = (w_B - 2\Gamma + 2)/(2\Gamma - 1)$. Tracking occurs when Γ is approximately constant and greater than 1. A potential with such features will exhibit a broad insensitivity to the initial conditions; the tracking solution is an attractor. As illustrated in Fig. 4, the tracking trajectory represents an idealization. For a wide range of initial conditions, the scalar field will either freeze or dump kinetic energy until it reaches and then joins the tracker line.

A few further examples of scaling fields borrow from assisted inflation. [18, 19] First, consider one field with many exponential potentials: $V = \sum_i V_i e^{-\alpha_i \phi/M_P}$. Suppose $i = 1, 2$ with $\alpha_1 > \sqrt{24\pi}$ and $\alpha_2 < \sqrt{24\pi}$. At late times ϕ dominates with $w \to -1 + \alpha_2^2/24\pi$, but at early times ϕ scales with the radiation then matter. Unless α_1 is sufficiently large, then the scalar field may contribute a non negligible fraction of the energy density. When this occurs, in a scenario called "early" dark energy, a host of new effects must be taken into account. We will return to early dark energy a bit later in these lectures. Second, consider many exponential potentials: $V = \sum_i V_i e^{-\alpha_i \phi_i/M_P}$. In this case, for all the fields that are actively rolling down their potentials, the ϕ_i's behave collectively as a single field with $\alpha_{\rm eff}^{-2} = \sum_i \alpha_i^{-2}$. For a large number of fields, for which none can drive accelerated expansion alone, the collective action is such that the equation of state is driven towards $w \to -1 + \alpha_{\rm eff}^2/24\pi$.

5. Cosmological Perturbations

An important facet of dark energy cosmology is the effect of dark energy perturbations on the evolution of inhomogeneities, which are observed in the cosmic microwave background and large scale structure. To follow these effects properly, we need to develop cosmological perturbation theory to include quintessence or, more generally, dark energy.

We start by writing down the relevant equations. A scalar field ϕ with potential V obeys the equation $\Box\phi = V'$, which reduces to $\phi'' + 2\mathcal{H}\phi' + a^2 V' = 0$. Here $' = d/d\tau$ where τ is the conformal time, $\mathcal{H} = a'/a$, and the prime on V means $dV/d\phi$. Next we perturb the metric, the Einstein Equations, and the scalar field equation of motion. In the conformal Newtonian gauge, the metric is written

$$ds^2 = a^2(\tau)\left[-(1+2\psi)d\tau^2 + (1-2\phi)d\vec{x}^2\right]$$

whereas in the synchronous gauge

$$ds^2 = a^2(\tau)\left[-d\tau^2 + (\delta_{ij} + h_{ij})dx^i dx^j\right]$$

$$h_{ij} = \int d^3k\, e^{i\vec{k}\cdot\vec{x}}\left[\hat{k}_i\hat{k}_j h + 6(\hat{k}_i\hat{k}_j - \frac{1}{3}\delta_{ij})\eta\right].$$

An outstanding reference for the connection between these two coordinate systems, and all things having to do with cosmological perturbations, is Ref. 20. To continue, we perturb the scalar field, $\phi \to \phi(\tau) + \delta\phi(\tau, \vec{x})$. Working in the synchronous gauge, the scalar field equation of motion is

$$\delta\phi'' + 2\mathcal{H}\delta\phi' + a^2 V''\delta\phi - \nabla^2\delta\phi = -\frac{1}{2}h'\phi'.$$

Pausing to inspect this equation, we see the scalar field responds like a damped, driven harmonic oscillator. The a^2V'' term acts like the mass term. Since ∇^2 is just the Laplacian in comoving coordinates, then high frequency perturbations to ϕ propagate at the speed of light. From the source term on the right hand side, we see that fluctuations of the scalar field are sourced when the background field is rolling and density perturbations are evolving.

We can build a fluid-like description of the scalar field fluctuations. The energy density, pressure, and momentum density perturbations are

$$\begin{aligned} \delta\rho &= \frac{1}{a^2}\phi'\delta\phi' + V'\delta\phi \\ \delta p &= \frac{1}{a^2}\phi'\delta\phi' - V'\delta\phi \\ (\rho + p)\theta &= \frac{k^2}{a^2}\phi'\delta\phi. \end{aligned}$$

It is useful to recognize that $\delta p/\delta\rho \neq w$, 1 or anything else simple. Only on small scales is $\delta p/\delta\rho = 1$. We can use these fluid variables, plus the background equations of motion for ϕ to determine a relationship between the pressure and energy density perturbations:

$$\delta p = \delta\rho + \frac{\theta}{k^2}\rho(w' + 3\mathcal{H}(1 - w^2)).$$

Then, the perturbed equations of motion can be recast into the form of conservation laws for the fluid variables:

$$\begin{aligned} \delta' &= -(1 + w)\left(\theta + \frac{1}{2}h'\right) - 3\mathcal{H}(\delta p/\delta\rho - w)\delta \\ \theta' &= -\mathcal{H}(1 - 3w)\theta - \frac{w'}{1 + w}\theta + \frac{1}{1 + w}\frac{\delta p}{\delta\rho}k^2\delta. \end{aligned}$$

These three above equations are precisely what is needed to evolve quintessence perturbations consistently with the inhomogeneous matter and radiation fluid. We only need to supply w and w', meaning that we can model any scalar field evolution.

What are the correct initial conditions for quintessence perturbations? The answer is that it almost doesn't matter. [21] The homogeneous solutions to the $\delta\phi$ equation, which depend on initial conditions, decay in general and so these solutions are subdominant to the inhomogeneous solutions. That is, fluctuations in the scalar field develop in response to the large scale gravitational potentials created by the other matter fields. It is only in the case that the initial perturbations are unrealistically large, or else if the scalar field tracks the radiation at early times, do the initial perturbations leave a detectable imprint. In any case, if inflation does endow quintessence with adiabatic perturbations, then $\delta_Q/(1+w_Q) = \delta_\gamma/(1+w_\gamma)$ and $\delta p_Q = (\dot{p}_Q/\dot{\rho}_Q)\delta\rho_Q$, where Q, γ indicates quintessence and photons, respectively.

The equations given above are sufficient to adapt one of the public Boltzmann codes, such as CMBfast [22] or CAMB [23], to simulate the cosmic microwave background anisotropy spectrum in a dark energy cosmology. These codes, which are both built upon an original code by Ma & Bertschinger [20], evolve the perturbations in the synchronous gauge. It is only when the temperature and polarization anisotropy are finally computed, or the effects of weak gravitational lensing are taken into account, that a conversion into the conformal-Newtonian gauge is made. As a word of warning, some codes have an option to turn off or ignore dark energy perturbations. In general, this is ill advised – the large scale perturbations are gauge dependent, and ignoring them can lead to gross inaccuracies. However, in a typical scenario in which the dark energy has a constant equation of state, and perturbations are evolved in the synchronous gauge, then if one restricts attention to small angular scales (or multipole moments $\ell \ll 100$) then one may safely ignore the scalar field perturbations.

A rite of passage for many cosmology students is to adapt one of these Boltzmann codes to compute the CMB pattern for a specific dark energy model. I encourage students of the summer school to take on an ambitious project, such as a study of the thawing models with $w(a) = -1 + (1 + w_0)(a/a_0)^N$ and $N > 1$, which have not been analyzed in detail.

6. Dynamical Modeling

A dynamical systems approach to studying dark energy helps reveal the attractor structure of certain models. Attractors, in which the field solution is driven towards special solutions such as scaling, as in the case of the exponential potential, or negative equation of state, as in the case of the tracker models, are of great interest. The reason is that such models might play a role in helping to solve the coincidence problem, or alleviate the challenges of fine tuning the quintessence field, as discussed earlier.

We consider a model universe containing Einstein gravity, a scalar field, and perfect fluid matter and radiation. The Lagrangian is

$$\mathcal{L} = \frac{R}{2\kappa^2} - \frac{1}{2}(\partial\phi)^2 - V(\phi) + \mathcal{L}_{R,M}$$

where we now set $\kappa = \sqrt{8\pi G}$. The equations of motion are $\ddot{\phi} + 3H\dot{\phi} + V' = 0$, $3H^2 = \kappa^2(\frac{1}{2}\dot{\phi}^2 + V + \rho_{R,M})$, and $2\dot{H} = -\kappa^2(\dot{\phi}^2 + \rho_{R,M} + p_{R,M})$. Following Ref. 4 we define dimensionless variables $x_1 \equiv \kappa\dot{\phi}/\sqrt{6}H$ and $x_2 \equiv \kappa\sqrt{V}/\sqrt{3}H$. Plugging these into our equations of motion, we find that $\Omega_{R,M} = 1-x_1^2-x_2^2$, $\Omega_\phi = x_1^2+x_2^2$ and $w_\phi = (x_1^2-x_2^2)/(x_1^2+x_2^2)$. Introducing a new time variable, $N = \ln a$, and a parameter $\lambda = -V'/\kappa V$, then the evolution equations for x_1 and x_2 are

$$\begin{aligned}\frac{dx_1}{dN} &= -3x_1 - \frac{\dot{H}}{H^2}x_1 + \frac{\sqrt{6}}{2}\lambda x_2^2 \\ \frac{dx_2}{dN} &= -\frac{\dot{H}}{H^2}x_1 - \frac{\sqrt{6}}{2}\lambda x_1 x_2 \\ \dot{H}/H^2 &= -3x_1^2 - \frac{3}{2}(1 + w_{R,M})(1 - x_1^2 - x_2^2).\end{aligned}$$

This may look like a complicated system, but it is actually quite simple in practice. For a given potential, one can determine $\lambda(\phi)$. One uses the evolution equations to solve for x_1 and x_2, using $w_{R,M} = \frac{1}{3}a_{eq}/(a_{eq} + a)$ and $d\phi/dN = \sqrt{6}x_1/\kappa$ to close the circle. Let's take a look at two cases.

In the case of an exponential potential, λ is a constant. Then we can use the above equations to find fixed points, where $dx_1/dN = dx_2/dN = 0$. It turns out that there are four. (1) $x_1 = x_2 = 0$, which corresponds to the trivial case of no scalar field, $\Omega_\phi = 0$. (2) $x_1 = \pm 1$ and $x_2 = 0$, which corresponds to a universe filled with a kinetic-dominated scalar field, $\Omega_\phi = 1$ and $w_\phi = 1$. (3) $x_1 = \lambda/\sqrt{6}$ and $x_2 = \pm\sqrt{1 - \lambda^2/6}$ which corresponds to a universe filled with a potential dominate scalar field, $\Omega_\phi = 1$ and $w_\phi = -1 + \lambda^2/3$. (4) $x_1 = \sqrt{3/2}(1 + w_{R,M})/\lambda$ and $x_2 = \pm\sqrt{3/2}(1 - w_{R,M})/\lambda$ which corresponds to the scaling solution, with $\Omega_\phi = 3(1 + w_{R,M})/\lambda^2$ and $w_\phi = w_{R,M}$. Next, we can carry out a stability analysis of these fixed points. That is, we perturb the solutions $x_{1,2} \to x_{1,2} + \delta x_{1,2}$ and solve for the eigenvalues of the matrix M where

$$\frac{d}{dN}\begin{pmatrix}\delta x_1 \\ \delta x_2\end{pmatrix} = M\begin{pmatrix}\delta x_1 \\ \delta x_2\end{pmatrix}.$$

Stable points have positive determinant and negative eigenvalues. Thus we find that fixed point (3) is stable provided $\lambda^2 < 3(1+w_{R,M})$ and (4) is stable provided $\lambda^2 > 3(1+w_{R,M})$. This confirms known results, which we obtained in an earlier section, and gives us confidence to proceed.

In general, the potential function $\lambda(\phi)$ varies with time. However, there may be models for which λ changes sufficiently slowly that the above analysis can be adapted. For example, try the potential $V = M^{4-n}\phi^n$ with $n > 0$ (so this is *not* the tracker potential). The shape parameter introduced to describe the tracker is $\Gamma = V''V/(V')^2 = 1 - 1/n$ which is less than unity, and so no tracking is possible. Next, the potential function is $\lambda = -V'/\kappa V = -n/\kappa\phi$. So if ϕ changes slowly then we might be able to adapt the fixed point (3) from the analysis of the exponential potential, whereby $\Omega_\phi = 1$ and $w_\phi = -1 + \lambda^2/3$. However, cosmic acceleration requires $w_\phi \ll -1/3$ or $\lambda^2 \ll 2$. This translates into $\phi \gg nM_P/4\sqrt{\pi}$. Just like in the case of chaotic inflation, the field needs to be perched high up on the potential for a self-consistent, accelerating solution. So we learn something from this technique.

Now we can consider the tracker potential, $V = M^{4+n}\phi^{-n}$ for $n > 0$. Here, $\lambda = n/\kappa\phi$ which we take to be positive. As the field evolves, $\dot{\phi} > 0$ and so $x_1 > 0$. Rather than looking for the usual fixed points, we can add an equation for λ. It is a good exercise to derive this for yourself. The answer is

$$\frac{d\lambda}{dN} = \sqrt{6}\lambda^2(1-\Gamma)x_1.$$

Since $\Gamma > 1$ for this potential, we see that $d\lambda/dN$ is negative, confirming that λ decreases and ϕ gets larger. Since the time rate of change of λ is sufficiently slow, we can adapt the results for the exponential potential. In particular, the system can evolve from stable fixed point (4), for which $\lambda^2 > 3(1+w_{R,M})$, to the other stable fixed point (3) with $\lambda^2 < 2$. These results help establish the attractor nature of the tracking solution, and the shift to "freezing" behavior as the field slows and becomes potential dominated.

7. Mass Varying Neutrinos

The apparent similarity of the milli-eV scale of dark energy and the mass splitting of neutrinos has not escaped the notice of theorists, and it has been asked whether the two numbers might have a common origin, linking two disparate sectors of particle physics. One realization of this idea is the mass varying neutrino scenario [24], where it is proposed that the neutrino mass depends on a cosmic field. This model ultimately does not work, but provides a useful lesson for model building: couplings to dark energy must not allow the dark energy to decay, lest it lose its "dark" and negative pressure nature.

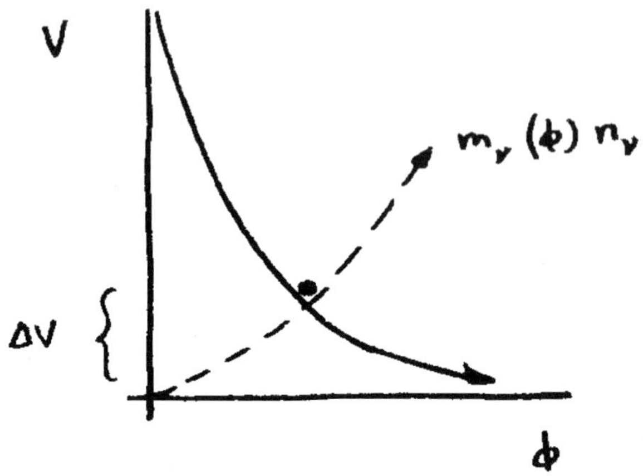

Fig. 5. The mass-varying neutrino potential.

The Lagrangian for this theory includes gravity, the usual radiation and matter fields, plus a scalar field coupled to a neutrino beyond the Standard Model:

$$\mathcal{L} = -\frac{1}{2}(\partial\phi)^2 - V(\phi) - m_\nu(\phi)\bar{\nu}\nu.$$

Because we are going to focus on the scalar field, and because we are going to assume that the neutrinos are non-relativistic, then we are justified to ignore the neutrino kinetic term. This means that the $\bar{\nu}\nu$ term is actually the neutrino number density. As a consequence, the scalar field sees an effective potential due to V plus $m_\nu(\phi)n_\nu$. If V is a runaway or racetrack potential which decreases at large ϕ, and m_ν grows with ϕ, as illustrated in Fig. 5, then the scalar field can find itself lifted up at a non-zero local minimum. If this potential energy is of the order $(10^{-3}\text{eV})^4$, then the interaction may explain dark energy.

The equation of motion of the neutrino fluid is obtained under the assumption that the interaction does not affect the conservation of neutrino number. This means $\rho_\nu = m_\nu(\phi)n_\nu$ and $n_\nu \propto a^{-3}$. The scalar field equation of motion is

$$\ddot{\phi} + 3H\dot{\phi} + V' = -Q(\phi)\rho_\nu$$

where $Q(\phi) = d\ln m_\nu/d\phi$. Now we make the important assumption that the scalar field sits at the local minimum of the effective potential, as the neutrino number density dilutes with the cosmic expansion. The effective potential is $V_{\rm eff} = V + m_\nu n_\nu$ and the condition for the minimum is $V'_{\rm eff} = V' + Q\rho_\nu = 0$. We can use this to simplify the ϕ equation of motion, whereby we obtain $\phi =$ constant plus a decaying term. Hence, the scalar field kinetic energy is found to be negligible compared to the potential energy. The energy density and pressure of the coupled system are therefore $\rho = V + m_\nu n_\nu$ and $p = -V$ so we find that the equation of state is $w = -V/(V + m_\nu n_\nu) = -1 + \Omega_\nu/(\Omega_\nu + \Omega_\phi)$. The conditions are right for cosmic acceleration.

The sound speed, however, reveals that the system is unstable. Since the scalar field is trapped at the minimum of the effective potential, the system is characterized by just a single degree of freedom, and may be regarded as a barotropic fluid. Hence, we can use the adiabatic sound speed, $c_a^2 = \dot{p}/\dot{\rho}$ to assess the stability. In detail,

$$c_a^2 = \frac{\dot{p}}{\dot{\rho}} = \frac{-V'\dot{\phi}}{V'\dot{\phi} + Q\dot{\phi}\rho_\nu + m_\nu \dot{n}_\nu}.$$

We can use the condition for the minimum of the potential to simplify the numerator: $V' = -Q\rho_\nu$. In the denominator, the first two terms thereby cancel, and the last term is just $-3H\rho_\nu$, which leaves

$$c_a^2 = -\frac{1}{3}\frac{\dot{\phi}}{H}\frac{d\ln m_\nu}{d\phi}.$$

Since $\dot{\phi}$ is positive, and m_ν is an increasing function of ϕ, then $c_a^2 < 0$ and the sound speed is complex. The perturbations grow large, upsetting the standard structure formation scenario. This is the adiabatic instability, reviewed and dissected in Refs. 25 and 26.

8. Phantom Energy

A natural question to ask is how negative might be the dark energy equation of state? A fluid with equation of state $w < -1$, dubbed a "phantom" [27] is consistent with current observational data. Such a fluid violates important energy conditions that are essential to proving the singularity theorems in general relativity. So, perhaps it is not a surprise that phantom energy leads to a future singularity, the Big Rip. [28]

To see how this all works, let us consider a simple model of a scalar field having a wrong-sign kinetic term. Indeed, this is a ghost but at the classical level it is stable assuming it has no interactions other than gravity. The scalar field equation of motion is $\ddot{\phi} + 3H\dot{\phi} - V' = 0$ meaning that the field will tend to run *up* a potential rather than down. The perturbation equations are also stable, such that fluctuations grow on large scales but free stream relativistically on small scales. Classically, the theory is ok.

A ghost model of phantom energy is unstable at the quantum level. Even if the only quantum interaction permitted is through the emission of gravitons, the sea of phantom energy will rapidly decay. Postulating a cutoff in the energy scale of gravitons, to restrict the available phase space for decay of the scalar field, can slow the decay timescale to an acceptable level, comparable to the age of the universe. Quantum aspects of a phantom field are discussed further in Refs. 29, 30, and 31.

The effect of $w < -1$ on the cosmic expansion is interesting. For simplicity, let us assume that w is a constant. In this case the evolution of the scale factor from a matter-dominated era over to a phantom-dominated era is given by

$$a(t) = \begin{cases} a_m (t/t_m)^{2/3} & t < t_m \\ a_m \left[-w + (1+w)(t/t_m)\right]^{2/(3+3w)} & t > t_m. \end{cases}$$

The scale factor above diverges in finite time, when $t - t_m = -t_m/(1+w)$. One can show that the expansion rate, $H = \frac{2}{3}\left[-wt_m + (1+w)t\right]^{-1}$, and the Ricci scalar both diverge. This is a real, physical singularity. One can also show that nearby geodesics move apart with an acceleration that ultimately diverges. Hence, any bound objects, whether gravitational, electromagnetic, or other, will ultimately be pulled apart by the runaway cosmic acceleration. It remains an open question whether the dark energy with $w < -1$ can be ruled out observationally.

9. Non-Standard Scalar Field Theories

There is much interest in scalar field theories with non-standard kinetic terms, as motivated by string theories, ghost condensate models, and galileon models of gravity. In the simplest form, a k-inflation [32] or k-essence [33] model (where "k" indicates that the kinetic energy is unusual) drives accelerated expansion with a field that rolls with constant velocity, $\dot{\phi}$ = constant. In the case of k-essence, the Lagrangian has the form $\mathcal{L} = P(\phi, X)$ where $X = -\frac{1}{2}(\partial\phi)^2$. With our mostly positive metric signature, $X = +\frac{1}{2}\dot{\phi}^2$ for a homogeneous scalar field. The stress-energy tensor is $T_{\mu\nu} = Pg_{\mu\nu} + P_{,X}\partial_\mu\phi\partial_\nu\phi$ so that the energy density and pressure are $\rho = 2XP_{,X} - P$, $p = P$. A function $P(X)$ with a minimum in the X direction leads to a pressure-dominated fluid with equation of state $w \to -1$. The equations of motion are then obtained from the familiar fluid conservation equation in a Robertson-Walker spacetime:

$$(P_{,X} + 2XP_{,XX})\dot{X} + (-P_{,\phi} + 2XP_{,X\phi})\dot{\phi} + 6HP_{,X}X = 0.$$

The perturbation equations are very similar to the case of quintessence with one difference:

$$\delta' = -(1+w)\left(\theta + \frac{1}{2}h'\right) - 3\mathcal{H}\left(\frac{\delta p}{\delta\rho} - w\right)\delta$$

$$\theta' = (3c_s^2 - 1)\mathcal{H}\theta + \frac{c_s^2 k^2}{1+w}\delta$$

$$\delta p = c_s^2\delta\rho + \frac{\theta}{k^2}\rho(w' + 3\mathcal{H}(1+w)(c_s^2 - w)).$$

Quintessence has the same equations, but with sound speed $c_s^2 = 1$. For k-essence $c_s^2 = (1 + 2XP_{,XX}/P_{,X})^{-1}$. Care must be taken to avoid negative values of c_s^2 as well as superluminal values – although the severity of the consequences depends on the scenario.

An interesting k-essence scenario was found in which the k-essence scales throughout the radiation and matter era, but is driven at late times towards an accelerating fixed point solution. [32–34] On the face of it, this sounds like a nice answer to the fine tuning and coincidence problems for dark energy. However, the achilles heel in this model is that the sound speed is shown to exceed the speed of light at some point during the matter era. [35] This does not mean that violations of causality and closed timelike curves follow automatically. It does mean that the superluminal modes emit gravitational Cerenkov radiation, but the emission rate contains two factors of G and so is probably too slow to have much of an effect on the scenario. Unless the k-essence field is also coupled to other matter fields, there may not be any damning consequence of $c_s^2 > 1$. [36]

It is worthwhile to take a closer look at the perturbations, in particular the perturbed Hamiltonian density, H_k, to see where there may be danger:

$$H_k = (P_{,X} + 2XP_{,XX})\frac{1}{2}(\dot{\delta\phi})^2 + P_{,X}\frac{1}{2}(\nabla\delta\phi)^2 - P_{,\phi\phi}\frac{1}{2}(\delta\phi)^2.$$

We define the coefficients $\xi_1 = P_{,X} + 2XP_{,XX}$, $\xi_2 = P_{,X}$, and $\xi_3 = -P_{,\phi\phi}$. The sound speed is $c_s^2 = \xi_2/\xi_1$. At the classical level, we only require that the sound speed is real. In general, positive energy requires all three coefficients to be positive. If ξ_3 is negative then long wavelength perturbations will grow at a power-law rate, so positivity of this coefficient is not essential. However, quantum stability of the system requires $\xi_{1,2} > 0$ or else there will be ghosts or a gradient instability whereby the k-essence energy can grow arbitrarily negative via quantum gravitational processes. Let us consider a few specific cases.

Phantom dark energy as a scalar field with a wrong-sign kinetic term has $P = -X - V$. This means $\xi_1 = \xi_2 = -1$. The sound speed is $c_s^2 = 1$ but there is a ghost and gradient instability. As mentioned earlier, processes like the decay of the vacuum to two ϕ particles plus other particles will rapidly destabilize the vacuum in a runaway process. [29–31]

A dilatonic ghost condensate model [37] of dark energy has $P = -X + (X/M^2)^2 e^{\lambda\kappa\phi}$. It is straightforward to check that $\xi_1 = -1 + 6Q$ and $\xi_2 = -1 + 2Q$ where $Q = Xe^{\lambda\kappa\phi}/M^4$. Also, one finds that $\xi_3 < 0$, although this is not a show-stopper. Hence, quantum stability requires $Q > 1/2$. Since $c_s^2 = (2Q-1)/(6Q-1)$ we are assured of a stable, real sound speed. And since $w = (Q-1)/(3Q-1)$, we find that $w \geq -1$. Despite the "$-X$" term in the Lagrangian, this is not much of a ghost!

A dynamical systems analysis of the dilatonic ghost condensate reveals interesting behavior. The equations of motion are

$$\left(3\frac{\dot\phi^2}{M^4}e^{\lambda\kappa\phi} - 1\right)\ddot\phi + 3H\left(\frac{\dot\phi^2}{M^4}e^{\lambda\kappa\phi} - 1\right)\dot\phi + \frac{3\lambda\kappa\dot\phi^4}{4M^4}e^{\lambda\kappa\phi} = 0$$

$$3H^2 = \kappa^2\left(-\frac{1}{2}\dot\phi^2 + \frac{3\dot\phi^4}{4M^4}e^{\lambda\kappa\phi} + \rho_{R,M}\right)$$

$$2\dot H = \kappa^2\left(\dot\phi^2 - \frac{\dot\phi^4}{M^4}e^{\lambda\kappa\phi} - \rho_{R,M} - p_{R,M}\right).$$

We introduce dimensionless variables $x_1 = \kappa\dot\phi/\sqrt{6}H$, $x_2 = \dot\phi^2 e^{\lambda\kappa\phi}/2M^4$, and $x_3 = \kappa\sqrt{\rho_R}/\sqrt{3}H$. In this language, the scalar field equation of state is $w_\phi = (1-x_2)/(1-3x_2)$, the sound speed is $c_s^2 = (1-2x_2)/(1-4x_2)$, the energy density fraction is $\Omega_\phi = 3x_1^2 x_2 - x_1^2$ and $\Omega_R = x_3^2$. The evolution equations are now

$$\frac{dx_1}{dN} = -x_1\left(\frac{6(2x_2-1) + 3\sqrt{6}\lambda x_1 x_2}{2(6x_2-1)}\right) + \frac{1}{2}x_1(3 - 3x_1^2 + 3x_1^2 x_2 + x_3^2)$$

$$\frac{dx_2}{dN} = x_2 \left(\frac{3x_2(4 - \sqrt{6}\lambda x_1) - \sqrt{6}(\sqrt{6} - \lambda x_1)}{1 - 6x_2} \right)$$
$$\frac{dx_3}{dN} = \frac{1}{2} x_3(-1 - 3x_1^2 + 3x_1^2 x_2 + x_3^2).$$

Inspection of these equations reveals three fixed points. (1) $(x_1, x_2, x_3) = (0, 1/2, 1)$ which corresponds to $\Omega_\phi = \Omega_M = 0$ and $\Omega_R = 1$. This is simply the radiation era. (2) $(0, 1/2, 0)$ which corresponds to $\Omega_\phi = \Omega_R = 0$ and $\Omega_M = 1$. This is the matter era. (3) $x_1 = -\frac{\sqrt{6}\lambda}{4}(1 - \sqrt{1 + 16/(3\lambda^2)})$, $x_2 = \frac{1}{2} + \frac{\lambda^2}{16}(1 + \sqrt{1 + 16/(3\lambda^2)})$, and $x_3 = 0$ which corresponds to $\Omega_\phi = 1$ with $-1 \leq w_\phi \leq -1/3$ for $0 \leq \lambda \leq \sqrt{2/3}$. So there is an accelerating fixed point. (We leave it as an exercise to the reader to determine the stability of these fixed points.)

10. Dark-Matter Coupled Quintessence

Models of quintessence that couple to a species of dark matter have been investigated, motivated in part by an effort to unify the dark sectors, but also with the idea in mind to develop a trigger that causes quintessence to dominate at late times. Here we give a snapshot of work on this topic. Consider a Lagrangian of the form

$$\mathcal{L} = P(\phi, X) + \mathcal{L}_M(\phi, g_{\mu\nu}, \psi_M)$$

where "M" represents the matter fields coupled to quintessence. It is presumed that the coupling is manifest in the conservation laws

$$\nabla_\mu T^\mu_\nu(\phi) = -QT(M)\nabla_\nu \phi, \quad \nabla_\mu T^\mu_\nu(M) = +QT(M)\nabla_\nu \phi,$$

where $Q = Q(\phi)$ controls the coupling to the stress-energy trace $T(M) = -\rho_M + 3p_M$. Expanding these equations, we find

$$\frac{d\rho_\phi}{dN} + 3\rho_\phi(1 + w_\phi) = -Q(\phi)\rho_M \frac{d\phi}{dN}$$
$$\frac{d\rho_M}{dN} + 3\rho_M(1 + w_M) = Q(\phi)\rho_M \frac{d\phi}{dN}.$$

It can be shown that the Lagrangian $P(\phi, X)$ takes on a specific form if solutions with constant w_ϕ and Ω_ϕ are to be obtained. To see this, let us define $\lambda = Q(1 + w_{\rm eff})/(\Omega_\phi(w_M - w_\phi)$ where $w_{\rm eff} = w_M \Omega_M + w_\phi \Omega_\phi$. The equation of motion for P obeys

$$\left(1 + \frac{2}{\lambda}\frac{d\ln Q}{d\phi}\right)\frac{d\ln P}{d\ln X} - \frac{1}{\lambda}\frac{d\ln P}{d\phi} = 1.$$

The solution is $P(\phi, X) = XQ^2 g$ where g is a function of $XQ^2 e^{\lambda\psi/Q}$ and $\psi = \int^\phi Q(\phi')d\phi'$. Now we have to choose a coupling function to proceed any further. Let's set Q to be constant, in which case $P = Xg(Y)$ where $Y = Xe^{\lambda\phi}$. Note that a canonical scalar field with an exponential potential has $P = X(1-C/Y)$ where C is a constant. The dilatonic ghost condensate has $P = X(-1 + Y/M^4)$. So both are amenable to coupling to matter.

The general scaling and coupled system has fixed points. The first corresponds to scalar field domination, with $\Omega_\phi = 1$ and $w_\phi = -1 + \lambda^2/3P_{,X}$. The second corresponds to scaling with $\Omega_\phi = -Q/w_\phi(Q + \lambda)$, $w_{\text{eff}} = -Q/(Q + \lambda)$, $w_\phi = -Q(Q + \lambda)/(Q(Q + \lambda) + 3(g + dg/d\ln Y))$. Both fixed points can give rise to acceleration. Hence, there are many possibilities to drive cosmic acceleration. But the most interesting features of this model may lie in the behavior of the species of dark matter and how it influences clustering on large scales. Further details of this model may be found in Refs. 4 and 38.

11. Early Dark Energy

Cosmological models in which dark energy scales during the radiation and early stages of the matter era are called "early dark energy" (EDE) or early quintessence [39] due to its early appearance. If the fraction of critical energy in dark energy is above a few percent, then there may be observational consequences. For example, during the radiation era, if dark energy is scaling with fixed ρ_{EDE}/ρ_R then the relationship between the expansion rate and the photon temperature is altered, which may have an adverse effect on Big Bang Nucleosynthesis (BBN). One can characterize the effect through an effective number of excess neutrino species:

$$\frac{\rho_{EDE}}{\rho_\gamma} = \frac{\rho_{EDE}}{\rho_R}\frac{\rho_R}{\rho_\gamma} = \Delta N_{\text{eff}} \times \frac{7}{8}\left(\frac{4}{11}\right)^{1/3}$$

or $\Delta N_{\text{eff}} \simeq 7\Omega_{DE}/(1 - \Omega_{EDE})$ where Ω_{EDE} is to be evaluated in the radiation era. Allowing the equivalent of one extra species, which is consistent with recent data [40], implies an upper limit $\Omega_{DE} \simeq 0.1$. During the epoch of recombination, the added presence of a cosmic scalar field that does not cluster in the same fashion as baryonic matter can have a noticeable effect on the temperature anisotropy pattern. As illustrated in Fig. 6, dark energy as well as EDE alters the angular diameter distance to recombination, which shifts the geometry of the acoustic peaks. EDE generically dampens oscillations of the baryon-photon plasma, thereby weakening the peak and

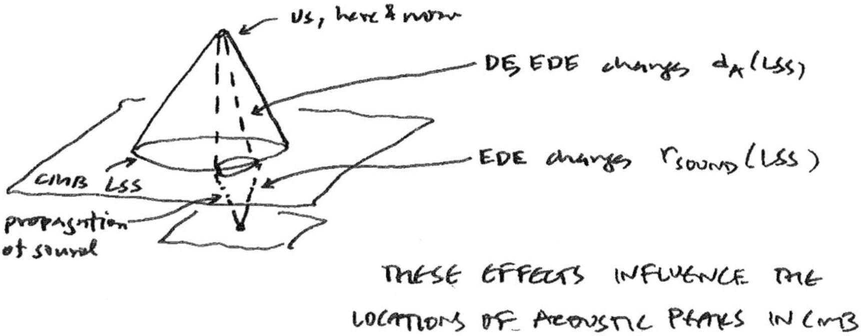

Fig. 6. The effects of early quintessence on the CMB.

trough amplitudes. EDE also dampens the growth of perturbations that enter the horizon before radiation-matter equality. This can be understood by examining the evolution of density perturbations: in the synchronous gauge,

$$\delta_M'' + 2\mathcal{H}\delta_M' = 4\pi G a^2(\delta\rho_M + \delta\rho_{EDE} + 3\delta p_{EDE}).$$

The added terms on the right hand side, which are evolved separately, contribute to slow the growth of δ_M. In most EDE models, the dark energy abundance is negligible during the epoch between recombination and the onset of cosmic acceleration, so the suppression of growth does not affect all length scales. Consequently, tighter limits can be placed on the abundance of EDE during recombination [41], on the order of $\Omega_{EDE} \simeq 0.02$.

12. Dark Energy & Inflation

It is reasonable to ask if dark energy and inflation are due to the same field. However, the disparity of the energy scales, 10^{15} GeV and 10^{-3} eV, makes this seem unlikely. Nevertheless, some toy models have been built that attempt to economize on the use of scalar fields. For example, quintessential inflation [42] uses a potential

$$V = \begin{cases} \lambda(\phi^4 + M^4) & \phi < 0 \\ \lambda M^8/(\phi^4 + M^4) & \phi \geq 0 \end{cases}$$

however neither inflation with $\lambda\phi^4$ nor tracking quintessence with that curvature are a very good fit to current data. A somewhat better model

[43] uses the amplitude and phase of a complex scalar field for the inflaton and quintessence, respectively. In this case, the potential is $V = \lambda(\phi^\dagger\phi - \sigma^2)^2 + M^4(1 + \cos(\arg\phi))$.

One of the most interesting possibilities is to join not merely the inflaton and quintessence, but also the Higgs. A generalization of recent models of Higgs inflation [44], one adds an additional scalar field to the non-minimal coupling to curvature. [45] The Higgs-Dilaton action is

$$S = \int d^4x\sqrt{-g}\left[\frac{1}{2}\left(\xi_\phi\phi^2 + 2\xi_H H^\dagger H\right)R + \mathcal{L}_{SM}^{\lambda\to 0} - \frac{1}{2}(\partial\phi)^2 - V(\phi, H)\right],$$

where $\mathcal{L}_{SM}^{\lambda\to 0}$ is the Lagrangian for the Standard Model with the Higgs self-coupling turned off, since the potential V carries the Higgs potential as well as an interaction with the quintessence field, ϕ. In this model, the scalar potential is $V = \lambda(H^\dagger H - \frac{\alpha}{2\lambda}\phi^2)^2 + \beta\phi^4$. It is a useful exercise for the ambitious student to transform this system from the Jordan frame to the Einstein frame. There, the student will discover that the direction in the $H - \phi$ field space corresponding to quintessence has a tracker-like potential.

13. Summary

A wide variety of dark energy models have been introduced and examined in these lectures, and yet there are many more fascinating models. One significant topic that was not covered is the possibility of couplings between quintessence and the Standard Model. Ref. 11 has shown that higher dimension operators consisting of gradients of the quintessence field are probably the safest interactions that keep the dark energy dark. Recent articles, Refs. 46 and 47, completed several months after the summer school, cover the most recent developments for a possible coupling between quintessence and electromagnetism.

Good luck, students! Hopefully your work will lead to breakthroughs that solve the cosmological constant problem and explain dark energy.

14. Exercises

- Exercise 1: Analyze Recent Supernova Data
  Download the most recent supernova magnitude-redshift data set. At the current time, this honor falls to the Union 2.1 data set (supernova.lbl.gov/union). Write a computer program to compare the observed

supernova magnitudes $M_{\rm obs}$ to the predictions of theory:

$$M_{\rm thy}(z) = 5\log_{10}[H_0 d_L(z)] + \mathcal{M}$$
$$d_L(z) = (1+z)\int_0^z dz'/H(z')$$
$$H(z) = H_0\sqrt{\Omega_M(1+z)^3 + (1-\Omega_M)}.$$

The above expression for $H(z)$ is specific to the ΛCDM model. Now you get to do M-theory. (joke) You should find the Ω_M that minimizes $\chi^2 = \sum_i (M_{\rm obs,i} - M_{\rm thy}(z_i))^2/(\sigma_{M,i})^2$. Treat $\mathcal{M}$ as a nuisance parameter and so, to be simplistic, for each value of Ω_M use the value of $\mathcal{M}$ that minimizes χ^2. What is the value of Ω_M? How well does the ΛCDM model fit the data? Refer to Table 36.2 in the review on statistics by the Particle Data Group [48] to find the value of $\Delta\chi^2$ that determines the 68% probability range for Ω_M.

- Exercise 2: Re-Analyze Recent Supernova Data
  Repeat Exercise 1, but this time allow for dark energy with a constant equation of state. Try restricting $-1 \le w \le 0$ at first. Then, drop the lower bound and let w be arbitrarily negative. You should use $H(z) = H_0\sqrt{\Omega_M(1+z)^3 + (1-\Omega_M)(1+z)^{3(1+w)}}$.
- Exercise 3: Re-Analyze Recent Supernova Data
  Repeat Exercise 1, but this time allow for dark energy with a time-varying equation of state. Assume that $w(a)$ satisfies the differential equation $dw/d\ln a = N(1+w)$ for $1 < N < 3$. You should use $H(z) = H_0\sqrt{\Omega_M(1+z)^3 + (1-\Omega_M)f(z)}$, where $\ln f(z) = 3\int_0^z d\ln(1+z)(1+w)$.
- Exercise 4: Scaling Field
  (a) Find the functional form of the scalar field potential $V(\phi)$ that corresponds to a constant equation of state w, for $-1 < w < 0$, where the scalar field ϕ evolves monotonically in the background of the matter-dominated era. (b) Generalize to the case for which the background has an equation of state w_B. (c) What happens to the scalar field equation of state when it comes to dominate the cosmic expansion?
- Exercise 5: Oscillating Field
  (a) Consider a scalar field oscillating in the potential $V(\phi) = \lambda|\phi|^N$. Assume that the oscillation period is much shorter than H^{-1}. Find the time-averaged equation of state w as a function of N. (b) Make a stability analysis to determine whether such a potential can be used to describe dark energy with $w < -1/3$.

- Exercise 6: Assisted Acceleration
  Show that the potential $V(\phi_1, \phi_2) = V_1 e^{-\kappa\lambda_1\phi_1} + V_2 e^{-\kappa\lambda_2\phi_2}$, where $\kappa = \sqrt{8\pi G}$, has an accelerating fixed point at which $w_{\rm eff} = -1 + \lambda_{\rm eff}^2/3$ and $\lambda_{\rm eff}^{-2} = \lambda_1^{-2} + \lambda_2^{-2}$.
- Exercise 7: Higgs-Dilaton
  Transform the Higgs-Dilaton Lagrangian into the Einstein frame. If the Higgs is stabilized at some vev, what is the resulting quintessence potential? (You will have to redefine the dilaton to put its kinetic term in canonical form. Feel free to make approximations where necessary!)

References

[1] R. R. Caldwell and M. Kamionkowski, Ann. Rev. Nucl. Part. Sci. **59**, 397 (2009) [arXiv:0903.0866 [astro-ph.CO]].
[2] J. Frieman, M. Turner and D. Huterer, Ann. Rev. Astron. Astrophys. **46**, 385 (2008) [arXiv:0803.0982 [astro-ph]].
[3] P. J. E. Peebles and B. Ratra, Rev. Mod. Phys. **75**, 559 (2003) [astro-ph/0207347].
[4] L. Amendola and S. Tsujikawa, "Dark Energy: Theory and Observations," Cambridge University Press, UK (2010).
[5] Y. B. Zeldovich, JETP Lett. **6**, 316 (1967) [Pisma Zh. Eksp. Teor. Fiz. **6**, 883 (1967)].
[6] S. Weinberg, Rev. Mod. Phys. **61**, 1 (1989).
[7] S. Nobbenhuis, Found. Phys. **36**, 613 (2006) [gr-qc/0411093].
[8] C. Wetterich, Nucl. Phys. B **302**, 668 (1988).
[9] S. M. Carroll, San Francisco, USA: Addison-Wesley (2004) 513 p
[10] R. R. Caldwell, R. Dave and P. J. Steinhardt, Phys. Rev. Lett. **80**, 1582 (1998) [astro-ph/9708069].
[11] S. M. Carroll, Phys. Rev. Lett. **81**, 3067 (1998) [astro-ph/9806099].
[12] R. D. Peccei, hep-ph/0009030.
[13] C. F. Kolda and D. H. Lyth, Phys. Lett. B **458**, 197 (1999) [hep-ph/9811375].
[14] R. R. Caldwell and E. V. Linder, Phys. Rev. Lett. **95**, 141301 (2005) [astro-ph/0505494].
[15] A. R. Liddle and R. J. Scherrer, Phys. Rev. D **59**, 023509 (1999) [astro-ph/9809272].
[16] I. Zlatev, L. -M. Wang and P. J. Steinhardt, Phys. Rev. Lett. **82**, 896 (1999) [astro-ph/9807002].
[17] P. J. Steinhardt, L. -M. Wang and I. Zlatev, Phys. Rev. D **59**, 123504 (1999) [astro-ph/9812313].
[18] A. R. Liddle, A. Mazumdar and F. E. Schunck, Phys. Rev. D **58**, 061301 (1998) [astro-ph/9804177].
[19] S. A. Kim, A. R. Liddle and S. Tsujikawa, Phys. Rev. D **72**, 043506 (2005) [astro-ph/0506076].

[20] C.-P. Ma and E. Bertschinger, Astrophys. J. **455**, 7 (1995) [astro-ph/9506072].
[21] R. Dave, R. R. Caldwell and P. J. Steinhardt, Phys. Rev. D **66**, 023516 (2002) [astro-ph/0206372].
[22] U. Seljak and M. Zaldarriaga, Astrophys. J. **469**, 437 (1996) [astro-ph/9603033].
[23] http://camb.info/
[24] R. Fardon, A. E. Nelson and N. Weiner, JCAP **0410**, 005 (2004) [astro-ph/0309800].
[25] N. Afshordi, M. Zaldarriaga and K. Kohri, Phys. Rev. D **72**, 065024 (2005) [astro-ph/0506663].
[26] R. Bean, E. E. Flanagan and M. Trodden, Phys. Rev. D **78**, 023009 (2008) [arXiv:0709.1128 [astro-ph]].
[27] R. R. Caldwell, Phys. Lett. B **545**, 23 (2002) [astro-ph/9908168].
[28] R. R. Caldwell, M. Kamionkowski and N. N. Weinberg, Phys. Rev. Lett. **91**, 071301 (2003) [astro-ph/0302506].
[29] S. M. Carroll, M. Hoffman and M. Trodden, Phys. Rev. D **68**, 023509 (2003) [astro-ph/0301273].
[30] J. M. Cline, S. Jeon and G. D. Moore, Phys. Rev. D **70**, 043543 (2004) [hep-ph/0311312].
[31] S. D. H. Hsu, A. Jenkins and M. B. Wise, Phys. Lett. B **597**, 270 (2004) [astro-ph/0406043].
[32] C. Armendariz-Picon, T. Damour and V. F. Mukhanov, Phys. Lett. B **458**, 209 (1999) [hep-th/9904075].
[33] C. Armendariz-Picon, V. F. Mukhanov and P. J. Steinhardt, Phys. Rev. D **63**, 103510 (2001) [astro-ph/0006373].
[34] M. Malquarti, E. J. Copeland and A. R. Liddle, Phys. Rev. D **68**, 023512 (2003) [astro-ph/0304277].
[35] C. Bonvin, C. Caprini and R. Durrer, Phys. Rev. Lett. **97**, 081303 (2006) [astro-ph/0606584].
[36] E. Babichev, V. Mukhanov and A. Vikman, JHEP **0802**, 101 (2008) [arXiv:0708.0561 [hep-th]].
[37] F. Piazza and S. Tsujikawa, JCAP **0407**, 004 (2004) [hep-th/0405054].
[38] L. Amendola, Phys. Rev. D **62**, 043511 (2000) [astro-ph/9908023].
[39] R. R. Caldwell, M. Doran, C. M. Mueller, G. Schafer and C. Wetterich, Astrophys. J. **591**, L75 (2003) [astro-ph/0302505].
[40] G. Hinshaw, D. Larson, E. Komatsu, D. N. Spergel, C. L. Bennett, J. Dunkley, M. R. Nolta and M. Halpern *et al.*, arXiv:1212.5226 [astro-ph.CO].
[41] C. Reichardt, R. de Putter, O. Zahn and Z. Hou, Astrophys. J. **749**, L9 (2012) [arXiv:1110.5328 [astro-ph.CO]].
[42] P. J. E. Peebles and A. Vilenkin, Phys. Rev. D **59**, 063505 (1999) [astro-ph/9810509].
[43] R. Rosenfeld and J. A. Frieman, JCAP **0509**, 003 (2005) [astro-ph/0504191].
[44] F. L. Bezrukov and M. Shaposhnikov, Phys. Lett. B **659**, 703 (2008) [arXiv:0710.3755 [hep-th]].

[45] J. Garcia-Bellido, J. Rubio, M. Shaposhnikov and D. Zenhausern, Phys. Rev. D **84**, 123504 (2011) [arXiv:1107.2163 [hep-ph]].
[46] V. Gluscevic, D. Hanson, M. Kamionkowski and C. M. Hirata, Phys. Rev. D **86**, 103529 (2012) [arXiv:1206.5546 [astro-ph.CO]].
[47] M. V. Romalis and R. R. Caldwell, arXiv:1302.1579 [astro-ph.CO].
[48] J. Beringer *et al.* [Particle Data Group Collaboration], Phys. Rev. D **86**, 010001 (2012).

Chapter 2

Cosmological Perturbations

J. Lesgourges

ITP, EPFL, 1015 Lausanne, Switzerland
Theory Division, CERN, 1211 Geneva 23, Switzerland
LAPTh (CNRS - Université de Savoie), BP 110, 74941 Annecy-le-Vieux Cdx, France
E-mail: Julien.Lesgourgues@cern.ch
www.cern.ch/Julien.Lesgourgues

We present a self-contained summary of the theory of linear cosmological perturbations. We emphasize the effect of the six parameters of the minimal cosmological model, first, on the spectrum of Cosmic Microwave Background temperature anisotropies, and second, on the linear matter power spectrum. We briefly review at the end the possible impact of a few non-minimal dark matter and dark energy models.

Contents

Purposes

Cosmology is progressing by leaps and bounds thanks to a spectacular amount of observational data. It provides crucial clues for particle physics, and more generally for high energy physics. A key role is played by Cosmic Microwave Background (CMB) and Large Scale Structure (LSS) data. Their interpretation relies on the *theory of cosmological perturbations*.

The aim of this course is to introduce this theory, staying at the linear level, and hopefully with a simple, self-contained and original approach. It is targeted at PhD students in astrophysics and cosmology, as well as researchers in particle physics willing to follow current developments in cosmology. We discuss in particular the effect of the parameters of the minimal cosmological model on the spectrum CMB temperature anisotropies, and on the matter power spectrum. We briefly review at the end the possible impact of non-minimal dark matter and dark energy models.

These notes are based on a cycle of five lectures of 90 minutes. Since this is not much for introducing linear cosmological perturbation theory, the lectures remained at a qualitative level, with more emphasis on physical ideas than on the algebra. These written notes contain roughly the same amount of information as the lectures. After reading these few pages, the reader may wish to improve his/her level of understanding, while sticking to the same structure and same notations. For this purpose, we suggest to read chapters 2, 5 and 6 of Ref. 1: they contain a similar presentation, still simple and qualitative, but more detailed. Optionally, the reader can also grab there a description of neutrino effects, which are completely neglected here.

For a quantitative description of perturbation theory, gauge transformations, CMB physics and linear or non-linear structure formation, the reader should refer to the specialized literature. In the bibliography, we list a very incomplete selection of outstanding references.

Sections 2 and 3 of these notes are illustrated with several figures that represent the qualitative evolution of cosmological perturbations. It would have been easy to produce them with a Boltzmann code, as we did in Ref. 1. We deliberately chose instead to include only scanned hand-drawings. The first reason is that drawings offer the possibility to exaggerate some effects for pedagogical purposes: they are often more illuminating than exact numerical plots. The second reason is that they better render the conviviality of the blackboard presentation at TASI.

1. Linear Cosmological Perturbations

1.1. *Classification*

We decompose the metric and stress-energy tensor of the universe into spatial averages and linear perturbations,

$$g_{\mu\nu}(t,\vec{x}) = \bar{g}_{\mu\nu}(t) + \delta g_{\mu\nu}(t,\vec{x}) \ , \tag{1}$$

$$T_{\mu\nu}(t,\vec{x}) = \bar{T}_{\mu\nu}(t) + \delta T_{\mu\nu}(t,\vec{x}) \ , \tag{2}$$

where $\bar{g}_{\mu\nu}$ stands for the metric of the homogeneous and isotropic Friedmann–Lemaître (FL) model. Being symmetric, the two perturbed tensors contain ten degrees of freedom each, describing different aspects of gravity. Bardeen showed in 1980 that they can be decomposed on the basis of scalars, vectors and tensors under spatial rotations (spatial rotations play a special role because they leave the FL background invariant). These three sectors are decoupled at first order in perturbation theory.

In the vacuum, scalar and vector perturbations vanish, while tensor perturbations can propagate if they have been excited: they account for gravitational waves, the only "real" (propagating) gravitational degrees of freedom in General Relativity (GR). In the presence of matter, scalars represent the response of the metric to an irrotational distribution of matter, and generalize Newton's theory of gravitation. Vectors represent the response of the metric to vorticity, and describe phenomena with no equivalent in Newton's theory, called "gravito-magnetism".

In minimal cosmological models, the vorticity of the various matter components decays with time, and vectors can be neglected. Tensors may play a small role in CMB anisotropies, that we will mention briefly in Sec. 2.8. They can be studied separately, since they decouple from the scalar sector at first order in perturbations. Hence this course will be essentially focused on scalar perturbations.

The four scalar components of both the metric and stress-energy perturbed tensors are contained in:

(1) the (00) term,
(2) the trace of the (ij) matrix,

(3) the irrotational part of the $(0i)$ vector,
(4) the traceless longitudinal part of the (ij) tensor.

For the perturbed metric $\delta g_{\mu\nu}$, these components correspond (in the same order) to:

(1) the generalized gravitational potential ψ,
(2) the local distortion ϕ of the average scale factor $a(t)$: the "local scale factor" is given by $(1-\phi)a$,
(3) the potential b such that $\delta g_{0i} = \partial_i b$,
(4) the potential μ of the metric shear: $\delta g_{ij} = (\partial_i \partial_j - \frac{1}{3}\delta_{ij}\Delta)\mu$.

For the perturbed stress-energy tensor $\delta T_{\mu\nu}$, these components (still in the same order) represent:

(1) the energy density perturbations $\delta\rho$ (we will usually refer to the relative perturbations $\delta \equiv \delta\rho/\bar{\rho}$),
(2) the pressure perturbations δp,
(3) the potential v of the irrotational component of the flux of energy, $\delta T^0_i = \partial_i v$ (v is sometimes called the velocity potential, since in the case of a fluid it is related to the bulk velocity),
(4) the potential s of the shear stress or anisotropic stress: $\delta T^i_j = (\partial_i \partial_j - \frac{1}{3}\delta_{ij}\Delta)s$.

It is equivalent to use as a variable the "velocity potential" v or the "velocity divergence" θ defined as

$$\partial_i \delta T^0_i \equiv (\bar{\rho}+\bar{p})\theta = \Delta v \; . \tag{3}$$

Similarly, we can use the function σ instead of s, with the definition

$$\left(\partial_i \partial_j - \frac{1}{3}\delta_{ij}\Delta\right)\delta T^i_j \equiv (\bar{\rho}+\bar{p})\sigma = \Delta(\Delta s) \; . \tag{4}$$

The function σ is usually called the anisotropic stress, although the true anisotropic stress is the component of δT^i_j deriving from the potential s. The factors $(\bar{\rho}+\bar{p})$ in the two previous equations are introduced in the definitions in order to obtain simple equations (some authors use alternative notations without these factors or with different ones). To summarize, we see that we can manipulate four degrees of freedom representing the scalar perturbations of matter fields, that can be chosen to be the density fluctuation, pressure perturbation, velocity divergence and anisotropic stress: $\{\delta, \delta p, \theta, \sigma\}$.

1.2. *Gauges*

In an idealised FL universe, there is only one time slicing compatible with the assumption of homogeneity. Instead, in a perturbed universe, there is an infinity of time slicings compatible with perturbation theory (i.e. such that on each slice, all quantities remain close to their average value).

The perturbation of any quantity in a given point is the difference between the true and the average quantity in this point. For instance, for the total energy density ρ,

$$\delta\rho(t, \vec{x}) = \rho(t, \vec{x}) - \bar{\rho}(t) \ . \tag{5}$$

While $\rho(t, \vec{x})$ is a locally, unambiguously defined quantity, $\bar{\rho}(t)$ depends on the choice of equal-time hypersurface going through the point $(t, \vec{x})$. With a different choice, $\rho(t, \vec{x})$ would be compared to the average performed on a different sheet, that would take a different value. Hence $\delta\rho(t, \vec{x})$ also depends on the choice of time slicing.

A gauge is a choice of time slicing. Gauge transformations are induced by coordinate transformations $x_\mu \mapsto x_\mu + \epsilon_\mu$ mapping the points of one time slicing to those of another time slicing. All coordinate transformations do not induce a valid gauge transformation: the condition that perturbations must still be linear after the transformation restricts ϵ_μ to be small in every point.

A naive study of the equations of motion of perturbed quantities would be plagued by the freedom to change the gauge without changing physical results: some solutions of the full equations would be "gauge modes" with no observable consequences. To deal with this issue, one can adopt one of two point of views:

- one can derive gauge-invariant quantities (i.e., non-trivial integro-differential combinations of the metric and stress-energy tensor components left invariant by a gauge transformation), and gauge-invariant equations of motions for these quantities. Note that there are four scalar degrees of freedom in $\delta g_{\mu\nu}$ and two scalar degrees of freedom in the four-vector field ϵ_μ inducing gauge transformations: namely, ϵ_0 and the potential e such that $\epsilon_i = \partial_i e$. Hence we can use gauge transformations to cancel two scalar degrees of freedom, and build up two independent gauge-invariant scalars. One way to define them is through the two Bardeen potentials Φ_A and Φ_H, defined by Bardeen (1980) as two integro-differential combinations of ψ, ϕ, b, μ.
- one can fix the gauge, i.e. introduce a condition such that the time

slicing is unique. Then the number of independent solutions to the equations will be the same as in the gauge-invariant formalism. In any case, one can show that truly observable quantities are always independent of the gauge. Obtaining them after solving equations in the gauge-invariant formalism or in one particular gauge should not make any difference in practice.

A convenient gauge choice for a pedagogical introduction to CMB and matter scalar perturbations is the so-called Newtonian gauge or longitudinal gauge, in which one imposes that non-diagonal scalar perturbations of the metric vanish: $b = \mu = 0$. This prescription can be showed to fix a unique time slicing. In this gauge, adopting units such that $c = 1$ and using proper time t, the line element reads:

$$ds^2 = -(1 + 2\psi)dt^2 + (1 - 2\phi)a^2 d\vec{l}^{\,2} , \tag{6}$$

where $d\vec{l}^{\,2}$ stands for the cartesian measure $(dx^2 + dy^2 + dz^2)$ for a flat FL model, or for $(1 \pm kr^2)^{-1}dr^2 + r^2(\sin\theta^2 d\theta^2 + d\varphi^2)$ for an open/closed FL model in spherical coordinates. We are still free to redefine time (by definition, a time redefinition leaves the time slicing invariant). Lots of results in cosmological perturbation theory look simpler when using conformal time η, defined up to a constant by $d\eta = dt/a$. In this course we fix the constant in such way that $\eta \longrightarrow 0$ at the vicinity of the initial singularity, when $\rho \longrightarrow \infty$ (this prescription would not work if we were studying cosmological inflation, but in this course, we are not). Conformal time is convenient because photons traveling in a flat unperturbed FL universe along geodesics crossing the origin of the system of coordinates obey to $dr = d\eta$ (this comes from $a\,dl = dt$, i.e. from $ds = 0$ with the restriction $d\theta = d\phi = 0$). Hence conformal time is a measure of time based on the comoving distance travelled by a photon,[a] and the comoving distance to a given object coincides with its "look-back conformal time". In this course we will use dots for derivatives with respect to proper time and primes for derivatives with respect to conformal time. The Hubble parameter (or expansion rate parameter) reads

$$H = \frac{\dot{a}}{a} = \frac{a'}{a^2} , \tag{7}$$

the Hubble radius is $R_H = 1/H$, and the condition that a Fourier mode of

[a]In a universe with non-zero spatial curvature, this remains true, provided that the comoving distance is defined not like r, but like $\chi \equiv \int (1 \pm kr^2)^{-1/2} dr$.

physical wavelength λ crosses the Hubble radius is

$$\lambda = R_H \quad \Leftrightarrow \quad \frac{2\pi}{k}a = \frac{1}{H} \quad \Leftrightarrow \quad k \sim aH = \frac{a'}{a} \; . \tag{8}$$

One advantage of the Newtonian gauge is that the gauge-invariant Bardeen potentials $\{\Phi_{\rm A}, \Phi_H\}$ reduce in this gauge to the metric perturbations $\{\phi, \psi\}$: the evolution of the Newtonian metric perturbations informs us directly on that of two gauge-invariant quantities. Other interesting properties of $\{\phi, \psi\}$ appear when writing the Einstein equations in the Newtonian gauge. The full Einstein equations linearized at first order in perturbations feature four equations relating scalar degrees of freedom. One of them, associated to the traceless longitudinal part of $\delta G^i_j = 8\pi G T^i_j$, gives (assuming a flat FL background)

$$\frac{2}{3}(k/a)^2(\phi - \psi) = 8\pi G \sum_x (\bar{\rho}_x + \bar{p}_x)\sigma_x \; , \tag{9}$$

where the index x runs over all the species contributing to the total stress-energy tensor. This means that when the universe contains only shearless components with $\sigma_x = 0$ (as would be the case in the presence of perfect fluids), the two metric perturbations are equal. Next, the Einstein equation $\delta G^0_0 = 8\pi G T^0_0$ gives (still assuming a flat FL background)

$$2a^{-2}\left[k^2\phi + 3\frac{a'}{a}\left(\phi' + \frac{a'}{a}\psi\right)\right] = -8\pi G \sum_x \bar{\rho}_x \delta_x \; . \tag{10}$$

The term on the right-hand side involves the total energy perturbation $\delta\rho_{\rm tot} = \sum_x \bar{\rho}_x\delta_x$. In the short scale (more precisely, sub-Hubble) limit, the term containing k^2 dominates the other terms in the square brackets, and we recover the Poisson equation

$$-\frac{k^2}{a^2}\phi = 4\pi G\, \delta\rho_{\rm tot} \; , \tag{11}$$

where the factor $-k^2/a^2$ represents the Fourier transform of the physical Laplace operator in an expanding universe. Note that in the Poisson equation one may have expected to see the generalized gravitational potential ψ instead of ϕ: however, in the sub-Hubble limit, the shear of individual components is usually either null or negligible, so that $\phi = \psi$.

1.3. *Equations of motion*

In the minimal cosmological scenario, the universe features several species with spatial fluctuations, described with different equations because of their distinct properties: cold dark matter (CDM) is non-relativistic and collisionless, neutrinos are ultra-relativistic and collisionless at the times of interest, baryons are non-relativistic and smoothly interpolating from a strongly coupled to decoupled regime, and finally photons are ultra-relativistic and interpolating between the same two regimes.

The equation of conservation of the total stress-energy tensor, $D_\mu T^\mu_\nu = 0$ (deriving from Bianchi identities), yields two scalar and two vector equations. The scalar ones are the conservation of energy equation and the Euler equation.

For an single component experiencing no interaction with other species (other than gravitational), the equation $D_\mu T^\mu_\nu = 0$ applies to the individual stress-energy tensor: it gives one continuity and one Euler equation for that component. The evolution of single component experiencing interactions is also given by the continuity and Euler equation, but with an extra source term accounting for stress-energy injection/leak caused by the interaction.

We have seen that the perturbations of each component x can be described by four variables $\{\delta_x, \delta p_x, \theta_x, \sigma_x\}$. Hence, in general, two equations of motion are not sufficient for closing the system. However:

- for a perfect fluid, microscopic interactions impose local thermodynamical equilibrium. The pressure is then isotropic,[b] with $\sigma_x = 0$. In addition, pressure perturbations obey to $\delta p_x = c_\mathrm{a}^2 \delta \rho_x$, where c_a is the adiabatic sound speed inferred from the equation of state of the fluid. If σ_x vanishes and δp_x is a function of $\delta \rho_x$, perturbations in the fluid are described by only two independent functions $\delta_x \equiv \delta\rho_x/\bar{\rho}_x$ and θ_x. If the collision term also vanishes or is specified, the two equations of motion (continuity and Euler) are sufficient for closing the system and computing the evolution of perturbations.
- for a decoupled or weakly interacting species, there are no such simplifications concerning the anisotropic stress and pressure perturbation. Hence, in general, the two equations inferred from stress-energy conservation are not sufficient. For such species, one has to use the more general Boltzmann equation, giving the evolution of each phase-space

[b]An anisotropic shear stress $\sigma \neq 0$ reflects the fact that in each given point, particles travel with different velocities (due to some intrinsic velocity dispersion and/or a superposition of several flows in phase space), leading to anisotropic pressure. This contradicts the assumption of a perfect fluid, in which local interactions result in a unique bulk velocity (after coarse-graining over microscopic scales), and erase anisotropic pressure.

distribution:

$$\frac{d}{d\eta} f_x = \sum_y C_{xy}[f_x, f_y] \ , \tag{12}$$

where the sum holds over the species y coupled with x. Each phase-space distribution can be decomposed into a background and perturbation part:

$$f_x(\eta, \vec{x}, \vec{p}) = \bar{f}_x(\eta, |\vec{p}|) + \delta f_x(\eta, \vec{x}, \vec{p}) \ , \tag{13}$$

where $\vec{p}$ stands for momentum, $|\vec{p}|$ for its modulus, and the background part does not depend on the direction of $\vec{p}$ by assumption of isotropy.

- fully decoupled CDM is a particular case of a collisionless species with negligible velocity dispersion (the word "cold" refers precisely to this last assumption). Hence, it behaves in the same way as a pressureless perfect fluid, although in reality it has no interactions and should not be called a fluid. Since the velocity dispersion is negligible, in a given point, all particles share the same velocity, imposed by gravitational flows (while for non-cold collisionless species, the velocity would get two contributions, one from gravitational flows, and one from the phase-space distribution function). Hence the anisotropic pressure vanishes (see the previous footnote). The pressure perturbation δp_x is also related to the velocity dispersion in a given point, anf can be neglected with respect to the density perturbation in the CDM case. Hence CDM is formally equivalent to a perfect fluid with no anisotropic stress and no pressure perturbation (or in other words, with a sound speed $c_s = 0$). In that case, the two equations of motion inferred from $D_\mu T^\mu_\nu = 0$ are sufficient, like for a fluid.

Gravitational interactions between species are accounted by the presence of metric perturbations in each equation of motion (more specifically, terms in $k^2\psi$ accounting for gravitational forces, and terms in ϕ' accounting for dilation effects, i.e. for local distortions of the scale factor with respect to a). Hence, in order to close the full system of equations, we still need two independent relations, to be chosen among the four scalar Einstein equations: they provide the value of $k^2\psi$ and ϕ' at each time, as a function of all matter fields.

1.4. *Initial conditions*

We wish to study the evolution of matter perturbations, starting from some early time at which all Fourier modes of interest (those which are observable in the CMB spectrum and in the matter power spectrum on linear or mildly non-scales) are still outside the Hubble radius. Indeed, super-Hubble modes experience a trivial evolution, unaffected by small scale interactions (Thomson or Coulomb scattering, usual gravitational force $\vec{\nabla}\phi$, etc.) Hence, the perturbations evaluated at some arbitrary time but on super-Hubble scales reflect directly the mechanism responsible for the formation of perturbations in the very early universe. In the standard cosmological model, these initial conditions can be inferred from inflation.

Typically, a good time for setting initial conditions is when the redshift $z = a_0/a - 1$ is of the order of 10^5: at this time, all comoving scales that are observable in the CMB and linear matter power spectrum still verify $k \ll aH$.

It is crucial to understand that, as long as the background cosmology is assumed to be of the FL type, the perturbed stress-energy momentum tensor δT^ν_μ must be diagonal on super-Hubble scales. Indeed, the background tensor $\bar{T}^\nu_\mu$ is diagonal, and of the form: $\mathrm{diag}(-\bar{\rho}, \bar{p}, \bar{p}, \bar{p})$. This can be showed to be the most general assumption compatible with homogeneity and isotropy. Let us assume that we Taylor-expand δT^ν_μ in powers of the variable $(k\eta)$. For any power-law scale factor, aH is given by $1/\eta$ times a factor of order one. Hence the limit $k\eta \ll 1$ represents precisely the super-Hubble limit. In the Taylor expansion, the zero mode should share the same properties as the background solution, and be diagonal. Higher order terms account for contributions to δT^ν_μ growing with time, and possibly becoming important around the time of Hubble crossing.

The total scalar perturbations $\delta\rho$ and δp are the only one preserving the diagonal form of δT^ν_μ: we conclude that they are the only one that do not vanish at order zero in the $(k\eta)$ expansion. Using stress-energy conservation equations, one can show that the part of δT^ν_μ associated with the velocity divergence is of order one in $(k\eta)$, while the part associated to the total anisotropic stress is of order two.

Suppose that the universe contains initially N uncoupled perfect fluids,[c] with N known sound speeds $c_x^2 = \delta p_x/\delta\rho_x$. There are $2N$ independent initial conditions, corresponding to possible initial values of each δ_x and each δ_x'. Importantly, in the $2N$-dimensional basis of IC's, one basis vector is very special, as we shall see below.

Before studying perturbations, one should have solved for the back-

[c]The discussion presented in this section could be generalized to N coupled species, not all of them being perfect fluids: the conclusions would not change qualitatively, and we restrict here to N uncoupled fluid for simplicity.

ground evolution: all background quantities should be known, including for instance the density $\bar{\rho}_x(\eta)$ and pressure $\bar{p}_x(\eta)$ of each species x. Now, let us assume that the real universe is perturbed initially by a single degree of freedom (one may say, by a single initial time shifting function). This is the case in single-field inflationary cosmology: during inflation, there is a single clock (the inflaton), and perturbations arise from a single time shifting function (the inflaton perturbation).

As long as we are dealing with super-Hubble modes, we can neglect microscopic interactions and say that the evolution in each point (in fact, in each Hubble patch) is still given by homogeneous cosmology, taking this shift function $\delta\eta(\vec{x})$ into account:

$$\forall x, \qquad \begin{aligned} \rho_x(\eta,\vec{x}) &= \bar{\rho}_x\left(\eta+\delta\eta(\vec{x})\right) = \bar{\rho}_x(\eta) + \bar{\rho}'_x(\eta)\,\delta\eta(\vec{x}) \\ p_x(\eta,\vec{x}) &= \bar{p}_x\left(\eta+\delta\eta(\vec{x})\right) = \bar{p}_x(\eta) + \bar{p}'_x(\eta)\,\delta\eta(\vec{x}) \end{aligned} \tag{14}$$

where in the last equalities, terms of order two or higher in $\delta\eta$ have been neglected. The above ansatz restricts a lot the choice of possible initial conditions. Indeed, for each uncoupled species, the energy conservation law for the background gives

$$\bar{\rho}'_x = -3\frac{a'}{a}(\bar{\rho}_x + \bar{p}_x)\ . \tag{15}$$

Combining this equation with the above ansatz, we get

$$\forall x, \qquad \frac{\delta\rho_x}{\bar{\rho}_x+\bar{p}_x} = -3\frac{a'}{a}\frac{\delta\rho_x}{\bar{\rho}'_x} = -3\frac{a'}{a}\,\delta\eta(\vec{x})\ . \tag{16}$$

Since the last term is independent of the index x, we see that for each pair of fluids,

$$\forall x, y, \qquad \frac{\delta\rho_x}{\bar{\rho}_x+\bar{p}_x} = \frac{\delta\rho_y}{\bar{\rho}_y+\bar{p}_y}\ . \tag{17}$$

Note that this famous relation is a consequence of Eqs. (14). It shows that in presence of such initial conditions, everything is fixed up to a single function of $\vec{x}$. Let us take the example of a universe containing only photons, baryons, cold dark matter and neutrinos. We can use the fact that for non-relativistic species $\bar{p}_x \ll \bar{\rho}_x$, while for ultra-relativistic ones $\bar{p}_x = \bar{\rho}_x/3$. Hence, if one function is known — for instance, $\delta_\gamma(\vec{x})$ at initial time — the others can be derived from

$$\delta_{\rm b} = \delta_{\rm cdm} = \frac{3}{4}\delta_\nu = \frac{3}{4}\delta_\gamma\ . \tag{18}$$

It is a simple exercise to prove that the ansatz of Eqs. (14) implies additional important relations:

- first,

$$\forall x, \qquad \delta p_x(\eta, \vec{x}) = \frac{\bar{p}'_x(\eta)}{\bar{\rho}'_x(\eta)}\, \delta\rho_x(\eta, \vec{x}) \ , \tag{19}$$

  i.e., as long as we are dealing with super-Hubble modes and Eqs. (14) are satisfied, each component features an adiabatic sound speed independent of x, obeying to $c_x^2(\eta) \equiv \bar{p}'_x(\eta)/\bar{\rho}'_x(\eta)$,
- second,

$$\delta p_{\rm tot}(\eta, \vec{x}) = c_{\rm s}^2(\eta)\, \delta\rho_{\rm tot}(\eta, \vec{x}) \ , \tag{20}$$

  where the squared total sound speed $c_{\rm s}^2$ can be easily expressed as a weighted average over the squared adiabatic sound speed of individual components. Hence, the total matter content resulting from the sum of all components also features an adiabatic sound speed.

For the reason expressed in the last item, such initial conditions are usually called *adiabatic initial conditions.* The word adiabatic refers to thermodynamics, but it is actually more helpful to think of adiabatic initial conditions as resulting from the assumption of a single inhomogeneous degree of freedom in the early universe: then Eqs. (14) stands for the true definition of adiabatic initial conditions, and Eqs. (17, 18, 19, 20) for its consequences.

More general initial conditions not obeying to Eqs. (14) can be expanded on different bases. A famous basis is formed by (i) the adiabatic mode; (ii) $(N-1)$ non-decaying isocurvature modes, getting their name from the property that for each of them, the total density (and spatial curvature) perturbations vanish asymptotically in the super-Hubble limit, while two species have opposite density perturbations compensating each other; (iii) N decaying modes that are irrelevant for most purposes.

It is clear from the previous discussion that non-adiabatic initial conditions should only be considered when assuming that primordial perturbations are generated by more than one degree of freedom (for instance, two inflaton fields, or one inflaton and one axion, etc.) The assumption of several degrees of freedom is necessary but not sufficient. The primordial universe may contain a mixture of adiabatic and isocurvature modes until a time at which all species are brought in thermal equilibrium. At that time, if we further assume that all chemical potentials vanish, the perturbations of each species can be inferred from those of temperature, $T(\eta, \vec{x}) = \bar{T}(\eta) + \delta T(\eta, \vec{x})$. Then, temperature plays precisely the role of a single time-shifting function. Any non-adiabatic initial condition is washed out and becomes irrelevant. Hence, isocurvature modes can be observable — for instance in the CMB — only under additional assumptions. For instance, one species carrying isocurvature perturbations might remain out of equilibrium at all times, or might feature a chemical potential with spatial fluctuations. There exist a few non-minimal, but still reasonable scenarii featuring isocurvature perturbations (with axions, curvatons etc.). Ultimately, the presence of isocurvature modes is to be tested with observations. Current CMB observations put strong limits on the amplitude of such modes, and prefer purely adiabatic initial conditions. We will restrict to this case in the rest of this course.

For adiabatic initial conditions, we found the relation (17) holding between density fluctuations. Also, if we do not consider the case of species with a non-zero anisotropic stress, we can assume that $\phi = \psi$ at initial time. By plugging Eqs. (18), (20) and $\phi = \psi$ into the Einstein equations, one is led to a second-order differential equation for ψ only. During the radiation dominated era, one can show that this equation has two solutions, one constant in time, and one decaying. The (00) Einstein equation then shows that the constant solution is related to density fluctuations through

$$-2\psi = -2\phi = \delta_{\text{tot}} \simeq \delta_\gamma = \text{constant} . \tag{21}$$

Hence, in the Newtonian gauge and for adiabatic initial conditions, super-Hubble metric fluctuations and density fluctuations are static. In fact, on super-Hubble scales, there can only be some time evolution when the universe changes of total equation of state (or equivalently, of expansion law). This is the case at the time of equality between radiation and matter. During matter domination and on super-Hubble scales, they freeze out again. Then, the relation $-2\psi = -2\phi = \delta_{\text{tot}}$ and Eq. (18) are still satisfied, but now $\delta_{\text{tot}} \simeq \delta_{\text{b}} = \delta_{\text{cdm}} = \frac{3}{4}\delta_\gamma$. This detail will become important when studying the Sachs-Wolfe effect in section 2.3.

1.5. *Power spectra and transfer functions*

The theory of cosmological perturbations is a stochastic theory: the fluctuations of a given quantity in a given point, $A(\eta, \vec{x})$, obey to a distribution of probability. As long as we stick to linear perturbation theory, there is a "linear transport of probability" from one time to another time. Let the probability of A in a given point at time η_1 be $\mathcal{P}_1(A)$. In the same point and at η_2, the linear evolution would have transformed each value A into αA, where α is the linear growth (or decrease) factor of A between η_1 and

η_2. So the probability of A at time η_2 is given by $\mathcal{P}_2(A) = \mathcal{P}_1(A/\alpha)$. This "linear transport of probability" implies that the linear evolution respects the shape of the probability distribution, and rescales all its statistical moments of order n by α^n. In particular, if the initial probability is Gaussian, the statistics will remain Gaussian at any later time, and the evolution of the system can be formulated as an evolution of the root mean square of all quantities. In summary, as long as we assume linear perturbations with Gaussian initial conditions, our goal is to solve for the evolution of the root mean squares of the fluctuations.

The equal-time 2-point correlation function of any quantity A in real space is given by

$$\langle A(\eta,\vec{x})A(\eta,\vec{x}')\rangle \equiv \xi(\eta,\vec{x},\vec{x}') \stackrel{SHI}{=} \xi(\eta,|\vec{x}'-\vec{x}|) \; , \qquad (22)$$

where the last equality holds as a consequence of Statistical Homogeneity and Isotropy (SHI, assumed to hold in a perturbed FL universe). Indeed, the correlation function should be invariant under spatial translations and rotations.

We can go to Fourier space and use the same letter to denote the Fourier transform of A. If A is real, $A(\eta,-\vec{k}) = A^*(\eta,\vec{k})$. It is easy to show that the previous equation (using the assumption of SHI) implies that the two-point correlation function in Fourier space reads

$$\langle A(\eta,\vec{k})A^*(\eta,\vec{k}')\rangle \stackrel{SHI}{=} \delta_D(\vec{k}'-\vec{k})P_{\rm A}(k) \; , \qquad (23)$$

where δ_D is the Dirac distribution. Here, statistical homogeneity is responsible for the fact that the two-point correlation vanishes for $\vec{k} \neq \vec{k}'$, and statistical isotropy for the fact that $P_{\rm A}$ only depends on the modulus $k = |\vec{k}|$. The function $P_{\rm A}(k)$ is usually called the power spectrum of A. Some authors prefer to refer to the "dimensionless power spectrum", defined as

$$\mathcal{P}_{\rm A}(k) \equiv \frac{k^3}{2\pi^2} P_{\rm A}(k) \; . \qquad (24)$$

The reason is that typical expressions for the average of various quantitites in real space consist in the convolution of the power spectrum with some window function $f(k)$, like in

$$\int \frac{d^3\vec{k}}{(2\pi)^3} P_{\rm A}(k) f(k) = \int \frac{4\pi k^2 dk}{(2\pi)^3} P_{\rm A}(k) f(k) = \int d\,\log\,k\;\mathcal{P}_{\rm A}(k) f(k) \; . \qquad (25)$$

Hence the dimensionless spectrum $\mathcal{P}_{\rm A}(k)$ stands for the weight of each logarithmic interval in the integral. The term "scale-invariant spectrum" refers to $\mathcal{P}_{\rm A}(k)$ being independent of k, i.e. $P_{\rm A}(k) \propto k^{-3}$.

We know that for adiabatic initial conditions, all perturbations are related to each other through Eqs. (18, 21). Hence, with Gaussian adiabatic initial conditions, specifying the primordial power spectrum of one quantity is sufficient for knowing everything about the system. For instance, if we assume that the power spectrum of the metric perturbation ψ is a given function $P_\psi(k)$, then we infer from Eqs. (18, 21) that the photon and baryon primordial spectra are given by $P_\gamma(k) = 4P_\psi(k)$ and $P_{\rm b}(k) = \frac{9}{16}P_\gamma(k) = \frac{9}{4}P_\psi(k)$.

By convention, the primordial spectrum is usually given for the variable $\mathcal{R}$, which represents the spatial curvature perturbation on one initial comoving hypersurface (i.e. an hypersurface orthogonal to the energy flux of the total cosmic fluid in each point). The advantage of using this quantity is that it is conserved on super-Hubble scales for adiabatic initial conditions (while ϕ and ψ get rescaled when the equation of state of the universe changes, e.g. at radiation-matter equality). In the Newtonian gauge, $\mathcal{R}$ relates to ψ and to the total density perturbation through

$$\mathcal{R} = \psi - \frac{1}{3}\frac{\delta\rho_{\rm tot}}{\bar{\rho}_{\rm tot} + \bar{p}_{\rm tot}} \,. \tag{26}$$

The power spectrum of a given quantity at some arbitrary time can be decomposed into two parts, one accounting for initial conditions, and one accounting for linear evolution with time:

$$\langle A(\eta, \vec{k}) A^*(\eta, \vec{k}')\rangle = \delta_D(\vec{k}' - \vec{k}) \left[\frac{A(\eta, \vec{k})}{\mathcal{R}(\vec{k})}\right]^2 P_{\mathcal{R}}(k) \,. \tag{27}$$

In the above equation, there is no time argument for $\mathcal{R}(\vec{k})$ since this quantity is conserved on super-Hubble scales. We just assume that $\mathcal{R}$ is evaluated at a time such that $k \ll aH$ for all modes of interest. In a FL universe, the equations of motion of all perturbations respect isotropy, and do not depend on the direction of the wavector $\vec{k}$. Hence the ratio between brackets in the last equation is a function of k, not $\vec{k}$. This function accounts for the linear evolution, independently of initial conditions. It is called the "transfer function" of A. In this course, we will denote transfer functions with the same letter as the perturbations themselves, but with the modulus of k as an argument, i.e.

$$A(\eta, k) \equiv \frac{A(\eta, \vec{k})}{\mathcal{R}(\vec{k})} \,. \tag{28}$$

In summary of this section, solving for cosmological perturbations (in a model with adiabatic Gaussian IC's) amounts in

- postulating a primordial spectrum, or calculating it within the framework of a model, for instance of inflation;
- solving the equations of motion of all perturbations, with quantities normalized initially to $\mathcal{R}(\vec{k}) = 1$, in such a way to obtain all transfer functions $A(\eta, k)$ and to infer the evolution of the various root mean squares.

2. CMB Temperature Anisotropies

From now on, we will assume for simplicity that the universe is flat.

2.1. *Photon scattering rate*

CMB physics consists is the study of electron, baryon and photon perturbations on cosmological scales, taking into account their gravitational coupling with collisionless species such as decoupled neutrinos and CDM. Electrons and baryons carry opposite electric charges and are coupled to each other through very efficient Coulomb scattering processes. Electrons and photons are coupled through Thomson scattering, which is the limit of Compton scattering when electrons are non-relativistic and photons carry a smaller energy than the rest mass of the electron. Then, the scattering process results mainly in a deflection of the photon, with a negligible transfer of energy between the two particles. The leading interaction between baryons and photons is the gravitational one.

In units such that $c = 1$, the Thomson scattering rate (with respect to conformal time) is given by $\Gamma = \sigma_T a n_e x_e$, where σ_T is the Thomson scattering rate, a is the scale factor, n_e is the total electron number density (scaling like a^{-3} due to dilution), and x_e is the ionized electron fraction. The product $a n_e$ scales like a^{-2}. The ionized fraction is close to one at high energy. Then, at the time of recombination between electrons and nuclei (around $z \sim 1080$), which falls at the beginning of matter domination, n_e drops abruptly to very small values. This causes Thomson scattering to become suddenly very inefficient, and photons to decouple from electrons.

Hence photon decoupling is the story of Thomson scattering becoming inefficient, while Coulomb scattering remains very strong. For that reason, one can describe electrons and baryons as a single tightly-coupled fluid. People often refer only to baryons for simplicity. At early time, the full system of (electrons)-baryons-photons is also tightly coupled, but later on, it splits progressively into two collisonless species, (electrons)-baryons on the one hand, photons on the other hand.

The thermodynamical description of recombination is very technical, due to the different energy level of atoms (in particular, of hydrogen). In order to understand CMB anisotropies, we only need to describe the main results of recombination studies at a very qualitative level.

- The free electron fraction $x_e(\eta)$ starts from one at high redshift. It decreases sharply at the recombination time (corresponding to $z \sim 1080$ or $T \sim 0.3$ eV), and freezes at a very small value (due to departure from thermal equilibrium). Around $z \sim 10$, star formation causes a reionization of the universe, and x_e goes up again to one.
- The Thomson scattering rate $\Gamma = \sigma_T a n_e x_e$ evolves like $a^{-2}x_e$. Before recombination, $\Gamma \gg \frac{a'}{a}$, and the universe is opaque. The sudden drop of x_e at recombination renders the universe transparent: $\Gamma \ll \frac{a'}{a}$. Due to the dilution factor coming from n_e, Γ remains much smaller than $\frac{a'}{a}$ even during reionization, and the universe keeps being transparent (this is why despite of reionization, most photons emitted at recombination do not interact anymore, and allow us to observe anisotropies on the last scattering surface).
- The optical depth $\tau(\eta) \equiv \int_\eta^{\eta_0} d\eta \Gamma(\eta)$ represents the opacity of the universe at a given time, when seen from today (when $\eta = \eta_0$). It tends to infinity when $\eta \longrightarrow 0$, falls below one at recombination and stabilize at a value of the order of 0.1 between recombination and reionization. After reionization it decreases smoothly and reaches zero today by definition.
- The visibility function $g(\eta) \equiv -\tau' e^{-\tau}$ gives the probability that a CMB photon seen today experienced its last scattering at time η. It starts from negligible values at high redshift (suppressed by the $e^{-\tau}$ factor). It has a narrow spike around the time of recombination, and then it falls again to negligible values due to the smallness of τ' between recombination and reionization. It develops a second smaller and wider spike around reionization. This function shows that most CMB photons did not interact between the last scattering surface and today, while a minority rescattered at reionization. The width of the recombination spike gives an indication on the thickness of the last scattering "surface".
- The diffusion length $\lambda_{\rm d}(\eta)$ is an important quantity for understanding the damping of temperature anisotropies on small scales. At any given

time, the comoving mean free-path (mfp) of photons is given by $r_{\rm mfp} = 1/\Gamma(\eta)$ (still in units where $c = 1$). If the photons experience a random walk analogous to Brownian motion in a gas, the comoving distance over which they travel between time $\eta_{\rm ini}$ and η can be approximated by

$$r_{\rm d} \sim \left[\int_{\eta_{\rm ini}}^{\eta_0} d\eta \; \Gamma \; r_{\rm mfp}^2 \right]^{1/2} = \left[\int_{\eta_{\rm ini}}^{\eta_0} d\eta \; \Gamma^{-1} \right]^{1/2} . \tag{29}$$

This physical diffusion length of photons is given in this approximation by $\lambda_{\rm d} \simeq a r_{\rm d}$. It grows quickly from very small to very large scales (comparable to the Hubble scale) around the time of recombination.

2.2. *Boltzmann equation*

Since photons decouple from baryons near the recombination time, we cannot describe them with fluid equations, and need to solve the Boltzmann equation

$$\frac{d}{d\eta} f = C\left[f, f_e\right] \tag{30}$$

at order one in perturbations. The right-hand side stands for the photon-electron coupling due to Thomson scattering. As explained in the previous section, electrons and baryons are so tightly coupled that it makes no difference to think of this term as a photon-electron or photon-baryon coupling term. Solving this equation is involved because the photon phase-space distribution f involves many arguments, $f(\eta, \vec{x}, \vec{p})$. Fortunately one can reduce the dimensionality of the problem. First, we notice that as long as photons are in thermal equilibrium with electrons (and hence with baryons), they are entirely described in any point by the local value of the equilibrium temperature $T(\eta, \vec{x})$. The phase space distribution is then of the Bose-Einstein form:

$$f(\eta, \vec{x}, \vec{p}) = \frac{1}{e^{\frac{p}{T(\eta,\vec{x})}} - 1} . \tag{31}$$

It can be expanded into a background part and a first-order perturbation, $f = \bar{f} + \delta f$, with:

$$\bar{f}(\eta, p) = \frac{1}{e^{\frac{p}{\bar{T}(\eta)}} - 1} , \qquad \delta f(\eta, \vec{x}, \vec{p}) = \frac{d\bar{f}}{d\log p} \, \frac{\delta T(\eta, \vec{x})}{\bar{T}(\eta)} . \tag{32}$$

We see that in the tightly-coupled regime we could replace the variable $f(\eta, \vec{x}, \vec{p})$ by the lower-dimensional variable $\Theta(\eta, \vec{x}) \equiv [\delta T(\eta, \vec{x})/\bar{T}(\eta)]$. The Boltzmann equation leads to an equation of motion for $\Theta(\eta, \vec{x})$.

At later times, when photons decouple, the shape of the Bose-Eisntein distribution of photons is preserved. This can be infered from the geodesic equation. This equation tells how the individual momentum p of photons evolve when they are decoupled and they travel in the perturbed universe. In the Newtonian gauge, it reads

$$\frac{d(a\,p)}{d\eta} = -a\,p\,\phi' - a\,\epsilon\,\hat{n}\cdot\vec{\nabla}\psi\ . \tag{33}$$

The left-hand side represents the time evolution of the product (ap) for a photon of momentum p traveling over a geodesic. In a perfectly homogeneous universe, the photon would only experience the average cosmological redshifting, $p \propto a^{-1}$, and the product (ap) would be conserved. Due to presence of perturbations in the universe, photons experience gravitational interactions and the product (ap) varies. The first term $-a\,p\,\phi'$ accounts for dilation, i.e. for the fact that locally, the expansion of the universe is a bit more advanced or delayed than the average (remember that $a(1+\phi)$ can be seen as the "local scale factor"). This means that the redshifting of the photon is also a bit more advanced or delayed locally. The second term accounts for the gravitational blueshifting of photons falling in gravitational potential wells (or redshifting of those leaving potential wells). The energy $\epsilon \equiv \sqrt{p^2+m^2}$ can be simply replaced by p for massless photons. In that case we can rewrite the geodesic equation as

$$\frac{d\ln(a\,p)}{d\eta} = -\phi' - \hat{n}\cdot\vec{\nabla}\psi\ . \tag{34}$$

The fact that p disappeared from the right-hand side is crucial: it shows that photons of different momenta traveling in the perturbed universe along a given geodesic all experience the same relative momentum variation. The consequence is that there can be no distortions of the Bose-Einstein shape of f. However, photons traveling along different geodesics and in different directions experience different redshifting. Hence the distribution function acquires a dependence on one extra argument, the direction $\hat{n}$ of propagation ($\hat{n} \equiv \vec{p}/p$):

$$f(\eta,\vec{x},\vec{p}) = \frac{1}{e^{\frac{p}{T(\eta,\vec{x},\hat{n})}}-1}\ . \tag{35}$$

We can perform the same decomposition in terms of background and perturbations as in Eq. (32). The difference is that $\Theta = \frac{\delta T}{T}$ is now a function of $(\eta,\vec{x},\hat{n})$. The Boltzmann equation can now be used to derive an equation of motion for $\Theta(\eta,\vec{x},\hat{n})$.

If we work in Fourier space, we can derive the equation of motion for the function $\Theta(\eta, \vec{k}, \hat{n})$. Because of the statistical isotropy of the FL universe, this equation does not depend explicitly on $\vec{k}$ nor $\hat{n}$, since there is no preferred direction: it depends only on the direction of propagation relatively to the considered wavenumber , i.e. on the product $(\vec{k} \cdot \hat{n})$. Hence the equation of motion can be written in terms of k and of the angle θ such that $(\vec{k} \cdot \hat{n}) = k \cos\theta$. The initial conditions for Θ do depend on the wavevector $\vec{k}$ (since each mode gets random initial conditions), but they also depend only on θ rather than $\hat{n}$, for a reason that will become clear in the paragraphs below. Hence we can entirely eliminate $\hat{n}$ from the problem, and solve the equation of motion for $\Theta(\eta, \vec{k}, \theta)$.

Finally, we can expand the temperature anisotropy with respect to θ using a Legendre transformation:

$$\Theta(\eta, \vec{k}, \theta) = \sum_l (-i)^l (2l+1) \Theta_l(\eta, \vec{k}) P_l(\cos\theta) \ . \tag{36}$$

Here the Θ_l's are the temperature anisotropy multipoles, and P_l the Legendre polynomials. It can be shown that the monopole Θ_0 is related to the photon density fluctuation δ_γ in a given point, the dipole Θ_1 to its velocity divergence θ_γ, and the quadrupole Θ_2 to its anisotropic stress σ_γ. The Boltzmann equation can be written as an infinite hierarchy of equations of motion for the coupled multipoles Θ_l.

Actually, the equation of motion for Θ takes a striking form when written in real space, before Fourier and Legendre expansions:

$$\Theta' + \hat{n} \cdot \vec{\nabla}\Theta - \phi' + \hat{n} \cdot \vec{\nabla}\psi = -\Gamma\left(\Theta - \Theta_0 - \hat{n} \cdot \vec{v}_e\right) \ . \tag{37}$$

We recall that Γ is the conformal Thomson scattering rate, Θ_0 the temperature monopole (i.e. the average of $\Theta(\eta, \vec{x}, \hat{n})$ over all directions $\hat{n}$), and $\vec{v}_e$ the bulk velocity of electrons, equal to that of baryons due to tight Coulomb interactions. We recall that the variable $\theta_{\rm b}$ is defined as the divergence of $\vec{v}_{\rm b} = \vec{v}_e$.

This equation is illuminating, since it shows that at early times, when photons, electrons and baryons form a tightly-coupled fluid, the fact that Γ is huge forces Θ to evolve in such way that the parenthesis on the right-hand side vanishes. When this is the case, Θ can only have two non-zero component: a monopole Θ_0, and a dipole equal to $\hat{n} \cdot \vec{v}_{\rm b}$. The condition on the dipole can be written equivalently as $\theta_\gamma = \theta_{\rm b}$.

This should remind us of the discussion of Sec. 1.3, when we noticed that a perfect fluid can be described in terms of δ_x, $c_{\rm s}^2$ and θ_x only, with a vanishing anisotropic stress σ_x. Here, we see concretely how this conclusion emerges from the Boltzmann equation in the tightly-coupled regime: Θ_2 (related to σ_γ) and all higher multipoles vanish; the photon perturbations can be described in terms of only two independent variables Θ_0 and Θ_1 (or δ_γ and θ_γ), like in a perfect fluid.

The fact that in the tightly-coupled limit the system is driven towards $\theta_\gamma = \theta_b$ simply shows that the interaction imposes a common bulk velocity to photons, baryons and electrons, as it should be the case in any tightly-coupled fluid (note that we are referring to bulk velocities, not to individual velocities of particles, which are still equal to c for interacting photons).

In the tightly-coupled regime, the dipole component of $\Theta(\eta, \vec{x}, \hat{n})$ is given by $\hat{n} \cdot \vec{v}_b$, with $\theta_b \equiv \vec{\nabla} \cdot \vec{v}_b$. Hence in Fourier space this component reads $i(\hat{n} \cdot \vec{k}) k^{-2} \theta_b = i(\cos\theta) k^{-1} \theta_b$. This justifies the fact that initial conditions for Θ in Fourier space depend on θ, but not on the two degrees of freedom of $\hat{n}$. As mentioned above, this property is preserved by the isotropic equations of motion, so that at all times we can study Θ as a function of the arguments $(\eta, \vec{k}, \theta)$ instead of $(\eta, \vec{k}, \hat{n})$.

2.3. *Temperature anisotropy in a given direction*

The map of temperature anisotropies that we observe today ($\eta = \eta_0$) in our location of the universe ($\vec{x} = \vec{o}$ with a proper choice of origin) when looking in a direction $\hat{n}$ is represented mathematically by

$$\frac{\delta T}{\bar{T}}(\hat{n}) = \Theta(\eta, \vec{o}, -\hat{n}) \tag{38}$$

(since in a direction $\hat{n}$, we see photons traveling towards $-\hat{n}$). Our scope is now to relate this quantity to perturbations on the point of the last scattering surface seen in the same direction $\hat{n}$. This can be done by integrating the Boltzmann equation along the corresponding line-of-sight.

A good starting point consists in computing the total derivative of the product $e^{\tau}(\Theta + \psi)$ along the trajectory of photons between the last scattering surface and the observer. The reason for choosing this product rather than just Θ will become clear in a few lines: the derivative of this term will be easy to simplify, using the Boltzmann equation.

The total derivative of an arbitrary function $\mathcal{F}$ of $(\eta, \vec{x}, \hat{n})$ along the trajectory of photons going in a direction $\hat{n}$ reads

$$\frac{d}{d\eta}\mathcal{F}(\eta, \vec{x}, \hat{n}) = \mathcal{F}' + \frac{dx_i}{d\eta}\frac{\partial \mathcal{F}}{\partial x_i} + \frac{dn_i}{d\eta}\frac{\partial \mathcal{F}}{\partial n_i} \,. \tag{39}$$

If $\mathcal{F}$ is of order one in perturbations, the first two terms on the right-hand side are also of order one. Instead the last term is of order two, since $\frac{dn_i}{d\eta}$ is of order one (this is clear from the fact that in an unperturbed universe, photons would travel in straight line with $\frac{dn_i}{d\eta} = 0$). Hence we can drop this term in first-order perturbation theory, and for $\frac{dx_i}{d\eta}$ we only need to keep the zero-th order contribution, that can be computed assuming a homogeneous universe:

$$\frac{dx_i}{d\eta} = \hat{n} \ . \tag{40}$$

The previous relation is just telling that photons are traveling in the direction $\hat{n}$ and at the velocity of light: hence, in units where $c = 1$, $d\vec{x}^2 = d\eta^2$. In summary, the total derivative of $\mathcal{F}$ is given at first order by

$$\frac{d}{d\eta}\mathcal{F}(\eta, \vec{x}, \hat{n}) = \mathcal{F}' + \hat{n} \cdot \vec{\nabla}\mathcal{F} \ . \tag{41}$$

Let us now replace the generic function $\mathcal{F}$ by

$$\mathcal{F}(\eta, \vec{x}, \hat{n}) = e^{-\tau(\eta)} \left(\Theta(\eta, \vec{x}, \hat{n}) + \psi(\eta, \vec{x})\right) \ . \tag{42}$$

The total derivative of this function reads

$$\frac{d}{d\eta}\left[e^{-\tau}(\Theta + \psi)\right] = e^{-\tau}\left(\Theta' + \psi' + \hat{n} \cdot \vec{\nabla}(\Theta + \psi)\right) - \tau' e^{-\tau}(\Theta + \psi) \ . \tag{43}$$

We now use the linearized Boltzmann equation (37) and the fact that $\tau' = -\Gamma$ to write the result as

$$\frac{d}{d\eta}\left[e^{-\tau}(\Theta + \psi)\right] = -e^{-\tau}\tau'(\Theta_0 + \psi + \hat{n} \cdot \vec{v}_{\rm b}) + e^{-\tau}(\phi' + \psi') \ . \tag{44}$$

Finally, using the definition of the visibility function g given in Sec. 2.1, we get

$$\frac{d}{d\eta}\left[e^{-\tau}(\Theta + \psi)\right] = g\left(\Theta_0 + \psi + \hat{n} \cdot \vec{v}_{\rm b}\right) + e^{-\tau}(\phi' + \psi') \ . \tag{45}$$

We can integrate this relation along the line of sight, i.e. along a straight line seen by the observer in a given direction $-\hat{n}$ (since the photons go in the direction $\hat{n}$), starting from an early time *before* recombination (such that $e^{-\tau(\eta_{\rm ini})} \simeq 0$) until the present time at which photons reach the observer (with by definition $e^{-\tau(\eta_0)} = 1$). The result reads

$$(\Theta + \psi)|_{\rm obs} = \int_{\eta_{\rm ini}}^{\eta_0} d\eta \left[g\left(\Theta_0 + \psi + \hat{n} \cdot \vec{v}_{\rm b}\right) + e^{-\tau}(\phi' + \psi')\right] \ , \tag{46}$$

where the notation $|_{\rm obs}$ means "evaluated at the observer location, along this line of sight", i.e. at the coordinate $(\eta_0, \vec{o}, \hat{n})$.

We can gain further intuition from this equation if we use the instantaneous decoupling approximation, in which all photons are assumed to decouple precisely at the time $\eta_{\rm dec}$. In this limit, we can replace the visibility function g by the Dirac function $\delta_D(\eta - \eta_{\rm dec})$, and $e^{-\tau}$ by the Heaviside function $H(\eta - \eta_{\rm dec})$. Note that the approximation $g(\eta) = \delta_D(\eta - \eta_{\rm dec})$ is correctly normalized, since the definition of g implies $\int d\eta\, g(\eta) = 1$. In the instantaneous decoupling limit, Eq. (46) reads:

$$(\Theta + \psi)|_{\rm obs} = (\Theta_0 + \psi + \hat{n} \cdot \vec{v}_{\rm b}))\,|_{\rm dec} + \int_{\eta_{\rm dec}}^{\eta_0} d\eta\, (\phi' + \psi')\ , \tag{47}$$

where the notation $|_{\rm dec}$ means "evaluated on the last scattering surface, along this line of sight", i.e. at the coordinate $(\eta_{\rm dec}, -r_{\rm dec}\hat{n}, \hat{n})$ (here $r_{\rm dec}$ is the comoving radius of the last scattering surface). Let us now give the interpretation of each term in this crucial equation.

First, $\Theta|_{\rm obs}$ is the temperature anisotropy measured by the observer in the direction $-\hat{n}$, while $\Theta_0|_{\rm dec}$ is the temperature anisotropy in the point of the last scattering surface seen in the same direction. If only these two terms where present, this relation would simply tell us that that the temperature anisotropy seen today in a given direction is equal to the intrinsic anisotropy in the point where the observed photons last scattered.

Second, the term $\hat{n} \cdot \vec{v}_{\rm b}|_{\rm dec}$ stands for the correction to this temperature coming from by the usual Doppler. Indeed, this correction is caused by the velocity of the baryon-photon fluid (that we assumed to be tightly coupled until $\eta_{\rm dec}$) projected along the line of sight.

Next, we expect a correction from gravitational effects. The redshifting and blueshifting of the photons traveling along gravitational potential fluctuations should affect the observed temperature anisotropy relatively to the intrinsic one. It turns out that if the gravitational potential was constant in time (but, of course, not in space), this effect would be conservative, and would only depend on $\psi|_{\rm obs} - \psi|_{\rm dec}$. This explains the second term on the left-hand and right-hand sides. But if ψ varies in time, the effect is not conservative anymore: intuitively, the amount of blueshifting and redshifting experienced by photons traveling across a potential well do not compensate each other if the gravitational well gets deeper between the time at which the photon enters and leaves the well. This explains the addition term $\int d\eta\, \psi$. A similar effect is caused by dilation effects along the line-of-sight, and contributes like $\int d\eta\, \phi$.

Finally, we can drop the second term on the left-hand side, because this term represents only a tiny *isotropic* correction to the observed anisotropies.

It is impossible to measure it with the CMB map only, because it is formally equivalent to a redefinition of the average temperature $\bar{T}$, but only by a tiny amount of the order of $10^{-5}\bar{T}$.

Let us write once more our result, dropping this unobservable correction, and grouping the terms in a particular way:

$$\Theta|_{\rm obs} = \underbrace{(\Theta_0 + \psi)|_{\rm dec}}_{\rm SW} + \underbrace{\hat{n}\cdot\vec{v}_{\rm b}|_{\rm dec}}_{\rm Doppler} + \underbrace{\int_{\eta_{\rm dec}}^{\eta_0} d\eta\,(\phi' + \psi')}_{\rm ISW} . \tag{48}$$

The first term is conventionally called the Sachs-Wolfe (SW) term, and includes the intrinsic temperature term Θ_0 and the "gravitational Doppler shift" term ψ at one point on the last scattering surface. The second term is the conventional Doppler term. The last term is called the Integrated Sachs-Wolfe (ISW) term and contains all non-conservative gravitational effects occurring in a universe with non-static metric fluctuations.

We can gain further insight on the Sachs-Wolfe term. We will try to find a simpler expression for $(\Theta_0 + \psi)|_{\rm dec}$, that applies at least for describing large angular patterns on CMB maps, i.e. maps smoothed over small scales. For instance, this expression would describe very well the map of the COBE satellite, which had limited angular resolution. In more precise terms, we wish to calculate the contribution to the Sachs-Wolfe term of large wavelengths, which are bigger than the Hubble radius at the time of recombination.

Let us first focus on the term $\Theta_0|_{\rm dec}$. We have seen that on super-Hubble scales, temperature anisotropies only have a monopole and a dipole component, related respectively to δ_γ and θ_γ. We can be more precise now. We know from thermodynamics that the local value of the photon density is proportional to the temperature to the power four. Taking the derivative of $\log\rho_\gamma = \log T^4$, we get $\delta_\gamma = 4\,\delta T/\bar{T}$, where on the right-hand side the temperature anisotropy is averaged over all directions $\hat{n}$: hence $\delta_\gamma = 4\Theta_0$.

Let us now focus on the term $\psi|_{\rm dec}$. We have seen in Sec. 1.4 that on super-Hubble scales and for adiabatic initial conditions, $\delta_\gamma = \frac{4}{3}\delta_{\rm b}$. Also, we know that decoupling takes place at the beginning of matter domination, and we mentioned at the very end of section 1.4 that for super-Hubble scales and during matter domination, one has $-2\phi = -2\psi = \delta_{\rm tot} = \delta_{\rm b} = \delta_{\rm cdm}$. Hence, $-2\psi = \frac{3}{4}\delta_\gamma$. Putting all these equalities together, we conclude that on the last scattering surface,

$$\Theta_0 + \psi = \frac{1}{4}\delta_\gamma + \psi = \left(-\frac{1}{4}2\frac{4}{3} + 1\right)\psi = \left(-\frac{2}{3} + 1\right)\psi = \frac{1}{3}\psi . \tag{49}$$

Moreover, on super-Hubble scales, we can neglect the Doppler term, which can be shown to be important only on sub-Hubble scales. We will see later that the integrated Sachs-Wolfe term plays a role on large scales, but only a small role, because ϕ and ψ are static during most of the evolution after decoupling. Hence we get an approximation for CMB anisotropies smoothed over small scales (that was first derived by Sachs and Wolfe in 1967):

$$\Theta|_{\text{obs, large scales}} \simeq \frac{1}{3}\psi|_{\text{dec}} = -\frac{1}{8}\delta_\gamma|_{\text{dec}} \,. \tag{50}$$

In this calculation, we have seen that the term ψ wins over the term Θ_0, leading to a minus sign in front of δ_γ in the above relation. This means that an over-density on the last scattering surface ($\delta_\gamma > 0$), corresponding to a potential well ($\psi < 0$), leads to a cold spot in the observed map ($\Theta < 0$). Conversely, a hot spot corresponds to an under-density. Hence, due to the "gravitational Doppler shift" effect accounted by the term ψ (often called the Sachs-Wolfe effect), the patterns that we observe on CMB map are inverted with respect to intrinsic fluctuations on the last scattering surface.

Equation (48) (and its large-scale approximation (50)) are important for pedagogical purposes, but they have no practical application: indeed, what we wish to calculate and to compare to observations is a theoretical prediction for the statistical properties of CMB anisotropies. Hence, we need to compute at least the CMB two-point correlation function.

2.4. *Spectrum of temperature anisotropies*

Definition. The map of CMB temperature anisotropies can be expanded in spherical harmonics:

$$\frac{\delta T}{\bar{T}}(\hat{n}) = \Theta(\eta_0, \vec{o}, -\hat{n}) = \sum_{lm} a_{lm} Y_{lm}(\hat{n}) \,. \tag{51}$$

Using the Legendre expansion of Θ introduced in Eq. (36), and some basic relations between Legendre polynomials and spherical harmonics, it is easy to express a_{lm} as a function of Θ_l:

$$a_{lm} = (-i)^l \int \frac{d^3\vec{k}}{2\pi^2} Y_{lm}(\hat{k}) \Theta_l(\eta_0, \vec{k}) \,, \tag{52}$$

where we recall that hats denote unit vectors: $\hat{k} \equiv \vec{k}/k$. In linear perturbation theory and assuming Gaussian initial conditions, both Θ_l and a_{lm} are Gaussian random variables. Using the orthogonality relation of spherical harmonics and the definitions given in Sec. 1.5, we can infer the two-point correlation function of the a_{lm}'s as a function of the power spectrum of Θ_l, or even better, of the primordial curvature power spectrum:

$$\langle a_{lm} a^*_{l'm'} \rangle = \delta^K_{ll'} \delta^K_{mm'} \left[\frac{1}{2\pi^2} \int \frac{dk}{k} \Theta^2_l(\eta_0, k) \mathcal{P}_\mathcal{R}(k) \right] . \tag{53}$$

Here $\delta^K_{ll'}$ represents the Kronecker symbol. The fact that $\langle a_{lm} a^*_{l'm'} \rangle$ vanishes for $l \neq l'$ or $m \neq m'$ comes out of the algebra, but physically, it is a consequence of the homogeneity of the universe — just like the fact that all power spectra in Fourier space are proportional to $\delta_D(\vec{k}' - \vec{k})$. Similarily, the fact that the quantity between brackets is a function of l but not of m is a consequence of isotropy — like the fact that in Fourier space, power spectra are functions of k but not $\vec{k}$.

The quantity between brackets is usually denoted C_l, and is called the power spectrum of temperature anisotropies in harmonic space, or the temperature harmonic power spectrum:

$$C_l \equiv \frac{1}{2\pi^2} \int \frac{dk}{k} \Theta^2_l(\eta_0, k) \mathcal{P}_\mathcal{R}(k) . \tag{54}$$

In a universe with linear and Gaussian perturbations, the C_l's encode all the information concerning the cosmological model describing our universe that is contained in the CMB temperature map.

It is worth coming back on the meaning of the averaging symbols $\langle ... \rangle$ in Eq. (53). Since the theory of cosmological perturbation is stochastic, the a_{lm}'s should be seen as random numbers, and the average is meant over many realizations of the theory. In a sense, "a given realization" means "a given universe", and the average holds over many universes, all obeying to the same cosmological model, which is encoded in the spectrum C_l.

However, we observe CMB anisotropies in our universe, i.e. in only one realization. CMB maps allow us to measure a definite value of each a_{lm}. Hence the squares $|a_{lm}|^2$ are not expected to be equal to C_l, even if we postulated the right model: there should be some scattering around C_l. This scattering limits the possibility to find the best theory matching the observations. However we can reduce considerably the scattering by noticing that for any fixed l, the statistical distribution of $|a_{lm}|^2$ is independent of m (as a consequence of isotropy, and as expressed by Eq. (54)). Hence, for an ideal full-sky CMB experiment, the best estimator of the underlying C_l's is the average between all observed coefficients $|a_{lm}|^2$ with fixed l,

$$C_l^{obs} \equiv \frac{1}{2l+1} \sum_{-l \leq m \leq l} |a^{obs}_{lm}|^2 . \tag{55}$$

In a typical universe, this quantity should be closer to the underlying C_l than a single $|a_{lm}^{obs}|^2$. The way to see this mathematically is to consider again the theoretical (stochastic) a_{lm}'s, and to define

$$\hat{C}_l \equiv \frac{1}{2l+1} \sum_{-l \leq m \leq l} |a_{lm}|^2 \; . \tag{56}$$

By performing averages in the same sense as in equation (53), one can easily show that for Gaussian a_{lm}'s,

$$\langle \hat{C}_l \rangle = C_l \qquad \text{and} \qquad \langle (\hat{C}_l - C_l)^2 \rangle = \frac{2}{2l+1} C_l^2 \; . \tag{57}$$

The first equality shows that the observed C_l's defined in Eq. (55) are unbiased estimators of the true underlying C_l's. The second equality gives the typical scattering between the theory and the observations. Since for larger l we perform an average over more values of m, the relative scattering decreases. This is the case with true data points, which are distributed qualitatively like in Fig. 1. This scattering is called cosmic variance. It can be seen as a theoretical error: because of cosmic variance, we cannot reconstruct the underlying model with infinite precision, even if we have infinitely precise observations. Cosmic variance is large for small l's, meaning that the shape of the true underlying C_l's will always be poorly known at low l.

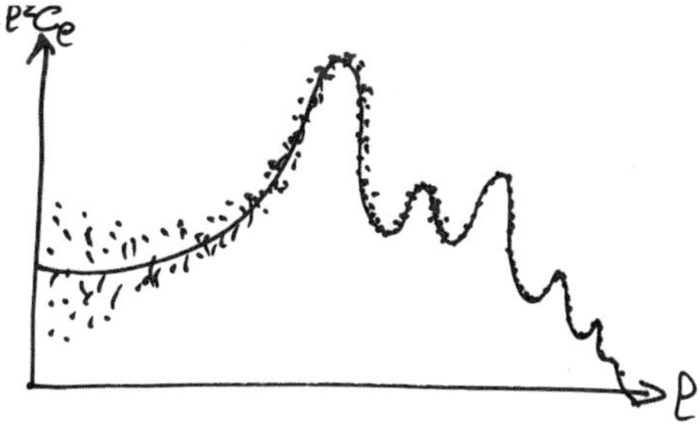

Fig. 1. Observed values of the temperature harmonic spectrum, C_l^{obs}, are expected to be scattered around the true underlying C_l's. The scattering, called cosmic variance, decreases with increasing l.

Line-of-sight integral in Fourier space. According to equation (54), for a given primordial spectrum, the shape of the CMB spectrum C_l depends on the square of the transfer function $\Theta_l(\eta_0, k)$. We would like to understand this shape at least qualitatively. In real space, we did learn a lot on the behavior of $\Theta(\eta, \vec{x}, \hat{n})$ by using the line-of-sight integral approach presented in section 2.3. A similar approach can be worked out in Fourier and harmonic space, i.e. for the variable $\Theta_l(\eta, k)$. We do not present here the intermediate steps, which can be found in Ref. 2. The final result shows many similarity with its real space counterpart, Eq. (46). It can be decomposed into:

$$\begin{aligned} \Theta_l(\eta_0, k) &= \int_{\eta_{\rm ini}}^{\eta_0} d\eta \; S_T(\eta, k) \; j_l(k(\eta_0 - \eta)) \;, \\ S_T(\eta, k) &\equiv \underbrace{g\,(\Theta_0 + \psi)}_{\rm SW} + \underbrace{\left(g\,k^{-2}\theta_{\rm b}\right)'}_{\rm Doppler} + \underbrace{e^{-\tau}\,(\phi' + \psi')}_{\rm ISW} \;. \end{aligned} \tag{58}$$

We see that $\Theta_l(\eta_0, k)$ is given by the convolution of spherical Bessel functions $j_l(x)$ with a function $S_T(\eta, k)$, called the temperature source function, which contains the usual three terms: Sachs-Wolfe, Doppler, and Integrated Sachs-Wolfe. Like in the previous section, we can use the instantaneous decoupling approximation, integrate the Doppler term by part, and write $\Theta_l(\eta_0, k)$ as:

$$\begin{aligned} \Theta_l(\eta_0, k) \simeq & \left[\Theta_0(\eta_{\rm dec}, k) + \psi(\eta_{\rm dec}, k)\right] j_l(k(\eta_0 - \eta_{\rm dec})) \\ & + k^{-1}\theta_{\rm b}(\eta_{\rm dec}, k) \; j_l'(k(\eta_0 - \eta_{\rm dec})) \\ & + \int_{\eta_{\rm dec}}^{\eta_0} d\eta \; \left[\phi'(\eta, k) + \psi'(\eta, k)\right] j_l(k(\eta_0 - \eta)) \end{aligned} \tag{59}$$

(note that in the second line, i.e. in the Doppler term, the prime stands for the derivative of the function $j_l(x)$ with respect to its argument, not with respect to conformal time). This approximate result can be plugged into Eq. (54) to obtain the final spectrum C_l. We see that each C_l can be decomposed into six terms: the power spectrum $C_l^{\rm SW}$ of the SW term, coming from the first line of Eq. (59) squared, that of the Doppler term, coming from the second line of Eq. (59) squared, that of the Integrated Sachs-Wolfe term, coming from the third line of Eq. (59) squared, and finally the three cross-spectra involving each pair of terms.

For large values of l, the spherical Bessel functions $j_l(x)$ and $j_l'(x)$ are very peaked near $x \simeq l$. Hence, for the Sachs-Wolfe and Doppler contributions to the spectrum C_l, the integral over k in Eq. (54) will pick up mainly modes with $k(\eta_0 - \eta_{\rm dec}) \simeq l$. This shows that the SW contribution to C_l is given by the product of the primordial spectrum with the squared transfer function $(\Theta_0 + \psi)$ at a given value of η and k:

$$C_l^{\rm SW} \sim \left[\Theta_0(\eta_{\rm dec}, k) + \psi(\eta_{\rm dec}, k)\right]^2 \mathcal{P}_{\mathcal{R}}(k) \ , \qquad k \simeq \frac{l}{(\eta_0 - \eta_{\rm dec})} \tag{60}$$

(for simplicity, we did not write numerical factors and powers of l or k in front of this expression). In other words, $C_l^{\rm SW}$ depends on the power spectrum of the perturbation $(\Theta_0 + \psi)$, evaluated at the time of decoupling, and for wavenumbers in the vicinity of $k = \frac{l}{(\eta_0 - \eta_{\rm dec})}$,

$$C_l^{\rm SW} \sim \langle |\Theta_0 + \psi|^2 \rangle_{(\eta,k) \simeq (\eta_{\rm dec},\, l/(\eta_0 - \eta_{\rm dec}))} \ . \tag{61}$$

We reached this result with mathematical arguments, but it has a very simple geometrical interpretation, illustrated in Fig. 2. The spectrum C_l encodes the correlation between structures on CMB maps seen under an angle $\theta = \pi/l$. This angle subtends a given physical scale on the last scattering surface, namely $\theta \times d_{\rm a}(z_{\rm dec})$, where $d_{\rm a}(z)$ is the angular diameter distance to objects of redshift z. Since the Sachs-Wolfe term $(\Theta_0 + \psi)$ contributes to the temperature map only at the time $\eta \simeq \eta_{\rm dec}$, the spectrum $C_l^{\rm SW}$ should depend on the power spectrum of the Sachs-Wolfe term $\langle |\Theta_0 + \psi|^2 \rangle$ at that time, and for a wavenumber such that

$$\frac{\lambda}{2} = \frac{\pi a(\eta_{\rm dec})}{k} = \theta \, d_{\rm a}(z_{\rm dec}) \ . \tag{62}$$

The reason for which λ has been divided by two is that for spherical harmonics, $\theta = \pi/l$ is the angle between a maximum and a minimum, while for a Fourier mode the distance between a maximum and a minimum is one half of the wavelength, $\frac{\lambda}{2} = \frac{\pi a}{k}$. Eq. (62) leads to

$$\frac{a(\eta_{\rm dec})}{k} = \frac{d_{\rm a}(z_{\rm dec})}{l} \ . \tag{63}$$

We recall that in a flat FL universe the angular diameter distance is given by

$$d_{\rm a}(z) = a(t(z)) \int_{t(z)}^{t_0} \frac{dt}{a} \ , \tag{64}$$

$t(z)$ being the proper time at which an object seen today with a redshift z emitted light. The conformal time $\eta(z)$ is defined similarly. In terms of conformal time,

$$d_{\rm a}(z) = a(\eta(z)) \int_{\eta(z)}^{\eta_0} d\eta = a(\eta(z)) \left[\eta_0 - \eta(z)\right] \ , \tag{65}$$

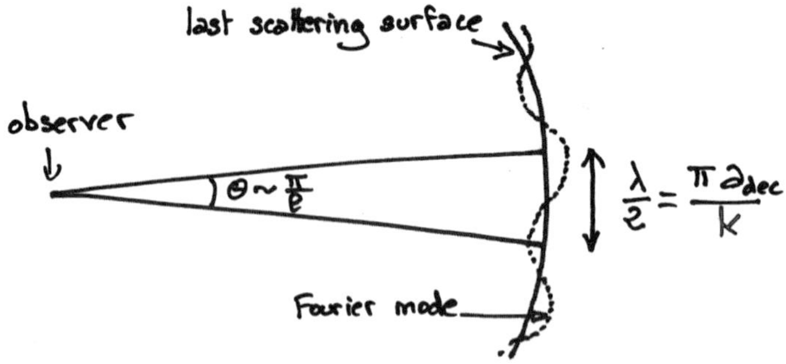

Fig. 2. A multipole l refers to structures seen on the last scattering surface under an angle $\theta = \pi/l$. These structures are seeded by Fourier modes with a half wavelength $\frac{\lambda}{2} = \frac{\pi a_{\rm dec}}{k}$.

and for the case of a point located on the last scattering surface,

$$d_{\rm a}(z_{\rm dec}) = a(\eta_{\rm dec})\ (\eta_0 - \eta_{\rm dec})\ . \tag{66}$$

Hence, Eq. (63) can be written as

$$\frac{1}{k} = \frac{(\eta_0 - \eta_{\rm dec})}{l}\ , \tag{67}$$

which is the same relation between k and l as in Eqs. (60, 61). In those equations, we implicitly performed a small-angle approximation. For large angles (small l's), it is inaccurate to say that a given angle/mutipole corresponds to a single Fourier mode on the last-scattering surface, and it is important to keep the spherical Bessel function of Eq. (58) and the integral over k of Eq. (54). In summary, Eqs. (60, 61) represent the instantaneous decoupling *and* small-angle limit of the true power spectrum $C_l^{\rm SW}$.

Using the same two limits, a similar discussion can be carried for the Doppler and Integrated Sachs-Wolfe power spectra. The Doppler term depends on the power spectrum of the baryon velocity divergence evaluated roughly at the same time and scale,

$$C_l^{\rm Doppler} \sim \langle|\theta_{\rm b}|^2\rangle_{(\eta,k)\simeq(\eta_{\rm dec},\ l/(\eta_0-\eta_{\rm dec}))}\ , \tag{68}$$

while the ISW term can be written approximately in terms of the integral

$$C_l^{\rm ISW} \sim \int_{\eta_{\rm dec}}^{\eta_0} d\eta\,(\eta_0 - \eta)\ \langle|\phi' + \psi'|^2\rangle_{(\eta,k)\simeq(\eta,\ l/(\eta_0-\eta))}\ . \tag{69}$$

Again, for simplicity, we did not write numerical factors and powers of l and k in front of these expressions.

In the next sections, we will infer the shape of the full spectrum C_l from that of the three power spectra appearing in Eqs. (61, 68, 69):

$$\begin{array}{llll} \text{SW} & : & \langle|\Theta_0+\psi|^2\rangle & \text{at } (\eta,k)\simeq(\eta_{\rm dec},\ l/(\eta_0-\eta_{\rm dec}))\ , \\ \text{Doppler} & : & \langle|\theta_{\rm b}|^2\rangle & \text{at } (\eta,k)\simeq(\eta_{\rm dec},\ l/(\eta_0-\eta_{\rm dec}))\ , \\ \text{ISW} & : & \langle|\phi'+\psi'|^2\rangle & \text{for all } (\eta,k)\simeq(\eta,\ l/(\eta_0-\eta))\ . \end{array}$$

Before entering into details, we can make a guess. Even if the primordial spectrum $\mathcal{P}_\mathcal{R}(k)$ is smooth, the temperature spectrum C_l should contain some structure. Indeed, as illustrated by Fig. 3, any primordial over-density

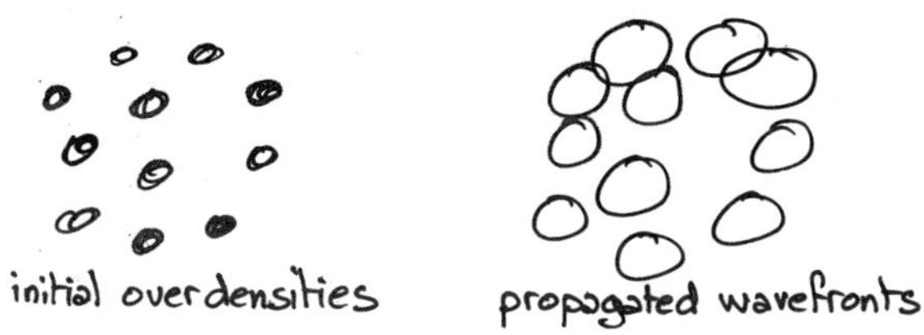

Fig. 3. Initial over-densities in the early universe propagate in the form of wavefronts. The distance travelled by any wavefront at a given time is given by the sound horizon in the photon-baryon fluid. Here, we represent initial over-densities as spherical patterns at a given scale, while in the real universe primordial over-densities result from a superposition of structures on all scales.

is expected to propagate. In Fig. 3, we represented naively the primordial over-densities with little dots, giving rise to spherical wavefronts at later time. In the real universe, wavefront patterns are less visible, because primordial over-densities result from a superposition of structures on all scales. However, there is always a characteristic scale in this problem: namely, the distance by which a wavefront travels between some time in the primordial universe and the time of photon decoupling. This distance, called the sound horizon at decoupling $d_{\rm s}(\eta_{\rm dec})$, obeys to

$$d_{\rm s} \equiv a \int_{t_{\rm ini}}^{t} \frac{c_{\rm s} dt}{a} = a \int_{\eta_{\rm ini}}^{\eta} c_{\rm s} d\eta\ , \tag{70}$$

where $c_{\rm s}$ is the sound speed in the photon-baryon fluid (in units of the speed of light). Two points on the last scattering surface separated by this distance should be partially correlated, since density waves have propagated from one point to the other. Hence, in angular space, the two-point correlation function of CMB anisotropies should exhibit a characteristic feature for angular scales corresponding to the sound horizon at decoupling,

$\theta \sim d_\mathrm{s}(\eta_\mathrm{dec})/d_\mathrm{a}(z_\mathrm{dec})$. Similarly, the harmonic power spectrum C_l should exhibit a feature at the corresponding scale, $l \sim \pi/\theta \sim \pi\ d_\mathrm{a}(z_\mathrm{dec})/d_\mathrm{s}(\eta_\mathrm{dec})$, and also for all the harmonics of this scale. We will get a confirmation of this in the next section.

2.5. *Acoustic oscillations*

As long as electrons, baryons and photons are tightly coupled, they form an effective single fluid in which density waves propagate at the sound speed

$$c_\mathrm{s}^2 = \frac{\delta p_\gamma + \delta p_\mathrm{b}}{\delta\rho_\gamma + \delta\rho_\mathrm{b}} \tag{71}$$

(the density and pressure of electrons is always negligible with respect to that of photons). The density fluctuation δ_x of each species $x = \gamma, b$ can be inferred from the local value of the equilibrium temperature. The fact that $\rho_\mathrm{b} \propto T^3$ and $\rho_\gamma \propto T^4$ implies $\delta_\gamma = \frac{4}{3}\delta_\mathrm{b}$, and tight coupling imposes $\theta_\gamma = \theta_\mathrm{b}$, as we already saw in Sec. 2.2. We can simplify the expression of the sound speed, using also the fact that $|\delta p_\mathrm{b}| \ll |\delta p_\gamma|$. The result reads

$$c_\mathrm{s}^2 = \frac{1}{3(1+R)}\ , \qquad R \equiv \frac{4\bar{\rho}_\mathrm{b}}{3\bar{\rho}_\gamma} \propto a\ . \tag{72}$$

It is possible to derive a simple equation of motion for the photon temperature fluctuation $\Theta_0(\eta, \vec{x})$ in the tightly-coupled regime:

$$\Theta_0'' + \frac{R'}{1+R}\Theta_0' + k^2 c_\mathrm{s}^2 \Theta_0 = -\frac{k^2}{3}\psi + \frac{R'}{1+R}\phi' + \phi''\ . \tag{73}$$

This equation follows from the combination of the continuity and Euler equations for photons and baryons. Given that R is proportional to the scale factor, we could replace R' by $(a'/a)R = aHR$. The second term on the left-hand side is a damping term, increasing with the contribution of baryons to the total energy of the fluid. The third term accounts for pressure forces in the effective fluid. The first term on the right-hand side accounts for the gravitational force, and the last two terms for dilation effects.

This equation would be that of a simple harmonic oscillator if R was a constant (no friction term, constant sound speed) and in absence of gravitational source terms. Then, the solution would be of the form

$$\Theta_0 = \Theta_\mathrm{ini}\ \cos(kc_\mathrm{s}\eta + \varphi)\ , \tag{74}$$

with two constants of integration $(\Theta_\mathrm{ini}, \varphi)$. We know that for adiabatic initial conditions and in the Newtonian gauge, photon density/temperature fluctuations should be constant in the super-Hubble limit, $k\eta \ll 1$: this fixes the phase to $\varphi = 0$. In the opposite limit, this solution corresponds to the propagation of acoustic oscillations. Actually, the limit between the constant and oscillatory regime is not set by the value of $k\eta$, but by that of $kc_s\eta$. In fact, the condition $kc_s\eta \ll 1$ is equivalent to $\lambda \gg d_s$, where λ is a physical wavelength ($\lambda = 2\pi a/k$), and d_s is the physical sound horizon, given in the case of a constant sound speed by:

$$d_s = a \int_{\eta_{\rm ini}}^{\eta} c_s d\eta \simeq a c_s \eta \tag{75}$$

(assuming $\eta \gg \eta_{\rm ini}$). Hence, the phase $kc_s\eta$ of the cosine stands for the ratio $2\pi d_s/\lambda$. Modes start oscillating when their wavelength becomes smaller than the sound horizon, and later on, the number of oscillations is given by the ratio between these two scales.

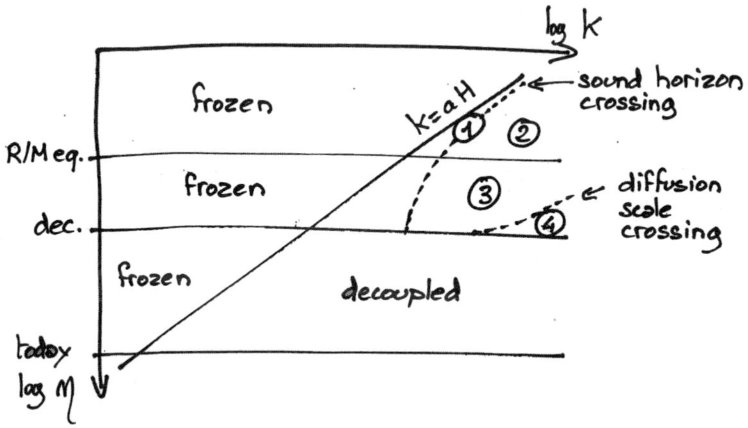

Fig. 4. Different regions in (k, η) space, corresponding to qualitatively different behaviors for photon (and baryon) perturbations.

In reality, R grows with time (and crosses one roughly around the time of decoupling). In addition, the gravitational source terms in Eq. (73) can play a role in some regimes. Let us describe qualitatively the evolution of Θ_0 in the different regions in (k, η) space shown in Fig. 4. In this figure, the horizontal axis corresponds to wavenumbers k (large wavelengths are on the left), and the vertical axis to conformal time, flowing from top to bottom. The super-Hubble and sub-Hubble regions are separated by the solid diagonal line corresponding to $k = aH$ (equivalent to $k\eta = 1$ during radiation domination). From top to bottom, the horizontal lines correspond to the time of equality between radiation and matter, to the time of photon decoupling, and to the time today. The upper dashed line separates wavelengths bigger/smaller than the sound horizon in the baryon-photon fluid before decoupling (after decoupling, this notion does not make sense anymore). As we have just seen, this limit corresponds to $\lambda = d_s$, or (up to a factor 2π) to $k = (a/d_s)$. At early times, $c_s = 1/\sqrt{3}$, and this condition reads $k\eta = \sqrt{3}$. Just before decoupling, R becomes large, c_s goes to zero, and the comoving sound horizon (d_s/a) becomes asymptotically constant, explaining the shape of the upper dashed line. Finally, the lower dashed line separates wavelengths bigger/smaller than the diffusion length defined in section 2.1: this line corresponds to $k = 1/r_d$, where r_d is given in first approximation by Eq. (29).

The evolution of Θ_0 in the super-Hubble region is trivial: as long as $k\eta \ll 1$ — and *a fortiori* $kc_s\eta \ll 1$ — the fluctuation Θ_0 is frozen, and remains approximately equal to its initial value.

The region marked with a ① in the figure corresponds to modes that are crossing the sound horizon before decoupling. This is precisely the region in which gravitational source terms are important. They shift the zero point of oscillations, and boost their amplitude (due to gravitational forces and dilution effects). This happens during a limited amount of time, because the metric fluctuations quickly decay inside the sound horizon during radiation domination, making the gravitational source terms negligible. An approximation for the zero-point of oscillations can be found by setting Θ_0'' and Θ_0' to zero in equation (73), and by keeping only the first gravitational term:

$$\Theta_0^{equilibrium} = -\frac{1}{3c_s^2}\psi = -(1+R)\psi \; . \tag{76}$$

Since the gravitational potential is non-zero on super-Hubble scales (we have seen that for adiabatic initial conditions $-2\psi = \delta_{tot}$), the equilibrium point is shifted away from zero on those scales. It reaches asymptotically zero on sub-sound-horizon scales.

Region ② corresponds to wavelengths smaller than the sound horizon during radiation domination. In this regime, the metric fluctuations have decayed, so the source term in Eq. (73) can be neglected. The friction term can also be neglected, because during radiation domination, $R \ll 1$. Finally the effective mass $k^2c_s^2$ is constant in time because $R \ll 1$ implies

$c_s^2 = 1/3$. Hence we are in the simple case discussed before, and the solution is proportional to $\cos(kc_s\eta)$, corresponding to stationary oscillations, symmetric around $\Theta_0 = 0$.

Region ③ refers to wavelengths smaller than the sound horizon during the intermediate stage between the time of equality and that of photon decoupling. In this region, the metric perturbations have decayed, but R cannot be neglected (baryons and photons contribute to the total energy density with the same order of magnitude). Hence the oscillator equation has a non-negligible friction term (increasing with time), and a time-varying effective mass (decreasing with time). The solution of the equation corresponds to damped oscillations. Physically, this damping is caused by the increasing inertia and decreasing pressure of the baryon-photon fluid when the energy density of non-relativistic baryons takes over.

Region ④ refers to modes with smaller wavelength than the diffusion scale in the photon-baryon fluid. We have defined this scale in section 2.1. At early times, in the tightly-coupled limit, the mean free path of particles in the fluid is negligible, and cosmologically interesting scales are all well above the diffusion length. At the approach of decoupling, the diffusion length suddenly increases, and encompasses most of sub-sound-horizon wavelengths. In this regime, the oscillator equaion (73) does not apply anymore, because we cannot describe baryons and photons in terms of a perfect fluid. Perturbations are then strongly damped, since diffusion tends to average out any small-scale fluctuation.

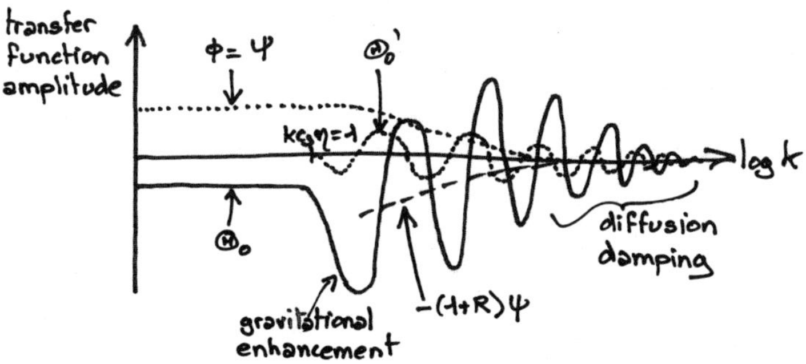

Fig. 5. Transfer functions at the time of decoupling.

After describing these different regions, we are ready for understanding the qualitative behavior of the various relevant transfer functions, evaluated at the time of decoupling. Figure 5 shows the transfer functions $\Theta_0(\eta_{\rm dec}, k)$ (solid line), $\psi(\eta_{\rm dec}, k)$ (upper dotted line) and $\Theta_0'(\eta_{\rm dec}, k)$ (middle dotted line) as a function of $\log k$. For simplicity, we neglect the role of the anisotropic stress generated by neutrinos (and to a lesser extent by photons near decoupling time). Hence we can assume $\phi = \psi$ at all times. For scales above the sound horizon ($kc_{\rm s}\eta \ll 1$), the transfer functions are constant: indeed we know that, on those scales and in the Newtonian gauge, density and metric fluctuations are frozen. Adiabatic initial conditions impose $-2\psi = \delta_{\rm tot} \simeq \delta_{\rm b} \simeq \frac{3}{4}\delta_\gamma \simeq 3\Theta_0$. The opposite sign of Θ_0 and ψ reflects the fact that an over-density corresponds to a temperature excess and a gravitational potential well (and vice-versa). We remember that transfer functions are all normalized to $\mathcal{R}(\vec{k}) = 1$ (see section 1.5): this corresponds to a negative δ_γ (and Θ_0) and to a positive ψ, like in the figure.

Because of the decay of metric fluctuations inside the Hubble radius, the ψ curve smoothly decreases and tends towards zero in the small wavelength limit. The behavior of Θ_0 is more complicated. Modes which are just crossing the sound horizon near $\eta = \eta_{\rm dec}$ are experiencing the boost caused by gravitational source terms in Eq. (73): this explains the first bump in the solid line. For smaller wavelengths, we see oscillatory patterns corresponding to acoustic oscillations. Smaller wavelengths crossed the sound horizon earlier, and had more time for oscillating before decoupling. The maxima observed at the time of decoupling correspond to modes that could experience 0.5, 1, 1.5, 2, ..., periods of oscillations before that time. The zero point of oscillations follows $-(1 + R)\psi$, represented on the figure with a dashed line: this zero point reaches zero well inside the sound horizon. The amplitude of the oscillations is maximal for the first oscillatory pattern, i.e. for modes that crossed the sound horizon very recently. The second oscillatory pattern is reduced by the fact that those modes stayed for a longer time inside the sound horizon during the matter dominated regime, and experienced more damping due to baryons. The third and higher oscillatory patterns are reduced even more by diffusion damping just before photon decoupling. Temperature fluctuations on very small wavelengths are completely suppressed by photon diffusion.

Figure 5 also shows qualitatively the behavior of the time derivative $\Theta_0'(\eta_{\rm dec}, k)$, which exhibits oscillations that are out of phase with respect to those of $\Theta_0(\eta_{\rm dec}, k)$. This will be important in a few paragraphs, when discussing the Doppler effect.

Now that we understand qualitatively the behavior of the metric and photon transfer functions, we can go back to the decomposition of the

CMB temperature spectrum C_l in three terms (Sachs-Wolfe, Doppler and integrated Sachs-Wolfe) discussed in section 2.5.

Sachs-Wolfe contribution. We have seen that the Sachs-Wolfe contribution to C_l is approximately given by the power spectrum of the combination $(\Theta_0 + \psi)$ at $\eta = \eta_{\rm dec}$, with a correspondence between k and l given by Eq. (67). We know that this power spectrum is given by the product of the primordial spectrum $\mathcal{P}_\mathcal{R}(k)$ (that we can choose to be scale-invariant in first approximation) by the square of the transfer function $(\Theta_0 + \psi)^2$. The qualitative behavior of the latter has no more secrets for us. We can pick up the solid and upper dotted lines in Fig. 5, add them up, and square the result. The result is shown in Fig. 6 as a function of $\log k$. For modes

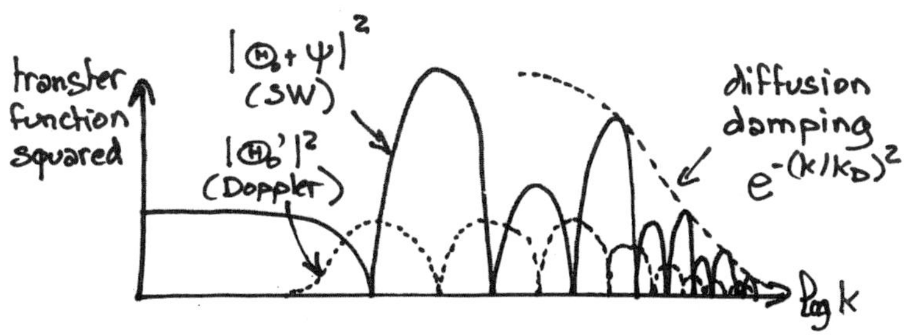

Fig. 6. Squared transfer functions at the time of decoupling.

with $kc_s^2\eta \ll 1$, we see a flat plateau: there, our previous calculation of the Sachs-Wolfe effect would apply (this part of the curve is actually called the Sachs-Wolfe plateau). We then observe a series of peaks. Due to the shift of the zero-point of oscillations given by $-(1+R)\psi$ for Θ_0, and hence by $-R\psi$ for the Sachs-Wolfe term $(\Theta_0 + \psi)$, there is an asymmetry between the first few odd and even peaks, with odd peaks being enhanced. Moreover the overall amplitude of the peaks is suppressed in the large k limit by diffusion damping. A bit of algebra would show us that the envelope of the peaks is given in first approximation by the function $\exp[-(k/k_{\rm d})^2]$, where $k_{\rm d}$ is the diffusion wavenumber, related to the diffusion comoving scale of Eq. (29) by $k_{\rm d} r_{\rm d} = 1$.

Doppler contribution. Next, we know that the C_l's receive a second contribution from the Doppler effect, related to the power spectrum of $\theta_{\rm b}$, which is equal to θ_γ until baryon and photons decouple from each other. It turns out that the photon velocity divergence θ_γ is itself related to the time derivative of the temperature fluctuation Θ_0. At a very qualitative level, we can infer the Doppler contribution from the shape of the transfer function $\Theta_0'(\eta_{\rm dec}, k)$. This contribution is null for scales above the sound horizon, since in this regime there are no oscillations and no significant dynamics in the fluid. On smaller scales, the Doppler contribution has oscillatory patterns, that are out of phase with respect to those of the Sachs-Wolfe term. The Doppler contribution is represented schematically as a dotted line in Fig. 6.

We can now sum up the Sachs-Wolfe and Doppler contributions. Also, we can transpose our results for power spectra as a function of k in terms of C_l's as a function of l. We have already seen in section 2.5 that there is a mapping betwwen the two, at least in the small-angle and instantaneous decoupling approximation, with a correspondence between values of k and l given by Eq. (67). The result is shows in Fig. 7. Note that the vertical

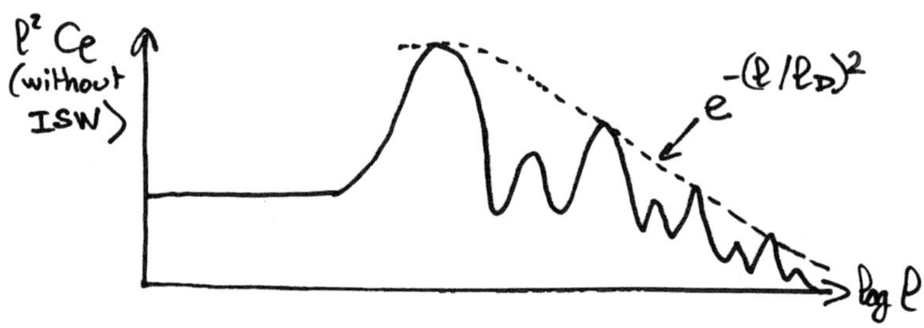

Fig. 7. Contribution to the C_l's from the SW and Doppler terms.

axis stands for $l^2 C_l$, for the following reason. If the primordial spectrum was scale invariant ($\mathcal{P}_\mathcal{R}(k) = $ constant) and the transfer functions were flat (as it is the case for large wavelengths/small l's), the quantity $l(l+1)C_l$ would also be flat and independent of l. This would follow from Eq. (60) if we had been more carefull in keeping all the factors. It is convenient to plot $l^2 C_l$ or $l(l+1)C_l$ instead of C_l, in order to display a roughly constant curve, just modulated by acoustic oscillations and diffusion damping.

Diffusion damping effect. In Fig. 7, we can identify all the features mentioned before: the flat Sachs-Wolfe plateau, the series of oscillations with enhanced odd peaks, and the exponentially decaying envelope of the peaks for large l. The envelope is now given by $\exp[-(l/l_{\rm d})^2]$, where $l_{\rm d}$ is the diffusion multipole, related to the diffusion angle by $l_{\rm d} = \pi/\theta_{\rm d}$. The diffusion angle is related to the diffusion scale $r_{\rm d}$ by the usual angular diameter distance relation,

$$\theta_{\rm d}\, d_{\rm a}(\eta_{\rm dec}) = \lambda_{\rm d} = a(\eta_{\rm dec}) r_{\rm d}(\eta_{\rm dec}). \tag{77}$$

At this point, we are almost done with the qualitative description of the CMB temperature spectrum C_l. We only missed the integrated Sachs-Wolfe contribution, and the effect of reionization.

Integrated Sachs-Wolfe contribution. We have seen that the integrated Sachs-Wolfe effect is given by an integral over $(\psi' + \phi')$ between photon decoupling and today. In fact, $(\psi' + \phi')$ remains vanishingly small in a large part of the space (k, η). In Fig. 8, we hatched all the regions in (k, η) space where metric fluctuations are expected to vary with time. Let us discuss these different regions.

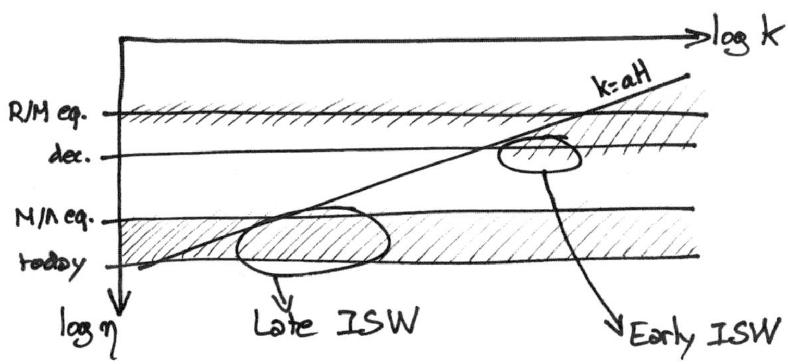

Fig. 8. Regions in (k, η) space where metric fluctuations are expected to vary with time (giving rise potentially to an ISW effect after decoupling).

Before photon decoupling, we know that metric fluctuations decay inside the sound horizon. Instead, in the Newtonian gauge, they remain frozen outside the Hubble radius, excepted near times at which the equation of state of the universe changes: namely, at the time of equality between radiation and matter.

Let us now discuss the variation of metric perturbations after photon decoupling (this is the relevant epoch for the ISW effect). Deep inside the matter dominated regime, one can show that metric fluctuations are static, even inside the Hubble radius (at least within linear perturbation theory). We will justify this result in section 3.2. Hence in Fig. 8 there are no hatches during matter domination on whatever scales. Note however that at the beginning of matter domination, it takes some time for sub-sound-horizon metric fluctuations to freeze around a constant value: hence the hatches continue below the line corresponding to the time of equality, and extend till the line corresponding to photon decoupling.

Later on, during the Λ (or dark energy) dominated regime, the equation of state of the universe changes again, so metric fluctuations vary on all scales, like at the time of equality. A simple calculation based on Einstein equations would show that metric fluctuations are damped during this stage.

In summary, contribution to the integrated Sachs-Wolfe effect can only come from two regions:

- just after photon decoupling, on sub-sound-horizon scales, and
- during Λ domination, on all scales.

These two distinct contributions are usually called the Early Integrated Sachs-Wolfe (EISW) and Late Integrated Sachs-Wolfe (LISW) effects. One can show that both effects decrease with wavelength, for geometrical reasons. Hence the EISW effect is maximal for scales crossing the sound horizon just at the time of photon decoupling, while the LISW effect is maximal for the largest observable scales today. In multipole space, this means that the EISW effect contributes mainly to the scale of the first peak, i.e. to $l \sim 200$, while the LISW contributes mainly to the smallest multipoles $l = 2, 3, 4$, etc. The two ISW contributions are drawn on Fig. 9. The EISW effect enhances the first peak, while the LISW effect tilts the Sachs-Wolfe plateau even if the primordial spectrum is exactly scale invariant.

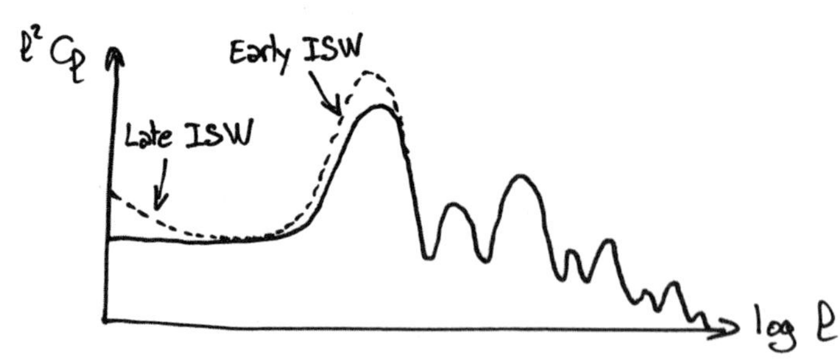

Fig. 9. ISW contribution to the temperature spectrum.

Reionization effect. The last effect that we omitted to describe is that of reionization. We have seen in section 2.1 that at small redshift ($z \simeq 10$), the reionization of the universe produces a small secondary bump in the visibility function, corresponding physically to a small probability for CMB photons to rescatter at late times. This rescattering will tend to smooth out any temperature anisotropy pattern. Hence, reionization lowers the overall amplitude of the C_l', but only a small amount (by approximately 15%). Note that the suppression of power is not uniform over the whole multipole range: smoothing effects cannot reach the largest observable scales (corresponding to the smallest values of l). Hence the effect of reionization is step-like shaped, and saturates for l of the order of 40 or so (as illustrated on Fig. 10).

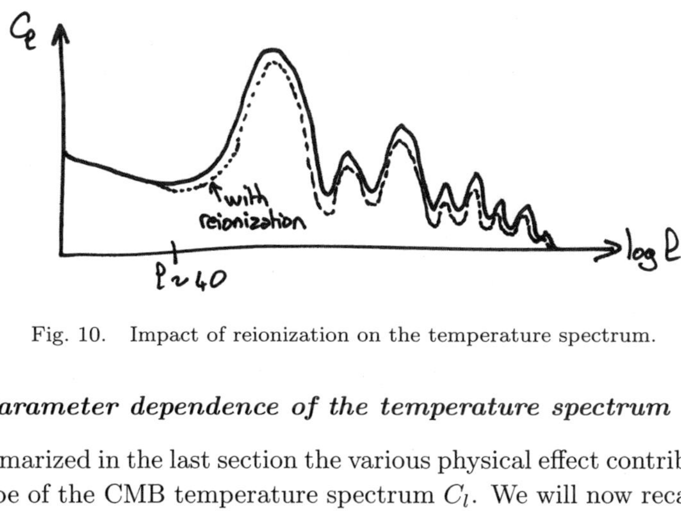

Fig. 10. Impact of reionization on the temperature spectrum.

2.6. *Parameter dependence of the temperature spectrum*

We summarized in the last section the various physical effect contributing to the shape of the CMB temperature spectrum C_l. We will now recapitulate the effect of the various cosmological parameters on the C_l's, within the framework of the minimal ΛCDM model.

This model assumes zero spatial curvature, and a power-law primordial spectrum of scalar perturbations:

$$\mathcal{P}_{\mathcal{R}}(k) = A_{\rm s}(k/k_*)^{n_{\rm s}-1} , \qquad (78)$$

where k_* is an arbitrary fixed pivot scale, $A_{\rm s}$ is the spectrum amplitude at this scale, and $n_{\rm s}$ is called the scalar tilt (the exponent is chosen to be $n_{\rm s} - 1$ rather than just $n_{\rm s}$ for historical reasons; with such notations, a scale-invariant spectrum corresponds to $n_{\rm s} = 1$).

We recall that the Hubble parameter today, H_0, can be expressed in terms of a dimensionless reduced Hubble parameter h:

$$H_0 \equiv 100\, h\, \mathrm{km.s^{-1}.Mpc^{-1}} \tag{79}$$

The physical energy density of a given component x today can be expressed in terms of a dimensionless parameter $\omega_x \equiv \Omega_x h^2$:

$$\rho_x^0 = \Omega_x \rho_{\mathrm{crit}}^0 = \Omega_x \frac{3H_0^2}{8\pi G} = \beta\, \omega_x \tag{80}$$

where $\beta = \frac{3(H_0/h)^2}{8\pi G}$ is a fixed number, with the dimension of an energy per volume.

The six free parameters of the minimal ΛCDM model can be chosen to be

$$\{A_{\mathrm{s}}, n_{\mathrm{s}}, \omega_{\mathrm{b}}, \omega_{\mathrm{m}}, \Omega_\Lambda, \tau_{\mathrm{reio}}\}, \tag{81}$$

where τ_{reio} is the optical depth to recombination, which is non-zero because of reionization in the recent universe. At first order, this is the only parameter that one needs to introduce for describing reionization. For a very accurate description, one should introduce other parameters specifying the full reionization history, but these extra parameters are very difficult to probe experimentally. Hence, they are usually not specified.

The above parameter basis specifies the baryon density ω_{b}, the total non-relativistic (baryons + CDM) matter density ω_{m}, and the fractional density of cosmological constant Ω_Λ. The photon density is implicitly assumed to match the measured value of the CMB temperature ($T = 2.725$ K implies $\omega_\gamma \sim 2.10^{-5}$). Neutrinos are assumed to be still relativistic today, with an abundance relative to photons given by the prediction of the standard neutrino decoupling model.[d] Finally, we did not include the parameter H_0 (or h) in the parameter basis (81). Given that we are assuming a flat universe, h can be inferred from other parameters in (81): $h = \sqrt{\omega_{\mathrm{m}}/(1-\Omega_\Lambda)}$.

The parameters basis (81) is just one particular choice. Many other bases would be valid, for instance, $\{A_{\mathrm{s}}, n_{\mathrm{s}}, \omega_{\mathrm{b}}, \omega_{\mathrm{cdm}}, H_0, \tau_{\mathrm{reio}}\}$). The choice of (81) is dictated by purely pedagogical considerations: we will show that the shape of the CMB spectrum can easily be related to these parameters.

We summarize in Fig. 11 the evolution of background densities in the minimal ΛCDM model. The normalization of the radiation density ρ_r is

[d]In this course, for simplicity, we do not discuss the effect of neutrinos; this effect is far from being negligible, but since we consider the abundance of neutrinos as fixed, and their mass as irrelevant, we can explain the effect of other free parameters without taking neutrinos into account. Neutrino effects are described in Ref. 1.

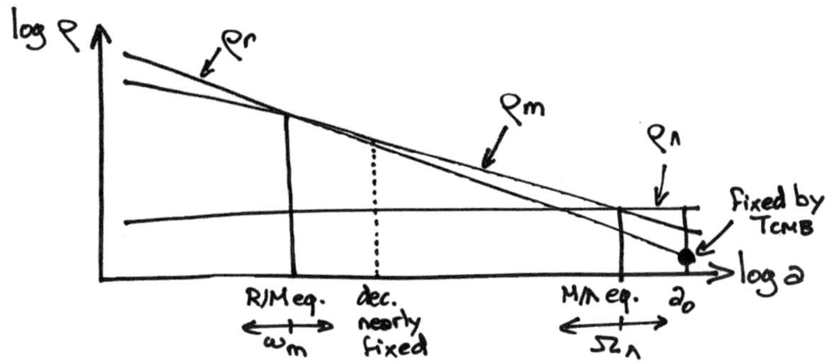

Fig. 11. In the minimal ΛCDM model, evolution of the background density of radiation, matter and cosmological constant as a function of the scale factor.

fixed by the CMB temperature and by the standard neutrino decoupling model. The normalization of the matter density $\rho_{\rm m}$ and of the cosmological constant density ρ_Λ is given respectively by the parameters $\omega_{\rm m}$ and Ω_Λ. The scale factor at radiation/matter equality is given by

$$\rho_r = \rho_{\rm m} \quad \Longrightarrow \quad \omega_r \left(\frac{a_0}{a_{\rm eq}}\right)^4 = \omega_{\rm m} \left(\frac{a_0}{a_{\rm eq}}\right)^3 \quad \Longrightarrow \quad \frac{a_{\rm eq}}{a_0} = \frac{\omega_r}{\omega_{\rm m}} . \tag{82}$$

Since ω_r is considered as fixed, the redshift of equality is controlled by $\omega_{\rm m}$ only. The scale factor at matter/Λ equality is given by

$$\rho_{\rm m} = \rho_\Lambda \quad \Longrightarrow \quad \omega_{\rm m} \left(\frac{a_0}{a_\Lambda}\right)^3 = \Omega_\Lambda h^2 \quad \Longrightarrow \quad \frac{a_\Lambda}{a_0} = \left(\frac{1-\Omega_\Lambda}{\Omega_\Lambda}\right)^{1/3} , \tag{83}$$

so it is controlled by Ω_Λ only. The scale factor at photon decoupling has a small (logarithmic) dependence on $\omega_{\rm b}$, that we can neglect — we will consider that decoupling takes place at a fixed temperature, and hence a fixed scale factor and redshift.

Among the various physical effects described in the previous section, we can identify eight independent leading effect. We summarize them in Table 1, and show by which parameters they are governed. Below, we give more details on these effects.

(C1) Since all peaks in multipole space correspond to the harmonics of a single correlation length in real space, the scale of the peak is controlled (in good approximation) by a single number $l_{\rm peak}$. It depends on the ratio of the sound horizon at decoupling by the angular

Table 1. Independent leading effects controlling the shape of the CMB temperature power spectrum C_l in the minimal ΛCDM model.

	Effect	Relevant quantity	Parameter
(C1)	Peak scale	$\theta_{\rm peak} = \frac{\pi}{l_{\rm peak}} \sim \frac{d_{\rm s}\vert_{\rm dec}}{d_{\rm a}\vert_{\rm dec}}$	$\leftarrow \omega_{\rm m}, \omega_{\rm b}$ $\leftarrow \Omega_\Lambda, \omega_{\rm m}$
(C2)	Odd/even peak amplitude ratio	$R\vert_{\rm dec}$	$\omega_{\rm b}$
(C3)	Overall peak amplitude	$\frac{a_{\rm dec}}{a_0}$	$\omega_{\rm m}$
(C4)	Damping enveloppe	$\theta_{\rm d} = \frac{\pi}{l_{\rm d}} = \frac{a_{\rm dec} r_{\rm d}\vert_{\rm dec}}{d_{\rm a}\vert_{\rm dec}}$	$\leftarrow \omega_{\rm m}, \omega_{\rm b}$ $\leftarrow \Omega_\Lambda, \omega_{\rm m}$
(C5)	Global amplitude	$\mathcal{P}_{\mathcal{R}}(k_*)$	$A_{\rm s}$
(C6)	Global tilt	$\frac{d \log \mathcal{P}_{\mathcal{R}}}{d \log k}$	$n_{\rm s}$
(C7)	Additional plateau tilting (LISW)	$\frac{a_\Lambda}{a_0}$	Ω_Λ
(C8)	Amplitude for $l \geq 40$ only	$\tau_{\rm reio}$	$\tau_{\rm reio}$

diameter distance to decoupling. The first quantity depends on the evolution prior to decoupling, and in particular on the expansion history and sound speed. Hence it depends on $\omega_{\rm m}$ (governing the time of equality) and $\omega_{\rm b}$ (governing $c_{\rm s}^2$ as a function of a). The second quantity depends on the expansion and geometry of the universe after decoupling, i.e. on Ω_Λ and H_0, or in our parameter basis Ω_Λ and $\omega_{\rm m}$.

(C2) When studying the SW contribution to the C_l's, we have seen that the asymmetry between the amplitude of odd and even peaks depends on the shift of the zero-point of acoustic oscillations by a term $-R\psi$. The value of the ratio R at decoupling is governed in our parameter basis by $\omega_{\rm b}$.

(C3) A shift in the time of radiation/matter equality affects the amplitude of all peaks for two reasons: it controls the duration of the intermediate stage between equality and decoupling, during which acoustic oscillations are damped by baryonic effects, and the EISW. Both effects go in the same direction. If equality takes place earlier, there is less time for damping (hence all peaks are increased). The metric fluctuations are also less stabilized at decoupling, and the EISW is larger (hence the first peak is increased even more).

(C4) Diffusion damping near the time of recombination controls the envelope of the peaks (the function $\exp[-(l/l_\mathrm{d})^2]$ suppresses them, starting essentially from the third one). It depends on the ratio of the damping scale at decoupling by the angular diameter distance to decoupling. The first quantity depends on the Thomson scattering rate prior to decoupling, and in particular on ω_m (governing the value of conformal time at equality) and ω_b (governing the ionization fraction as a function of a). The second quantity depends on the expansion and geometry of the universe after decoupling, i.e. on Ω_Λ and H_0, or in our parameter basis Ω_Λ and ω_m. The parameter dependence of effects (C1) and (C4) could be thought to be similar: in fact, it is not, because the sound horizon and the diffusion scale depend on very different combinations of ω_m and ω_b.

(C5) The global amplitude of the C_l's depends trivially on that of the primordial spectrum, fixed by A_s.

(C6) The global slope of the C_l's depends trivially on the tilt of the primordial spectrum, fixed by n_s.

(C7) The LISW effect tilts the Sachs-Wolfe plateau (on top of the effect of n_s). The plateau is more lifted at small l's if Λ domination is longer, i.e. if metric fluctuations decay during a larger amount of time. Hence this effect is enhanced by large values of Ω_Λ.

(C8) The C_l amplitude is suppressed if reionization takes place early, i.e. if τ_reio is large, but without affecting the largest scales (small l's).

We see that in the framework of the minimal ΛCDM model, six parameters control eight distinct physical effects with different impacts on the C_l's. This suggests that an accurate enough measurement of the temperature spectrum is sufficient for fixing the six parameters of the cosmological model describing our universe, at least if the data is consistent with ΛCDM. This conclusion is roughly correct, but must be refined with some words of caution. Indeed, we remember that for small l's, cosmic variance is large, so that the average C_l's of the "true model" describing our universe cannot be measured precisely, even in the case of an ideal experiment. However, two of the previous effects (C1) - (C8) can only affect the smallest multipoles:

- effect (C7) affects only the Sachs-Wolfe plateau,
- a combination of effects (C5) and (C8), corresponding to the product $e^{-2\tau_\mathrm{reio}}A_\mathrm{s}$, controls the global amplitude for $l \gg 40$, but a variation of both τ_reio and A_s with the previous product being kept fixed would only affect the smallest multipoles.

Hence, the measurement of the CMB temperature spectrum is sufficient for constraining the six parameters of the ΛCDM model, but with a relatively large error bar for Ω_Λ and for a particular combination of $\tau_{\rm reio}$ and $A_{\rm s}$. In the next section, we will say a few words on the measurement of CMB polarization, which allows to better constrain reionization: polarization data allow to remove the degeneracy between $\tau_{\rm reio}$ and $A_{\rm s}$. Instead, the error bar on Ω_Λ can only be improved by combining CMB data with other cosmological probes (e.g. supernovae luminosity or large scale structure data).

2.7. *A quick word on polarization*

The Bolztmann equation presented in Eq. (37) was a bit over-simplified. We did as if the only degree of freedom describing photons with a black-body spectrum was their temperature. In fact, photons are described by more degrees of freedom, called the Stokes parameters, involving also their polarization. In Eq. (37), the Thomson scattering rate was integrated over polarization parameters, but in reality some polarized correction terms are present.

Well before decoupling, photons remain unpolarized on average. Indeed, we have seen in section 2.2 that, throughout the tight-coupling regime and in the frame comoving with the fluid (i.e. such that $\theta_{\rm b} = \theta_\gamma = 0$), the photon temperature is isotropic is every point. This isotropy (resulting from frequent interactions) implies that photons acquire no net polarization patterns when they scatter over electrons.

Instead, at the approach of decoupling, when Thomson scattering becomes inefficient, the photon temperature is no longer isotropic in the frame comoving with the electrons. This means that a given electron will scatter simultaneously some hotter photons coming from one direction, and some colder photons coming from another direction. What is important for polarization is the quadrupolar component of the the temperature distribution in each point. This component starts from zero and grows at the approach of decoupling. When it becomes significant, photon scattering leads to a net linear polarization. Hence, today, CMB photons have a different polarization amplitude and orientation in each direction of the sky.

The map of CMB temperature anisotropies is a scalar map: it can be represented with a one-dimensional color code. The map of CMB polarization can be represented with sticks of different size and orientation in different points of the map. Roughly speaking, the size and orientation of the sticks can be related to the magnitude and orientation of the quadrupole anisotropy in each point of the last scattering surface. Since these quadrupolar patterns reflect variations of temperature in the region of the last scattering surface, it is clear that there is a non-zero correlation between temperature and polarization maps. Still, both maps contain some independent information.

In general, a vector field can be decomposed into a gradient and a curl component, like the electric and magnetic fields. Similarly, CMB polarization maps can be expanded in two scalar maps called, by analogy, E-type and B-type polarization maps.

One can show that primordial scalar perturbations can produce both temperature anisotropies and E-type polarization anisotropies. If the early universe only features Gaussian scalar perturbations on cosmological scales, all the information contained in CMB maps is encoded in the temperature power spectrum C_l^{TT}, the E-type polarization power spectrum C_l^{EE}, and the cross-correlation power spectrum C_l^{TE}. The later really contains additional independent information, because polarization patterns are only partially correlated with temperature patterns (in mathematical terms, $[C_l^{TE}]^2 \leq C_l^{TT} C_l^{EE}$).

We will mention in the next section the possibility that a significant amount of tensor perturbations are produced in the primordial universe. Such perturbations can also generate B-type polarization. If this is the case, all the information is encoded in C_l^{TT}, C_l^{EE}, C_l^{TE} and C_l^{BB}. We did not include in this list the cross-correlation spectra C_l^{TB} and C_l^{EB}, because the parity symmetry imposes that they should vanish (at the level of primary anisotropies).

Mapping the CMB polarization is interesting because it contains additional information with respect to temperature anisotropies.

First, reionization has a very distinct effect on the polarization spectrum. The small fraction of photons rescattered around the time of reionization acquire additional polarization that will show up in the form of a peak in the small-l branch of C_l^{EE} (see Fig. 12) and C_l^{TE}. We have seen in the previous section that with temperature data only, cosmic variance limits our ability to measure $\tau_{\rm reio}$, and *a fortiori* to constrain additional parameters describing the reionization history. The effect of reionization in the small-l branch of C_l^{EE} and C_l^{TE} is so big that despite of cosmic variance, $\tau_{\rm reio}$ can be well measured using polarization data.

So the effect of reionization, described before as (C8), is very different

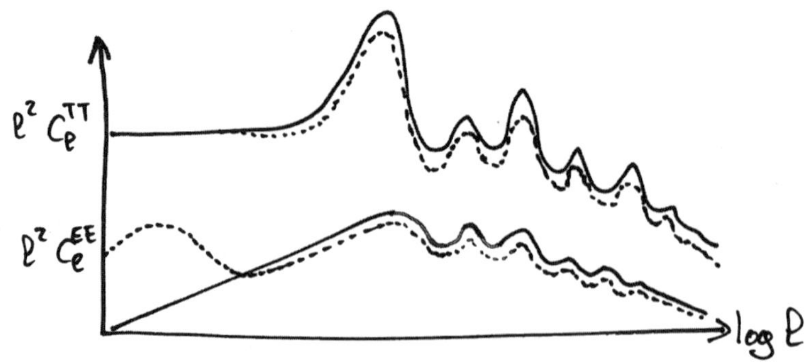

Fig. 12. Typical shape of the temperature and E-type polarization power spectra in a ΛCDM model without (solid lines) or with (dotted lines) reionization of the universe at small redshift.

for temperature and polarization. Instead, the other effects (C1) - (C7) have a qualitatively similar impact on C_l^{EE} and C_l^{TE}. Still, measuring polarization is interesting, because temperature anisotropies feature a particular combination of the SW, Doppler and ISW effects, while polarization provides information on the quadrupole Θ_2 on the last scattering surface, which is correlated with the SW and Doppler terms, but probing a different combination of them. Hence the role of polarization measurements is to remove some of the parameter degeneracies appearing in the analysis of temperature data.

Note that in the full Boltzmann equation, the equation of motion of temperature and polarization degrees of freedom are coupled. In the last sections, we neglected such a coupling. In fact, the impact of polarization on the evolution of temperature anisotropies is very small. Hence our qualitative discussion of the various effects affecting CMB temperature remains valid.

2.8. *A quick word on tensors*

We have seen in section 1.1 that tensor modes are related to the traceless divergenceless components of the metric and stress-energy tensor, obeying to

$$\sum_i \delta T_{ii} = 0 \qquad \text{and} \qquad \forall j, \;\; \sum_i \partial_i \delta T_{ij} = 0 \; . \tag{84}$$

Since these relations impose four constraints on the six degrees of freedom of the 3×3 symmetric matrix δT_{ij} (or δg_{ij}), there are two tensor degrees of freedom in $g_{\mu\nu}$ and $T_{\mu\nu}$.

The two degrees of freedom in δg_{ij} are the two degrees of polarization of gravitational waves, that can propagate even in the vacuum. In the presence of matter with a non-diagonal stress tensor δT_{ij}, gravitational waves can be seeded by the tensor components δT_{ij}.

If tensor perturbations are sufficiently large at recombination or later, they can generate CMB temperature and polarization anisotropies: one can show that they interact with CMB photons, and produce effects similar to the SW and ISW ones. As mentioned in the previous section, gravitational waves can even seed B-type polarization.

The stress-energy tensor is diagonal for perfect fluids, and negligible for pressureless components like CDM. Hence, during radiation and matter domination, tensor perturbations can only be seeded by decoupled neutrinos and/or decoupled photons. However, neutrino and photon tensor modes are far too small for generating a detectable amount of CMB anisotropies. However, in the early universe, other mechanisms may produce gravitational waves on scales sufficiently large to be observable in the CMB. The most famous one is related to inflation. During inflation, quantum fluctuations of the metric can excite primordial tensor perturbations at a level that will produce a detectable signature in CMB maps. The amplitude of this signal is directly proportional to the energy scale of inflation. If this scale is large enough, tensors can contribute to C_l^{TT}, but only on small l's, because gravitational waves decay quickly inside the Hubble radius. Hence, one way to detect tensors would be through a small distortion of the C_l^{TT} spectrum shape for $l \leq 100$ (i.e. for scales that are equal or larger than the Hubble radius at the time of decoupling).

However, if the tensor signal is small with respect to the error bars associated to cosmic variance, it will remain undetectable. In that case, further sensitivity could be obtained by measuring C_l^{BB}: in absence of tensors, C_l^{BB} would vanish, so even if the tensor amplitude is very small it can still dominate the C_l^{BB} signal. Unfortunately, this is true only up to some extent, because secondary anisotropies (in particular, those generated by weak lensing) produce a non-zero C_l^{BB} that could mask the primary tensor anisotropy spectrum.

It is very challenging for CMB experiments to reach the sensitivity level required for detecting a C_l^{BB} signal, even for that generated by weak lensing. Current limits on the tensor primordial spectrum (and on the energy scale of inflation) mainly come from the observation of C_l^{TT} at small l by WMAP. The sensitivity of Planck to tensors will also mainly come from temperature. Future experiments dedicated to CMB polarization will improve the sensitivity to B-type polarization and obtain more precise bounds, until they reach the theoretical limit set by the lensing contamination.

3. Matter Power Spectrum

3.1. *Definition*

The total energy perturbation in the universe can be expanded as

$$\delta\rho_{\rm tot} = \delta\rho_\gamma + \delta\rho_{\rm b} + \delta_{\rm cdm} + \delta\rho_\nu \, (+\, \delta\rho_{\rm de} + ...) \tag{85}$$

where $\delta\rho_{\rm de}$ refers to possible Dark Energy (DE) perturbations, and the three dots for extra relics. In the minimal ΛCDM model, only the first four components are present.

Many Large-Scale Structure (LSS) observables are related to the power spectrum of $\delta\rho_{\rm tot}$, at different wavenumbers and redshifts. This is the case of the galaxy and of the halo correlation function, of the cluster mass function, of CMB lensing, of the cosmic shear spectrum, etc.

All these observations probe the power spectrum during matter or Λ/DE domination, when photons are subdominant and $\delta\rho_\gamma$ can be neglected. If neutrinos are still ultra-relativistic today, $\delta\rho_\nu$ can also be neglected (in this course, we do not have time to discuss the impact of small neutrino masses, for which we refer the reader to Ref. 1). If the acceleration of the universe is caused by a cosmological constant, there is no term $\delta\rho_{de}$. More generally, most Dark Energy models would predict a negligible amount of DE perturbations. In summary, in a wide category of cosmological scenarii including ΛCDM, we can use $\delta\rho_{\rm tot} \simeq \delta_{\rm m} \equiv \delta\rho_{\rm b} + \delta_{\rm cdm}$. Hence, in the context of LSS observations, it is customary to refer to the power spectrum of the non-relativistic matter fluctuation $\delta_{\rm m}$, defined as

$$\delta_{\rm m} = \frac{\delta\rho_{\rm m}}{\bar{\rho}_{\rm m}} = \frac{\delta\rho_{\rm b} + \delta\rho_{\rm cdm}}{\bar{\rho}_{\rm b} + \bar{\rho}_{\rm cdm}} \, , \tag{86}$$

which is indistinguishable from the ratio $\delta\rho_{\rm tot}/\bar{\rho}_{\rm m}$. Only in models with large dark energy perturbations or modifications of Einstein gravity, the two quantities might be different, and special care about the definition of the matter power spectrum is needed.

The matter power spectrum $P(z,k)$ of $\delta_{\rm m}$ is defined like in section 1.5:

$$\langle \delta_{\rm m}(z,\vec{k})\delta^*_{\rm m}(z,\vec{k}') \rangle = \delta_D(\vec{k}-\vec{k}') \; P(z,k) \; . \tag{87}$$

Here we used the redshift as a time variable, but we could have indifferently used proper or conformal time. We have seen in section 1.5 that for Gaussian initial conditions and as long as perturbations are linear, the power spectrum at a given time can be written as the product of the primordial spectrum by the square of the relevant transfer function, in our case $\delta_{\rm m}(z,k)$. Sticking to the same conventions as in section 1.5 and assuming a power-law primordial spectrum like in section 2.6, this gives

$$P(z,k) = \frac{2\pi^2}{k^3} A_{\rm s} \left(\frac{k}{k_*}\right)^{n_{\rm s}-1} \delta^2_{\rm m}(z,k) \ . \tag{88}$$

Hence, by studying qualitatively the evolution of the transfer function $\delta_{\rm m}(z,k)$, we will get some insight on the cosmological information encoded in the matter power spectrum.

3.2. *Transfer function evolution*

CDM dominated universe. In order to simplify the presentation, let us first assume that we leave in a ΛCDM universe with a negligible amount of baryons: $\Omega_{\rm b} \ll \Omega_{\rm cdm}$ and $\delta_{\rm m} \simeq \delta_{\rm cdm}$. In section 2.5, we wrote the master equation governing the evolution of photon perturbations during the tightly-coupled regime. Similarly, by combining the continuity and Euler equation of CDM perturbations, we can write here a master equation for $\delta_{\rm cdm}$, actually valid in all regimes:

$$\delta''_{\rm cdm} + \frac{a'}{a}\delta'_{\rm cdm} = -k^2\psi + 3\phi'' + 3\frac{a'}{a}\phi' \ . \tag{89}$$

In an expanding universe, the clustering rate depends on the expansion rate: expansion increases distances, weakens gravitational forces, and slows down clustering processes. In the above equation, this is accounted by the second term, often called the Hubble friction term. On the right-hand side, the first term represents gravitational forces, and the last two terms account for dilation effects.

On super-Hubble scales, we have seen that in the Newtonian gauge, adiabatic ICs predict constant density fluctuations $\delta_{\rm cdm}$. To be precise, ϕ and $\delta_{\rm cdm}$ vary on super-Hubble scales only when the total equation of state of the universe changes (i.e. around the time of radiation/matter equality, and during Λ domination). Instead, they remain constant on those scale during the radiation and matter dominated regime.

Inside the Hubble radius, we can neglect dilation terms, and replace the gravitational potential term:

$$\delta''_{\rm cdm} + \frac{a'}{a}\delta'_{\rm cdm} - \frac{3}{2}\left(\frac{a'}{a}\right)^2 \Omega_{\rm cdm}(a)\, \delta_{\rm cdm} = 0 \ , \tag{90}$$

where $\Omega_{\rm cdm}(a)$ is the fraction of the critical density coming from CDM at a given value of time (or of the scale factor). This equation is often called the Mészáros equation. It can be obtained by combining Eq. (89) with the (00) component of the Einstein equation (or its Poisson limit) and the Friedmann equation. The careful reader might have noticed something suspicious: the gravitational force term $k^2\psi$ has been eliminated in favor of $\delta_{\rm cdm}$, while the first Einstein equation (or its Poisson limit) involves the total density fluctuation. Hence, shouldn't Eq. (90) feature also δ_γ, $\delta_{\rm b}$ and δ_ν? Actually, it turns out that the above equation is a really good approximation for the CDM equation of evolution in all regimes, under our assumption $\Omega_{\rm b} \ll \Omega_{\rm cdm}$. It applies even during radiation domination, when photon fluctuations are potentially large. In this course, we do not have enough time for justifying this point, but we refer the reader to Weinberg [3, 4] for a detailed explanation.

During radiation domination, the Friedmann equation gives $a \propto \eta$, and $\Omega_{\rm cdm}$ is much smaller than one. Hence we can neglect the last term in the Mészáros equation, and find that the two solutions are $\delta_{\rm cdm}$ = constant and $\delta_{\rm cdm} \propto \log \eta$. Hence CDM fluctuations grow logarithmically. During matter domination, $a \propto \eta^2$ and $\Omega_{\rm cdm} \simeq 1$, so the solutions are $\delta_{\rm cdm} \propto \eta^2$ and $\delta_{\rm cdm} \propto \eta^{-3}$. Then, CDM fluctuations grow quadratically with η. (These are only asymptotic solutions, but the Mészáros equation can actually be solved analytically at all times). During Λ domination, the function $a(\eta)$ is more complicated, and $\Omega_{\rm cdm}$ decreases. With a bit of work, one can show that $\delta_{\rm cdm}$ grows at a smaller rate than during matter domination (i.e. slowlier than η^2), and that the reduction of the growth rate does not depend on k.

In summary, during radiation domination, $\delta_{\rm cdm}(\eta, k)$ is constant on super-Hubble scales and grows logarithmically on sub-Hubble scales. A more precise calculation would show that up to a numerical factor of order one, $\delta_{\rm cdm}$ is given on sub-Hubble scales by $\delta_{\rm cdm}(\eta, k) = \log(k\eta)$. During matter domination, $\delta_{\rm cdm}(\eta, k)$ is still constant on super-Hubble scales, and grows like η^2 on sub-Hubble scales. Finally, during Λ domination, it grows more slowly. These different behaviors are reported in Fig. 13.

This simple discussion is sufficient for understanding the shape of the matter power spectrum at different time. Like in a cartoon, Fig. 14 shows

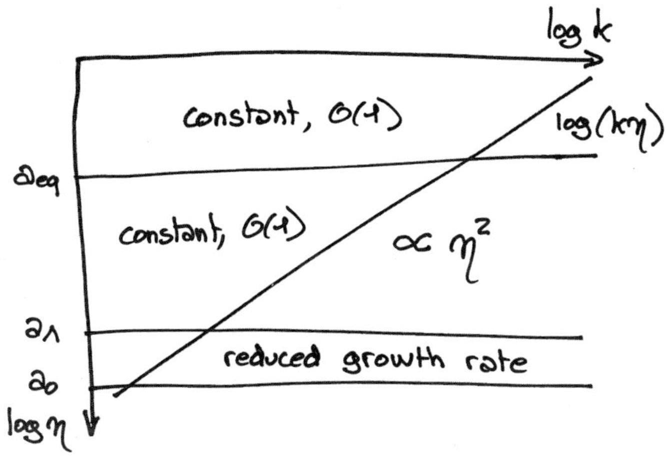

Fig. 13. Qualitative evolution of the transfer function $\delta_{\rm cdm}(\eta, k)$ (normalized as usual to $\mathcal{R}(k) = 1$ at initial time) in different regimes: during radiation domination, matter domination, Λ domination, and on super/sub-Hubble scales.

this shape at four different times: at some initial time when all relevant modes are super-Hubble; at radiation/matter equality; at matter/Λ equality; and today. Let us comment these plots. We first need to define the comoving wavenumbers corresponding to wavelengths crossing the Hubble radius at the time of radiation/matter equality, of matter/Λ equality, and today:

$$k_{\rm eq} = a_{\rm eq} H_{\rm eq} \,, \qquad k_\Lambda = a_\Lambda H_\Lambda \,, \qquad k_0 = a_0 H_0 \,. \tag{91}$$

We can now review the evolution of $P(k)$ with respect to time, following the same steps as in Fig. 14.

(1) At initial time, if we assume a scale-invariant spectrum with $n_{\rm s} = 1$, we see from Eq. (88) that $P(k) \propto k^{-3}$, with an amplitude given by $A_{\rm s}$.
(2) During radiation domination, modes grow logarithmically inside the Hubble radius, like $\log(k\eta)$. At equality, super-Hubble modes ($k_{\rm eq}\eta \ll 1$) are still shaped like at initial time, while sub-Hubble modes ($k_{\rm eq}\eta \gg 1$) have been enhanced by a factor $[\delta_{\rm cdm}(\eta_{\rm eq}, k)/\delta_{\rm cdm}(\eta_{\rm ini}, k)]^2 \simeq [\log(k\eta_{\rm eq})]^2$. The two asymptots of $P(k)$ are then given by k^{-3} for $k \ll k_{\rm eq}$ and $k^{-3}[\log(k)]^2$ for $k \gg k_{\rm eq}$.
(3) At the end of matter domination, when $a = a_\Lambda$, modes still outside the Hubble radius keep being shaped like at initial time. This concerns all

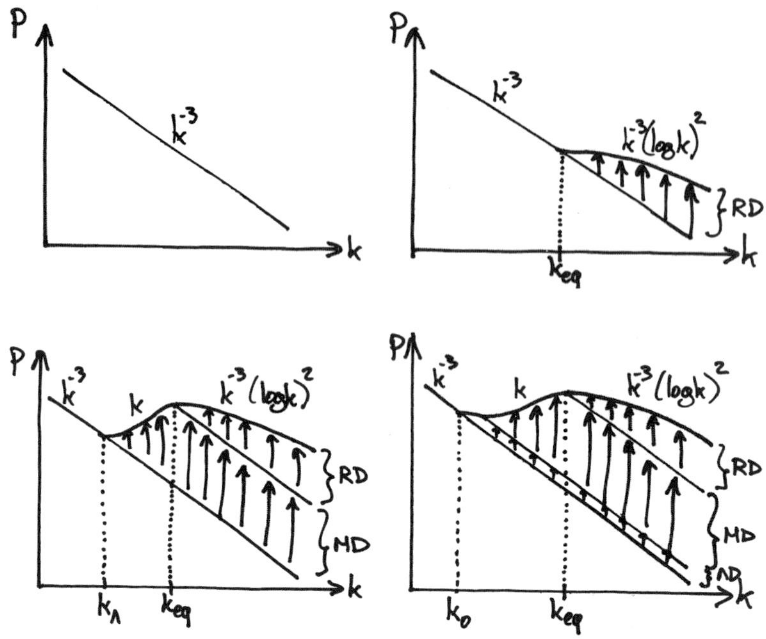

Fig. 14. Shape of the matter power spectrum $P(k)$ (log-log scale) at four different times: *(upper left)* when initial conditions are imposed (and all wavenumbers are super-Hubble); *(upper right)* at radiation/matter equality (arrows show the logarithmic growth during radiation domination); *(lower left)* at matter/Λ equality (lower set of arrows show the growth during matter domination); *(lower right)* today (lower set of arrows show the growth during Λ domination).

modes with $k \ll k_\Lambda$. Modes $k \gg k_{\rm eq}$ have been amplified during matter domination by a factor $[\delta_{\rm cdm}(\eta_\Lambda, k)/\delta_{\rm cdm}(\eta_{\rm eq}, k)]^2 \simeq (\eta_\Lambda/\eta_{\rm eq})^4$. This factor does not depend on k and preserves the shape of the power spectrum on those scales. Finally, intermediate modes entering the Hubble scale during matter domination have been amplified by $(\eta_\Lambda/\eta_*)^4$, where η_* is their time of Hubble crossing, given approximately by $\eta_* = 1/k$. Hence they have been amplified by $(k\eta_\Lambda)^4$. Putting all these informations together, we see that the spectrum has three branches, scaling respectively like:

- $P(k) \propto k^{-3}$ for $k < k_\Lambda$,
- $P(k) \propto k^{-3}k^4 = k$ for $k_\Lambda < k < k_{\rm eq}$,
- $P(k) \propto k^{-3}(\log k)^2$ for $k > k_{\rm eq}$.

(4) During Λ domination, $\delta_{\rm cdm}(k,\eta)$ grows more slowly than η^2, but it still grows at the same rate for all sub-Hubble modes. So the shape of the power spectrum today is unaltered by this stage, and given by:

- $P(k) \propto k^{-3}$ for $k < k_0$,
- $P(k) \propto k^{-3}k^4 = k$ for $k_\Lambda < k < k_{\rm eq}$,
- $P(k) \propto k^{-3}(\log k)^2$ for $k > k_{\rm eq}$.

We do not enter into details for the small range of modes obeying $k_0 < k < k_\Lambda$.

All this discussion was carried under the assumption of a scale-invariant primordial spectrum. If $n_{\rm s} \neq 1$, the above shape should simply be rescaled by $k^{n_{\rm s}-1}$, and the three branches of the power spectrum are given by:

- $P(k) \propto k^{n_{\rm s}-4}$ for $k < k_0$,
- $P(k) \propto k^{n_{\rm s}}$ for $k_\Lambda < k < k_{\rm eq}$,
- $P(k) \propto k^{n_{\rm s}-4}(\log k)^2$ for $k > k_{\rm eq}$.

This closes the presentation of the shape of $P(z,k)$ in the limit $\Omega_{\rm b} \ll \Omega_{\rm cdm}$. What remains to be seen is the impact of a non-negligible baryon fraction on the power spectrum.

Baryon corrections. We must define a new important time in the evolution of the universe: the baryon drag time. The photon decoupling time was defined through the maximum of the photon visibility function $g(\eta)$. But there are many more photons than baryons in the universe. Hence, for some amount of time after photon decoupling, baryons keep tracking photon perturbations: in other words, the baryon population decouples later than the photon population (seen as a whole).

Until the baryon drag time, we know that $\delta_{\rm b} = \frac{3}{4}\delta_\gamma$. After that time, baryons do not experience significant Thomson scattering anymore. They only feel gravity, and collapse in gravitational potential wells.

We know qualitatively the behavior of the baryon transfer function $\delta_{\rm b}(\eta, k)$ just before baryon drag, since in the CMB section we have studied the behavior of the photon transfer function. We know that $\delta_{\rm b} = \frac{3}{4}\delta_\gamma$ is constant on super-sound-horizon scales, experiences stationary oscillations on sub-sound-horizon scales during radiation domination, and finally damped oscillations on sub-sound-horizon scales during matter domination. Moreover, we expect that at any time — and particularly at the approach of decoupling — the relation $\delta_{\rm b} = \frac{3}{4}\delta_\gamma$ breaks on very small wavelengths comparable to the mean free path of baryons in the imperfect baryon-photon fluid, but here we will not discuss such small scales.

The behavior of the CDM transfer function $\delta_{\rm cdm}(\eta, k)$ before baryon drag is still given by the Mészáros equation: deep inside the sound horizon, baryons are much less clustered than CDM, so that CDM is almost self-gravitating.

In the left plot of Fig. 15, we sketch the qualitative behavior of the individual power spectrum of baryons and CDM just before baryon drag ($P_{\rm b}(k)$ and $P_{\rm cdm}(k)$ are defined as in Eq. (88), with $\delta_{\rm m}$ replaced either by $\delta_{\rm b}$ or $\delta_{\rm cdm}$).

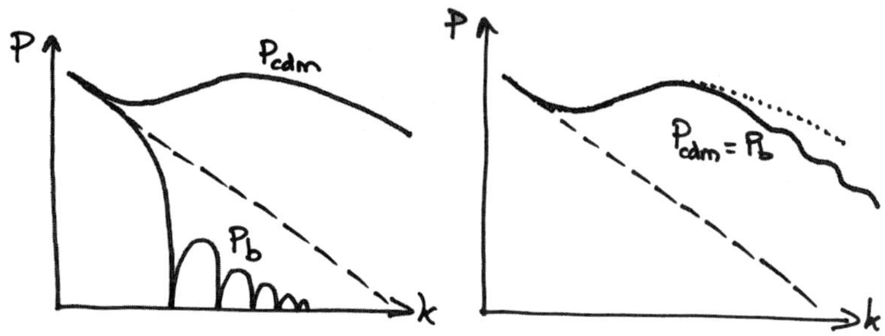

Fig. 15. *(Left)* Shape of the baryon and CDM power spectra just before the baryon drag time, compared to the shape of the primordial spectrum (dashed line). *(Right)* Shape of the matter power spectrum well after the baryon drag time, when baryons and CDM perturbations have reached gravitational equilibrium, compared to the shape of the primordial spectrum (dashed lines) and to the spectrum of a CDM-dominated universe (dotted line).

After baryon drag, baryons collapse in potential gravitational wells. Since $\delta_{\rm b}$ grows, CDM will start to feel the gravity of baryons. Finally, because CDM and baryons are two collisionless species feeling the same gravitational forces, they will equilibrate with $\delta_{\rm b} = \delta_{\rm cdm}$. The gravitational potential is then related to this common value by the (00) Einstein equation (or its Poisson sub-Hubble limit). This is not true however on very small scales, for which the baryon pressure cannot be neglected, but here we do not discuss such small scales.

Hence, after a quick relaxation period, the two power spectra of baryons and CDM are equal to each other, $P_{\rm b}(k) = P_{\rm cdm}(k) = P(k)$. In order to relate this common power spectrum to the individual power spectrum of baryons and CDM before baryon drag, one must follow a matching process described in details in Einsteinstein and Hu [5]. Intuitively, the matching depends on the relative weight of baryons and CDM, i.e. on the ratio $\Omega_{\mathrm{b}}/\Omega_{\mathrm{cdm}}$. In the limit $\Omega_{\mathrm{b}} \ll \Omega_{\mathrm{cdm}}$, what we wrote at the beginning of this section applies. In the limit $\Omega_{\mathrm{b}} \gg \Omega_{\mathrm{cdm}}$, the power spectrum is very suppressed with respect to the CDM-dominated case, with a much more negative slope on average, and some large oscillations corresponding to photon-baryon acoustic waves before decoupling. Finally, if Ω_{b} is a bit smaller than Ω_{cdm} but not negligible (as it is the case in our universe), the power spectrum departs slightly from a pure CDM one, with a smooth step-like suppression plus small oscillatory patterns, which are the smoking gun in the recent universe of baryon-photon acoustic oscillations happening before photon decoupling. This is illustrated on the right plot of Fig. 15. These oscillations are called Baryon Acoustic Oscillations (BAO).

LSS observations can only probe the power spectrum on scales much smaller than the radius of the observable universe, i.e. than the current Hubble radius. For this reason, the first of the three branches described above is unobservable.[e] We can only measure the second and third branches, behaving respectively like $P(k \ll k_{\mathrm{eq}}) \propto k^{n_{\mathrm{s}}}$ and, in first approximation, $P(k \gg k_{\mathrm{eq}}) \propto [k^{n_{\mathrm{s}}-4}(\log k)^2]$, plus the titling and superimposed oscillations coming from baryons.

3.3. *Parameter dependence*

The discussion presented in the previous section allows us to understand which effects and which parameters impact the shape of the matter power spectrum in the minimal (flat) ΛCDM model, parametrized by

$$\{A_{\mathrm{s}}, n_{\mathrm{s}}, \omega_{\mathrm{b}}, \omega_{\mathrm{m}}, \Omega_{\Lambda}, \tau_{\mathrm{reio}}\} \tag{92}$$

(see Sec. 2.6 for details on this parametrization). We can already notice that the reionization optical depth is relevant for the CMB spectrum but not for $P(k)$, since it only impacts the scattering rate of photons in the recent universe. Other effects are described in Table 2. Below, we give

[e]The first unobservable branch is actually gauge-dependent: it behaves like $k^{n_{\mathrm{s}}-4}$ in the Newtonian gauge, but not in other gauges. Note that truly observable quantities are always gauge-invariant. We have been a bit uncareful when saying that LSS data probes the power spectrum of δ_{m}. In fact, different LSS observations probe different quantities, each of them being gauge-invariant. However, well inside the Hubble region, all these quantities coincide with each other and with the spectrum of δ_{m} computed in an arbitrary gauge (up to small corrections that may become important in the future, but not with current experimental sensitivities).

Table 2. Independent leading effects controlling the shape of the matter power spectrum $P(k)$ in the minimal ΛCDM model.

	Effect	Relevant quantity	Parameter
(P1)	scale of the maximum	$k_{\rm eq}$	$\omega_{\rm m}, \Omega_\Lambda$
(P2)	slope for $k \gg k_{\rm eq}$ and BAO amplitude	$\Omega_{\rm b}/\Omega_{\rm cdm}$	$\omega_{\rm b}, \omega_{\rm m}$
(P3)	BAO scale	$r_{\rm s}(\eta_{\rm drag})$	$\omega_{\rm b}, \omega_{\rm m}$
(P4)	Global amplitude	amplitude of primordial spectrum and duration of ΛD	$A_{\rm s}, \Omega_\Lambda$
(P5)	Global tilt	tilt of primordial spectrum	$n_{\rm s}$

more details on these effects.

(P1) The time of equality determines the scale $k_{\rm eq}$ of the power spectrum peak. More precisely, if this scale is expressed in units of $h\,{\rm Mpc}^{-1}$, which is the usual convention, then one can show that the scale of the maximum depends on both $z_{\rm eq}$ and $\Omega_{\rm m}$ (i.e. on our parameter basis on $\omega_{\rm m}$ and $\Omega_\Lambda = 1 - \Omega_{\rm m}$).

(P2) The baryon abundance (relative to CDM) is crucial at the matching time: a high baryon abundance leads to more suppression of the power spectrum for $k \geq k_{\rm eq}$, and to more pronounced BAOs.

(P3) We have seen that the scale of acoustic oscillations is set by the sound horizon at a given time. Since photons decouple at $\eta_{\rm dec}$, the scale of oscillations on the last scattering surface is set by $d_s(\eta_{\rm dec})$, corresponding to the comoving scale $r_s(\eta_{\rm dec}) = d_s(\eta_{\rm dec})/a(\eta_{\rm dec})$. Similarly, since baryons decoupled at $\eta_{\rm drag}$, the scale of BAOs depends on $d_s(\eta_{\rm drag})$ at baryon drag, corresponding to the comoving scale $r_s(\eta_{\rm drag}) = d_s(\eta_{\rm drag})/a(\eta_{\rm drag})$. As explained in Sec. 2.6, the sound horizon at a given time depends on $\omega_{\rm b}$ and $\omega_{\rm m}$. In the case of $d_s(\eta_{\rm drag})$ the dependence on $\omega_{\rm b}$ is even stronger because the time of baryon drag itself depends strongly on the baryon abundance.

(P4) The global amplitude depends of course on the primordial spectrum amplitude, i.e. on $A_{\rm s}$. Also, we have seen that during Λ domination, the growth rate of fluctuations is reduced with respect to matter domination, but is independent of k for all sub-Hubble scales. Hence, Ω_Λ also affects the global amplitude of the power spectrum.

(P5) The global tilt depends of course on the primordial spectrum tilt, i.e. on n_s.

This discussion shows that in principle, a precise measurement of the matter power spectrum today (or at a given redhsift) would allow to measure independently ω_b, ω_m, n_s, A_s and Ω_Λ (assuming a flat ΛCDM universe). In practice, given the limited precision of current data sets, some of the effects described above are degenerate with each other, and matter power spectrum observations are mainly useful in combination with CMB observations, since they bring independent and complementary information.

Future experiments will use all the discriminating power of the matter power spectrum. They will perform accurate measurements of $P(k, z)$ at different redshifts. This is crucial for at least two reasons:

- First, the effect of Ω_Λ in (P5) is redshift dependent: at different redshifts, fluctuations have spent more or less time during the Λ dominated regime, and the power spectrum amplitude has been more or less affected. Hence, the comparison of the amplitude at different redshifts allows to measure Ω_Λ, or more generally to test the compatibility of the data with a cosmological constant (rather than some dynamical Dark Energy model).
- Second, BAOs appear at a fixed comoving wavenumber k_{BAO} in Fourier space (related to $r_s(\eta_{\mathrm{drag}})$), but if we measure it in different LSS datasets at different redshifts (corresponding to different shells in real space), this scale will be seen under different angles.[f] By comparing the BAO angle at different redshifts, one can reconstruct the angular diameter distance, and therefore the expansion history at different redshifts. This is another way of measuring Ω_Λ or testing the Λ model versus DE models.

4. Effects of Dark Matter on CMB and Matter Spectrum

The existence of Dark Matter (DM) was first postulated by astronomers in order to account for the velocity of stars in the Milky Way (Oort, 1932), and that of galaxies in the Coma cluster (Zwicky, 1933). Later on, it has

[f] Here we are assuming that the BAO scale is measured in each redshift shell transversally, i.e. in the direction orthogonal to the line of sight. Real experiments probe the BAO scale both transversally and longitudinally, so the situation is a bit more subtle than in this simplified discussion.

been firmly established that if general relativity applies to galactic scales and above, some form of DM halo must be present in most galaxies and clusters.

Most of the dark matter is cold. If DM is made of collisionless and stable particles, the only DM property that can be probed by cosmological perturbations is its velocity distribution.

At the level of homogeneous cosmology, the DM phase-space distribution f only depends on momentum p. The number of particles in a given momentum bin is given by $dp\,p^2 f(p)$. If $p^2 f(p)$ is very peaked near $p = 0$ or the DM mass is very large, DM particles have a negligible velocity dispersion. In this case, they can be represented effectively as a pressureless fluid, as explained in Sec. 1.3. We recall that they are not forming a real fluid, due to the absence of interactions, but since their velocity dispersion is negligible, all particles have approximately the same bulk velocity in each point (described by the divergence $\theta_{\rm cdm}$). In that case the DM component is called Cold Dark Matter (CDM).

If $p^2 f(p)$ is not very peaked near $p = 0$ or the DM mass is not very large, a significant fraction of DM particles may have a velocity p/m comparable to the speed of light, or at least to that of astrophysical objects. Typically, if the velocity dispersion is of the order of 10^{-3} (in units of c) or larger, it cannot be neglected when studying the CMB or LSS spectra. In that case, the velocity of individual particles results from two physical mechanisms: the dynamics imposed by gravitational forces, plus an intrinsic velocity distributed stochastically, according to the phase-space distribution. The notion of "velocity dispersion" refers to the second component, since the first effect tends to impose the same bulk velocity to all particles in a given point. DM with a large velocity dispersion cannot be described by fluid equations like CDM. A famous example of DM particles with a large velocity dispersion is neutrinos. The velocity dispersion is then fixed by the neutrino mass, and by their Fermi-Dirac distribution. Such DM is called Hot Dark Matter. Possible DM species with a smaller (but still significant) velocity dispersion are called Warm Dark Matter.[g]

Since CDM is a very generic limit, and neutrinos are know to exist since the 1950's, the two most popular DM models until the 1990's were CDM and HDM. CDM can be motivated by many possible extensions of the standard

[g]The limit between HDM and WDM is usually placed in the following way: both types of particles have a sizable velocity dispersion, but most HDM particles become non-relativistic after radiation/matter equality, while most WDM particles become non-relativistic before that time.

model of particle physics (supersymmetry, Peccei-Quinn symmetry, etc.), predicting some early decoupled stable particles, with a very small velocity dispersion. HDM was motivated at that time by massive neutrinos with masses of the order of ~ 10 eV.

These two models lead to very different predictions for the matter power spectrum. We have discussed before the case of CDM. A study of perturbations in a HDM (or ΛHDM) model would show that for $k \geq k_{\rm eq}$, the power spectrum $P(k)$ would decrease with a much steeper slope. The physical explanation is that the velocity dispersion of HDM particles prevents them from clustering on small scales: hence the growth rate of $\delta_{\rm m}$ inside the Hubble radius is considerably reduced in the HDM model.

In the 90's, the first estimates of the matter power spectrum $P(k)$ (or of some related quantities) on galaxy scales and cluster scales showed clearly that the HDM scenario was ruled out. Today, using more accurate LSS data, it is firmly established that most of the DM must be cold (or eventually, warm, but with such a small velocity dispersion that WDM would be indistinguishable from CDM at least on the largest observable scales). Still, we cannot exclude the presence of a subdominant hot or warm component. In fact, since we know that neutrinos have a small mass (of the order of ~ 0.1 eV), we do expect that a few percents of DM is hot. Since the signature of this sub-leading component has not yet been detected, the minimal model is still ΛCDM.

In the ΛCDM case, cosmological perturbations can only measure one DM-related parameter: its relic density, parametrized by $\omega_{\rm cdm}$. This is better than nothing: the value of $\omega_{\rm cdm}$ can be used to constrain some fundamental parameters governing DM production, within the framework of a postulated model.

Measuring $\omega_{\rm cdm}$ with the CMB. Going back to Sec. 2.6, we see that $\omega_{\rm cdm}$ can be measured "indirectly" using CMB observations: effect (C2) measures $\omega_{\rm b}$, effect (C3) measures $\omega_{\rm m}$, and $\omega_{\rm cdm}$ is given by the difference between these two parameters.

There is no more "direct" effect of CDM that could be probed with the CMB. A priori, we could think that photons are affected by gravitational forces generated by CDM. This gravitational interaction could only be relevant before photon decoupling, and around the time of sound horizon crossing for each given wavelength (we have seen in Sec. 2.4 that well inside the sound horizon, photons do not feel gravitational forces anymore, because pressure forces are much larger). But around sound horizon crossing, photons feel mainly the gravitational forces generated by themselves, by neutrinos, and to a lesser extent by baryons. In fact, in this regime, CDM and other components are effectively decoupled [3].

Hence the CMB probes the CDM abundance only indirectly, as a "missing mass", exactly like in astrophysics: it measures the total matter density through the time of equality, the baryon density through the ratio between odd and even peaks, and the difference must be attributed to a non-relativistic component not interacting with the photons: Dark Matter.

In this discussion, we concentrated only on effects (C2) and (C3). Indeed, effects (C1) and (C4) are also sensitive to $\omega_{\rm b}$ and $\omega_{\rm cdm}$, but only in combination with other parameters. So, unlike (C2)+(C3), they do not bring a robust and nearly model-independent evidence for DM.

Measuring $\omega_{\rm cdm}$ with the matter power spectrum. The overall shape of $P(k)$ for $k \geq k_{\rm eq}$ brings very strong support for CDM (or eventually WDM with a small enough velocity dispersion). Indeed, a universe with only baryonic matter, or baryonic plus hot dark matter, would have a radically different power spectrum (with a very steep slope for $k \geq k_{\rm eq}$, and eventually huge BAOs in the purely baryonic case). More precisely, data on the matter power spectrum allows to measure the ratio $\omega_{\rm b}/\omega_{\rm cdm}$ through effect (P2) of Sec. 3.3. Hence the matter power spectrum probes the CDM abundance independently of the CMB. This probe is more direct, since $P(k)$ is really sensitive to the gravitational clustering of CDM inside the Hubble radius.

Signatures of DM annihilation or decay in the CMB. Depending on the underlying particle physics model, DM particles could either annihilate or decay. Of course, the annihilation or decay rate should be very small, since we need a stable dark matter population. Still, a small annihilation or decay rate could play a role in the recent thermal history of the universe, and in particular in the evolution of the ionized fraction of atoms, $x_e(z)$, to which the CMB is sensitive. Indeed, DM annihilation/decay products may include photons that could directly ionize hydrogen atoms, or particles that could reheat the inter-galactic medium. These mechanisms tend to increase x_e at a given time. Their efficiency depends on the mass of DM particles, on their cross-section in the case of annihilation, and on their lifetime in the case of decay.

Before recombination and photon decoupling, the ionization fraction is anyway very close to one. In the minimal cosmological model, x_e drops abruptly after recombination, freezes out when particles leave chemical equilibrium, and increases again to one at the time of reionization, due to ionizing photons emitted at star formation. DM annihilation or decay could potentially modify this picture and play two roles:

- at high redshift (typically $600 < z < 1100$), it could slow down the drop of x_e after recombination, and increase its freeze-out value. This would appear in the CMB as a shift in the time of decoupling (impacting effect (C1) and (C4) of Sec. 2.6), as a broadening of the peak of the visibility function $g(\eta)$ (i.e. the last scattering "surface", which is in fact more like a shell, would be thicker, impacting (C4)), and as an increase of the optical depth to recombination (impacting (C8), but not on the same scales as reionization).
- at small redshift (typically $z < 40$), the concentration of DM in halos could enhance its annihilation/decay rate, and annihilation/decay products could reionize atoms and increase x_e. This would be equivalent to some extra reionization: it could modify the reionization time or history, and impact (C8).

The recent literature shows that for realistic models, CDM annihilation could play these two roles: hence, interesting bounds on DM annihilation can be derived from the CMB. These bounds are complementary from those inferred from cosmic ray data, or direct detection experiments.

DM decay could play the second role, but the bounds on the DM lifetime inferred in this way from the CMB are not as strong as those from cosmic ray data.

These effects had to be mentioned in this section, because they are part of the interplay between DM physics and cosmological perturbation theory. We do not discuss here bounds on DM annihilation/decay from cosmic ray data, since they have little to do with the evolution of linear cosmological perturbations.

What else could we learn on DM from CMB or LSS data? In the simplest cosmological scenario, cosmological perturbations only probe the relic abundance of DM particles. If we are lucky, they may provide some hint of DM annihilation. There are other properties of DM that can be tested with cosmological perturbations, even if at the moment, everything is consistent with the minimal ΛCDM picture:

- the small mass of neutrinos implies a small fraction of HDM, with distinct effects on the CMB and LSS power spectrum [1] (effects on $P(k)$ are larger: a small neutrino mass implies a step-like suppression of the power spectrum on small scales, by at least 4% if we assume the smallest possible neutrino masses that are consistent with neutrino oscillations experiments).

- a non-negligible velocity dispersion of the dominant DM component (like in the WDM model) or of a fraction of it (in the case of several DM species) would also induce a suppression of the small-scale matter power spectrum, at a scale and with a shape depending on the mass and momentum distribution of the warm species. However, such a feature is clearly not observed in the matter power spectrum on linear scales. Such a signature could only be found on smaller scales, affected by non-linear gravitational clustering.
- instead of being treated as collisionless, DM could experience small interactions with itself or with other species (photons, neutrinos baryons). All these possibilities require a case-by-case analysis, and usually predict observable features similar to WDM. Both the CMB and the matter power spectrum can be used to put very strong bounds on a possible DM interactions.

In conclusion of this section, it appears that CMB and LSS observations bring some crucial information concerning the abundance and the nature of Dark Matter. But as long as everything is compatible with the minimal ΛCDM scenario, CMB and LSS data are only telling us what DM cannot consist of, rather than what it is.

5. Effects of Dark Energy on CMB and Matter Spectrum

In this course, we assumed for simplicity that the (at least apparent) acceleration of the universe is caused by a cosmological constant. The possibility that acceleration would be caused by a dynamical component, called generically Dark Energy (DE), is one of the most investigated questions in cosmology. Of course, the acceleration could also be caused by some alternative theory of gravity, different from General Relativity on very large scales. Even in that case, everything could be reformulated in terms of a dark energy component (by moving new terms from the left to the right hand-side of the Einstein equation). Hence, measuring DE parameters is a rather generic way to look for deviations from a simple cosmological constant.[h]

[h]Note that alternatively, the observed apparent acceleration might be the consequence of a departure from the homogeneous Friemann-Lemaître model, but we will not discuss such a possibility (which is very constrained and maybe already excluded by observations) in this course.

Possible dark energy parameters. At the background level, the stress-energy tensor of any species compatible with the assumption of a homogeneous and isotropic FL universe must be diagonal, involving only two function $\bar{\rho}_{\rm de}(a)$ and $\bar{p}_{\rm de}(a)$, related to each other by an equation of conservation of energy. Hence these two functions are not independent of each other, at least as long as DE is not interacting with other species. Even if in general a DE model has no reasons to obey to an equation of state, one can always define the function $w(a) \equiv \bar{p}_{\rm de}(a)/\bar{\rho}_{\rm de}(a)$.

A cosmological constant corresponds to $w(a) = -1$, i.e. to a constant density $\bar{\rho}_{\rm de} = -\bar{p}_{\rm de}$, parametrized by the value of $\Omega_\Lambda = \Omega_{\rm de} \equiv \bar{\rho}_{\rm de}/\rho_{\rm crit}$ today.

A simple extension of this model is the wCDM scenario, in which one assumes a linear and constant equation of state $\bar{p}_{\rm de} = w\bar{\rho}_{\rm de}$, leading to $\bar{\rho}_{\rm de} \propto a^{-3(1+w)}$ by virtue of energy conservation. This model has two independent DE parameters, $\Omega_{\rm de}$ and w.

However, w could in principle vary with time. In a purely phenomenological approach, after confronting the wCDM model to observations, one can test a model with one more free parameter, accounting for the variation of w today. Then, there are three DE parameters, for instance $\Omega_{\rm de}$, $w(a_0)$ and $\frac{dw}{da}(a_0)$.

At the level of perturbations, different DE models with a given background evolution could differ through their pressure perturbation, or their anisotropic stress. First, let us stress that DE cannot be a perfect fluid. If this was the case, DE would have an adiabatic sound speed obeying to

$$c_a^2 \equiv \frac{\bar{p}'}{\bar{\rho}'} = \frac{(w\bar{\rho})'}{\bar{\rho}'} = w + \frac{w'}{3\frac{a'}{a}(1+w)} \tag{93}$$

where in the last equality we have used energy conservation. Given observational constraints on w and w' today, the squared adiabatic sound speed of dark energy must be negative. If the actual squared sound speed $c_s^2 \equiv \frac{\delta p_{\rm de}}{\delta\rho_{\rm de}}$ was negative, then DE perturbations would be unstable, and would explode exponentially, in contradiction with observations. This proves that for DE, $c_s^2 \neq c_a^2$. The stability of DE perturbations imposes $c_s^2 > 0$. On the other hand, the absence of superluminal propagation in conventional physics imposes an upper bound, $c_s^2 \leq 1$.

In phenomenological studies of DE models, the default assumptions concerning DE perturbations is that *(i)* pressure perturbations are related to density perturbations through $c_s^2 = 1$, and that *(ii)* anisotropic stress is negligible. This is justified by the fact that DE could be a scalar field (often called quintessence). If this is the case, one can easily show that the sound speed is indeed equal to one,[i] while anisotropic stress only appears at second order in perturbations.

A DE model with $c_s^2 = 1$ cannot lead to sizable DE perturbations. Indeed, $c_s^2 = 1$ implies that pressure forces in the fluid resist to gravitational collapse. Astrophysicists like to formulate this problem in terms of the Jeans instability. In this language, $c_s^2 = 1$ means that the Jeans length in the fluid is as large as the Hubble radius. Then DE perturbations remain as small as in the primordial universe (of the level of 10^{-5}), and can never leave an observable signature, even during DE domination.

Only if c_s^2 is assumed to be very small, DE perturbations can grow and impact LSS observables. Indeed, if c_s^2 goes to zero, the Jeans length can be arbitrarily small, and perturbations on cosmological scales can be amplified by the Jeans instability. In that case, one needs to postulate whether DE perturbations feature anisotropic stress or not. In a phenomenological approach, one may try to parameterize the anisotropic stress using some effective parameter (like a viscosity parameter). One should stress that a very small sound speed can be assumed, but these models sound artificial: $c_s^2 \ll 1$ is not a generic prediction of "realistic" DE models (if any such model exist).

Probing DE models with CMB and LSS data. If the DE sound speed is not very small, DE models can only be probed through their impact on the background evolution, which affects indirectly the evolution of metric and matter perturbations.

Since the background expansion at late time impacts the angular diameter distance, the CMB is sensitive to DE through effects (C1) and (C4) of Sec 2.6, both depending on $d_a(z_{\rm dec})$. However, (C1) and (C4) are also sensitive to many other quantities, so this probe is only useful in combination with other ones: in the best case it may remove parameter degeneracies.

The LISW effect (C7) probes the duration of the DE dominated era, and could bring one bit of information on DE models. In practise, it is however impossible to infer precise constraints from (C7) due to the large cosmic variance affecting the C_l's on the scale of the Sachs-Wolfe plateau.

[i] Actually, the quantity $\delta p_{\rm de}/\delta\rho_{\rm de}$ is gauge dependent. Everything becomes clearer if one decides to call "sound speed" the ratio $\delta p_{\rm de}/\delta\rho_{\rm de}$ calculated *in the frame comoving with the DE fluid*, i.e. in a gauge such that the velocity divergence $\theta_{\rm de}$ vanishes. If the sound speed of a given component derives from some fundamental principle, it should take a simple form precisely in this frame. Actually, it is only when this condition is imposed that one finds $\delta p_{\rm de} = \delta\rho_{\rm de}$ for a scalar field.

We have seen in Sec 3.3 that LSS data can be useful especially if the matter power spectrum is measured accurately at several distinct redshifts. Then, both the evolution of the power spectrum amplitude (in other words, of the linear growth factor) and of the BAO apparent scale encode information on the expansion rate at several redshifts, and therefore on w or even $w(a)$. This is why measurements of $P(k, z)$ with future ambitious programs, like for instance the Large Synoptic Survey Telescope or the Euclid satellite, are expected to provide the most accurate tests of DE models.

In summary, the CMB is not very useful for testing DE models, but can bring a few independent pieces of information; while LSS data at different redshifts is sensitive to the evolution of the background DE density. LSS data should anyway be used in combination with other probes of homogeneous cosmology (like supernovae luminosity) in order to get a maximum amount of information on DE.

The situation that we depicted would change radically if:

- DE would account for a significant fraction of the total energy density even at early times ($z \gg 1$), like in "early dark energy" models: then, DE could affect the CMB more strongly than assumed in the previous discussion;
- large DE perturbations would develop as a consequence of a very small sound speed $c_s^2 \ll 1$: then the total power spectrum (including DE fluctuations) may depart significantly from the ΛCDM one.

Both possibilities are still open, but no such effects are indicated by current data, and there are no solid theoretical motivations for our universe to fall in one of these two cases.

6. Conclusions and Bibliography

We hope that this short — but very dense — introduction to cosmological perturbations was clear enough, and that the reader will feel like going further.

Some readers might regret that, after exposing in details the effect of cosmological parameters on the CMB and LSS spectra (in Secs. 2.6, 3.3, 4 and 5), we did not show concrete bounds from current experiments. There are three reasons for this. First, the duration of the lectures and the length of these notes would have exploded. Second, such information would be outdated very quickly by new data releases. Third, it is actually a very good exercise (and an efficient way to digest this course) to look at published bounds on cosmological parameters (inferred from CMB data, or LSS data, or from both), and to figure out by oneself whether everything is consistent with these notes: we encourage the reader to check that the best and worst constrained parameter combinations can be related to the physical effects described here.

We promised a bibliography in the introduction. We already mentioned Ref. 1, in which one can find a similar but more detailed presentation, plus a discussion of neutrino effects. We have cited in the core of the text three seminal papers (Refs. 2, 3, 5) providing explicit calculations on precise topics: the line-of-sight integral in Fourier space [2], the justification of the Mészáros equation and of the fact that CDM behaves effectively as a self-gravitating species during radiation domination [3], and the matching between CDM and baryon perturbations around the time of baryon drag, in view of computing analytically baryon effects in the matter power spectrum [5].

Specialists of the field refer regularly to a concise paper of Ma & Bertschinger [6], containing a remarkably clear presentation of gauge issues and of all equations of motion in two gauges (including the Newtonian one).

Many recent textbooks cover most topics in cosmology, including linear perturbations, CMB and large scale structure. We only mention in the reference list those of S. Dodelson [7], V. Mukhanov [8] and S. Weinberg [4], all three outstanding, but there exist many excellent alternatives. For a presentation more focused on the CMB, we recommend the thesis of W. Hu [9] and the book of R. Durrer [10].

Acknowledgements

I warmly thank all TASI 2012 organizers — and particularly Elena Pierpaoli — for giving me the opportunity to lecture in such an inspiring atmosphere, and for patiently collecting and processing the lecture notes. I am grateful to the TASI students for their interest and feedback.

References

[1] J. Lesgourgues, G. Mangano, G. Miele and S. Pastor, *Neutrino Cosmology* (Cambridge Univ. Press, Cambridge, 2013).
[2] U. Seljak and M. Zaldarriaga, *Astrophys.J.* **469**, 437 (1996).
[3] S. Weinberg, *Astrophys.J.* **581**, 810 (2002).
[4] S. Weinberg, *Cosmology* (Oxford Univ. Press, Oxford, 2008).

[5] D. J. Eisenstein and W. Hu, *Astrophys.J.* **496**, p. 605 (1998).
[6] C.-P. Ma and E. Bertschinger, *Astrophys.J.* **455**, 7 (1995).
[7] S. Dodelson, *Modern cosmology* (Academic Press, San Diego, CA, 2003).
[8] V. Mukhanov, *Physical Foundations of Cosmology* (Cambridge Univ. Press, Cambridge, 2005).
[9] W. T. Hu, *Wandering in the Background: A CMB Explorer*, astro-ph/9508126, (1995).
[10] R. Durrer, *The Cosmic Microwave Background* (Cambridge Univ. Press, Cambridge, 2008).

Chapter 3

Implications of the Higgs Boson and the LHC for the MSSM

A. Pierce

Department of Physics, University of Michigan,
Ann Arbor, MI 48109, USA
E-mail: atpierce@umich.edu

These lectures were presented at the TASI 2012 summer school to a mixture of graduate students in particle theory and cosmology. They serve as an elementary introduction to the Minimal Supersymmetric Standard Model (MSSM) and discuss the implications of a 125 GeV Higgs boson for this theory. Some familiarity with the Standard Model is assumed.

Contents

1. Introduction

As recently as 2008, direct experimental knowledge of the Higgs boson was limited. All that was known was that its mass lay above the LEP bound [1] of 114 GeV. But progress has been rapid. At the beginning of this decade, the Tevatron and the Large Hadron Collider (LHC) quickly began to rule out large swathes of potential Higgs boson masses, and at the time these lectures were delivered, there were tantalizing hints of a Higgs boson at 125 GeV [2, 3]. Just a couple of weeks later, these hints were solidified by a discovery of a "Higgs-like" state near 125 GeV by both the ATLAS and CMS experiments [4, 5]. Around the same time, the Tevatron experiments also presented evidence for a Higgs-like state decaying to b quarks [6]. Working under the assumption that this state is in fact a Higgs boson, what are the implications for supersymmetry?

These lectures begin with an introduction to supersymmetry in two parts. First, we present a discussion of why we are interested in weak scale supersymmetry in the first place. With our appetites whetted, we move to overview the basic structure of the Minimal Supersymmetric Standard Model (MSSM). The treatment is a quick and dirty one. The goal is to give non-experts some appreciation for the structure of the MSSM, how to look for it, and how a 125 GeV Higgs boson fits into the picture. Because of our interest in the Higgs signal, we analyze the MSSM Higgs potential in some detail, giving much of the rest of the model short shrift. The motivated student will want to consult some of the several excellent reviews of this subject for more details. An incomplete list of recent pedagogical resources includes lecture notes, Refs. 7–13, as well as textbooks, Refs. 14–19.

2. Why Supersymmetry?

Before targeting supersymmetry, it is illuminating to first ask: why go beyond the Standard Model (SM) at all?

There are experimental reasons to extend the Standard Model. Some new particle must play the role of the Dark Matter which makes up the majority of mass in our universe. In addition, the Standard Model alone cannot explain the asymmetry between matter and antimatter in our universe. Moreover, if an epoch of inflation occurred early in the history of the universe, as is suggested by data from the Cosmic Microwave Background (CMB), there must be new physics to drive that inflation. All these observations warrant new physics. Unfortunately, they do not indicate the energy scale at which the new physics arises. It could be near the weak scale M_W (in which case we are on the cusp of probing it), or be stubbornly out of reach.

There are also purely aesthetic reasons to go beyond the Standard Model. For example, the Standard Model may have too many free parameters for comfort, some of which display strange patterns. Perhaps some semblance of order might be restored if the seemingly disparate forces could be satisfyingly unified into a single force, but presumably again at much higher energies, far removed from M_W. And there is the puzzle puzzle of why such disparate values of fermion masses are present within the Standard Model (e.g., the electron mass is six orders of magnitude smaller than that of the top quark). But again, there is no guarantee that the explanation — and the attendant new physics — should lie near the weak scale.

Of all the motivations to go beyond the Standard Model, both experimental and aesthetic, there is only one that firmly points to new physics at the weak scale: why is gravity so much weaker than the other forces? Equivalently, how can we explain the hierarchy $M_W << M_{pl}$? As we will discuss below, if the Standard Model model remains unmodified to high scales, the chasm between the weak scale, $M_W \sim 100$ GeV and the scale associated with gravity, $M_{pl} \sim 10^{18}$ GeV is unexpected. Quantum corrections attempt to drag the Higgs boson mass up to the highest energy scale at which the Standard Model remains valid. If this is the Planck scale, then we must ask why the Higgs boson stubbornly remains light in the face of these extreme quantum pressures. This is known as the fine–tuning or the hierarchy problem. The simplest solution is that the Standard Model is not, in fact, valid up to very high energies, and new physics arrives close to the weak scale to tame these quantum corrections.

But not just any new physics can play Siegfried and Roy to these quantum lions. But as we discuss below, supersymmetry is an example of the kind of physics that fits the bill, and this is one of the main reasons it is touted as an extension to the SM. In the MSSM the couplings of new particles conspire (because of the symmetry) so that the quantum pressures from these new particles precisely cancel those that would have dragged the

Higgs mass Planckward. But not without a price: supersymmetry posits the doubling of the Standard Model particles, an no such partners have been observed as yet.

2.1. *Supersymmetry isn't crazy*

This price may seem awfully high, and might cause one to dismiss supersymmetry out of hand. Here we present a historical analogy (which we believe first appeared in Ref. 20) that helps illustrate why such a doubling might not be so outlandish. Suppose, armed with knowledge of classical electromagnetism, one set about trying to calculate the correction to the electron's rest mass due to its electrical self energy. The result is parametrically

$$\begin{aligned}\Delta E_{elec} &\sim -\frac{\alpha}{r_e}, \\ &= -1.4\,\mathrm{MeV}\left(\frac{\mathrm{fm}}{\mathrm{r_e}}\right)\end{aligned} \tag{1}$$

where the r_e represents a "radius of the electron". It is not *a priori* clear what value one should choose for this electron radius. With detailed knowledge about the lack of electron substructure from LEP, one might choose something like something like $(\mathrm{TeV})^{-1}$ or even smaller. (This is a bit artificial, because these experimental limits depend upon on a knowledge of Quantum Field Theory, and we are imagining a strictly classical calculation, but let us see where this takes us.) Plugging in this value to Eq. (1), one would find a contribution to the *electron mass* (every electron has it after all) of -10 GeV. Comparing to its physical value, one would be forced to conclude that nature has engaged in a delicate balancing act:

$$m_e c^2 = .511\,\mathrm{MeV} = -10\ \mathrm{GeV} + 10.000511\ \mathrm{GeV} = \Delta\mathrm{E_{elec}} + \mathrm{m_0 c^2} \tag{2}$$

where $m_0 c^2$ represents all other contributions to the electron mass (including, e.g., the bare value).

But this calculation is incorrect — there is a sense in which the classical electron is in fact "large" on the TeV scale. Well before we get to these energies, the existence of the positron becomes important. As first calculated by Victor Weisskopf, once its contributions to the self-energy are taken into account, we find:

$$m = m_e^{bare}\left(1 + \frac{3\alpha}{4\pi}\log\left(\frac{\frac{\hbar}{m_e c}}{R}\right) + \cdots\right) \tag{3}$$

Here $\frac{\hbar}{m_e c} \approx 4 \times 10^{-13}$ cm, and R is a new length scale that tells us where this improved calculation breaks down. Even for $R \sim 1/M_{pl}$ this is only a 10% correction.

Thus the existence of partner particles (here the positron postulated by Dirac in 1928, and found by Andersen in 1932) was crucial to the cancellation. Furthermore, these particles are intimately tied to a symmetry (here the Poincaré symmetry that allows the formulation of a chiral symmetry). So, a symmetry forced nature to "double" the number of particles, and the new partner particles softened the self-energy. Is it possible that this story repeats itself in nature, with supersymmetry playing the role of the doubling symmetry?

2.2. *How would this apply to the Standard Model?*

The fermions of the Standard Model (thanks to the chiral symmetry discussed above) do not require a delicate balancing act in their mass budget. However, the situation is different for scalars, and in particular, the Standard Model Higgs boson, whose vacuum expectation value sets the electroweak scale. Consider the Standard Model Higgs potential:

$$V(H) = -\mu^2|H|^2 + \lambda|H|^4, \tag{4}$$

with

$$\langle H\rangle = \frac{1}{\sqrt{2}}\begin{pmatrix}0\\ v\end{pmatrix} \qquad v = \sqrt{\frac{\mu^2}{\lambda}} \qquad v = 246\,\text{GeV}. \tag{5}$$

Just as the classical electron received large corrections to its mass electrical interactions, the Higgs field receives large corrections to its mass squared parameter μ^2 via, e.g., its coupling to the top quark.

The diagram shown in Fig. 1 contributes a value

$$\Delta(\mu^2)^{1-loop}_{top} = -\frac{3\lambda_t^2}{8\pi^2}\Lambda^2, \tag{6}$$

where λ_t is the Yukawa coupling to the top quark, and Λ corresponds to the scale where our calculation breaks down. If, for example, Λ were identified with M_{pl} — as would be appropriate if the Standard Model were valid all the way up to scales were gravity becomes strong — then we have an equation of the form:

$$\mu_0^2 + \Delta\mu^2 \sim M_W^2 \tag{7}$$

where $\Delta\mu^2$ is about 10^{32} times M_W^2! This makes the conspiracy of Eq. (2) seem tame. Such a fine-tuning is said to be "unnatural". The desire for

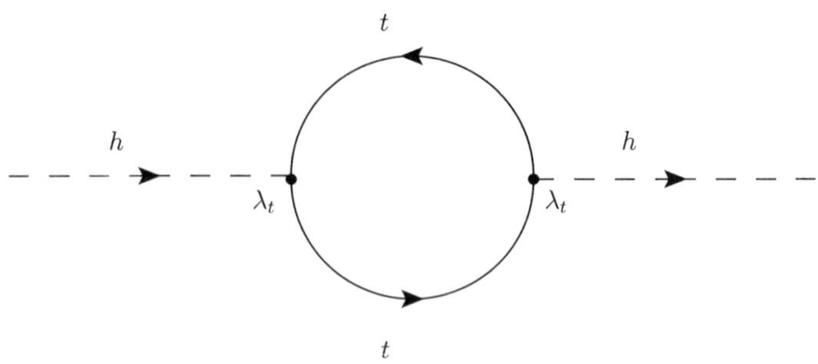

Fig. 1. Leading correction to the Standard Model Higgs mass squared parameter from the top quark.

naturalness in the physics of the electroweak scale is the main motivator for new physics at the LHC, for the way to avoid an unnatural theory is to posit that Λ is not M_{pl} but instead a much lower scale, of order a TeV. Then $\Delta\mu^2$ will be the same size as M_W.

Perhaps we can gain insight on fine-tuning by considering an example outside the realm of particle physics: the Landau-Ginzburg theory for superconductors. There the free energy takes a form

$$F = \alpha|\phi|^2 + \beta|\phi|^4, \tag{8}$$

with α having leading temperature dependence

$$\alpha(T) = \alpha_0(T - T_C), \tag{9}$$

where T_c is the critical temperature corresponding to the phase transition. For $T < T_c$, minimization of the free energy leads to a non-zero vacuum expectation value of ϕ. This non-zero value is the "order parameter" associated with the transition to the superconducting state. Furthermore, below the transition point, the size of the order parameter is proportional to the distance from the critical temperature. One would expect α_0 to be related to the fundamental mass scales in the theory. If the Standard Model is unchanged up to the Planck scale, the incredible smallness of the electroweak scale relative to the Planck scale is like having a superconducting system tuned to a temperature extraordinary close to the transition temperature. Moreover, the Higgs vacuum expectation is pegged there in spite of the quantum fluctuations (cf. thermal fluctuations in the superconductor case). In the case of the superconductor, it is clear what is doing the fine-tuning — it is the experimentalist, who is diligently keeping her sample within a hair's breadth of the critical temperature. But what physics is doing that job for the electroweak theory?

2.3. *A bit more on naturalness*

This discussion about naturalness and fine-tuning may seem abstract. Given its importance for determining whether we will, in fact, see new physics at the LHC, it would be useful to know whether or not there are examples where naturalness is a useful guiding principle. A lucid discussion of these points appears in Ref. 21.

One example is the mass splitting between the charged and neutral pion. This mass difference receives a contribution from the the the electromagnetic interaction, and is logarithmically divergent. This divergence is effectively cut-off by the ρ-meson mass, so that the size of the contribution to the mass splitting is "natural". The mass of the ρ was not predicted in this way, but it is indeed a natural system.

Another, perhaps the most compelling, historical example [22] relates to the discovery of the charm quark. If one calculates in an effective theory that has only u, d, s quarks and weak interactions, the contribution to the neutral kaon mass splitting is quadratically sensitive to the cutoff of the effective theory Λ:

$$\frac{\Delta M_K}{M_{K_L^0}} = \frac{G_F f_K}{6\pi^2} \sin^2\theta_C \Lambda^2, \tag{10}$$

with $\Delta M_K \equiv M_{K_L^0} - M_{K_S^0}$, G_F the Fermi constant, and $f_K = 114$ MeV the Kaon decay constant. The experimentally observed value [23] of the left hand side is 7×10^{-15}. Requiring a natural theory imposes that the right-hand-side not supersaturate this value. This is the condition that the contribution calculated in the effective theory need not be finely tuned against other contributions. Here this implies that the cutoff $\Lambda < 2$ GeV. So, naturalness indicates that new physics should arise before this scale, and in fact it does: the charm quark has mass $m_c = 1.2$ GeV. Its existence is the foundation of the Glashow-Illiopoulos-Maini (GIM) mechanism.

Whether the weak scale is in fact natural is an experimental question — one that we should be able to answer soon.

3. A Shortcut to Building the MSSM

We have foreshadowed that the doubling of particles in the MSSM will soften the ultraviolet behavior and reduce the fine-tuning. This results from a cancellation between loops of fermions (like the top quark) and bosons (like its superpartner, the stop). It is the supersymmetry itself that assures this cancelation. To appreciate why this is so, we first need to understand something about the interactions in these theories.

We take a relatively streamlined approach to quickly get to the structure of the MSSM, glossing over important details and stating some results without proof. For details, consult Ref. 7, or other lecture notes referred to in the introduction.

3.1. *Some basics of supersymmetric theories*

In supersymmetric theories, particles can be packaged into "supermultiplets" that serve as building blocks for the theory. Particles within a supermultiplet share gauge charges but differ in spin. Using these building blocks, we can find a compact presentation of the MSSM. Since supersymmetry relates fermions and bosons and supermultiplets contain both, they allow us to keep supersymmetry manifest throughout the construction. For the MSSM we will focus on two multiplet types[a]:

- Chiral multiplets, which contain a complex scalar and a Weyl fermion.
- Vector multiplets, which contain a vector field and a fermion (gaugino).

We will first introduce these multiplets, and then describe how to use them to construct the supersymmetric Lagrangian of the MSSM.

Precisely because supersymmetry relates bosons and fermions, in each multiplet the bosonic and fermionic degrees of freedom must match. In the chiral multiplet, two real bosonic degrees of freedom are bundled as a complex scalar, ϕ. The chiral superfield Φ can be written as

$$\Phi(y) = \phi(y) + \sqrt{2}\theta\psi(y) + \theta^2 F(y), \tag{11}$$

with ψ a Weyl Fermion. F is an auxiliary field (whose purpose we discuss below), and

$$y^\mu \equiv x^\mu - i\theta\sigma^\mu\bar{\theta}. \tag{12}$$

Each SM matter field (i.e. quark or lepton) gets promoted to a chiral multiplet and will be identified with these ψ — meaning that the MSSM will

[a]Here, for the most part we will ignore gravity, which requires a multiplet of its own. If the gravitino, the superpartner of the graviton is the lightest superpartner is the LSP, there can be important impacts on collider phenomenology. For a brief overview, see Ref. 7.

have scalar partners for each field. The names of scalar superpartners are found by prepending an "s" (for scalar) to the Standard Model particle name, e.g. "squark" and "slepton". (The names of fermionic superpartners are found by appending the suffix "ino", e.g. "Higgsino" or "gaugino".) The θ that appears here is a two component Grassman (anti-commuting) spinor. The product of ψ and θ is bosonic, as it should be — everything on the RHS of Eq. (11) should have the same statistics. Because θ is anti-commuting and only has two independent components, $\theta^n = 0$, for $n \geq 3$. A consequence is that Taylor expansions (about $\theta = 0$) quickly truncate. A Weyl fermion has two degrees of freedom, so this balances against the two bosonic degrees of freedom in ϕ. From studies of QED, four component fermions (and Dirac matrices flying fast and furiously) may be more familiar, but the Standard Model electroweak gauge group treats left-handed and right-handed fermions differently: it is a chiral gauge theory. When dealing with such theories — in contrast to the vectorlike theories of QCD and QED — it is more natural to use two component notation (and it is Pauli spin matrices do the flying).[b]

We now return to discuss the auxiliary field F. Off-shell a Weyl fermion has four, rather than two degrees of freedom. How, then, does the accounting work in the superfield to preserve the symmetry between bosons and fermions? The answer is to introduce a bosonic auxiliary field with two components of its own. It is non-propagating, and on-shell it will just be a function of the other scalar fields in the theory. Off-shell, it provides two additional bosonic degrees of freedom, and supersymmetry can be manifest, even off-shell.

The vector multiplet is a bit more complicated, but in a particular gauge [25], its components can be taken to be:

$$V = -\theta\sigma^{\mu}\bar{\theta}A^{a}_{\mu}(x) + i\theta^{2}\bar{\theta}\,\lambda^{a\,\dagger}(x) - i\bar{\theta}^{2}\theta\lambda^{a}(x) + \frac{\theta^{2}\bar{\theta}^{2}}{2}D^{a}(x). \tag{13}$$

Here, A_μ corresponds to a vector field (e.g., gluon in QCD), λ corresponds to its fermionic superpartner (e.g. a gluino) and D corresponds to another non-propagating auxiliary field, which again helps the degrees of freedom balance off-shell. a is the gauge index which runs over the adjoint representation of the gauge group in question. Another useful object is the (chiral)

[b]The conversion between these two languages can be a little tricky at first if you are not used to it. The recent treatise by Dreiner, Haber and Martin [24] is a comprehensive and useful reference on two-component notation, and includes a nice appendix on how to go back and forth to four-component notation.

superfield that packages the gauge field strength

$$\mathcal{W}^a = -i\lambda^a(y) + \theta D^a(y) - \sigma^{\mu\nu}\theta F^a_{\mu\nu}(y) - \theta^2\sigma^\mu D_\mu \lambda^a(y). \tag{14}$$

The degrees of freedom here are the same as those enumerated in V above (after our gauge fixing), but both packagings (i.e. V and $\mathcal{W}^a$) are useful for constructing parts of the Lagrangian. It is useful to think of the relationship between V and $\mathcal{W}^a$ as the supersymmetrization of the relationship between A_μ and its field strength $F_{\mu\nu}$.

To help us write down supersymmetric theories compactly, it turns out that Grassman integration is quite useful. For a Grassman variable η, we have

$$\int d\eta = 0 \quad \int \eta\, d\eta = 1. \tag{15}$$

Thus, an integral of a Grassman variable picks out the coefficient of that variable. It is useful to introduce the notation $\int d^2\theta$ — which selects out the coefficient of θ^2 — and $\int d^4\theta$ — which selects out the coefficient of $\theta^2\bar{\theta}^2$.

We present a few facts without proof (again, refer to, e.g., Refs. 7, 14 for discussion) that will allow us to construct supersymmetric Lagrangians in an efficient, compact way:

(1) The product of chiral superfields is itself a chiral superfield.
(2) The supersymmetric variation of the F component of a chiral superfield is a total derivative, hence when integrated over d^4x is invariant.
(3) The supersymmetric variation of the $\theta^2\bar{\theta}^2$ component of *any* superfield is a total derivative.

So, working with the basic building blocks of chiral and vector superfields, we can identify a recipe to construct the MSSM.

(i) Identify all the left-handed chiral superfields Φ_i that you need based on the particle content of the theory.
(ii) We want to generate the kinetic terms for the fields contained in the Φ_i. These arise from "Kähler potential interactions." Taking

$$\int K(\Phi_i, \Phi_i^\dagger)\, d^4\theta,$$

selects out a $\theta^2\bar{\theta}^2$ term, which is supersymmetric (by (3) above). The function K is known as the Kähler potential. A minimal Kähler potential $\Phi_i^\dagger\Phi_i$ will generate appropriate canonical kinetic terms for a theory of chiral superfields. If gauge interactions are added, this should be modified as

$$\int \Phi_i^\dagger e^V \Phi_i \, d^4\theta.$$

This generates kinetic terms for the fields in Φ along with the interactions of these fields with the gauge fields dictated by gauge invariance. (This requires starting with Eqs. (11), (12), and (13) and doing a Taylor expansion). It also generates interactions between scalars proportional to the gauge couplings once the auxillary field D is eliminated. These interactions are required by supersymmetry and are known as D terms. They will figure prominently in our discussion of the Higgs boson potential.

(iii) We can construct another supersymmetric object using the $\mathcal{W}^a$ field strength superfield of Eq. (14). For each gauge group write down:

$$\int \mathcal{W}^a(y)\mathcal{W}^a(y)\, d^2\theta.$$

This is supersymmetric because it is the θ^2 component of a chiral superfield. (Remember the product of two chiral superfields is itself a chiral superfield, see (1) above). This provides the kinetic terms for the gauge fields (and gauginos).

(iv) Finally, choose a *superpotential* $W(\Phi_i)$. (Too bad that everything is called W.) This superpotential is a polynomial function of chiral superfields. Here we use symmetry and renormalizability as guiding principles. Notably, supersymmetry imposes that the W is a function of the *chiral* superfields only, and not "anti-chiral" superfields, $\Phi^\dagger$. We say that the superpotential is a *holomorphic* function. Then (because it is an F term of a chiral superfield) the supersymmetric variation of $\int W(\Phi_i) d^2\theta$ will be a total derivative. So, the action that results from this construction will be supersymmetric (again, see (1)&(2) above).

All told, the interactions found by expanding out the expressions in (ii), (iii) and (iv) represent a supersymmetric Lagrangian.

3.2. *How to supersymmetrize the Standard Model*

The above prescription allows us to construct supersymmetric Lagrangians. Let us now turn to the MSSM itself. First we identify the superfield content.

Table 1. The Higgs and matter superfields of the minimal supersymmetric standard model.

Field	SU(3)	SU(2)$_L$	U(1)$_Y$
Q	3	2	1/6
U^c	$\bar{3}$	1	-2/3
D^c	$\bar{3}$	1	1/3
L	1	2	-1/2
E^c	1	1	1
H_u	1	2	1/2
H_d	1	2	-1/2

There is a vector superfield associated with each gauge group of $SU(3)_C \times SU(2)_L \times U(1)_Y$ — each of which contains a gaugino, λ_i, and a gauge field A^μ. There is a chiral superfield for each matter and Higgs field. It is worth noting that none of the known fermions of the Standard Model can play double duty as as a gaugino — none of the SM fermions are in the appropriate adjoint representation. Furthermore, one cannot build a model where the superparter of the neutrino is a Higgs boson. Although the quantum numbers in this case are basically correct, in these models one would get (much) too large neutrino masses. With these restrictions in mind, we are led to the particle content of Table 1. Note that the electric charge is given as $Q = T_3 + Y$.

The Higgs sector with its two superfields H_u, H_d warrants additional explanation. In the Standard Model, there is a single complex scalar doublet. So, we might have thought that one could promote this scalar to a superfield and be done with it. However, there are a number of independent arguments that indicate that this is insufficient. First, the supersymmetrization of the Higgs field introduces a new chiral fermion beyond the Standard Model content (a Higgsino). The Standard Model content (without this new fermion) is by itself non-anomalous [26]. This means that the presence of a single Higgsino introduces a gauge anomaly from the triangle diagram containing the Higgsino. This renders the theory sick at the quantum level. The introduction of two Higgs multiplets with opposite hypercharge eliminates this objection. A separate (also anomaly-based) argument is that adding a single doublet would introduce a global $SU(2)$ anomaly [27] — only theories with even numbers of doublets are non-anomalous. The addition of a second Higgs multiplet obviates this problem.

A separate line of argument is that we need to provide masses to all the fermions in the theory. In the SM, the Higgs boson is in a sense overly efficient. The Yukawa portion of the SM Lagrangian reads:

$$\mathcal{L} = \lambda_d \overline{Q_L} \cdot \phi \, d_R - \lambda_u \epsilon^{ab} \overline{Q_L}_a \phi_b^\dagger u_R + H.c. \tag{16}$$

The presence of the $\phi^\dagger$ presents problems for a naive supersymmetrization. Recall, to maintain supersymmetry, the superpotential $W(\Phi^i)$ can only contain superfields (and not their complex conjugates). Again, the introduction of a second Higgs supermultiplet solves the problem. We write

$$W_{Yukawa} = \lambda_u Q H_u U^c + \lambda_d Q H_d D^c + \lambda_\ell L H_d E^c \tag{17}$$

with contraction over relevant indices implied. The presence of the second Higgs multiplet also allows us to write

$$W_\mu = \mu H_u H_d. \tag{18}$$

This term is important because it allows a (Dirac) mass for the fermionic component of the Higgsino multiplet. These last two equations, along with the Kähler potential and gauge kinetic terms specify the supersymmetric interactions of the MSSM.

3.3. *Interactions and cancellation of divergences*

Let us write down a superpotential with three superfields, $\{Q, U^c, H_u\}$:

$$W = \lambda_u Q H_u U^c. \tag{19}$$

Assume there is also a minimal Kähler potential

$$K = Q^\dagger Q + H_u^\dagger H_u + U^{c\dagger} U^c. \tag{20}$$

What interactions result from the superpotential of this theory? Returning to our discussion in 3.1, the prescription is to pick out the θ^2 terms in the expansion of the superfields in Eq. (19). One possibility is Yukawa interactions (see Fig. 2) containing two fermions (which each come with a single power of θ) and a single scalar field (no θs). This does not yield just a single Yukawa interaction — rather you get three different ones, all with strength λ_u. Each superfield gets a turn being the scalar. These interactions — and the equality of the couplings — are forced by the supersymmetry of the theory.

In addition, this superpotential generates interactions solely involving scalars, the second line of Fig. 2. Each of these vertices comes with strength $|\lambda_u|^2$. These interactions arise when Grassman integration picks out two scalar fields (no θs) and a single auxilary field, F (with its θ^2). The F

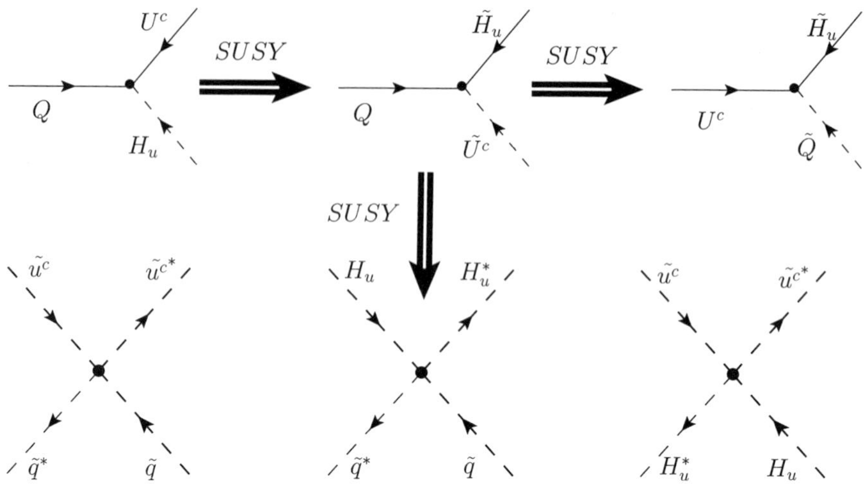

Fig. 2. Vertices generated by the superpotential interaction of Eq. (19).

can be eliminated using the equation of motion. The Lagrangian contains an FF^* term (arising from the Kähler potential) and no F term involving derivatives (it is not a propagating field); so the Euler-Lagrange equations are particularly simple: $\partial\mathcal{L}/\partial F = 0$. This equation sets F^* equal to a simple function of the scalar fields. The result is the interactions in Fig. 2.

The relationship between the couplings of the Yukawa interactions in the top line of Fig. 2 and the quartic scalar interactions in the second line of Fig. 2 is what enforces cancellation of corrections to the Higgs boson mass squared in supersymmetry. In a supersymmetric theory, a diagram with scalars running around the loop cancels against a diagram with fermions running around the loop because a closed fermion loop comes with an extra (-1).

3.4. *Supersymmetry breaking*

Supersymmetry cannot be an exact symmetry. If it were, scalars and fermions with identical quantum numbers that would be degenerate. The absence of a 511 keV charged scalar is a pretty good clue that supersymmetry is at best a broken symmetry. We will return to how this occurs momentarily, but first let us ask what supersymmetry breaking means for the ultraviolet behavior of this theory. After all, we imposed supersymmetry in an effort to soften the quadratic divergence of the Higgs boson mass parameters. Does adding supersymmetry breaking undo all this hard work? Not necessarily. In fact, it tells us something about how much supersymmetry breaking we expect in nature.

To understand this, consider a toy model where there is a single superfield Φ with superpotential

$$W_{toy} = \frac{\lambda}{3}\Phi^3 + \frac{M}{2}\Phi^2. \tag{21}$$

To this supersymmetric theory, we add a scalar mass.

$$\mathcal{L}_{\not{SUSY}} = \mu^2|\phi|^2, \tag{22}$$

where ϕ is the scalar component of Φ. The absence of an analogous fermion mass means this breaks supersymmetry. To fully compute the self-energy of the ϕ at one-loop there are actually three diagrams. But to understand how the quadratic divergence is softened, it is sufficient to examine the two diagrams of Fig. 3. The one on the left arsies from integrating out the F-term.

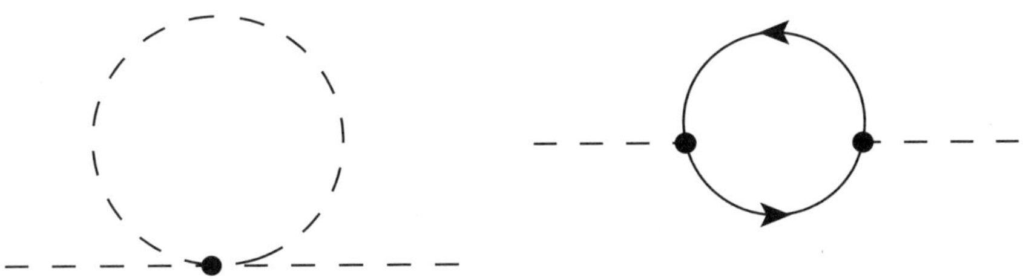

Fig. 3. Diagrams contributing to ϕ self energy in the toy model of Eq. (21).

They yield

$$\begin{aligned}\text{Fig. 1} &= \frac{\lambda^2}{16\pi^2}\left(2\int\frac{d^4k}{(k^2+M^2+\mu^2)} - \int\frac{d^4k\,Tr[(\not{k}-M)^2]}{(k^2+M^2)^2}\right),\\ &= \frac{2\lambda^2}{16\pi^2}\int\frac{d^4k\,\mu^2}{(k^2+M^2+\mu^2)(k^2+M^2)},\\ &= \frac{\lambda^2\mu^2}{8\pi^2}\log\frac{\Lambda}{M},\end{aligned} \tag{23}$$

where we have cut off the integral at energy Λ.[c] So, while the theory with supersymmetry breaking has a UV divergence, it is only logarithmic in

[c]Even for $\Lambda \sim M_{pl}$, this logarithm is manageable, much like in the case of the chiral symmetry of the electron.

nature. Furthermore, we note that the coefficient is directly proportional to the size of supersymmetry breaking. This is the crux of the argument for superpartners at the weak scale. If supersymmetry breaking is $\mathcal{O}$(TeV) or less, then the radiative corrections will not vastly exceed the weak scale.

3.5. *Soft supersymmetry breaking parameters*

We demonstrated that supersymmetry breaking scalar masses at the weak scale will not spoil the solution to the hierarchy problem. In fact, there are a variety of supersymmetry breaking terms that do not spoil the UV behavior [28]

- $m^2|\phi|^2$: soft scalar masses
- $B\phi_i\phi_j$: a mass term between scalars of the same "holomorphicity". Due to gauge invariance (and R-symmetry) this turns out to only be relevant for the combination H_uH_d
- $A_{ijk}\phi_i\phi_j\phi_k$: trilinear scalar interactions, or "A-terms"
- $m_\lambda\lambda\lambda$: gaugino masses.

Since the MSSM in the supersymmetric limit is completely controlled by Standard Model parameters, it is variation in the supersymmetry *breaking* parameters that control the variations in phenomenology.

Where do these Supersymmetry breaking terms come from? One might have thought that it could be possible to somehow spontaneously break supersymmetry within the MSSM itself. However, there is a theorem due to Dimopoulos and Georgi [29] that shows that this approach will lead to a squark with mass less than 10 MeV, which is well excluded. For an elementary review of the proof of this theorem, see Ref. 30.

We then are led to a picture where supersymmetry breaking occurs in a so-called hidden-sector quite separate from the MSSM, and this supersymmetry breaking is then subsequently communicated (or "mediated") to the visible sector, which is what we call the MSSM. The details of the mediation mechanism determines how we perceive supersymmetry breaking; how supersymmetry is broken primordially is less important for the details of the superpartner spectrum.

One example is so called gravity mediation, wherein supersymmetry is supposed to be broken primordially in another sector, and that supersymmetry breaking is communicated to the visible sector by Planck suppressed operators. Another example is gauge mediation wherein supersymmetry is broken and communicated to "messenger" fields that possess Standard

Model gauge quantum numbers. The Standard Model gauge interactions subsequently communicate the supersymmetry breaking to the visible sector.

One consequence of all the allowed supersymmetry breaking terms enumerated above is a proliferation of parameters. A careful accounting, see, e.g. Ref. 10, yields 105 physical parameters beyond the Standard Model. These parameters cannot be arbitrary. If chosen willy-nilly, loop diagrams involving superpartners will lead to dangerous flavor changing neutral currents (FCNCs), yielding unacceptably large contributions to processes like $b \to s\gamma$, $\mu \to e\gamma$, and $K \to \bar{K}^0$ oscillations. This is known as the supersymmetry flavor problem. If the supersymmetry breaking masses are nearly universal, this substantially mitigates the contributions to these processes.

In part motivated by this, and in part by a desire for simplicity, a toy spectrum called the constrained minimal supersymmetric standard model (CMSSM) has been widely adopted wherein one posts a single universal gaugino mass $m_{1/2}$, a single universal scalar mass m_0, and vanishing trilinear terms, all at the GUT scale. The additional parameters of this ansatz are $\tan\beta \equiv v_u/v_d$ and the sign of the μ parameter. Experimental results are often quoted in terms of this parameter space. It should be emphasized that this is a toy model and not the result of an underlying theory.

Gauge mediation, on the other hand, solves the supersymmetry flavor problem by design — the only flavor violation will be proportional to that already present in the standard model. It faces its own challenges, however, including finding a path to an attractive cosmology [31].

3.6. *Unification of couplings*

With the MSSM structure in place we are equipped to discuss one of the main pieces of "evidence" for the MSSM: unification [32, 33]. Can g_Y, g and g_3 all derive from a single unified gauge group at the high scale? To quantitatively discuss this question, it is useful to recall the renormalization group equation for a general (not necessarily supersymmetric) gauge theory. At one loop it is

$$\frac{dg}{dt} = \frac{b^{1-loop}}{16\pi^2} g^3, \tag{24}$$

with $t \equiv \ln\mu$ and

$$b^{1-loop} = -\frac{11}{3}C_2(G) + \frac{2}{3}\sum_f T_f(R) + \frac{1}{3}\sum_s T_s(R). \tag{25}$$

Here, $C_2(G)$ is the quadratic Casimir of the gauge group in question. The quadratic Casimir is defined as $(T^a)_r^s(T^a)_s^t \equiv C_2(R)\delta_r^t$ for a representation R, with the T^a in that representation. For our purposes, we only need $C_2(G)$ the quadratic Casimir for the adjoint representation; $C_2(G) = N$ for SU(N), $N > 2$, and $C_2(G) = 0$ for a U(1) factor. This contribution is due to the gauge field self-interactions. The vanishing of this quantity for a U(1) (which is abelian and lacks gauge self-interactions) is consistent with this interpretation. The $T(R)$ are the so-called Dynkin indices of a representation R, $Tr(T^aT^b) \equiv T(R)\delta^{ab}$. Note $T(R) = \frac{1}{2}$ for a fundamental of SU(N), $N > 2$, and is equal to the charge squared, Q^2, for a U(1) factor. For an adjoint of SU(N), $N > 2$, we have $T(adj) = N$. The solution to Eq. (24) is

$$\frac{1}{\alpha_{high}} = \frac{1}{\alpha_{low}} - \frac{b^{1-loop}}{2\pi} \log\left(\frac{\mu_{high}}{\mu_{low}}\right). \tag{26}$$

To check for unification, we can use the experimentally measured values of $\alpha_i(M_Z)$ and ask whether they are consistent with a universal value of α_{GUT}.

Contributions from the SU(3) gauge sector and the three generations of Q, U^c and D^c yield $b_{SM}^{SU(3)} = -7$. Similarly, contributions from SU(2) gauge fields, the Higgs doublet, and the three generations of Q and L yield $b_{SM}^{SU(2)} = -19/6$. Finally, carefully accounting for multiplicity of generations, doublets, and color, gives $b_{SM}^{Y} = 41/6$.

For supersymmetric theories, Eq. (25) simplifies.

$$b_{SUSY}^{1-loop} = -3C_2(G) + \sum_{r,matter} T_r, \tag{27}$$

where we have used the fact that there is a fermionic representation in the adjoint (gauginos) and that all matter representations have both fermions and scalars. With MSSM field content, it is straightforward to compute $b_{SU(3)}^{MSSM} = -3$, $b_Y^{MSSM} = 10$, and $b_{SU(2)}^{MSSM} = 1$. If we assume the superpartners all lie at the weak scale (more on this below), we can use these supersymmetric beta functions to evolve the measured values of gauge couplings at the weak scale up to high scales and check for unification.

Before checking unification, we must deal with a subtlety: b^Y and g^Y are not properly normalized to be embedded in a grand unified theory. To ensure proper normalization, we want Tr(generator^2) to be equal for any gauge generator in the unified group, where the trace runs over all particles in a representation. Consider a $\bar{5}$ of SU(5)

$$\bar{5} = \begin{pmatrix} d^c \\ d^c \\ d^c \\ \begin{pmatrix} \nu \\ e^- \end{pmatrix} \end{pmatrix}. \tag{28}$$

For color generators $SU(3)_C$, we have $Tr(T^a T^b) = T(R)\delta^{ab}$, with $T(R) = \frac{1}{2}$ coming from the the (anti-)fundamental down-quark. On the other hand, for hypercharge, $Tr(Y^2) = 3 \times (1/3)^2 + 2(1/2)^2 = 5/6$. Since $5/6 \neq 1/2$ we need a new normalization for our $U(1)$. We cannot arbitrarily change the physical combination of couplings and quantum numbers, so we must take

$$g_Y Y = g_1 Y', \tag{29}$$

such that

$$Y' = \sqrt{\frac{3}{5}} Y \qquad g_1 = \sqrt{\frac{5}{3}} g_Y. \tag{30}$$

We are now (finally) in a position to check unification. At one loop, we find (imposing a universal GUT coupling at the high scale) Eq. (26) implies

$$\frac{b_3 - b_2}{b_2 - b_1} = \left(\frac{\alpha_3^{-1} - \alpha_2^{-1}}{\alpha_2^{-1} - \alpha_1^{-1}} \right) \Bigg|_{M_Z}. \tag{31}$$

Using experimentally determined values [23] of the couplings on the RHS at M_Z, we find $RHS = .718 \pm 0.003$. On the other hand, substituting our beta functions above for the LHS we find $LHS_{SM} = .528$ and $LHS_{MSSM} = .714$. The Standard Model does not appear to unify. The MSSM seems plausibly consistent with unification. We did not take into account (important) two-loop effects, or threshold effects (from both the weak scale and GUT scale) arising from mass splittings within grand unified multiplets. Finally, we have assumed all the superpartners lie right at M_Z, and apparently, they do not. So, the precise agreement demonstrated here for the MSSM should be somewhat taken with a grain of salt, but it is reasonable to say that unification plausibly occurs in the MSSM once percent level threshold corrections are taken into account.

This unification of couplings is perhaps the single-most compelling piece of evidence (aside from the hierarchy problem itself) for finding supersymmetry at the weak scale. It is possible that it is a coincidence, but it is striking. Without this evidence, it is unlikely that supersymmetry would have received quite so much attention from the theoretical community.

4. Basics of Supersymmetry for Colliders

The goal of this section is to give a broad brush appreciation of the kinds of signatures that the MSSM can produce and are searched for at colliders.

4.1. *Where is SUSY?*

Building on the arguments of Sec. 2, we can ask a bit more precisely where we expect supersymmetry to show up. Consider the one-loop correction to the up-type Higgs $(mass)^2$ parameter. Interactions proportional to the top quark Yukawa coupling (see Fig. 1) would be quadratically divergent. However, when supplemented by diagrams involving the stop squark, they are softened to:

$$\Delta m_{H_u}^2 = -\frac{3\lambda_t^2}{8\pi^2}\left(\tilde{m}_{Q_3}^2 + \tilde{m}_{u_3}^2 + |A_t|^2\right)\log\left(\frac{\Lambda}{m_{\tilde{t}}}\right), \tag{32}$$

with $\tilde{m}$ representing soft scalar masses, and A_t a trilinear coupling for the top squarks. The soft supersymmetry breaking parameters appear explicitly (just as in our toy model) out in front — the correction to the Higgs mass is quadratically sensitive to these. Furthermore, there is a logarithmic enhancement, cutoff at some scale Λ. This is the scale at which the soft masses explicitly "become soft". The value of Λ depends on the supersymmetry breaking and mediation mechanism and can lie anywhere between 100 TeV and the Planck scale.[d] All told, this equation suggests that the third generation superpartners should be quite close to a TeV.

Also, somewhat less directly (but importantly), there is a correction to the stop mass coming from the gluino

$$\Delta m_{\tilde{t}}^2 = \frac{8\alpha_s}{3\pi}|M_3|^2\log\left(\frac{\Lambda}{m_{\tilde{t}}}\right). \tag{33}$$

A too-large gluino mass will feed into the stop mass, and then quickly feed into the Higgs mass parameter via Eq. (32). The combination of these facts argue that a natural supersymmetric theory wants both gluinos and stops to be "somewhat light". It is less clear where all the other superpartners must lie (though naturalness considerations also suggest light Higgsinos, more on this below). It might be that all superpartners are fairly close in mass — naturalness considerations alone, however, would allow the superpartners of the first two generations to be quite a bit heavier.

[d] In the case of gauge mediation, Λ corresponds to the messenger masses. For gravity mediation, the Planck scale would be appropriate; so the logarithm largely cancels the loop factor that appears out in front.

Table 2. R Charges of superfields and components.

Field	R	Field	R
Vector Superfield V	0	Chiral Superfield Φ	q
gauge field: v_μ	0	scalar field: A	q
gaugino: λ	1	fermion: ψ	q-1
auxiliary D	0	auxiliary F	q-2

So the above arguments suggest that stops and gluinos (at least) should be produced at the LHC if the theory is to avoid fine-tuning. As the limits on gluinos and stops increase, this means that the possibility of a natural theory of supersymmetry is becoming more and more squeezed. But before we are in a position to talk about how to see gluinos, stops, and other superpartners, we need to understand how they behave in the detector. Some will immediately decay, but — at least in the most common version of the MSSM — the lightest will not. To understand why this is so requires an examination of a new kind of symmetry possible in supersymmetric theories: the "R-symmetry".

4.2. *R-symmetries and the fate of the lightest superpartner*

Under an R-symmetry, the Grassman θ parameters that enter the superfield expansion transform:

$$\theta \to e^{i\alpha}\theta \;\; ; \;\; R[\theta] = 1, \tag{34}$$

$$\bar{\theta} \to e^{-i\alpha}\bar{\theta} \;\; ; \;\; R[\bar{\theta}] = -1. \tag{35}$$

An immediate consequence is that R-symmetry treats different components of a super-multiplet differently, see Table 2. The R charge of the superfield is equal to the combined R-charge of the component fields with the R-charges of the θ's that multiply the component fields in the superfield expansion.

The Kähler potential term $\int d^4\theta\, Q^\dagger e^V Q$ is invariant upon imposing $R[V] = 0$. What condition does R-invariance impose on the superpotential? Since $\int d^2 theta$ effectively strips away two θ's, we can write $R[d^2\theta] = -2$, and if $R[W] = 2$ the Lagrangian will be invariant, see Sec. 3.1.

Now suppose we write down a superpotential that respects this symmetry. An example is the superpotential of the MSSM

$$W = \lambda_u Q U^c H_u + \lambda_d Q D^c H_d + \lambda_\ell L E^c H_d + \mu H_u H_d, \tag{36}$$

with $R_Q = R_L = R_{H_u} = R_{H_d} = 1$, $R_{U^c} = R_{D^c} = 0$.

What does soft supersymmetry breaking do to this symmetry? We can use the spurion language (discussed in detail in J. Thaler's lectures at this school) to encapsulate the effects of supersymmetry breaking. We parameterize the effects of supersymmetry breaking by giving a vacuum expectation value to the F term of a spurion field Z:

$$Z = m\theta^2. \tag{37}$$

Effectively this Z field has an R-charge of 2 due to the presence of the two θs in its expansion. The spurion field is a shorthand device that allows tracking of the effects of supersymmetry breaking. The actual dynamics that gives rise to the supersymmetry breaking may be (much) more complicated than a single field.

This spurion allows us to write down scalar soft masses, for example, for squarks:

$$\int d^4\theta\, Q^\dagger Q Z^\dagger Z \to m^2 \tilde{q}^* \tilde{q}. \tag{38}$$

The θ^2 and $\bar{\theta}^2$ from the Z and $Z^\dagger$ saturate the $d^4\theta$ integral, and only the lowest components of the Q superfields are relevant. This term is R-invariant; the non-zero R charge of Z is compensated by the opposite R-charge of $Z^\dagger$. So, scalar soft masses do not violate R. On the other hand, the scalar trilinear couplings $A_{ijk}\phi_i\phi_j\phi_k$ are given as, e.g.,

$$\int d^2\theta Q U^c H_u Z \to m\tilde{q}\tilde{u^c}h. \tag{39}$$

Note that the superpotential of Eq. (36) absent Z was R-invariant. So, this means the inclusion of this term breaks R by two-units — the two units present in Z. Similarly, gaugino masses can be written down with the help of the spurion and the field strength chiral superfield $\mathcal{W}$

$$\int d^2\theta Z\mathcal{W}\mathcal{W} = m\lambda\lambda. \tag{40}$$

This will also break R by two units. So soft-terms break the $U(1)_R$ that we had present in our original theory. But the symmetry is not broken entirely — since the soft supersymmetry breaking terms only break R by two units, the R-symmetry is broken down to a residual parity, which we can denote as R_p. This residual parity will have important implications.

It turns out that the precise R_p as derived from the charge assignment in Eq. (36) is not so useful because the Higgs fields acquire vacuum expectation values, and so this naive parity is spontaneously broken. A more useful symmetry that is preserved is one where we have $\theta \to -\theta$, $H_{u,d} \to H_{u,d}$, and $(Q, U^c, D^c, L, E^c) \to -(Q, U^c, D^c, L, E^c)$. This parity symmetry, a combination of fermion number and matter parity, is the R-parity of the MSSM. With this choice, all Standard Model fields (including the Higgs scalars) are even, and all superpartners are odd.

Conservation of this symmetry is crucial for determining the collider phenomenology. In particular, it enforces

- Superpartners must be produced in pairs. We start in a R-parity even state, and end in an R-parity even state.
- The lightest superpartner is stable, as it has no R-odd final state to decay into. This means that super partners will exit the detector as missing energy.

This symmetry forbids terms that would otherwise be allowed by the combination of gauge symmetry and renormalizability:

$$W_{\Delta L} = \mu' L H_u + \lambda_{ijk} L_i L_j E_k^c + \lambda' Q_i D_j L_k, \tag{41}$$

$$W_{\Delta B} = \lambda'' U_i^c D_j^c D_k^c. \tag{42}$$

(Alternately, the presence of all these terms preclude an R-parity). These terms, if large, are a phenomenological disaster. They can mediate large flavor changing neutral currents as well as rapid proton decay. Note that the potential existence of these interactions can be viewed as a step backwards from the Standard Model. There, renormalizability and gauge invariance are sufficient to strongly suppress these kinds of processes. For example, if the Standard Model were valid up to the Planck scale, one might guess that the leading contribution to proton decay would be dimension-six operators suppressed by two powers of M_{pl}. One attitude frequently taken is that successful supersymmetric phenomenology motivates the imposition of an R-symmetry, and that this symmetry, amongst other things, gives rise to an excellent Dark Matter candidate (see lectures by R. Essig and S. Profumo at this school).

4.3. *Who are we producing anyway? Cast of characters: Mass eigenbasis*

After electroweak (and super-) symmetry breaking, the mass eigenstates do not align with the superfields. Fields with identical electric charge but different electroweak quantum numbers are expected to mix. In particular, amongst neutral fields, we expect mixing between the superpartners of the

Higgs fields (known as Higgsinos) and the superpartners of the neutral $SU(2)$ gauge boson (the wino, $\tilde{w}^0$) and the $U(1)_Y$ gauge boson, known as the bino ($\tilde{B}^0$). These fields are collectively known as neutralinos. All told, have the symmetric matrix (written in the basis $\{\tilde{B}^0, \tilde{w}^0, \tilde{H}_d, \tilde{H}_u\}$):

$$M_{neutralino} = \begin{pmatrix} M_1 & 0 & -\frac{1}{2}g_Y v_d & -\frac{1}{2}g_Y v_u \\ 0 & M_2 & -\frac{1}{2}g v_d & -\frac{1}{2}g v_u \\ -\frac{1}{2}g_Y v_d & -\frac{1}{2}g v_d & 0 & -\mu \\ -\frac{1}{2}g_Y v_u & -\frac{1}{2}g v_u & -\mu & 0 \end{pmatrix}. \quad (43)$$

The M_1 and M_2 are Majorana gaugino mass terms that arise from supersymmetry breaking, see Sec. 3.5. The μ terms arise directly from the superpotential Eq. (18). The terms in the upper-right hand block arise from terms in the Lagrangian like $\mathcal{L} \ni g'\tilde{B}\tilde{H}_u\langle H_u\rangle$. Their presence is required by the combination of gauge interactions and supersymmetry. Note that these interactions are just Yukawa couplings: they have two fermions ($\tilde{B}$ and $\tilde{H}_u$ in our example) that are married together once the Higgs field H_u takes on its expectation value. Because of the additional (super)symmetry, this *a priori* independent Yukawa coupling is related to a gauge coupling. This mass matrix can be diagonalized by an orthogonal matrix. The mass eigenstates are $\chi^0_{1,2,3,4}$ with $M^{diag}_{neutralino} = N^T M_{neutralino} N$.

A similar story obtains for the charged sector. The mass matrix for the charginos is

$$M_{chargino} = \begin{pmatrix} M_2 & \sqrt{2}\sin\beta M_W \\ \sqrt{2}\cos\beta M_W & \mu \end{pmatrix}, \quad (44)$$

where we have traded the M_W for the product of the Higgs vacuum expectation value (vev) and gauge coupling. Here the columns are labelled by $\tilde{w}^+$ and $\tilde{H}_u^+$and rows are labelled by $\tilde{w}^-$ and $\tilde{H}_d^-$. The diagonal elements are present due to supersymmetry breaking gaugino mass M_2 and supersymmetric Higgsino mass μ. The off-diagonal terms are again due to the Yukawa couplings that arise upon supersymmetrization of gauge interactions. The end result is that the wino and Higgsinos marry off to form a pair of Dirac fermions, denoted $\chi^\pm_{1,2}$. Because this 2×2 matrix is not symmetric, two different rotation matrices to do the singular-value decomposition of this matrix, $M^{chargino}_{diag} = U M_{chargino} V^\dagger$:

$$\begin{pmatrix} \chi_1^+ \\ \chi_2^+ \end{pmatrix} = V \begin{pmatrix} \tilde{w}^+ \\ \tilde{H_u}^+ \end{pmatrix}, \quad \begin{pmatrix} \chi_1^- \\ \chi_2^- \end{pmatrix} = U \begin{pmatrix} \tilde{w}^- \\ \tilde{H_d}^- \end{pmatrix}. \quad (45)$$

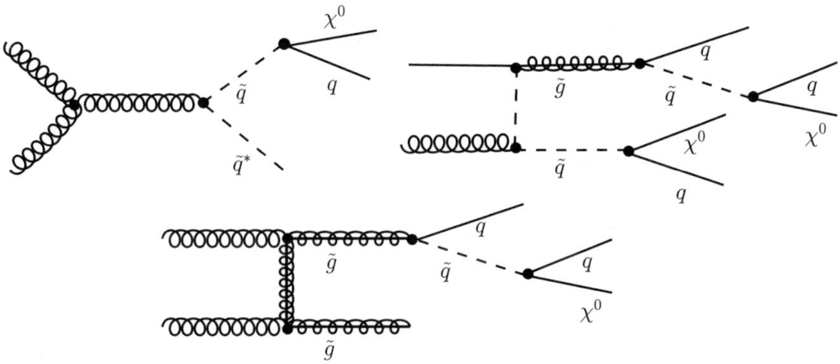

Fig. 4. Production of colored superpartners. Clockwise from upper-left are representative diagrams for squark pair production, squark-gluino associated production and gluino pair production. Their decays are shown to quarks and the lightest supersymmetric particle, here assumed to be the neutralino. The result is a final state with jets and missing energy.

4.4. *Production modes*

With the cast of characters in place, let us examine the production modes that are most relevant for the LHC. Because the LHC is a hadron collider, there are plenty of gluons and quarks in the initial parton distribution functions. The result is that if colored states are kinematically accessible, they are likely to dominate production. Pair production of colored particles is shown in Fig. 4. To set the scale, production cross sections are shown in Table 3. Its apparent that the processes that are being searched for make up a tiny fraction of the total event rate at the LHC. So, to identify these rare processes, care must be taken to cut away the dominant Standard Model backgrounds. Details of these experimental searches are discussed elsewhere at this school.

4.5. *Classic SUSY signals*

We are finally ready to discuss a few of the classic signatures of supersymmetry. As discussed above, because the LHC is a hadron collider, it is likely that the dominant production of superpartners will be colored ones, as shown in Fig. 4. The colored particles will then decay down to the lightest supersymmetric particle, which is stable, as described in Sec. 4.2. It leaves the detector, carrying energy with it. This "missing energy" signature is a hallmark of supersymmetry searches. After quarks hadronize

Table 3. Order of Magnitude Production Cross Sections at LHC, data taken from Ref. 34, 35.

Process	LHC 8	LHC 14
Total	.1b	.1 b
$b\bar{b}$ ($p_T > 30$ GeV)	.3 μb	1 μb
$t\bar{t}$	200 pb	800 pb
$gg \to h$	15 pb	50 pb
gluino ($m_{\tilde{g}} = 500$ GeV)	4 pb	30 pb
gluino ($m_{\tilde{g}} = 750$ GeV)	200 fb	3 pb
gluino ($m_{\tilde{g}} = 1$ TeV)	20 fb	400 fb

to form jets, the final state will be jets and missing energy. This search is an excellent all-purpose search for supersymmetry, and in much of the parameter space of simplified models (e.g., the CMSSM described above), it provides the strongest bounds.

Another classic signature is the production of like-sign dileptons. In the Standard Model, most pair production processes (e.g. $t\bar{t}$, WW) and decays (e.g. Z decay) typically lead to opposite sign leptons. In supersymmetry things can be different. At the LHC, two initial state u-quarks can form two u-squarks via the t-channel exchange of a gluino. The two u squarks can then "cascade" decay down to the LSP. Because the squarks have the same sign, it is possible that like-sign W's (and hence leptons) can be produced (see Fig. 5). Two squarks of the same sign can also be produced as a result of gluino pair production (see Fig. 5). The reason is that the gluino does not carry quark number. Dominant standard model backgrounds include $t\bar{t} + W/Z$, $W^{\pm}W^{\pm}qq, W^{\pm}Z$.

The last classic signal we will mention is the trilepton search. This can be produced via the direct production of charginos and neutralinos (Fig. 6). The leptons can arise either through the decays of gauge bosons, or if the sleptons are relatively light.

As of yet, not excesses have been seen in any of these channels. The latest supersymmetry searches at ATLAS and CMS can be found at Ref. 36, 37. The purpose of these notes is not to review the present searches in any detail (see lectures by V. Sanz at this school for some related case studies). Depending on the details of the superpartner spectrum, current limits can exclude superpartners up to masses of nearly a TeV. It is possible to soften these limits somewhat in cases of specific designer spectra. Nevertheless, given our desire to have superpartners near the weak scale to tame the quadratic divergence in the Higgs boson mass parameter, it is fair

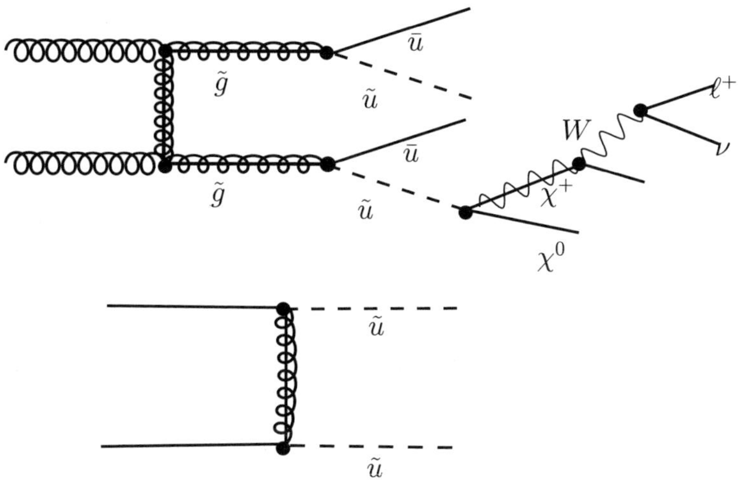

Fig. 5. Processes leading to same sign leptons and missing energy. At top, gluino pair production can lead to production of same signed quarks in their decay. At bottom, direct production of same sign squarks is shown. Subsequent decay of the squarks can lead to same sign leptons.

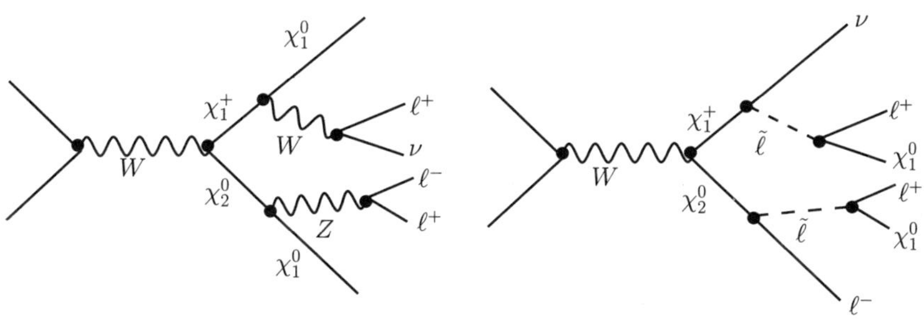

Fig. 6. Production of charginos and neutralinos leading to trilepton signals. Decays may proceed via gauge bosons or potentially via sleptons. The result is a final state with three leptons and missing energy.

to say things are becoming somewhat uncomfortable for the most natural instantiations of supersymmetry.

4.6. *Not so classic signals*

Our guiding principle in our discussion of collider phenomenology was the existence of a R-parity. While there is strong motivation for this symmetry, there is no guarantee of its existence. For example, it is possible to avoid proton decay without an imposing the full R-symmetry. For example, conservation of baryon number (to forbid $W_{\Delta B}$) is enough to ensure proton stability. Actually, the absence of either $W_{\Delta L}$ or $W_{\Delta B}$ is enough.[e] Recently, there has been a resurgence in activity in examining models with such a reduced symmetry — in part because the collider phenomenology differs dramatically from that described above, and it allows a weakening of the constraints on the supersymmetric parameter space. Typically R parity violation is a small perturbation, so that all heavy superpartners that are produced will decay down to the lightest superpartner, which (since it has no other choice) uses the R-parity violating couplings to decay. This last decay degrades the missing energy signal. Instead, typically, events have a large multiplicity. For example, if the superpotential contains the $U^cD^cD^c$ operator, the lightest neutralino might decay to three quarks.

Another direction of interest is to examine compressed spectra [39, 40]. The missing energy observed at colliders depends largely on mass splittings, and not as directly on the overall mass scale. Thus, if the superpartner spectrum is compressed, the missing energy signals will be degraded, this somewhat softens the limits on the superpartners.

4.7. *Split supersymmetry*

We spent a long time arguing for a scale of supersymmtery breaking $\tilde{m} \approx M_W$ on the basis of naturalness. However, it should be noted that there is an example where the idea of naturalness fails spectacularly. If one were to guess the size of the cosmological constant (CC) — completely unprotected by any symmetry — the guess would be $\Lambda_{CC}^4 \sim (\text{cutoff})^4$. If the Standard Model is valid up to the Planck scale, this estimate is off by 120 orders of magnitude: the true value of the CC is $(\text{meV})^4$. Even if there were supersymmetry right at the TeV scale (which would cutoff contributions to the CC above this scale), we would be wrong by sixty(!) orders of magnitude. What are we to make of this? Perhaps the CC, which after all depends (in part) on a quantum theory of gravity, differs fundamentally from the Higgs mass parameter, whose fine-tuning we were so worried about? Maybe one day, with a fuller understanding of String Theory, we will understand why the CC is so small. Of course, this need not be so, in which case, the spectacular failure of naturalness might give one pause

[e]This conclusion holds as long as the gravitino is not too light [38].

about the likelihood of observing supersymmetry. And even if this is so, we might worry that the solution to the CC problem does not commute with the hierarchy problem. If it does not, we might worry that a universe that has the right cosmological constant might look fine-tuned in the Higgs sector [41, 42].

But if the Higgs sector is fine-tuned is there *any* reason to talk about supersymmetry at all? Suppose, e.g., that string theory "likes" supersymmetry — perhaps it is necessary for the ultimate stability of the theory — but the scale of supersymmetry breaking has nothing to do with the hierarchy problem (or at best solves it incompletely). Then, at what scale should we see superpartners? Absent the fine-tuning motivation, it turns out there is not really any scale strongly preferred for the scalar masses.[f] The fermionic superpartners are another story. They are responsible for the improvement in gauge coupling unification with respect to the Standard Model. The reason is that the new scalar superpartners (all the squarks and sleptons) come in complete SU(5) multiplets. Carefully examining our discussion in 3.6, we see that complete GUT multiplets do not contribute to the LHS of Eq. (31). So, if we are to maintain successful unification of couplings these fermions should be "somewhat close" to the TeV scale. Unfortunately, this unification argument is not strong enough to say they should live below 1 TeV, and the difference between 1 TeV and 10 TeV is everything when it comes to the LHC. An independent argument for relatively light gauginos is that they make up an excellent Dark Matter candidate. Again, a successful cosmological history typically indicates gaugino masses near a TeV, but the argument is not sharp enough to say exactly where they would lie. For example, a pure wino (with very small admixture of other neutralinos) can give the right thermal relic abundance of Dark Matter if its mass is somewhat above 2 TeV, but a proper admixture of bino and Higgsino could be a couple hundred GeV.

The end result of this line of reasoning is a "split spectrum" wherein the scalar can be (quite heavy), but fermions (gauginos and Higgsinos) are near the weak scale. How large the splitting can be is model dependent. One especially simple scenario involves a splitting of approximately one-loop factor between the scalars and fermions, a scenario discussed in Ref. 44, 45, and recently revisited in Ref. 46, 47. A similar spectrum was also motivated from a "top-down" approach from a string compactification [48].

---

[f]Except perhaps by some indirect cosmological arguments, see, e.g., Ref. 43. Also, the 125 GeV Higgs mass may favor scalar mass that are not *too* heavy.

How can one test this scenario? One can look for the Dark Matter in direct [42, 49] or indirect [50–52] detection experiments. Direct production of charginos and neutralinos at the LHC may also be possible. Perhaps one of the most interesting searches is for gluinos. Because the squarks are potentially heavy, the lifetime of the gluinos can in principle be quite long. If the squarks are sufficiently heavy, the gluinos transit macroscopic distances in the detector before decaying. In fact, it is possible the gluinos hadronize, eventually coming to a stop [53] in the detector prior to decaying. Decays out of synch with the beam (perhaps even occurring when the LHC is off!) would be a striking signature indeed. Thus far, no such decays of long-lived hadrons have been observed, and limits of near a TeV have been placed, see Ref. 54–58. However, it should be noted that of parameter space of theories with the heaviest scalars (and most striking gluino signatures) can give too heavy Higgs boson masses. It may be that gluino lifetimes are only long enough to allow slightly displaced vertices (if at all).

5. Electroweak Symmetry Breaking in the MSSM

But the absence of supersymmetric signatures is not the most exciting data we have received from the LHC to date. That honor falls on the discovery of a 125 GeV Higgs boson. What are the implications of this discovery for supersymmetry? Before we can answer this, we need to understand a bit about the Higgs potential in supersymmetry. As alluded to above, the MSSM is a two Higgs doublet model (2HDM). For the two Higgs doublets of hypercharge $Y = \pm 1/2$, we have scalar multiplets

$$\begin{pmatrix} H_u^+ \\ H_u^0 \end{pmatrix}, \qquad \begin{pmatrix} H_d^0 \\ H_d^- \end{pmatrix}. \tag{46}$$

However, supersymmetry restricts the form of the potential with respect to a general 2HDM. In particular, some quartics are not allowed, and others are fixed to values related to the gauge couplings. This is to be expected — any time you impose a symmetry on a theory, there are relationships that you would not otherwise have anticipated. After imposing these constraints, we find

$$V(H_u, H_d) = m_2^2|H_u|^2 + m_1^2|H_d|^2 - (\mu B(H_u \cdot H_d) + c.c.)$$
$$+ \frac{g^2}{2}\left(H_d^* \frac{\tau}{2} H_d + H_u^* \frac{\tau}{2} H_u\right)^2 + \frac{g_Y^2}{2}\left(\frac{1}{2}|H_u|^2 - \frac{1}{2}|H_d|^2\right)^2. \tag{47}$$

Note, $m_{1,2}^2 \equiv \mu^2 + m_{H_{d,u}}^2$ with $m_{H_{d,u}}^2$ the relevant soft supersymmetry breaking scalar $(mass)^2$ parameters. The τ correspond to the Pauli matrices.

There are a variety of different contributions to this potential. Some are supersymmetry preserving contributions arising from the superpotential ($\propto \mu^2$), some arise from supersymmetry breaking, and others (those proportional to gauge couplings) derive from the D-terms. The $(B\mu H_u H_d + h.c.)$ term breaks not only supersymmetry (again softly), but an additional global symmetry that takes $H_u \to e^{i\alpha} H_u, H_d \to e^{i\alpha} H_d$. So, we expect there to be an additional massless Goldstone boson in the limit where this term vanishes.

After doing some rearranging, we can rewrite the D-term contributions as:

$$V_D = \frac{\bar{g}^2}{8}(|H_u|^2 - |H_d|^2)^2 + g^2 |H_u^\dagger H_d|^2 \tag{48}$$

where we have defined $\bar{g^2} \equiv g^2 + g_Y^2$. It is worthwhile reemphasizing that these quartic coupling of the Higgs fields are related to the gauge couplings because of supersymmetry.

Let us explore the minimization of this potential. Using an $SU(2)_L$ gauge transformation, we can set, say, $H_u^+ = 0$. That is, we can rotate the vev into one component of the doublet. It is a straightforward exercise to check that

$$\frac{\partial V}{\partial H_u^+} = 0 \text{ at } \mathrm{H_u^+} = 0 \Rightarrow \mathrm{H_d^-} = 0. \tag{49}$$

This is good news, as it means that electric charge will not be broken at the minimum of the potential (no charged component gets a vacuum expectation value). Since only neutral components will get vevs, we can rewrite the potential as a function of v_u and v_d, where

$$H_u = \begin{pmatrix} 0 \\ v_u \end{pmatrix} \quad H_d = \begin{pmatrix} v_d \\ 0 \end{pmatrix}. \tag{50}$$

In principle these v_u and v_d could be complex. The potential simplifies as:

$$V(v_u, v_d) = m_1^2 |v_d|^2 + m_2^2 |v_u|^2 - \mu B(v_u v_d) + c.c. + \frac{\bar{g}^2}{8}(|v_d|^2 - |v_u|^2)^2. \tag{51}$$

Note that we can use a "Peccei-Quinn" rotation, under the H_u and H_d fields rotate by the same phase to make μB real and positive. From here we can explicitly check that the vevs are in fact real, i.e. spontaneous breaking of

CP does not occur. We define

$$v_u = |v_u|e^{i(\alpha+\beta)}, \tag{52}$$

$$v_d = |v_d|e^{i(\alpha-\beta)}. \tag{53}$$

Examining the potential as a function of the phases, we see that there is only a single term (the one proportional to μB) that depends on the overall phases of the vevs α. This contribution to the potential is minimized for $\alpha = 0$. The remaining relative phase β can be removed by a $U(1)_Y$ gauge transformation — recall the two doublets have opposite hyper charge.

From this potential, we should be able to derive the mass spectrum of the five physical Higgs boson states (there are eight degrees of freedom in two complex scalar doublets, and three are eaten by the $W^{\pm}$ and Z). These states are denoted $h^0, H^0, H^{\pm}$ and A^0.

But before doing this, note that there is a special direction where the quartic coupling vanishes. The existence of these so-called "flat-directions" is a generic feature of supersymmetry gauge theories. Since the coupling in question is generated by the D-terms, this direction is known as D-flat. Its existence means we need to take care that the potential will not be unbounded from below. To determine the condition needed to prevent this instability, we set

$$v_u = v_d = \phi_{flat}. \tag{54}$$

The potential in this direction is

$$V(\phi_{flat}) = \phi_{flat}^2(m_1^2 + m_2^2 - 2\mu B). \tag{55}$$

This implies

$$\mu B < \frac{m_1^2 + m_2^2}{2}. \tag{56}$$

On the other hand, to get EWSB (at all!), we need a negative mass eigenvalue at the origin ($v_u = v_d = 0$). One way to ensure this is the case is to require a negative determinant for the relevant mass matrix (recall $\lambda_1\lambda_2 = \det\mathcal{M}$, where λ_i are the eignenvalues).

$$\det\begin{pmatrix} \frac{\partial^2 V}{\partial v_u^2} & \frac{\partial^2 V}{\partial v_u \partial v_d} \\ \frac{\partial^2 V}{\partial v_u \partial v_d} & \frac{\partial^2 V}{\partial v_d^2} \end{pmatrix} < 0 \Rightarrow \det\begin{pmatrix} m_2^2 & -\mu B \\ -\mu B & m_1^2 \end{pmatrix} \tag{57}$$

$$\Rightarrow (\mu B)^2 > m_1^2 m_2^2. \tag{58}$$

Note that if $m_1^2 = m_2^2$, you cannot simultaneously satisfy Eqs. (56) and (58). This should not be viewed as a problem. In fact, renormalization group evolution rapidly suppresses $m_{H_u}^2$ (and drives it negative) due to its large coupling to the top quark. The ability to start with universal soft supersymmetry breaking masses, and see $m_{H_u}^2$ driven negative is viewed as a success of EWSB within the supersymmetric framework.

Now we want to move on to discuss the spectrum of the Higgs bosons. In much of the MSSM parameter space, there is a Higgs boson that is approximately Standard Model like. We will try to work in a way that makes both its identity and its properties as explicit as possible. So, rather than making a full analysis of the Higgs potential in the H_u, H_d basis (as is often done), we work in a basis that makes it manifest that one CP-even scalar, traditionally h^0, likely has properties that closely mimic those of the Standard Model Higgs boson. We closely follow the presentation of Ref. 59, see also Ref. 60, 61 in another context. This basis (dubbed the Runge basis in Ref. 59 for no good reason), involves a rotation away from the H_u, H_d basis to a basis where one field contains the full vacuum expectation value (vev). In a sense it is this field that is *the* Higgs boson. Denote vacuum expectation values $\langle H_d \rangle \equiv v_d$, $\langle H_u \rangle \equiv v_u$ and $v_u/v_d \equiv \tan\beta$, and

$$\Phi_{vev} = \frac{v_d}{v} H_d^c + \frac{v_u}{v} H_u, \Phi_\perp = -\frac{v_u}{v} H_d^c + \frac{v_d}{v} H_u. \tag{59}$$

Note that we have taken the conjugate of H_d, $H_d^c \equiv i\sigma^2 H_d^*$, to make sure that we are adding fields with identical hypercharge.

So, we can write:

$$\begin{aligned} \Phi_{vev} &= \begin{pmatrix} G^\pm \\ \frac{1}{\sqrt{2}}(v + h_v^0 + iG^0) \end{pmatrix} \\ \Phi_\perp &= \begin{pmatrix} H^\pm \\ \frac{1}{\sqrt{2}}(H_v^0 + iA^0) \end{pmatrix}, \end{aligned} \tag{60}$$

with $G^\pm, G^0$ the Goldstone bosons who get eaten to make $W^\pm, Z^0$ massive. Here we also clearly see five extra degrees of freedom $(H^\pm, h_v^0, H_v^0, A^0)$. Although $H^\pm, A^0$ are mass eigenstates, h_v^0 and H_v^0 are not. h_v^0 is by construction aligned with the vacuum expectation value. It is this field that is responsible for giving the gauge bosons their masses. In the Standard Model, there is a single Higgs field (that must perforce be aligned with the vev), so we expect that h_v^0 will have identical couplings to the gauge bosons as a Standard Model Higgs. We expect any misalignment between h_v^0 and the mass eigenstate h^0 will result in non-Standard Model like couplings.

We can rewrite the original Higgs potential as

$$\begin{aligned} V = &(\mu^2 + \sin^2\beta\, m_{H_u}^2 + \cos^2\beta\, m_{H_d}^2 - \sin 2\beta B\mu)|\Phi_{vev}|^2 \\ &+ (\mu^2 + \cos^2\beta\, m_{H_d}^2 + \sin^2\beta\, m_{H_u}^2 + \sin 2\beta B\mu)|\Phi_\perp|^2 \\ &+ \left(\frac{\sin 2\beta}{2}(m_{H_u}^2 - m_{H_d}^2) - \cos 2\beta B\mu\right)(\Phi_{vev}^\dagger \Phi_\perp + c.c) \\ &+ \frac{\bar{g}^2}{8}\left[\cos 2\beta\left(|\Phi_\perp|^2 - |\Phi_{vev}|^2\right) + \sin 2\beta(\Phi_{vev}^\dagger\Phi_\perp + c.c.)\right] \\ &+ \frac{g^2}{2}|\Phi_{vev}\cdot\Phi_\perp|^2. \end{aligned} \tag{61}$$

As usual, we can examine the minimum by taking derivatives and forcing them to be equal to zero. Letting

$$\phi_i = \left\{G^\pm, H^\pm, G^0, h_v^0, H_v^0, A^0\right\}, \tag{62}$$

we have:

$$\left.\frac{\partial V}{\partial \phi_i} \stackrel{!}{=} 0\right|_{\{h=v,\phi_i\neq h=0\}\equiv minimum}. \tag{63}$$

It is a straightforward exercise to check that only

$$\left.\frac{\partial V}{\partial h_v} \stackrel{!}{=} 0\right|_{minimum}, \quad \left.\frac{\partial V}{\partial H_v} \stackrel{!}{=} 0\right|_{minimum}, \tag{64}$$

yield non-trivial conditions:

$$h_v: \ \frac{\bar{g}^2}{8}v^2\cos^2 2\beta + \mu^2 + m_{H_d}^2\cos^2\beta + m_{H_u}^2\sin^2\beta - B\mu\sin 2\beta = 0, \tag{65}$$

$$H_v: \ -\frac{\bar{g}^2}{8}v^2\cos 2\beta\sin 2\beta + \frac{1}{2}(m_{H_u}^2 - m_{H_d}^2)\sin 2\beta - B\mu\cos 2\beta = 0. \tag{66}$$

Taking linear combinations (and using $M_Z^2 = \bar{g}^2 v^2/4$), we find:

$$M_Z^2 + \mu^2 = \frac{m_{H_d}^2 - \tan^2\beta m_{H_u}^2}{\tan^2\beta - 1}, \tag{67}$$

$$\frac{B\mu}{\sin 2\beta} = \mu^2 + \frac{1}{2}\left(m_{H_d}^2 + m_{H_u}^2\right). \tag{68}$$

These two conditions should be familiar to MSSM aficionados, even though the route we have taken may not be.

Before moving on to discuss the phenomenology of the Higgs bosons, it is worthwhile to pause and consider the implications of this equation for fine-tuning. The weak scale (i.e. M_Z) is determined in terms of μ, $m_{H_d}^2$, and $m_{H_u}^2$. We already pointed out that the absence of fine-tuning will require relatively light stops and gluinos. Here, we see clearly why: if they are heavy, it puts pressure on $m_{H_u}^2$ and in turn M_Z. There is a new lesson, too. The μ term cannot be too large. If it is, there is already fine tuning at the *tree* level in Eq. (67). So, it seems that a natural spectrum requires that Higgsinos not be too heavy, either.

From here, to get the Higgs boson masses, we just need to take the potential, and

(1) Expand in terms of $\{v, h_v^0, H^\pm, H_v^0, A^0\}$.
(2) Use conditions (65) and (66) above.
(3) Pick off terms quadratic in the fields, i.e. the mass terms.

The results are particularly simple for A^0 and $H^\pm$ because they do not mix with other fields. The pseudoscalar Higgs boson's mass is given by

$$M_A^2 = (m_{H_u}^2 - m_{H_d}^2)\sec 2\beta - M_Z^2. \tag{69}$$

The interested reader can check that this expression is proportional to $B\mu$ — the A is a Goldstone boson of the PQ breaking in the limit that $B\mu$ vanishes. The charged Higgs mass is given by

$$M_{H^\pm}^2 = M_A^2 + M_W^2. \tag{70}$$

For CP-even neutral Higgs bosons,we have to diagonalize the 2×2 h_v^0 and H_v^0 mass matrix given by

$$\mathcal{M}_{hH}^2 = \begin{pmatrix} M_Z \cos^2 2\beta & -M_Z^2 \sin 2\beta \cos 2\beta \\ -M_Z^2 \sin 2\beta \cos 2\beta & (m_{H_u}^2 - m_{H_d}^2) - M_Z^2 \cos^2 2\beta \end{pmatrix}. \tag{71}$$

The $(1,1)$ entry of the mass matrix (corresponding to the h_v^0 state) has $(mass)^2 = M_Z^2 \cos^2 2\beta$. A theorem from linear algebra states that the smallest eigenvalue of a matrix is smaller than the smallest diagonal element of that matrix. Trigonometry then indicates a Higgs mass less than M_Z. Furthermore, this is saturated as $(m_{H_u}^2 - m_{H_d}^2) \to \infty$, where $h^0 \equiv h$. But the Higgs boson is (apparently) at 125 GeV. How robust is this prediction? After all, we have not yet taken supersymmetry breaking into account. We do not expect loop effects to cancel once supersymmetry is broken [62–64].

There are many ways to compute these effects. They include

(1) A direct diagrammatic calculation
(2) A renormalization group analysis
(3) Computation of the effective potential. (For a textbook treatment, see Sec. 10.6 of Ref. 15).

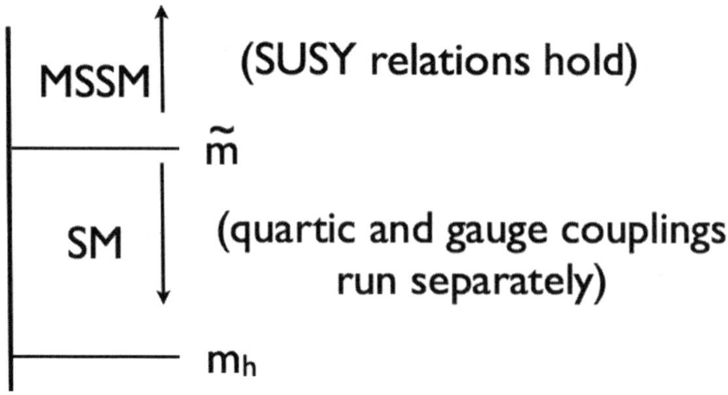

Fig. 7. Matching in a calculation of the mass of the Higgs boson.

We concentrate on the second of these — in some ways it offers the most physical insight. We take the scale of supersymmetry breaking $\tilde{m} >> M_Z$, we can imagine matching the full Minimal Supersymmetric Standard Model onto the Standard Model at that scale. Above $\tilde{m}$ supersymmetry enforces the equality of the quartics and the gauge couplings. Because the Higgs quartic is tied to gauge couplings in the supersymmetric regime, at scales above the supersymmetry breaking scale $\tilde{m}$, we have

$$\begin{aligned} m_h^2 &= 2\lambda v^2 \\ &= 2\left(\frac{1}{8}(g^2 + g_Y^2)\right) v^2. \end{aligned} \tag{72}$$

Below $\tilde{m}$ the quartic is no longer tied to the gauge coupling — the symmetry no longer connects them, and they are free to evolve independently. For scales not *too* far from $\tilde{m}$ we have (with $t \equiv \log Q$)

$$\lambda(Q) = \lambda(\tilde{m}) + \frac{\partial \lambda}{\partial t}(t - t_0) + \ldots \tag{73}$$

$$= \lambda(\tilde{m}) - \beta_\lambda \log\left(\frac{\tilde{m}}{Q}\right). \tag{74}$$

Here β_λ is the beta-function for the quartic in the Standard Model. The leading contribution (arising from the top quark Yukawa coupling) is

$$\beta_\lambda = -\frac{3y_t^4}{8\pi^2}. \tag{75}$$

Setting $Q = m_t$, and using Eq. (72), we find

$$m_h^2 = M_Z^2 \cos^2 2\beta + \frac{3g^2 m_t^4}{4\pi^2 M_W^2} \log\left(\frac{\tilde{m}}{m_t}\right) \tag{76}$$

For a Higgs boson mass of 125 GeV, this indicates a supersymmetry breaking scale of approximately 1 TeV.

This is the core of the argument that supersymmetry is now at least somewhat fine tuned. The above line of reasoning indicates pressure for the top squarks to be heavy. On the other hand, the heavier the stops get, the more fine-tuned the Higgs mass squared parameter is. So, even if limits from the LHC had not been strengthening the bounds on superpartners, the Higgs boson mass would have been a clue that the superpartners are likely somewhat heavy.

It should noted that the above formula is a simplification. In particular, it neglects the possible contribution of the trilinear scalar term $A_t \tilde{q}_3 \tilde{u}_3^c h$ to the Higgs mass. Indeed, pushing A_t to be somewhat large (its effects are maximized at $A_t \sim \sqrt{6}\tilde{m}_t$) can have an important effect on the Higgs boson mass. This does not come for free: while (finite) diagrams with the A-term can push the effective value of the quartic (and hence the Higgs mass) higher, the same A terms will contribute a radiative correction to the Higgs mass squared parameter, and thus the fine-tuning, see Eq. (32).

In fact, if the supersymmetry breaking scale is not "too high", then one is forced to consider relatively large A-terms . What implications does this have for the underlying theory? It is interesting to note that in its simplest forms of so-called "gauge mediated supersymmetry breaking" do not communicate large A-terms. However, A terms can be generated via renormalization group evolution from a high mediation scale (where the messenger particles lie) down to the TeV scale. As recently reviewed in Ref. 65, this indicates that gauge mediated versions of the MSSM will necessarily possess relatively heavy squarks, high mediation scales, or both.

6. What Does 125 GeV Mean?

6.1. *How uncomfortable is 125 GeV?*

So, is 125 GeV "natural"? It is borderline. The Higgs boson could have been found at 160 GeV. To achieve such a large value would have required extraordinarily fine-tuning in an supersymmetric theory. In fact, a theory that is the MSSM above $m_{GUT} \simeq 10^{16}$ GeV and the Standard Model below predicts a Higgs boson closer to 140 GeV [66]. But nature didn't choose that path. On the other hand, nature could have chosen to place m_h right near M_Z, with minimal contribution to the quartic from supersymmetry breaking. In this case radiative corrections would have been minimal. But that ship sailed some time ago. 125 GeV is tantalizingly close to M_Z, but just far enough away to give us pause. Where does that leave us?

- One approach (see below) is to ask for new physics to affect the Higgs boson mass. If the low energy theory is not just the Higgs sector of the MSSM, it is possible that there are new contributions to mass of the Standard Model-like Higgs boson.
- The alternative is to accept some (or a lot, see Sec. 4.7) of tuning. Supersymmetry was designed to get rid of a tuning of some thirty-two orders of magnitude. How uncomfortable would we be with a 1% accident in the fundamental theory of our universe? How about a part in 10^4? At what point do we stop thinking of supersymmetry as a solution to a hierarchy problem?
- Supersymmetry is "just around the corner", but the Higgs mass is achieved through particularly large A-terms. There a designer spectrum that gives rise to a large Higgs, and we have eluded the LHC for now, but won't for much longer.

6.2. *Higgs properties implications*

At present, there is no strong deviation of the observed "Higgs boson" properties from that of the Standard Model Higgs boson. This is already quite an interesting statement for supersymmetry. In the MSSM, it would have been possible (via h_v^0-H_v^0 mixing) to modify the properties of the Higgs boson significantly. Though it is early days, the apparent absence of deviations indicates that we may lie in the so-called "decoupling regime", where $M_A \to \infty$ and the other Higgs bosons decouple from the theory.

The only deviation worth mentioning at present — and it is a mild one whose significance is unclear — is a slight enhancement in the rate $R_{\gamma\gamma}$: the number of Higgs bosons decaying to two photons is higher than expected from the Standard Model by roughly $1.8 \pm .4$ (ATLAS [67]) and 1.6 ± 0.5 (CMS [68]). This is tantalizing because the decay to two photons is induced in the Standard Model via loops of W-bosons and to a lesser extent top quarks. This raises the possibility that there could be new particles with weak scale masses (very roughly at the few 100 GeV scale or below if they are to have a large effect) that could also be running around the loop, and modifying the branching ratio. This has been a subject of much theoretical speculation in a variety of theoretical frameworks. The best suspect that could conceivably be responsible for modifying this branching ratio within the MSSM is the stau (the superpartner of the τ lepton). Because it has large $|\text{charge}| = 1$ it has a large coupling to photons, and because it lacks color, it would not disturb the $gg \to h$ rate, which might have shown up, e.g. in the overall rate for the process $gg \to ZZ^*$ (this disfavors the stops as the culprit). Depending on the stau trilinear coupling, the staus can in fact cause the branching ratio to go in either direction. A recent analysis of this can be found in Ref. 69, 70; it is too early to draw conclusions from the experimental data.

6.3. *Beyond the MSSM?*

The MSSM has a curious feature. In its superpotential, there is a term,

$$W = \mu H_u H_d \tag{77}$$

with positive mass dimension. What should the size of this μ term be. From our considerations of electroweak symmetry breaking in the previous section, we see that, phenomenologically, it must be of order M_Z to realize electroweak symmetry breaking without fine-tuning. This presents a puzzle. After all, terms in the superpotential are, by construction *supersymmetric.* They have nothing to do with the scale of supersymmetry breaking.

A natural extension to the MSSM is to introduce a single electroweak singlet superfield, S. The part of the superpotential that gives rise to the Yukawa couplings remains unchanged, but the piece that previously gave rise to the μ term is replaced:

$$W_\mu = \mu H_u H_d \to W_{NMSSM} = \lambda S H_u H_d + \kappa S^3. \tag{78}$$

If S gets a vacuum expectation value, an effective μ term is generated with $\mu = \lambda \langle S \rangle$. It might appear that no progress has been made — all that has been done is to replace an unexplained value of μ with an unexplained value of $\langle S \rangle$. However, it is reasonably straightforward to arrange for $\langle S \rangle$ to vanish in the supersymmetric limit, in which case it is plausible that μ_{eff} is of order the weak scale. One possibility is to have a negative $(\text{mass})^2$ term for S coming from supersymmetry breaking that drives the vev.

Interestingly, this superpotential also gives a novel contribution to the mass of the Higgs boson. In particular, the first term contributes at tree level

$$\Delta m_h^2\big|_{NMSSM} = \frac{\lambda v^2}{2} \sin 2\beta \tag{79}$$

with $v = 246$ GeV. This term gives the possibility of raising the Higgs boson mass without a large supersymmetry breaking contribution. It should be noted that this term is maximized precisely only where the tree-level MSSM (D-term) contribution is minimized. So, this is no panacea when it comes to getting a heavy Higgs boson. Furthermore, the maximum value of λ is limited if one imposes that there is no Landau pole (and no new physics) below the grand unified scale: $\lambda < 0.7$. Because one extraordinarily nice feature of the MSSM was its unification, hitting a Landau pole before the couplings had a chance to unify would be a pity. As recently reviewed [71], large values of λ and small values of $\tan\beta$ can allow relatively light stops, with masses of roughly 500 GeV.

7. Conclusion

No direct hints for supersymmetry have been found. The recent discovery of a Higgs boson, however, at 125 GeV gives us some important tea leaves to read. Because supersymmetry relates the Higgs boson mass to the Z-boson mass, the discovery of a "light" Higgs boson, i.e. near 100 GeV, is perhaps encouraging. On the other hand, 125 GeV is not 91 GeV. What should we make of the discrepancy?

Perhaps supersymmetry is broken more strongly than would would have expected based on naive fine-tuning arguments. One interesting possibility is one whether the fermionic superpartners are found near the TeV scale, but the superpartners are parametrically a loop heavier, near the 100 TeV scale. This can realize the Higgs boson mass rather simply, but raises sharp questions about naturalness. Supersymmetry will have softened the hierarchy problem, but the theory at the weak scale will be tuned to a part in a 10^5 or so. From a bottom-up perspective, this is puzzling, but perhaps there are motivations from a more fundamental theory.

Another possibility is that the theory at the weak scale is not quite the MSSM, but rather has some twist, e.g. the NMSSM or beyond. These extensions can help raise the Higgs boson mass, but one still must address the absence of direct production of superpartners.

Of course it is possible that supersymmetry is not important for physics at the weak scale at all.

The upgrade of the LHC will have a great deal to say about whether there are in fact any superpartners near the TeV scale. Information from experiments searching for weakly interacting massive particles (WIMPs) via either direct or indirect detection are also rapidly approaching the sensitivities relevant for the MSSM. It is an exciting time, and if I were to give similar lectures a few years from now, my bet is that they would look very different.

Acknowledgments

This work was supported in part by the DOE Office of Science under Grant DE-SC0007859 and NSF CAREER Grant NSF-PHY-0743315. Thanks to the Berkeley Center for Theoretical Physics for their hospitality while these notes were completed. Thanks to the students of TASI 2012 for good questions and a stimulating atmosphere. These notes also owe a debt to Nima Arkani-Hamed, Savas Dimopoulos, Lawrence Hall, Gordy Kane, Hitoshi Murayama, Jay Wacker, and James Wells, who have impacted my thinking on these topics through courses, collaborations, and conversations. Thanks also to Jack Kearney for a reading of this manuscript. All errors, are of course, my fault.

References

[1] R. Barate *et al.*, *Phys. Lett.* **B565**, 61 (2003).
[2] G. Aad *et al.*, *Phys. Lett.* **B710**, 49 (2012).
[3] S. Chatrchyan *et al.*, *Phys. Lett.* **B710**, 26 (2012).
[4] S. Chatrchyan *et al.*, *Phys. Lett.* **B716**, 30 (2012).
[5] G. Aad *et al.*, *Phys. Lett.* **B716**, 1 (2012).
[6] T. Aaltonen *et al.*, *Phys. Rev. Lett.* **109**, p. 071804 (2012).
[7] S. P. Martin, *A Supersymmetry primer*, in *Perspectives on Supersymmetry II*, ed. G. L. Kane 2010, pp. 1–153.
[8] S. P. Martin, TASI 2011 lectures notes: two-component fermion notation and supersymmetry, arXiv:1205.4076.
[9] M. Drees, An Introduction to supersymmetry, hep-ph/9611409.
[10] M. Dine, Supersymmetry phenomenology (with a broad brush), hep-ph/9612389.
[11] J. D. Lykken, Introduction to supersymmetry, hep-th/9612114.
[12] J. A. Bagger, Weak scale supersymmetry: Theory and practice, hep-ph/9604232.
[13] H. Murayama, Supersymmetry phenomenology, hep-ph/0002232.
[14] J. Wess and J. Bagger, *Supersymmetry and supergravity* (Princeton University Press, 1992).
[15] M. Drees, R. Godbole and P. Roy, *Theory and phenomenology of sparticles: An account of four-dimensional N=1 supersymmetry in high energy physics* (World Scientific Publishing, 2004).
[16] S. Weinberg, *The quantum theory of fields. Vol. 3: Supersymmetry* (Cambridge University Press, 2000).

[17] P. Binetruy, *Supersymmetry: Theory, experiment and cosmology* (Oxford University Press, 2006).

[18] M. Dine, *Supersymmetry and string theory: Beyond the standard model* (Cambridge University Press, 2007).

[19] J. Terning, *Modern supersymmetry: Dynamics and duality* (Oxford University Press, Oxford, 2006).

[20] H. Murayama, Supersymmetry, in *Tokyo 1994, Proceedings, Physics with high energy colliders*, 1994.

[21] G. F. Giudice, *Naturally Speaking: The Naturalness Criterion and Physics at the LHC*, in *Perspectives on LHC Physics*, eds. G. L. Kane and A. Pierce 2008, pp. 155–178.

[22] M. Gaillard and B. W. Lee, *Phys. Rev.* **D10**, p. 897 (1974).

[23] J. Beringer *et al.*, *Phys. Rev.* **D86**, p. 010001 (2012).

[24] H. K. Dreiner, H. E. Haber and S. P. Martin, *Phys. Rept.* **494**, 1 (2010).

[25] S. Ferrara, J. Wess and B. Zumino, *Phys. Lett.* **B51**, p. 239 (1974).

[26] D. J. Gross and R. Jackiw, *Phys. Rev.* **D6**, 477 (1972).

[27] E. Witten, *Phys. Lett.* **B117**, 324 (1982).

[28] L. Girardello and M. T. Grisaru, *Nucl. Phys.* **B194**, p. 65 (1982).

[29] S. Dimopoulos and H. Georgi, *Nucl. Phys.* **B193**, p. 150 (1981).

[30] C. Csaki, *Mod. Phys. Lett.* **A11**, p. 599 (1996).

[31] J. Fan, M. Reece and L.-T. Wang, *JHEP* **1109**, p. 126 (2011).

[32] S. Dimopoulos, S. Raby and F. Wilczek, *Phys. Rev.* **D24**, 1681 (1981).

[33] P. Langacker and M.-x. Luo, *Phys. Rev.* **D44**, 817 (1991).

[34] T. Han, Collider phenomenology: Basic knowledge and techniques, hep-ph/0508097.

[35] W. Beenakker, R. Hopker, M. Spira and P. Zerwas, *Nucl. Phys.* **B492**, 51 (1997).

[36] Atlas experiment, public results, susy searches:, `https://twiki.cern.ch/twiki/bin/view/AtlasPublic/SupersymmetryPublicResults`.

[37] Cms supersymmetry physics results:, `https://twiki.cern.ch/twiki/bin/view/CMSPublic/PhysicsResultsSUS`.

[38] K. Choi, E. J. Chun and J. S. Lee, *Phys. Rev.* **D55**, 3924 (1997).

[39] S. P. Martin, *Phys. Rev.* **D75**, p. 115005 (2007).

[40] T. J. LeCompte and S. P. Martin, *Phys. Rev.* **D85**, p. 035023 (2012).

[41] N. Arkani-Hamed and S. Dimopoulos, *JHEP* **0506**, p. 073 (2005).

[42] G. Giudice and A. Romanino, *Nucl. Phys.* **B699**, 65 (2004).

[43] A. Arvanitaki, C. Davis, P. W. Graham, A. Pierce and J. G. Wacker, *Phys. Rev.* **D72**, p. 075011 (2005).

[44] G. F. Giudice, M. A. Luty, H. Murayama and R. Rattazzi, *JHEP* **9812**, p. 027 (1998).

[45] J. D. Wells, *Phys. Rev.* **D71**, p. 015013 (2005).

[46] A. Arvanitaki, N. Craig, S. Dimopoulos and G. Villadoro, Mini-Split, arXiv:1210.0555.

[47] L. J. Hall, Y. Nomura and S. Shirai, Spread Supersymmetry with Wino LSP: Gluino and Dark Matter Signals, arXiv:1210.2395.

[48] B. S. Acharya, K. Bobkov, G. L. Kane, J. Shao and P. Kumar, *Phys. Rev.* **D78**, p. 065038 (2008).

[49] A. Pierce, *Phys. Rev.* **D70**, p. 075006 (2004).
[50] A. Arvanitaki and P. W. Graham, *Phys. Rev.* **D72**, p. 055010 (2005).
[51] A. Masiero, S. Profumo and P. Ullio, *Nucl. Phys.* **B712**, 86 (2005).
[52] D. S. Alves, E. Izaguirre and J. G. Wacker, Higgs, Binos and Gluinos: Split Susy Within Reach, 1108.3390.
[53] A. Arvanitaki, S. Dimopoulos, A. Pierce, S. Rajendran and J. G. Wacker, *Phys. Rev.* **D76**, p. 055007 (2007).
[54] Searches for Heavy Long-Lived Sleptons and R-hadrons with the ATLAS detector in pp collisions at sqrt(s) = 7 TeV, ATLAS Collaboration, ATLAS-CONF-2012-075, ATLAS-COM-CONF-2012-127.
[55] V. Khachatryan *et al.*, *Phys. Rev. Lett.* **106**, p. 011801 (2011).
[56] S. Chatrchyan *et al.*, *JHEP* **1208**, p. 026 (2012).
[57] S. Chatrchyan *et al.*, *Phys. Lett.* **B713**, 408 (2012).
[58] G. Aad *et al.*, *Eur. Phys. J.* **C72**, p. 1965 (2012).
[59] J. D. Wells, Lectures on Higgs Boson Physics in the Standard Model and Beyond, arXiv:0909.4541.
[60] G. Yeghian, Upper bound on the lightest Higgs mass in supersymmetric theories, hep-ph/9904488.
[61] B. A. Dobrescu and K. T. Matchev, *JHEP* **0009**, p. 031 (2000).
[62] H. E. Haber and R. Hempfling, *Phys. Rev. Lett.* **66**, 1815 (1991).
[63] J. R. Ellis, G. Ridolfi and F. Zwirner, *Phys. Lett.* **B257**, 83 (1991).
[64] Y. Okada, M. Yamaguchi and T. Yanagida, *Prog. Theor. Phys.* **85**, 1 (1991).
[65] P. Draper, P. Meade, M. Reece and D. Shih, *Phys. Rev.* **D85**, p. 095007 (2012).
[66] L. J. Hall and Y. Nomura, *JHEP* **1003**, p. 076 (2010).
[67] Observation and study of the Higgs boson candidate in the two photon decay channel with the ATLAS detector at the LHC, ATLAS Collaboration, ATLAS-CONF-2012-168.
[68] Evidence for a new state decaying into two photons in the search for the standard model Higgs boson in pp collisions, CMS Collaboration, CMS-PAS-HIG-12-015.
[69] M. Carena, S. Gori, N. R. Shah and C. E. Wagner, *JHEP* **1203**, p. 014 (2012).
[70] M. Carena, S. Gori, N. R. Shah, C. E. Wagner and L.-T. Wang, *JHEP* **1207**, p. 175 (2012).
[71] L. J. Hall, D. Pinner and J. T. Ruderman, *JHEP* **1204**, p. 131 (2012).

Chapter 4

Astrophysical Probes of Dark Matter

S. Profumo

Department of Physics and Santa Cruz Institute for Particle Physics
University of California, Santa Cruz
CA 95064, United States of America
E-mail: profumo@ucsc.edu

What is the connection between how the dark matter was produced in the early universe and how we can detect it today? Where does the WIMP miracle come from, and is it really a "WIMP" miracle? What brackets the mass range for thermal relics? Where does $\langle\sigma v\rangle$ come from, and what does it mean? What is the difference between chemical and kinetic decoupling? Why do some people think that dark matter cannot be lighter than 40 GeV? Why is $b\bar{b}$ such a popular annihilation final state? Why is antimatter a good way to look for dark matter? Why should the cosmic-ray positron fraction decline with energy? How do you calculate the flux of neutrinos from dark matter annihilation in a celestial body, and when is it independent of the dark matter pair-annihilation rate? How does dark matter produce photons? — Read these lecture notes, do the suggested 10 exercises, and you will find answers to all of these questions (and to many more on what You Always Wanted to Know About Dark Matter But Were Afraid to Ask).

Contents

Introduction

These are lecture notes for a series of four lectures on astrophysical probes of dark matter, I delivered in June 2012 at the 2012 Theoretical Advanced Study Institute in Elementary Particle Physics (TASI) Summer School at the University of Colorado, Boulder. The aim of this set of lectures was neither to present in detail particle dark matter models nor to focus on the technical aspects of experiments and on the (alas, necessary!) understanding of astrophysical backgrounds; rather, I tried to convey and to work out order-of-magnitude estimates that can be applied to a variety of particle dark matter models and physical situations. Practice with these estimates might be perhaps appreciated by the theory-minded scholar who is not necessarily keen on the fine print of experimental setups or the complications of astrophysical processes.

The first lecture is devoted to an introduction to the cosmology of particle dark matter, specifically in connection with the indirect detection of dark matter: what are the relevant/expected energy scales, particle products and rates? The second lecture discusses the thermal (or non-thermal) processes associated with the production of dark matter in the early universe, a few exceptions to the "standard lore", and the process of kinetic decoupling. The third lecture introduces indirect detection of dark matter, and details on charged cosmic rays produced by dark matter in the Galaxy. Finally, the fourth lecture discusses gamma rays and high-energy neutrinos from dark matter. Ten simple exercises are scattered through the lectures. The interested Reader might find them useful practice for the material being discussed.

The reference list is vastly incomplete, and it primarily egotistically contains self-citations: I consider it part of a good scientific education to go the extra mile and find relevant papers on given topics of interest. This will therefore be Exercise #11! Enjoy!

Lecture 1: Particle Dark Matter: Zeroth-order Lessons from Cosmology

The fundamental, particle nature of dark matter can be probed with the detection of photons, neutrinos or charged cosmic rays that are produced or affected by dark matter as an elementary particle. The key processes are:

(a) the *pair annihilation* of dark matter particles (which we shall generically indicate here with the symbol χ), producing Standard Model (SM) particles in the final state, $\chi + \chi \to \text{SM}$;
(b) the *decay* of dark matter particles into SM particles, $\chi \to \text{SM}$;
(c) the *elastic scattering* of dark matter particles off of SM particles, $\chi + \text{SM} \to \chi + \text{SM}$.

Other processes might exist, but are less common in the literature and in model building, and we will not entertain them here. It is important to note that none of the processes (a-c) listed above is bound to necessarily occur in any particle dark matter model: for example (b) will not occur if the dark matter is absolutely stable, and (a) and (c) can be highly suppressed or even not occur at all if the coupling of the dark matter sector to the SM is suppressed, or if the dark matter sector is somehow "secluded".

There exist, however, reasons to be optimistic with respect to the prospect of detecting non-gravitational signatures from dark matter: firstly, some of the best motivated (from a theoretical standpoint) extensions to the SM encompassing a dark matter candidate χ predict a coupling of χ to SM particles that would entail processes (a)–(c) or (b) at some level; secondly, there exist "phenomenological" reasons, chiefly the so-called WIMP miracle (to be reviewed in what follows), that imply the occurrence of some of the processes above for models where the observed abundance of dark matter is connected to its particle nature.

There are three key ingredients to understand indirect dark matter detection at a qualitative level, and to be able to make predictions:

(1) *Production rates* of the relevant SM particles ("messengers"); this is related to the pair-annihilation or decay rate of the dark matter particle;
(2) *Energy scale* of the SM messengers: this is set by the mass of the dark matter particle (or by its momentum, for processes of type (c) above);
(3) *Annihilation products*: this largely model-dependent ingredient details on which SM particles are produced by the dark matter.

The rate $\Gamma_{e^\pm,\bar{p},\gamma,\nu,\ldots}$ for a given SM messenger (i.e. the flux of such particle species per unit time from a unit volume V containing dark matter particles) is generically the product of three factors: the number of dark matter particle pairs (or of dark matter particles, for decaying χ) in the volume V times the pair annihilation (respectively, the decay) rate, times the flux of SM particles per annihilation (decay) event. In formulae:

$$\Gamma_{\mathrm{SM,\ ann}} = \left(\int \frac{\rho_{\mathrm{DM}}^2}{m_\chi^2}\mathrm{d}V\right) \times (\sigma v) \times (N_{\mathrm{SM,\ ann}})\,,$$
$$\Gamma_{\mathrm{SM,\ dec}} = \left(\int \frac{\rho_{\mathrm{DM}}}{m_\chi}\mathrm{d}V\right) \times \left(\frac{1}{\tau_{\mathrm{dec}}}\right) \times (N_{\mathrm{SM,\ dec}})\,.$$

Interestingly for the present discussion, many key quantities (ρ_{DM}, m_χ, $\sigma v, \ldots$) are potentially connected to how the dark matter was produced in the early universe. The dark matter production mechanism in the very early universe is therefore an outstanding starting point both for model building and for eye-balling the relevant indirect detection techniques and setting constraints.

The one quantity we need to know from cosmology which is important to have in mind is the average dark matter density in the universe,

$$\bar{\rho}_{\mathrm{DM}} \simeq 0.23 \cdot \rho_{\mathrm{crit}} = \Omega_{\mathrm{DM}} \cdot \frac{3H_0^2}{8\pi G_N}\,.$$

It is useful in many social (as well as anti-social) situations to have on the tip of your tongue the value of this latter quantity both in "astronomical"[a] and in "particle physics" units:

$$\rho_{\mathrm{crit}} \simeq 3 \times 10^{10}\ \frac{M_\odot}{\mathrm{Mpc}^3} \simeq 10^{-6}\ \frac{\mathrm{GeV}}{\mathrm{cm}^3}\,.$$

From the "astronomical" units, we learn for example that clusters of galaxies, the largest bound dark matter structures in the universe, have typical over-densities of 10^5, since they approximately host hundreds to thousands of galaxies, whose mass is in the $\sim 10^{12}\ M_\odot$ range; from the "particle physics" units, we learn that in our particular location in the Milky Way, where $\rho_{\mathrm{DM}} \sim 0.3\ \mathrm{GeV/cm^3}$, the over-density is a factor of a few larger than in a typical cluster.

A successful framework for the origin of species in the early universe is the paradigm of thermal decoupling (see e.g. [1] and [2]). This framework

[a]Particle physicist: always a good idea to talk to astronomers; for example, I found my wife that way!

encompasses for example the successful predictions of recombination and of Big Bang nucleosynthesis, and it describes in detail the process of cosmological neutrino decoupling. In short, thermal decoupling consists of the relevant particle interaction rate Γ, initially, at high temperatures, much larger than the Hubble expansion rate H, falling to a "freeze-out" point where $\Gamma \sim H$; after this point in time/temperature, the particle species simply "redshift" their momenta and number densities.

In natural units, we can think of $\Gamma = n \cdot \sigma$, with n a particle density and σ an interaction cross section. As statistical mechanics kindly teaches us, the equilibrium number density of a particle of mass m in a thermal bath of temperature T has two asymptotic regimes:

$$n_{\rm rel} \sim T^3 \quad \text{for} \quad m \ll T,$$

$$n_{\rm non-rel} \sim (mT)^{3/2} \exp\left(-\frac{m}{T}\right) \quad \text{for} \quad m \gg T.$$

The right-hand side of $\Gamma \sim H$, i.e. $H(T)$, comes from general relativity, instead, and specifically from Friedmann's equation,

$$H^2 = \frac{8\pi G_N}{3}\rho.$$

In the radiation dominated epoch (i.e. $T \gtrsim 1$ eV),

$$\rho \simeq \rho_{\rm rad} = \frac{\pi^2}{30} \cdot g \cdot T^4,$$

with g the number of relativistic degrees of freedom (2 for photons). To a decent degree of approximation, and recalling that $M_P = 1/\sqrt{8\pi G_N}$, in order to eyeball when thermal decoupling occurs you can take $H \simeq T^2/M_P$.

Let's put all of this in practice, and estimate the temperature of neutrino freeze-out. We estimate the relevant scattering cross section in the Fermi four-fermion contact interaction approximation, and we take $E \sim T_\nu$, so that $\sigma \sim G_F^2 T_\nu^2$ (where $G_F \sim 10^{-5}$ GeV^{-2} is Fermi's constant). We thus have

$$n \cdot \sigma = H \quad \rightarrow \quad T_\nu^3 G_F^2 T_\nu^2 = T_\nu^2/M_P.$$

We thus have

$$T_\nu = (G_F^2 M_P)^{-1/3} \simeq (10^{-10} \times 10^{18})^{-1/3} \text{ GeV} \sim 1 \text{ MeV}.$$

Among various things to be happy about, we cheerfully verify that the T_ν we found indeed satisfies $T_\nu \gg m_\nu$, which we have implicitly assumed for the form of $n(T)$, and we learn that neutrinos are *hot relics*, not because they are particularly attractive, but because they freeze out when relativistic.

Now let's calculate the relic density for a dark matter particle χ. Let me introduce the notation $Y = n/s$ where n is a number density and s is the entropy density. In an iso-entropic universe, $s \cdot a^3 =$ constant, where a is the universe's scale factor. $Y \sim na^3$ is thus a "comoving" number density. If no entropy is produced, $Y_{\rm today} = Y_{\rm freeze-out}$. In the case of hot relics, like SM neutrinos,

$$Y_{\rm freeze\ out} = \frac{\rho_\nu(T_\nu)}{m_\nu \cdot s(T_\nu)}$$

and

$$n_{\rm today} = s_{\rm today} \times Y_{\rm freeze-out}.$$

For example, for the SM neutrinos, the fraction of the universe's critical density times h^2 (where h is today's Hubble constant in units of 100 km/s/Mpc – in practice $h^2 \simeq 0.5$) in a SM neutrino species is

$$\Omega_\nu h^2 = \frac{\rho_\nu}{\rho_{\rm crit}} h^2 \simeq \frac{m_\nu}{91.5\ \rm eV}.$$

While the normalization depends on the relevant cross section, it is a general fact that *a hot relic's thermal relic abundance scales linearly with the relic's mass.* For a weakly interacting dark matter particle, requiring that the thermal dark matter density be less or equal than the observed matter density leads to the so-called Coswik-McClelland limit on the mass of a hot dark matter relic.

> **Exercise #1:** Calculate $T_{\rm f.o.}$ for the $p\bar{p}$ annihilation reaction (you can use $\sigma \sim m_\pi^{-2}$) and estimate the relic proton/antiproton density; is this a hot relic problem? Compare what you find with the observed "baryon asymmetry".

A *cold relic* is one for which the freeze-out temperature is much lower than the mass of the particle, which thus decouples in the non-relativistic regime. An illustrative example of a cold relic is a "heavy" neutrino, with a mass $m_N \gg 1$ MeV. In this case, the appropriate asymptotic form for the equilibrium number density is the non-relativistic limit

$$n \sim (m_\chi T)^{3/2} \exp\left(-\frac{m_\chi}{T}\right).$$

Again, the condition $n\sigma \sim H$ yields

$$n_{\rm f.o.} \sim \frac{T_{\rm f.o.}^2}{M_P \cdot \sigma}. \tag{1}$$

Let us call $m_\chi/T \equiv x$; when dealing with cold relics, we are thus working in the $x \gg 1$ regime. We can re-cast the condition $n \cdot \sigma \sim H$ as

$$\frac{m_\chi^3}{x^{3/2}} e^{-x} = \frac{m_\chi^2}{x^2 \cdot M_P \cdot \sigma}.$$

We thus need to solve

$$\sqrt{x} \cdot e^{-x} = \frac{1}{m_\chi \cdot M_P \cdot \sigma} \sim \frac{1}{10^2 \cdot 10^{18} \cdot 10^{-6}} \sim 10^{-14}, \tag{2}$$

where I've substituted for the nominal values of an "electro-weak interacting" relic, with $\sigma \sim G_F^2 m_\chi^2$ and $m_\chi \sim 10^2$ GeV. Numerically, for the range $10^{-10}...10^{-20}$ for the right-hand side of Eq. (2), the resulting $x_{\rm f.o.} \simeq 20...50$. Now,

$$\Omega_\chi = \frac{m_\chi \cdot n_\chi(T = T_0)}{\rho_c} = \frac{m_\chi\, T_0^3}{\rho_c} \frac{n_0}{T_0^3},$$

with $T_0 = 2.75$ K $\sim 10^{-4}$ eV. Since for an iso-entropic universe $aT \sim$const,

$$\frac{n_0}{T_0^3} \simeq \frac{n_{\rm f.o.}}{T_{\rm f.o.}^3}$$

we have

$$\Omega_\chi = \frac{m_\chi\, T_0^3}{\rho_c} \frac{n_{\rm f.o.}}{T_{\rm f.o.}^3} = \frac{T_0^3}{\rho_c} x_{\rm f.o.} \left(\frac{n_{\rm f.o.}}{T_{\rm f.o.}^2}\right) = \left(\frac{T_0^3}{\rho_c\, M_P}\right) \frac{x_{\rm f.o.}}{\sigma},$$

where I used Eq. (1) in the last step. The equation above can be then cast, plugging in the numbers for the various constants, as

$$\left(\frac{\Omega_\chi}{0.2}\right) \simeq \frac{x_{\rm f.o.}}{20} \left(\frac{10^{-8}\ {\rm GeV}^{-2}}{\sigma}\right), \tag{3}$$

a relation that many refer to as "miraculous". Often, Eq. (3) is quoted with the thermally-averaged product of the cross section times velocity $\langle \sigma v \rangle$ (we will understand why this is, and what a thermal average is, in the next lecture), instead of the simple cross section σ. Since $v \sim c/3$ for $x \sim 20$, one has

$$\langle \sigma v \rangle \sim 10^{-8}\ {\rm GeV}^{-2} \left(3 \times 10^{-28}\ {\rm GeV}^2\ {\rm cm}^2\right)\ 10^{10}\ \frac{\rm cm}{\rm s} = 3 \times 10^{-26}\ \frac{{\rm cm}^3}{\rm s}.$$

Exercise #2: Convince yourself that $v \sim c/3$ for $x \sim 20$.

The $\langle \sigma v \rangle \sim 3 \times 10^{-26}$ cm^3/s is a "magic" number definitely worth keeping in mind!

Is the "magic number" we just found unique and peculiar to the electroweak scale? Not at all! Let us remind ourselves which ingredients we used to get the "right" relic density:

(i) the condition for having a cold relic, $m_\chi \cdot \sigma \cdot M_P \gg 1$;
(ii) a cross section $\sigma \sim 10^{-8}$ GeV^{-2}.

Now, suppose that the cross section be

$$\sigma \sim \frac{g^4}{m_\chi^2},$$

with g some coupling. Condition (ii) essentially enforces that

$$g^2 \sim \frac{m_\chi}{10 \text{ TeV}}$$

independently of which scale m_χ is at. Now go back to condition (i): this reads

$$M_P \gg \frac{1}{m_\chi \cdot \sigma} = \frac{m_\chi^2}{m_\chi \cdot (g^2)^2} \sim \frac{10^8 \text{ GeV}^2}{m_\chi}$$

if condition (ii) holds. Therefore, condition (i) implies, with condition (ii), that $m_\chi \gg 0.1$ eV. Therefore, thermal freeze-out giving the "right" recic abundance is not peculiar to the electroweak scale, as reiterated recently in the literature (see e.g. the "WIMPless" miracle of Ref. [3]). However, since

$$\sigma_{\text{EW}} \sim G_F^2 T_{\text{f.o.}}^2 \sim G_F^2 \left(\frac{E_{\text{EW}}}{20}\right) \sim 10^{-8} \text{ GeV}^{-2}$$

the electroweak scale is quite a "natural" place (whatever natural means) for the miracle to occur!

Is there any *upper* limit to the particle dark matter mass in the cold thermal relic scheme? Indeed there is! The coupling constant g cannot be arbitrarily large (a condition that can also be rephrased in terms of a unitarity limit in the partial wave expansion [4]; note that caveats to the unitarity argument do exist, and this limit can be evaded! (I suggest you read Ref. [4] and think about how to do that)). Roughly,

$$\sigma \lesssim \frac{4\pi}{m_\chi^2},$$

which implies

$$\frac{\Omega_\chi}{0.2} \gtrsim 10^{-8} \text{ GeV}^{-2} \cdot \frac{m_\chi^2}{4\pi}.$$

Therefore, demanding $\Omega_\chi \lesssim 0.2$ implies

$$\left(\frac{m_\chi}{120 \text{ TeV}}\right)^2 \lesssim 1,$$

or $m_\chi \lesssim 120$ TeV.

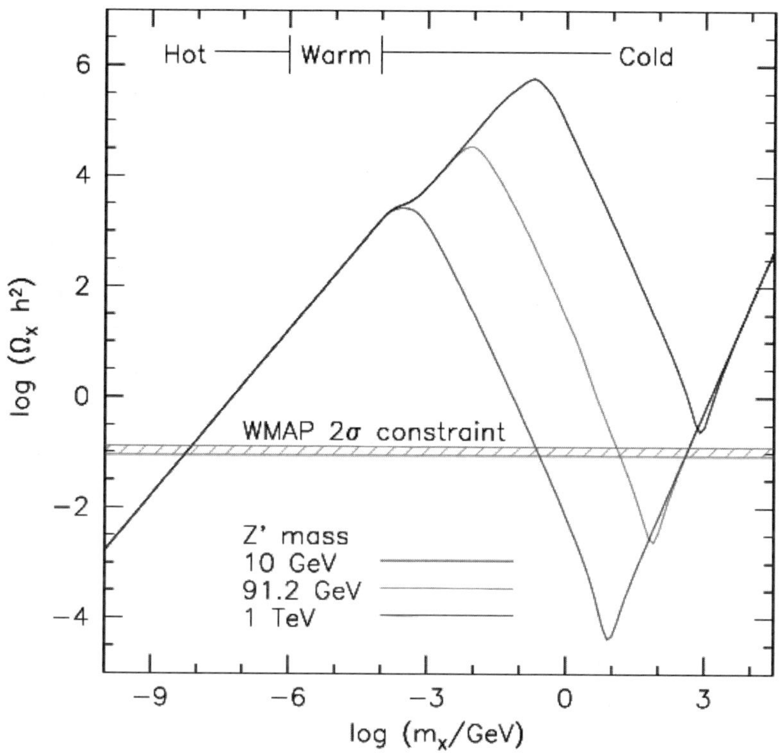

Fig. 1. The thermal relic density of a relic that pair-annihilates with the cross section of Eq. (4), for three values of $m_{Z'}$. From Ref. [6].

Is there, similarly, a *lower* limit in the cold thermal relic scheme? We commented above on the general limit, for arbitrarily low cross sections, $m_\chi \gg 0.1$ eV. But suppose now we have in mind a particle that interacts via electroweak interactions, for example, again, a massive neutrino with $\sigma \sim G_F^2\, m_\chi^2$. In this case

$$\Omega_\chi h^2 \sim 0.1 \frac{10^{-8}}{\text{GeV}^{-2}} \cdot \frac{1}{G_F^2\, m_\chi^2} \sim 0.1 \left(\frac{10\ \text{GeV}}{m_\chi}\right)^2 .$$

This implies that $m_\chi \gtrsim 10$ GeV for weakly interacting massive particles (WIMPs) – a limit known in the literature as the *Lee-Weinberg limit* [5].

Figure 1, from Ref. [6], illustrates the thermal relic density of a weakly interacting massive particle as a function of the particle's mass. The cross

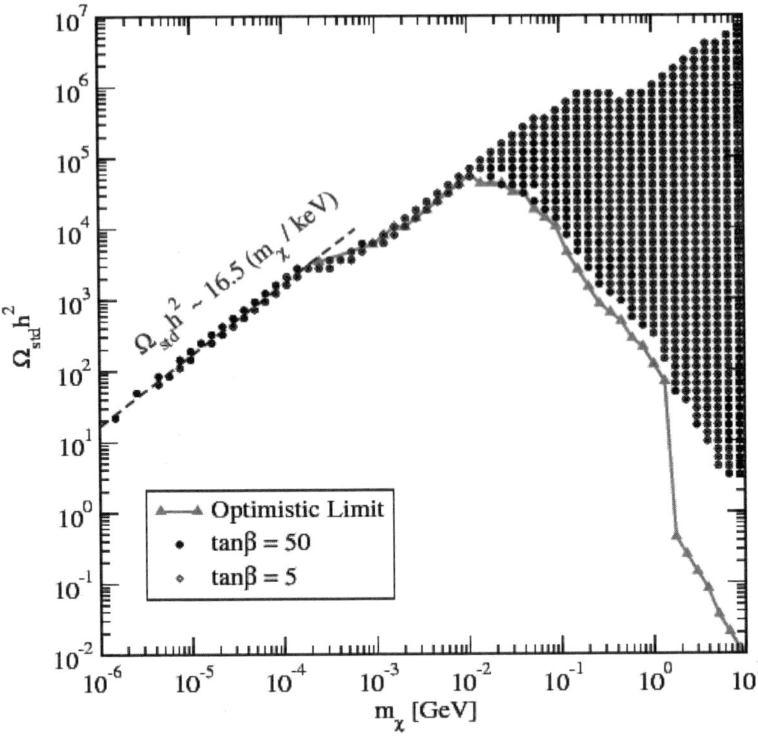

Fig. 2. The thermal relic density of the lightest neutralino in the MSSM, as a function of the neutralino mass. From Ref. [7].

section is assumed to be of the form

$$\sigma \sim \frac{m_\chi^2}{(s - m_{Z'}^2)^2 + m_{Z'}^4}, \tag{4}$$

with s the total center of mass energy squared. The mass of the mediator Z' is taken to be 10 GeV, 91.2 GeV (the Z mass) and 1 TeV. The asymptotic hot and cold relic behaviors are clearly visible and match the predictions we made above: a hot relic density scales linearly with mass, a cold relic with $m_\chi \gg m_{Z'}$ has $\Omega \sim 1/\sigma \sim m_\chi^2$, and a cold relic in the regime where $m_\chi \ll m_{Z'}$ has $\Omega \sim m_{Z'}^4/m_\chi^2$.

In theories with a more complicated dark matter pair-annihilation cross section than what appears in Eq. (4), for example for the lightest neutralino in the minimal supersymmetric extension of the Standard Model (MSSM), various effects blur the simple connection between mass and cross section/relic density, resulting in a much larger spread of results. The general feature of a lower limit of about 1-10 GeV is, however, rather resilient, as shown in Fig. 2, from Ref. [7], where I scanned generously over the relevant MSSM parameter space.

The last comment above leads to the general question of how light can a WIMP be. So far, we assumed an iso-entropic universe. Suppose at some point after a WIMP has frozen out and is thus decoupled from the universe's thermal bath, the entropy density changes from $s \to \gamma \cdot s$, $\gamma > 1$ from e.g. decaying relics (such as relic gravitinos, moduli, ...) or from a first order phase transition (see e.g. Ref. [8]). Then

$$Y_{\rm today} \to \frac{Y_{\rm today}}{\gamma} \quad \text{and} \quad \Omega_\chi \to \frac{\Omega_\chi}{\gamma}.$$

For a sufficiently large γ, the relic abundance of almost any over-abundant relic WIMP can be "diluted" enough to match the observed dark matter density. For example, in supersymmetry the lightest neutralino can be almost arbitrarily light as long as it is bino-like (to prevent an excessively light associated chargino) and if sfermions are sufficiently heavy to suppress energy loss mechanisms in stars (for example $e^+e^- \to \chi\chi$). The additional requirement is of course that the entropy injection happen at a temperature smaller than the neutralino freeze-out, which sets a (weak) constraint on the neutralino mass [7]: entropy injection cannot happen too close to the era of Big Bang Nucleosynthesis to prevent disrupting the successful predictions of the abundance of light elements.

Lecture 2: WIMP Relic Density, A Closer Look

The dark matter literature is flooded with the symbol $\langle\sigma v\rangle$, short for the *zero-temperature thermally-averaged pair-annihilation cross section times velocity* – but do we really understand what this symbol indicates, and where it comes from? How is the "thermal average" defined? What is v? is it a relative velocity? but a relative velocity is not Lorentz-invariant! etc...

The starting point for a closer look at the WIMP relic density is the Boltzmann equation, that can be symbolically cast as [1]:

$$\hat{L}[f] = \hat{C}[f], \tag{5}$$

where $f = f(\vec{p}, \vec{x}, t)$ is the phase space density, $\hat{L}$ is the *Liouville operator* describing the change in time of the phase space density, and $\hat{C}$ is the *collision operator* describing the number of particles per phase-space volume lost or gained per unit time. For those (like me) who need a refresher on the Liouville operator, its non-relativistic form reads

$$\hat{L}_{\rm NR} = \frac{\rm d}{{\rm d}t} + \frac{{\rm d}\vec{x}}{{\rm d}t}\vec{\nabla}_x + \frac{{\rm d}\vec{v}}{{\rm d}t}\vec{\nabla}_v,$$

while in its covariant form it reads

$$\hat{L}_{\rm cov} = p^\alpha \frac{\partial}{\partial x^\alpha} - \Gamma^\alpha_{\beta\gamma}\, p^\beta\, p^\gamma \frac{\partial}{\partial p^\alpha}.$$

In a homogeneous and isotropic cosmology (also known in the trade's slang as a Friedman-Robertson-Walker universe),

$$f(\vec{x}, \vec{p}, t) \to f(|\vec{p}|, t) \quad \text{or, equivalently :} \quad f(E, t).$$

Also, $\hat{L}$ simplifies to

$$\hat{L}[f] = E\frac{\partial f}{\partial t} - \frac{\dot{a}}{a}|\vec{p}|^2\, \frac{\partial f}{\partial E}.$$

We are interested in particle number densities, defined by

$$n(t) = \sum_{\rm spin} \int \frac{{\rm d}^3 p}{(2\pi)^3} f(E, t).$$

We will thus take Eq. (5) and consider (calling g the number of spin degrees of freedom)

$$\int L[f] \cdot g \frac{{\rm d}^3 p}{(2\pi)^3} = \frac{{\rm d}n}{{\rm d}t} + 3H \cdot n$$

where we introduced $H = \dot{a}/a$ and integrated by parts using

$$\frac{1}{a^3}\frac{\mathrm{d}}{\mathrm{d}t}\left(a^3 \cdot n\right) = \frac{\mathrm{d}n}{\mathrm{d}t} + 3H \cdot n.$$

Cleaning up the right-hand side of the Boltzmann equation is a bit messier and I recommend the classic paper by Gondolo and Gelmini, Ref. [9]. For definiteness, let us consider a process of the type $1+2 \to 3+4$ where we are interested in the number density of species 1, and where we assume that species 3 and 4 are in thermal equilibrium. The right-hand side of Eq. (5) can then be cast as:

$$g_1 \int \hat{C}[f_1]\frac{\mathrm{d}^3p}{(2\pi)^3} = -\langle \sigma \cdot v_{\text{Møl}}\rangle \left(n_1 n_2 - n_1^{\text{eq}} n_2^{\text{eq}}\right), \tag{6}$$

where $n_{1,2}$ are the number densities, while $n_{1,2}^{\text{eq}}$ indicate the equilibrium number densities, and where

$$\sigma = \sum_f \sigma_{12\to f}$$

indicates the invariant, unpolarized total cross section for processes $1+2 \to$ any final state f in thermal equilibrium, and, finally, where

$$v_{\text{Møl}} = \frac{\sqrt{(p_1 \cdot p_2)^2 - m_1^2 m_2^2}}{E_1\ E_2}. \tag{7}$$

A couple of comments:

(1) Notice that $v_{\text{Møl}}\ n_1\ n_2$ is a Lorentz invariant;
(2) Notice that in the rest frame of 1 (or 2; what we can think of as the "lab frame"), $v_{\text{Møl}} \to v_{\text{rel}} = |\vec{v}_1 - \vec{v}_2|$, where e.g. $\vec{v}_1 = \vec{p}_1/E_1$ etc.

Last ingredient: the thermal average: this is defined by the expression:

$$\langle \sigma \cdot v_{\text{Møl}}\rangle = \frac{\int \sigma \cdot v_{\text{Møl}}\ e^{-E_1/T} e^{-E_2/T}\ \mathrm{d}^3p_1\ \mathrm{d}^3p_2}{\int e^{-E_1/T} e^{-E_2/T}\ \mathrm{d}^3p_1\ \mathrm{d}^3p_2}. \tag{8}$$

Exercise #3: Evaluate the denominator of Eq. (8) for $m_1 = m_2$.

The diligent Reader who carried out the exercise proposed above found that the denominator of Eq. (8), for $m_1 = m_2 = m$ (the pair-annihilation case relevant for us) reads

$$\int e^{-E_1/T} e^{-E_2/T}\ \mathrm{d}^3p_1\ \mathrm{d}^3p_2 = \left(4\pi m^2 T K_2\left(\frac{m}{T}\right)\right)^2,$$

where K_2 is the modified Bessel function of the second order. The numerator reads, instead,

$$\int \sigma \cdot v_{\text{Møl}} e^{-E_1/T} e^{-E_2/T} \, \mathrm{d}^3 p_1 \, \mathrm{d}^3 p_2 = \int_{4m^2}^{\infty} \sigma(s - 4m^2)\sqrt{s} K_1 \left(\frac{\sqrt{s}}{T} \right) \mathrm{d}s, \quad (9)$$

a "convolution" of the cross section evaluated at $s - 4m^2$, where s is the center-of-mass total energy, and a temperature-dependent thermal kernel. This form is key to understand important caveats to, e.g. the magic relation that implies that $\langle \sigma v \rangle = 3 \times 10^{-26}$ cm^3/s gives the correct thermal relic density. From now on, I will suppress the subscript Møl and intend always that $v \to v_{\text{Møl}}$.

Caveats to the Standard Story

A classic paper that discusses departures from the vanilla calculation of the WIMP relic density described in the previous lecture is Ref. [10] by Griest and Seckel, "Three exceptions in the calculation of relic abundances". The three memorable exceptions are:

(1) Resonances
(2) Thresholds
(3) Co-annihilations

Resonant annihilation through a particle with the right quantum numbers and a mass $m_A \simeq 2m_\chi$, found for example in the so-called "funnel" region of the (soon to be gone, thanks LHC!) minimal supergravity/constrained MSSM model, can be relevant either if $m_\chi \gtrsim m_A/2$ or if $m_\chi \lesssim m_A/2$. In the first case, the cross section peaks at $s = m_A^2$ and is thus most relevant at temperatures $T_{\text{res}} \simeq m_A^2/(6m_\chi)$.

Exercise #4: Show that $\langle s \rangle \simeq 4m_\chi^2 + 6m_\chi T$.

If $T_{\text{res}} \simeq T_{\text{f.o.}}$ the resonance is extremely important at freeze-out, and hence for the thermal relic density; the pair annihilation cross section today, the one that we care about for indirect dark matter detection rates, will however be potentially much lower! In this case, the freeze-out cross section might be close to $\langle \sigma v \rangle = 3 \times 10^{-26}$ cm^3/s, but (potentially) the $T = 0$ cross section $\langle \sigma v \rangle_0 \ll 3 \times 10^{-26}$ cm^3/s.

If $m_\chi \lesssim m_A/2$, in the integral of Eq. (9) the cross section is always maximal for $T = 0$, the resonance can be (or not) subdominant at freeze-out, but we are in the (lucky, if you want a signal, or unlucky if you want to hide it!) circumstance where the $T = 0$ cross section$\langle\sigma v\rangle_0 \gg 3 \times 10^{-26}$ cm^3/s. It must be noted that this discussion has model-dependent caveats: for example, in supersymmetry the lightest neutralinos are Majorana particles, and in a purely s-wave annihilation (at $T = 0$) a pair of neutralinos are in a CP-odd state. Therefore, for example, the pair annihilation via a CP-even particle (such as a neutral CP-even Higgs) cannot contribute to the $T = 0$ pair annihilation cross section!

Thresholds affect the relation between the freeze-out and $T = 0$ pair-annihilation cross sections in an obvious way: the cross section in Eq. (9) suddenly increases as, e.g. $s > 4m_t^2$, where m_t is the particle in which pair our annihilating particle can go, $\chi\chi \to \bar{t}t$ (think e.g. of $m_\chi \lesssim m_t$ where t is the Standard Model top quark). Thresholds therefore always imply $\langle\sigma v\rangle_{\rm f.o.} > \langle\sigma v\rangle_0$.

Co-annihilation occur for particles whose freeze-out process is tangled with that of other particle species with a close enough mass so that the two freeze-out episodes are inter-connected. A necessary condition is that this second co-annihilating species 2 have a mass such that at freeze-out the Boltzmann suppression of its equilibrium number density is not dramatic. In formulae, we want $m_2 - m_1 \lesssim T_{\rm f.o.}$. In this case the relevant cross section is an "effective" cross section that includes the appropriately Boltzmann-weighed contribution from (N) co-annihilating particles, i.e., with obvious notation (if this is not obvious see the extensive and clear discussion of Ref. [11])

$$\langle\sigma v\rangle \;\to\; \langle\sigma_{\rm eff} v\rangle = \frac{\sum_{ij}\sigma_{ij}\exp\left(-\frac{\Delta m_1+\Delta m_2}{T}\right)}{\sum_{i=1}^{N} g_i \exp\left(-\frac{\Delta m_i}{T}\right)}. \tag{10}$$

Note that the denominator counts the effect of the additional degrees of freedom, suitably weighed.

Coannihilation comes in two varieties, that I like to call "*parasite*" and "*symbiotic*". If the additional degrees of freedom annihilate "less efficiently" than the particle whose number density we are interested in, then the coannihilating particles will have a parasitical effect, and produce a smaller effective pair-annihilation cross section. This is the typical case with, for example, the lightest Kaluza-Klein excitation of universal extra dimensions (UED; for a review see my own review! Ref. [12]). UED has a very compressed spectrum of particles (Kaluza-Klein modes of Standard Model particles, whose mass differs from the compactification scale by loop corrections or corrections of the order of the Standard Model particle mass) above the stable dark matter particle candidate. This large collection of particles brings a ton of additional effective degrees of freedom which outweighs the additional contribution to the pair-annihilation cross section, rendering UED a prototypical example of parasitical coannihilation.

An example of "symbiotic" co-annihilation is provided by a nimble particle that annihilates efficiently and doesn't carry a big bag of degrees of freedom. An example is co-annihilation of the lightest neutralino of the MSSM with the scalar partner of the tau lepton, or "stau" (but other equally good examples are co-annihilation with charginos or stops or other sfermions). Unfortunately, collider searches indicate that the days of the "stau coannihilation region" may be counted [13].

So far we have dealt with exceptions to the left-hand side of the Boltzmann equation (or of its short-hand version $\Gamma = n \cdot \sigma = H$). What happens if we fiddle around with the right-hand side instead, i.e. with H, the expansion history of the universe? To avoid self-promoting my own papers again, I will point you to the following example: a cosmology with a "quintessence" field that provides a dynamical dark energy term, whose impact on the relic density of WIMPs was first studied by Salati in Ref. [14].[b]

Let ϕ be the quintessence field, a spatially homogeneous real, scalar field. The field energy density and pressure are

$$\rho_\phi = \frac{1}{2}\left(\frac{\mathrm{d}\phi}{\mathrm{d}\phi}\right)^2 + V(\phi) \tag{11}$$

$$P_\phi = \frac{1}{2}\left(\frac{\mathrm{d}\phi}{\mathrm{d}\phi}\right)^2 - V(\phi) \tag{12}$$

An example of a suitable potential that exhibits the desired "tracking" behavior (for appropriate initial conditions), i.e. that tracks dynamically the dominant energy density component, is

$$V(\phi) = M_P^4 \exp\left(-\frac{\lambda\phi}{M_P}\right).$$

The field's equation of state $w = P_\phi/\rho_\phi$ moves from $w = -1$ in the "kination" phase, where the kinetic energy term dominates, to $w = +1$ in the "cosmological constant" phase, where V dominates. Tracking helps explaining the coincidence problem $\Omega_\Lambda = \Omega_\phi \sim \Omega_M$, although fine-tuning is not eliminated (as it creeps back in via the initial conditions).

[b]See also Ref. [15]. Couldn't resist.

Noticing that $\rho_\phi \sim a^{-3(1+w)}$, in the kination phase $\rho \sim a^{-6}$ and therefore the universe is kination-dominated as sufficiently early times, with

$$H \sim \frac{T^2}{M_P}\frac{T}{T_{\rm KRE}},$$

where $T_{\rm KRE}$ stands for the temperature of kination-radiation equality. To be relevant for the relic density of a particle species decoupling at $T = T_{\rm f.o.}$, kination must dominate before and at freeze-out, hence $T_{\rm KRE} > T_{\rm f.o.}$. However, to avoid disrupting Big Bang Nucleosynthesis, we must also require that $T_{\rm KRE} < T_{\rm BBN} \sim 1$ MeV.

In a kination-dominated universe, freeze-out works, schematically, exactly as we described in the previous section, and

$$\Omega_\chi^{\rm quint} = \frac{T_0^3}{M_P \cdot \rho_c} x_{\rm f.o.} \left(\frac{n_{\rm f.o.}}{T_{\rm f.o.}^2}\right) \tag{13}$$

but now the freeze-out condition reads

$$n_{\rm f.o.}\langle\sigma\, v\rangle \sim \frac{T^2}{M_P}\frac{T}{T_{\rm KRE}}.$$

We therefore have that

$$\frac{n_{\rm f.o.}}{T_{\rm f.o.}^2} \sim \frac{1}{M_P\ \langle\sigma\, v\rangle}\frac{T_{\rm f.o.}}{T_{\rm KRE}}.$$

To first order, the enhancement factor of the thermal relic density in the presence of quintessence above the standard thermal relic density is thus

$$\frac{\Omega_\chi^{\rm quint}}{\Omega_\chi^{\rm standard}} \sim \frac{T_{\rm f.o.}}{T_{\rm KRE}} \lesssim \frac{m_\chi}{20}\frac{1}{T_{\rm BBN}} \sim 10^4 \frac{m_\chi}{100\ {\rm GeV}}.$$

With more accurate calculations the enhancement factor is found to be potentially as large as 10^6 [14, 15].

The dark matter production mechanism can naturally be non-thermal, or it can arise from an asymmetry (as is probably the case for the origin of baryonic matter). Suppose a particle species ψ, with $m_\psi > m_\chi$ is produced in the early universe with an abundance Ω_ψ, and that ψ decays to χ, which is the stable dark matter particle, at a temperature when χ is out of equilibrium. The relic density that χ will then inherit (up to contribution from the decay of other particle species and from thermal production etc.) is simply

$$\Omega_\chi \simeq \Omega_\psi \frac{m_\chi}{m_\phi},$$

where the $\simeq$ sign indicates that additional effects (such as some entropy production in the decay process) can enter.

As you calculated in Ex. #1, the thermal abundance of protons and antiprotons is almost ten orders of magnitude smaller than the observed abundance. If an asymmetry is present, then the observed proton density can be inherited entirely from the asymmetry itself. The same could hold for the dark matter sector. Many have envisioned the possibility of explaining the coincidence $\Omega_{\rm DM} \simeq 5\Omega_{\rm B}$ (where DM is dark matter and B is baryonic matter) by postulating that perhaps $n_{\rm DM} \sim n_{\rm B}$ and that $m_{\rm DM} \sim 5$ GeV, with some mechanism that couples an asymmetry in one sector to the other sector, or many variants on this theme. It is interesting to ask the question of whether any indirect dark matter detection signal could exist if dark matter originates from an asymmetry. In principle, if no anti-dark-matter is generated (and if primordial pair-annihilation got rid of all of it), then it's very hard to get any indirect signals; however, if dark matter and anti-dark matter oscillate into each other (for example because of a $\Delta N_\chi = 2$ operator, such as a mass term, say $m_M \chi\chi$) then oscillations can populate the anti-dark matter content of the universe and residual annihilation occur [16]. Unless a symmetry in the theory explicitly prohibits such operators, oscillations generically occur, since after all we only need

$$\tau_{\rm universe} \sim 10^{17}\ {\rm s} \lesssim 10^{17}\ {\rm s}\left(\frac{10^{-41}\ {\rm GeV}}{m_M}\right), \quad \text{i.e.} \quad m_M \gtrsim 10^{-41}\ {\rm GeV}.$$

Of course many other interesting and witty dark matter production mechanisms exist: I recommend you explore at least the classic ones that pertain to sterile neutrinos and to axions (see e.g. Ref. [1]).

Kinetic Decoupling

The dark matter thermal history in the very early universe is important not only for the eventual relic density, but potentially also for the formation of structure, especially for (cold) WIMPs. In the early universe, elastic scattering processes such as $\chi f \leftrightarrow \chi f$, where f is a Standard Model fermion, keep the dark matter particle in kinetic equilibrium even *after* chemical decoupling (i.e. when $\Gamma_{\chi\chi\leftrightarrow X} \ll H$). The reason is that the target densities for the processes that keep the dark matter in chemical versus kinetic equilibrium are vastly different after chemical decoupling:

$$\begin{aligned} \chi\chi \leftrightarrow ff \quad &\rightarrow \Gamma = n_{\rm non-rel} \cdot \sigma \\ \chi f \leftrightarrow \chi f \quad &\rightarrow \Gamma = n_{\rm rel} \cdot \sigma \end{aligned}$$

with $n_{\rm non-rel}$ exponentially suppressed! Let us know estimate the kinetic decoupling temperature of a WIMP. We shall assume a default electroweak cross section

$$\sigma_{\chi f \leftrightarrow \chi f} \sim G_F^2 T^2.$$

We have to account for the fact that the WIMP is non-relativistic after chemical decoupling and that momentum transfer between the WIMP and the thermal bath becomes "inefficient", in a sense to be made quantitative with a simple estimate. The typical momentum transfer per collision is $\delta p \sim T$, while the WIMIP momentum in the non-relativistic regime satisfies the relation

$$\frac{p^2}{2m} \sim T$$

so that $p \sim \sqrt{m_\chi T}$. Momentum transfer is a stochastic process, so it takes

$$N = \left(\frac{\delta p}{p}\right)^2 \sim \frac{T^2}{m_\chi T} = \frac{T}{m_\chi}$$

collisions to establish kinetic equilibrium.

To calculate kinetic decoupling, we thus want to compare

$$n_{\rm rel} \cdot \sigma_{\chi f \leftrightarrow \chi f} \left(\frac{\delta p}{p}\right)^2 \sim T^3 \cdot G_F^2 T^2 \cdot \frac{T}{m_\chi} \sim H \sim \frac{T^2}{M_P}.$$

We thus find

$$T_{\rm k.d.} \sim \left(\frac{m_\chi}{M_P \cdot G_F^2}\right)^{1/4} \sim 30 \text{ MeV } \left(\frac{m_\chi}{100 \text{ GeV}}\right)^{1/4}.$$

What does this imply for structure formation? Roughly, the cutoff scale of the matter power spectrum will correspond to the size of the horizon at kinetic decoupling, so

$$M_{\rm cutoff} \sim \frac{4\pi}{3}\left(\frac{1}{H(T_{\rm kd})}\right)^3 \rho_{\rm DM}(T_{\rm kd}) \sim 30\ M_\oplus \left(\frac{10 \text{ MeV}}{T_{\rm kd}}\right)^3.$$

More precisely, the cutoff scale is set by the largest of the free-streaming versus acoustic damping scale, see e.g. [17] for a nice review. For typical WIMPs, we thus find that these "protohalos" (which correspond to the first structures that gravitationally collapse in the early universe) have a mass comparable to the Earth mass (i.e. roughly $10^{-6}\ M_\odot$). In specific theories, the range of variations ca however be very significant [18]. This fact has potentially important consequences for indirect detection, as it feeds in the

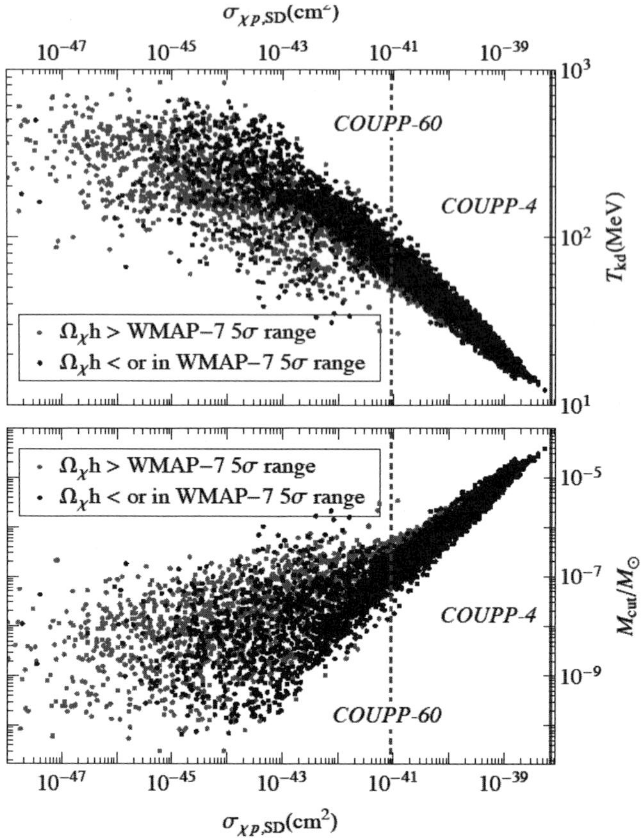

Fig. 3. The correlation between the spin-dependent dark matter-proton cross section and the kinetic decoupling temperature (upper panel) and the small scale cutoff mass (lower panel) in the MSSM. From Ref. [19].

problem of calculating the *boost factor* (i.e. the additional contribution to the count of pairs in a halo on top of the smooth halo component) from substructure, and for the dark matter "small-scale" problem.

Is there any way to probe the size of the dark matter small-scale cutoff? In Ref. [19] we pointed out that if $f = q$ (i.e. the particles the dark matter scatters off of are quarks), the processes relevant for kinetic decoupling are exactly the same as those participating in dark matter direct detection. Therefore, in principle, one could correlate $M_{\rm cutoff}$ with $\sigma_{\rm direct\ det}$. There is, however, an important caveat: usually, $T_{\rm kd} < T_{\rm QCD}$, the latter symbol indicating the temperature corresponding to the QCD confinement phase

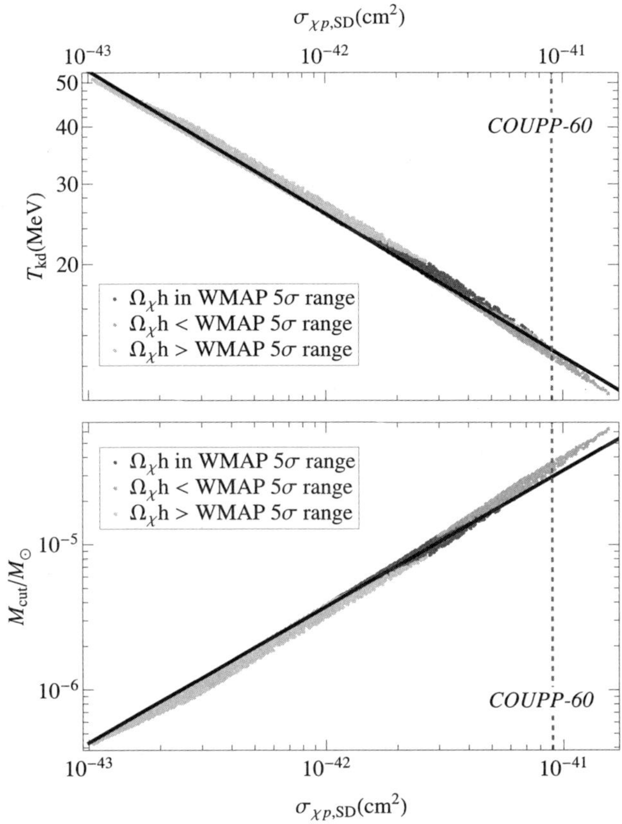

Fig. 4. The correlation between the spin-dependent dark matter-proton cross section and the kinetic decoupling temperature (upper panel) and the small scale cutoff mass (lower panel) in UED. From Ref. [19].

transition. After confinement, the number density of hadrons is negligible compared to light leptons, and the latter dominate and control kinetic decoupling. If, however, one has a dark matter theory where "quark-lepton universality" holds, then a correlation is expected, and indeed is found — both for UED and for supersymmetric dark matter [19] as illustrated in Figs. 3 and 4.

Lecture 3: Indirect Dark Matter Detection

What have we learned thus far about annihilation processes, that we could use to detect non-gravitational signals from particle dark matter?

— $\langle\sigma v\rangle \sim 3 \times 10^{-26}$ cm^3/s is a good "magic" benchmark;
— the magic number above is not WIMP-specific, and is independent of mass, to first order;
— there exist numerous caveats, both from the particle physics side (coannihilation, resonances, thresholds,...) and from the cosmology side (quintessence, non-thermal production, asymmetry,...) to that magic number.

What about dark matter decay? If dark matter is unstable with a lifetime well in excess of the age of the universe, the decay products would also be a great way to detect non-gravitational signals from dark matter! From a theoretical standpoint, a GUT-scale or Planck-scale dark matter number-violating operators should be generic. For example for a dimension-5 operator,

$$\Gamma_5 \sim \frac{1}{M^2} m_\chi^3$$

the resulting lifetime

$$\tau_5 \sim 1 \text{ s} \left(\frac{1 \text{ TeV}}{m_\chi}\right)^3 \left(\frac{M}{10^{16} \text{ GeV}}\right)^2$$

would potentially result in an impact on BBN, but would produce an excessively short-lived dark matter candidate.

For a dimension-6 operator things look more interesting,

$$\Gamma_6 \sim \frac{1}{M^4} m_\chi^5$$

with a lifetime

$$\tau_6 \sim 10^{27} \text{ s} \left(\frac{1 \text{ TeV}}{m_\chi}\right)^5 \left(\frac{M}{10^{16} \text{ GeV}}\right)^4,$$

which, turns out, to be a very interesting lifetime range to explain the Pamela positron excess and for searches for dark matter with gamma rays, as we shall see later on.

For definiteness, let me however concentrate on dark matter annihilation. Let us now try to corner the key ingredients to make predictions for indirect searches for dark matter. First and foremost: what do we know about the dark matter particle mass, which sets the energy scale for the particles produced in an annihilation event? We saw earlier that for WIMPs a useful lower limit is provided by the Lee-Weinberg bound at about 10 GeV, while unitarity constrains WIMPs, on the large mass end, to be lighter than a few 100 TeV. I think this is a reasonable range to keep in mind, if one is wed to the notion of a weakly interacting dark matter particle.

There exists, however, a number of theoretical prejudices that have populated this field for a long time. A historically interesting one has it that WIMPs must be heavier than 40 GeV. Interestingly, this prejudice somehow managed to distort how certain dark matter search experiments were optimized! It is worthwhile then to see where this prejudice comes from.

The first tenet of the "WIMPs must be heavier than 40 GeV" prejudice is that WIMPs are supersymmetric neutralinos. Browsing papers on dark matter (especially on *astro-ph*) the confusion between WIMPs and neutralinos is not unheard of. The second tenet is that there exists one universal soft supersymmetry breaking mass for all three gauginos at the GUT scale. If $M_1(M_{\rm GUT}) = M_2(M_{\rm GUT})$, where M_1 is the soft supersymmetry breaking scale associated with the $\mathrm{U}(1)_Y$ gaugino and M_2 that of the SU(2) gaugino, renormalization group evolution (and the assumption of a "desert" between the electroweak and the GUT scale) implies that $M_1(M_{\rm EW}) =\simeq 0.4 \times M_2(M_{\rm EW})$. Now, LEP2 constrains the chargino mass to be above about half its center of mass energy, or $m_{\tilde{\chi}_1^\pm} \gtrsim 100$ GeV. But one of the charginos (the wino-like, in susy slang) has a mass very close to $M_2(M_{\rm EW})$, therefore implying that the lightest, bino-like neutralino $m_{\tilde{\chi}_1^0} \simeq M_1(M_{\rm EW}) \gtrsim 0.4 \times 100\ \mathrm{GeV} = 40$ GeV. Amazing. Of course, GUT-scale universality, renormalization group evolution etc. are all model-dependent ingredient, not to mention the assumption that the dark matter is a neutralino...

What do we know about the annihilation final state? *Presque rien*, almost nothing. If the dark matter particle is a Majorana fermion, then the pair annihilation into a fermion-antifermion final state is "p-wave suppressed": $\chi\chi \to f\bar{f}$ requires a helicity flip, and thus the matrix element squared is proportional to the square of the fermion mass, $|M|^2 \propto m_f^2$ (in just the same way as for charged pion decay – a Majorana pair in a $l = 0$ wave is in a CP-odd state). As a result, pair-annihilation of Majorana dark matter into light fermions is highly suppressed. If $m_\chi < m_{\rm top}$ and if the annihilation channel $\chi\chi \to$ bosons (such as W^+W^- or hh) is suppressed, the dominant annihilation final states are $b\bar{b}$ and $\tau^+\tau^-$. This explains the otherwise surprising popularity of these two final states in the literature on dark matter indirect detection.[c] Note that besides $m_b \gg m_\tau$, the bottom quark final state wins by an additional factor 3 from color. There exist, however, circumstances where the $\tau\tau$ final state can be boosted, for example with a light scalar tau in supersymmetry (in the so-called stau coannihilation region).

In UED the situation is entirely different: the particle that is usually the stable lightest Kaluza-Klein excitation is the $n = 1$ mode of the hyper-charge gauge boson, or $B^{(1)}$. The matrix element squared for pair annihilation into a fermion-antifermion pair is proportional to the fourth power of the fermion's hyper-charge, $|M|^2 \propto |Y_f|^4$, thus up-type quarks ($Y_{u_L} = 4/3$) and charged leptons ($Y_{e_R} = 2$) are the preferred annihilation modes.

If the dark matter lives in an SU(2) multiplet (for example, higgsinos and winos in supersymmetry) everything is fixed by gauge interactions. For wino-like dark matter, the preferred final state is W^+W^-. The lightest neutralino is quasi-degenerate with the lightest chargino, with mass splittings on the order of a fraction of a GeV, and both lie at a scale close to M_2, the corresponding soft supersymmetry breaking mass. Coannihilation plays obviously a very significant role, and coannihilation is of the "symbiotic" type (charginos pair-annihilate quite efficiently). The resulting pair-annihilation cross section is, approximately [20]

$$\langle \sigma v \rangle_{\tilde{W}} \simeq \frac{3g^4}{16\ \pi\ M_2^2}$$

and the resulting thermal relic abundance is

$$\Omega_{\tilde{W}} h^2 \simeq 0.1 \left(\frac{M_2}{2.2\ \text{TeV}} \right)^2 ,$$

implying that thermal winos must weigh about 2.2 TeV. Higgsinos, instead, come in a set of two neutralinos and a chargino with again small mass splittings and at a mass scale around μ. The dominant annihilation final states are W^+W^- and ZZ, and the pair annihilation and thermal relic densities are given by

$$\langle \sigma v \rangle_{\tilde{H}} \simeq \frac{g^4}{512\ \pi\ \mu^2} \left(21 + 3\tan^2\theta_W + 11\tan^4\theta_W \right)$$

[c] If you are a graduate student at UCSC who happened to work on dark matter, and I sit on your thesis committee, I will ask you the reason why $b\bar{b}$ is such a popular final state during your thesis defense. Be ready.

and

$$\Omega_{\tilde{H}} h^2 \simeq 0.1 \left(\frac{\mu}{1\ \mathrm{TeV}}\right)^2$$

indicating that thermal higgsinos like to weigh about a TeV.

Indirect Detection: Warm up lap

Time to charge ahead on indirect detection. The name of the game is to get "enough" number counts, in symbols:

$$N = \phi_\chi \cdot A_{\mathrm{eff}} \cdot T_{\mathrm{exp}},$$

where:

- ϕ_χ indicates the relevant dark matter-induced event rate, such as the flux of a certain type of Standard Model particle, and has units $\mathrm{cm}^{-2} \cdot \mathrm{s}^{-1}$;
- A_{eff} is an effective area: good to have in mind some numbers here. For example, in the business of gamma-ray telescopes, the Fermi Large Area Telescope (LAT) has an effective area of about 1 m^2, while the top-of-the-line atmospheric Cherenkov (ground based) telescopes, such as H.E.S.S. or MAGIC or VERITAS have effective areas on the order of 10^5 m^2; the relevant numbers for the two key antimatter satellites are $\sim$0.01 m^2 for Pamela and $\sim$ 0.1 m^2 for AMS-02; finally, if you're asked to quote a number for high-energy neutrino telescopes, mention IceCube and mumble 1 km^2.
- T_{exp} indicates the relevant "exposure time": for satellites this is on the order of a year, which as you know is exactly $\pi \times 10^7$ s; for typical ground-based telescopes you can perhaps count on about 100h, or about 10^5 s, while balloon experiments have a typical exposure time of the order of a week, or about 10^6 s.

To detect a signal we need to fulfill two basic conditions:

(i) have some signal events, i.e. $\phi_\chi \cdot A_{\mathrm{eff}} \cdot T_{\mathrm{exp}} \gg 1$
(ii) have enough signal-to-noise, for example requesting $N_{\mathrm{signal}} > (\#\sigma)\sqrt{N_{\mathrm{background}}}$.

I like to classify astrophysical probes of dark matter into three categories:

(1) **Very indirect**: this category includes effects induced by dark matter on astrophysical objects or on cosmological observations;

(2) **Indirect**: I include in this category probes that don't "trace back" to the annihilation event, as their trajectories are bent as the particles propagate: charged cosmic rays;
(3) **Not-so-indirect**: neutrinos and gamma rays – we will discuss these guys in the next lecture.

1. Effects on Astrophysical Objects: folks have thought about an amazing variety of possibilities, including:

- Solar Physics (dark matter can affect the Sun's core temperature, the sound speed inside the Sun,...)
- Neutron Star Capture, possibly leading to the formation of black holes (notably e.g. in the context of asymmetric dark matter, see e.g. [21])
- Supernova and Star cooling (see the excellent book by Georg Rafelt [22])
- Protostars (e.g. WIMP-fueled population-III stars, available also in Swedish [23])
- Planets warming

Given the relevance of global warming to the general public (and to funding agencies), let's make an estimate of this latter effect. The "capture probability" for WIMPs is roughly

$$n_{\rm nucleons} \cdot \sigma_{\chi-N} \cdot R_{\rm planet} \lesssim 10^{-4},$$

where for the right-hand side I've used $n_{\rm nucleons} \sim N_A/{\rm cm}^3$, the current upper limit on the spin-dependent WIMP-nucleon cross section $\sigma_{\chi-N} \gtrsim 10^{-37}$ cm^2 and the radius of Uranus, $R \sim 3 \times 10^9$ cm – the choice of Uranus is motivated by an anomalous heat observed in the planet, of about 10^{14} W. Now, the power produced by dark matter assuming that all of the dark matter mass is converted to heat is

$$W \sim (\text{capture probability}) \cdot \pi R_{\rm planet} \cdot \rho_{\rm DM} \cdot v_{\rm DM} \lesssim 10^{12}\ {\rm W}$$

which tells us that we fall short by a couple orders of magnitude of explaining Uranus' anomalous heat. Too bad.

Exercise #5: Estimate the heat produced by dark matter annihilation in the Earth and compare with the accuracy of geothermal models (see also the much, much more refined discussion in [24]); how large should the dark matter-nucleon scattering cross section to cause global warming concerns?

1. Effects on Cosmology (cont'd): lots of work here, spanning effects on Big Bang Nucleosynthesis, on the cosmic microwave background, on reionization, on structure formation and many more. I don't even have time to give you a laundry list of all this! Go browse the arXiv and have fun!

2. Charged Cosmic Rays: here, the dark matter "source term" is unfortunately tangled with effects of propagation and energy losses of charged cosmic rays on their way to our human detectors.

Do we expect enough cosmic rays from dark matter annihilation or decay? The ballpark energy density of cosmic rays in the Milky Way is

$$\epsilon_{\rm CR} \sim 1\ \frac{\rm eV}{\rm cm^3}.$$

Let's estimate the energy density in cosmic rays dumped by dark matter annihilation in the Galaxy:

$$\epsilon_{\rm DM} \sim m_\chi \cdot \langle\sigma v\rangle \cdot n_{\rm DM}^2 \cdot T_{\rm MW},$$

with $m\chi \sim 100$ GeV, $\rho_{\rm DM} \sim 0.3$ GeV/cm^3, $\langle\sigma v\rangle \sim 3\times10^{-26}$ cm^3/s, and the Milky Way age $T_{\rm MW} \sim 10\times10^9$ yr, I get

$$\epsilon_{\rm DM} \sim 10^{-2}\ \frac{\rm eV}{\rm cm^3}.$$

Exercise #6: Improve on the estimate above using a Navarro-Frenk-White dark matter density profile and integrating over an appropriate cosmic-ray "diffusion region", e.g. a cylindrical slab of half-height 1 kpc and radius 20 kpc.

Exercise #7: Same as Ex.#6, but for a decaying dark matter particle, find $\epsilon_{\rm DM}(\tau)$ where τ is the dark matter lifetime. Do you expect to get interesting limits on τ from this calculation? If yes, please mention me and these lecture notes in the acknowledgements of your forthcoming paper.

The estimate above indicates that the contribution of annihilating dark matter to cosmic rays is, at best, subdominant to the observed cosmic ray energy density. In effect, models of Galactic cosmic rays decently match observation, so this is in some sense good news for dark matter model building... As a result, it is key in this business to target under-abundant species, namely either heavy nuclei or antimatter (for example positrons

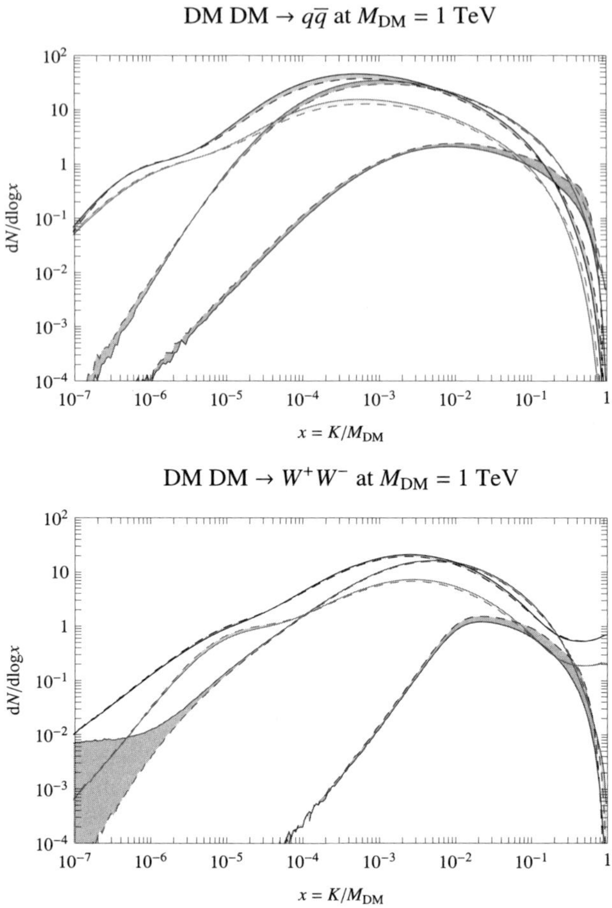

Fig. 5. The differential photon (red lines), neutrino (black lines), $e^{\pm}$ (green lines), $\bar{p}$ (blue lines) yield from dark matter pair-annihilation into a $q\bar{q}$ pair (top) and W^+W^- (bottom). From Ref. [25].

(e^+), antiprotons ($\bar{p}$), antideuterons $\bar{D}$,...). Unfortunately, it is quite hard to produce heavy nuclei from dark matter annihilation (that results, in its hadronic part, in a couple of high-energy jets only). Antimatter, on the other hand, is promising; typical dark matter models (exceptions are certain flavors of asymmetric dark matter) are democratic in producing as much matter as antimatter in the annihilation or decay final products.

Figure 5 illustrates the final yield of several particle species resulting from $\chi\chi \to q\bar{q}$ (left) and from $\chi\chi \to W^+W^-$ (I took these two nice figures from Ref. [25]). The red lines indicate photons, the black lines neutrinos, while the green and blue lines indicate $e^{\pm}$ and $\bar{p}$, respectively. All of these particle species primarily originate from the hadronization and cascade decays of jets initiated by the final state q and $\bar{q}$, or directly from the prompt decay modes of the W (notice the green and black lines getting "horizontal" at $x = 1$, where x is the particles' kinetic energy normalized by the dark matter mass).

In cosmic rays, antimatter is primarily produced by spallation processes, such as

$$p + p \to p + p + \bar{p} + p$$

where one of the protons in the initial state is a high-energy particle, and the second one is typically an H^+ nucleus in the interstellar medium gas, and baryon number conservation forces you to produce at least four nucleons in the final state. The process has a relatively large threshold (if you need a special relativity refresher carry out the two-lines calculation), $E_p \gtrsim 7.5$ GeV. Now, the spectrum of cosmic rays observed in the Galaxy falls steeply with energy,

$$\frac{\mathrm{d}N_{\text{cosmic-ray protons}}}{\mathrm{d}E} \sim E^{-2.7},$$

so compared to the maximal flux of cosmic-ray protons, observed at $E \sim 0.1$ GeV, antiprotons will be under abundant, at 0.1 GeV, by about a factor

$$\frac{\bar{p}}{p} \sim \left(\frac{0.1}{7.5}\right)^{2.7} \sim 10^{-5}.$$

This is in fact in remarkable agreement with what is observed, see Fig. 6, left, from Ref. [26].

There are therefore two effects that make antiprotons an interesting probe of dark matter (that, as Fig. 5 shows, tends to produce low-energy antinucleons): on the one hand there are few "beam" particles to produce cosmic-ray antiprotons, since the cosmic-ray proton spectrum falls steeply, and on the other hand the typical kinetic energy inherited by the final state antiproton will be on the same order as the threshold for the process (indeed, Fig. 6, left shows that the $\bar{p}/p$ ratio peaks right around 10 GeV!) a much higher energy than the typical anti nucleon produced by dark matter. These two effects are even more drastic for anti-deuterons (i.e., bound states of $\bar{p}$ and $\bar{n}$), for which the key astrophysical background comes from the reaction

$$\bar{D}: \quad p + p \to p + p + \bar{p} + p + \bar{n} + n$$

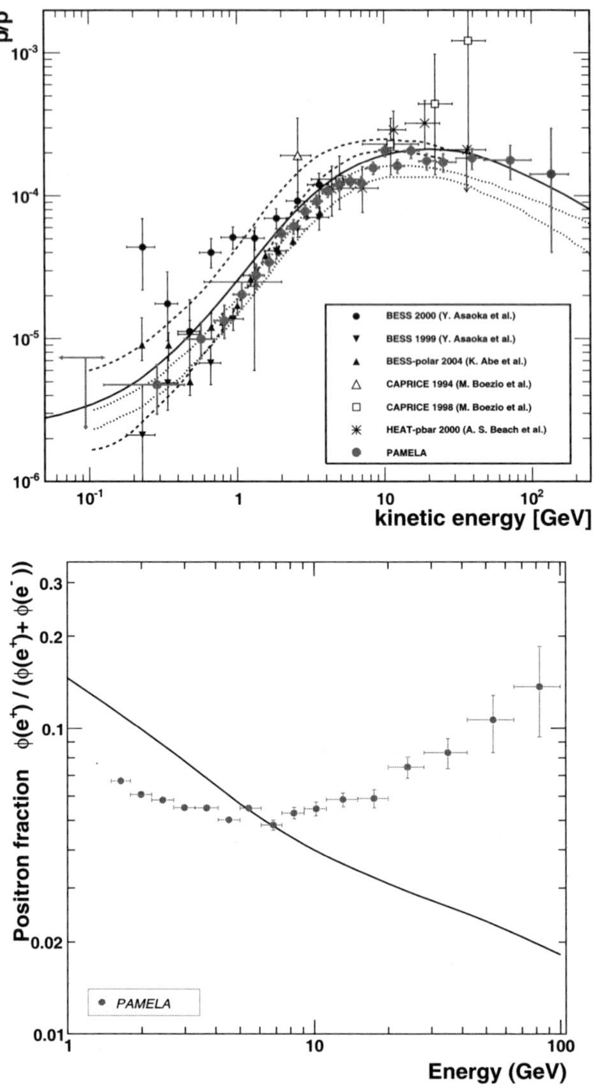

Fig. 6. The cosmic-ray antiproton to proton ratio (top, from Ref. [26]) and the positron fraction (bottom, from Ref. [27]) as measured by the Pamela experiment.

that has a threshold of about 17.5 GeV. In addition, $\bar{D}$ have a hard time loosing energy by elastic scattering (tertiary population) since the deuteron binding energy is very low, and when hit $\bar{D}$ tend to disintegrate rather than lose energy! There is a smart idea out there (the proposed satellite is called

GAPS [28]) to target specifically low-energy antideteurons and to detect them via the peculiar de-excitation X-rays that an atom capturing a $\bar{D}$ would produce.

How to deal with charged cosmic rays

How do we model cosmic-ray transport? The most successful framework is provided by the so-called diffusion models (adequate for cosmic-ray energies $E_{\rm CR} \lesssim 10^{17}$ eV). Let us indicate the differential (in energy) number density of cosmic rays with

$$\frac{{\rm d}n}{{\rm d}E} = \psi\left(\vec{x}, E, t\right).$$

The master equation of cosmic-ray diffusion models looks something like this:

$$\frac{\partial}{\partial t}\psi \;=\; D(E)\Delta\psi \;+\; \frac{\partial}{\partial E}\left(b(E)\;\psi\right) \;+\; Q\left(\vec{x}, E, t\right), \tag{14}$$

The first term on the right-hand side describes diffusion, the second one energy losses, and the third includes all possible sources. As always in life and in science, it is possible and easy to add complications – an incomplete list of popular ones and of the associated recipes is:

- *Cosmic-ray convection*; recipe: add: $\frac{\partial}{\partial z}(v_c \cdot \psi)$.
- *Diffusive re-acceleration*; recipe: add: $\frac{\partial}{\partial p}\; p^2\; D_{pp}\frac{\partial}{\partial p}\frac{1}{p^2}\psi$.
- *Fragmentation and decays*; recipe: add: $-\frac{1}{\tau_{\rm f,d}}\psi$.

When dealing with partial differential equations, we all learned in kindergarten that it is crucial to define boundary conditions. A popular choice is free-escape at the boundaries of a "diffusive region", whose geometry, for obvious reasons, is typically chosen to be a cylindrical slab, with

$$R \sim \mathcal{O}(1) \times 10\text{ kpc},$$

$$h \sim \mathcal{O}(1) \times 1\text{ kpc}.$$

The diffusion coefficient (that in certain models can depend also on position - it more than likely does in reality!) has a dependence on energy (a descendent of the fact that the Larmor radius scales with the particle's momentum!) that can be schematically cast as

$$D(E) \;\sim\; D_0\left(\frac{E}{E_0}\right)^{\delta}, \qquad E_0 \sim \text{GeV},\;\; D_0 \sim \text{few}\;\times\;10^{28}\;\frac{\text{cm}^2}{\text{s}}, \qquad \delta \sim 0.7.$$

The parameters entering cosmic ray diffusion are tuned self-consistently to reproduce key observational data, such as stable pure secondary to primary ratios as a function of energy (classic example: boron to carbon, B/C) or unstable secondary to primary ratios, such as ^{10}Be/^{9}Be. For example, this latter ratio constrains quite severely the height of the diffusion region.

What are the relevant time-scales for the diffusion equations? Two key quantities are the diffusion and the energy loss time scales:

$$\tau_{\rm diff} \sim \frac{R^2}{D_0} \cdot E^{-\delta}, \qquad \tau_{\rm loss} \sim \frac{E}{b(E)},$$

where R is the linear size of the diffusion region, or the relevant time/distance scale for which we want to calculate the typical associated diffusion length (for example, to infer which diffusion length corresponds to the energy loss time scale, we would plug in $R \sim c/\tau_{\rm loss}$). The steady-state diffusion equation (14) can then be re-written as

$$0 = -\frac{\psi}{\tau_{\rm diff}} - \frac{\psi}{\tau_{\rm loss}} + Q,$$

implying that

$$\psi \sim Q \cdot \min[\tau_{\rm diff},\ \tau_{\rm loss}]. \tag{15}$$

Let's see if this makes sense and consider cosmic ray protons and primary and secondary electrons and positrons:

- If the primary sources of cosmic-ray protons are supernova remnants, and if the injected particles are accelerated via a Fermi mechanism, we expect

  $$Q \sim E^{-2}.$$

  Energy losses for protons in the GeV-TeV range are relatively inefficient, and typically $\tau_{\rm diff} \ll \tau_{\rm loss}$, therefore Eq. (15) would predict

  $$\psi \sim E^{-2} \cdot E^{-\delta} \sim E^{-2.7}$$

  which is in great agreement with observation!
- For primary electrons, let us suppose that again $Q \sim E^{-2}$ – for example because the acceleration site is the same as for cosmic-ray protons (not such an unreasonable assumption...). At high energy ($E_e \gg$ GeV) the dominant energy loss mechanisms are inverse Compton scattering (i.e. the process of a high-energy electron up-scattering an ambient photon – the inverse of the classic Compton scattering where a high-energy photon up-scatters an electron at rest!) and synchrotron. Both have

the same dependence on energy, $\propto E^2$, and the resulting energy loss term reads

$$b_e(E) \simeq b^0_{\rm IC}\left(\frac{u_{\rm ph}}{1\ {\rm eV/cm^3}}\right)\cdot E^2 \ +\ b^0_{\rm sync}\left(\frac{B}{1\ \mu{\rm G}}\right)\cdot E^2,$$

where, in units of 10^{-16} GeV/s, the constants

$$b^0_{\rm IC} \simeq 0.76, \qquad b^0_{\rm sync} \simeq 0.025$$

and $u_{\rm ph}$ corresponds to the background radiation energy density and B to the ambient magnetic field. Depending on the size and geometry of the diffusion region, Eq. (15) predicts a break between a low-energy regime where

$$\psi_{\rm primary,\ low-energy} \sim\ Q\cdot\tau_{\rm diff} \sim E^{-2}\cdot E^{-\delta} \sim E^{-2.7}$$

and a high-energy regime where

$$\psi_{\rm primary,\ high-energy} \sim\ Q\cdot\tau_{\rm loss} \sim E^{-1}\cdot\frac{E}{E^2} \sim E^{-3}.$$

The general prediction is thus of a *broken power-law* with a break corresponding to $\tau_{\rm loss} \sim \tau_{\rm diff}$. This indeed (both directly and indirectly, i.e. from measurements of the secondary radiation produced by cosmic ray $e^{\pm}$) matches observation again!

- For secondary electrons and positron, produced e.g. by the decay of charged pions produced by cosmic-ray proton collisions with protons in the interstellar medium, the source term corresponds to the $Q_p \sim E^{-2.7}$ spectrum found above. The $e^{\pm}$ spectrum after diffusion and energy losses will then follow the same fate as that of primary particles discussed above: a broken power-law, with a hard low-energy spectrum $\psi_{secondary,\ low-energy} \sim E^{-3.4}$ and a softer high-energy tail due to energy losses, $\psi_{secondary,\ high-energy} \sim E^{-3.7}$. The key point is that, independently of the value of δ (that, remember, tunes the energy dependence of the diffusion coefficient) and of the primary injection spectrum and of energy, the ratio of secondary to primary species is

$$\frac{\psi_{e+}}{\psi_{e-}} \sim\ E^{-\delta}.$$

The prediction of a declining secondary-to-primary ratio was recently found to be at odds with the observed local positron fraction (see the right panel of figure 6, from Ref. [27]), a fact that spurred much speculation about the nature of the additional positrons responsible for the upturn in the secondary-to-primary ratio.

There are a couple of special limits in which one can get a simple solution to Eq. (14) that are worth remembering because they apply to certain interesting physical situations:

(1) **No diffusion**: this case corresponds to physical situations where the energy loss time-scale is much shorter than the diffusion time-scale: the cosmic rays effectively loose their energy before diffusing. In this case, the asymptotic, steady-state (i.e. after time transients) solution to the diffusion equation (14) is

$$\psi(\vec{x}, E) \propto \frac{1}{b(E)} \int \mathrm{d}E' \; Q(\vec{x}, E').$$

There are numerous circumstances where this is a relevant approximation: for example, when the system is very large and its physical size is much larger than the typical diffusion length associated with the energy-loss time (for example clusters of galaxies), or when the system is such that the energy loss term is very large (for example, a medium with a very dense radiation fields off of which $e^\pm$ can efficiently inverse-Compton scatter, or with large magnetic field inducing large synchrotron radiation energy losses).

(2) **Burst-like injection from a point source at time t in the past**: in this case, the relevant (spherically symmetric) solution, neglecting energy losses, is

$$\psi \propto Q \cdot \exp\left(-\left(\frac{r}{r_{\rm diff}}\right)^2\right), \tag{16}$$

where r is the source distance,

$$r_{\rm diff} \simeq \sqrt{D(E) \cdot t}.$$

This second case, a burst-like injection, is especially important in connection with Galactic pulsars as sources of high-energy $e^\pm$, a potential explanation to the anomalous rising positron fraction found by Pamela [27] and recently confirmed by the Fermi-LAT [29], as pointed out by many including Yours Truly, who once again is not shy to self-promote his own papers [30].

What are the requirements on the age and distance of a pulsar that contributes to the Pamela positron anomaly? It is easy to give general arguments: first, the pulsar age must be much smaller than the energy loss time scale for energies as large as about $E_e \sim 100$, in order to have at least some energetic $e^{\pm}$ around! This implies the condition

$$T_{\rm psr} \ll \tau_{\rm loss} = \frac{E}{b(E)}$$

$$\text{for } E = 100 \text{ GeV}, \tau_{\rm loss} \sim \frac{100}{10^{-16} \cdot 100^2} \text{ s} \sim 10^{14} \text{ s} \sim 3 \text{ Myr}.$$

Now, to avoid the exponential suppression of Eq. (16) we must have

$$\sqrt{D(E) \cdot T_{\rm psr}} \gg \text{ distance}$$

$$\rightarrow \quad \text{distance} \ll (3 \times 10^{28} \cdot 100^{0.7} \cdot 10^{14})^{1/2} \text{ cm} \sim 10^{22} \text{ cm} \sim 3 \text{ kpc}.$$

So our candidate pulsar is younger than about a mega-year and closer than a few kilo-parsec. One would also like the pulsar to have enough power injected in electron-positron pairs, but this condition, for such a nearby object, is usually fulfilled. Interestingly, several pulsar candidates exist within the desired age and distance, including possibly a handful of the newly discovered radio-quiet gamma-ray pulsars detected by Fermi-LAT (see e.g. Ref. [31]).

As some of you might be aware of, the dark matter annihilation explanation to the Pamela positron fraction anomaly gathered quite a bit of attention (in fact on the order of 10^3 publications entertain this possibility!). Dark matter as a source of the observed excess high-energy positrons faces various issues, including:

- there is no evidence for an associated antiproton excess, thus the dark matter must preferentially pair-annihilate into non-hardonic final states (it must be "leptophilic");
- diffuse secondary radiation from internal bremsstrahlung and inverse Compton is not observed;
- the needed pair-annihilation rate,

  $$\langle \sigma v \rangle \sim 10^{-24} \frac{\text{cm}^3}{\text{s}} \cdot \left(\frac{m_\chi}{100 \text{ GeV}}\right)^{1.5}$$

  is very large for thermal production, and generically leads to unseen gamma-ray or radio emission;
- a XIII century monk pointed out that "*entia non sunt multiplicanda præter necessitatem*", and the pulsar explanation works just fine to explain the excess positrons.

Despite these difficulties, theorists from all over the world (including myself, of course) have proposed models that circumvent all difficulties and show proof that Pamela might have perhaps detected the first non-gravitational signs of dark matter, providing more and more empirical evidence in favor of the so-called "Redman's theorem" [32]:

> *Any competent theoretician can fit any given theory to any given set of facts*

What's next? Well, myself and many others are quite eagerly awaiting results from the AMS-02 payload, successfully deployed and operational on the International Space Station since May 2011. An independent measurement of the positron fraction over an extended energy range and with much larger statistics, measurements of various cosmic ray species which will be key to a better understanding of cosmic ray propagation in the Galaxy, and possibly additional information on e.g. anisotropies of the positron arrival direction don't quite keep me awake as much as my newborn son (actually, not even nearly), but still...

Lecture 4: Not-So-Indirect Detection: Neutrinos and Gamma Rays

The tiny neutral ones

Detecting neutrinos (from an Italian made-up word that indicates the "tiny neutral one", with an English made-up plural form) is hard. In fact, despite building km^3 size detectors, only two astrophysical neutrino sources have been observed so far: the Sun and Supernova 1987A! The flip side of the coin is that astrophysical backgrounds are evidently quite low (albeit of course cosmic rays produce copious "atmospheric" neutrinos as they hit the atmosphere...) if one is to use neutrinos to search for dark matter. The key idea is that dark matter particles can accrue in celestial bodies until large enough densities start fueling a steady rate of annihilation yielding high-energy neutrinos. Neutrinos are pretty much the only thing produced by dark matter annihilation that can escape the core of a celestial body without losing much energy at all, and get all the way to our km^3 size detectors. The best bets are the Sun and the Earth, with the former, turns out, much better than (although somewhat complementary to) the latter. Let us know estimate this process for capture and annihilation in the Sun.

The dark matter capture rate in the sun is, roughly

$$C^{\odot} \sim \phi_\chi \cdot \left(\frac{M_\odot}{m_p}\right) \cdot \sigma_{\chi-p},$$

with the dark matter flux

$$\phi_\chi \sim n_\chi \cdot v_{\rm DM} = \frac{\rho_{\rm DM}}{m_\chi} \cdot v_{\rm DM}$$

the ratio $M_\odot/m_p$ estimating the number of target nucleons in the Sun, and the dark matter-nucleon interaction cross section $\sigma_{\chi-p}$ being bound by current experimental limits:

$$\sigma_{\chi-p}^{\text{spin dependent}} \lesssim 10^{-39}\ \text{cm}^2,$$
$$\sigma_{\chi-p}^{\text{spin independent}} \lesssim 10^{-44}\ \text{cm}^2.$$

Plugging in the relevant numbers, I find

$$C^{\odot} \sim \frac{10^{23}}{\text{s}} \left(\frac{\rho_{\rm DM}}{0.3\ \text{GeV/cm}^3}\right) \cdot \left(\frac{v_{\rm DM}}{300\ \text{km/s}}\right) \cdot \left(\frac{100\ \text{GeV}}{m_\chi}\right) \cdot \left(\frac{\sigma_{\chi-p}}{10^{-39}\ \text{cm}^2}\right).$$

We are interested in the number of dark matter particles in Sun: let's call this number N and write down a differential equation that describes the time evolution of $N(t)$:

$$\frac{\mathrm{d}N}{\mathrm{d}t} = C^{\odot} - A^{\odot}[N(t)]^2 - E^{\odot}N(t),$$

There are various elements I introduced in the otherwise self-explanatory equation above:

- $E^{\odot}$ describes the "evaporation" of dark matter particles, something that happens if the particles have a (thermal) velocity comparable with the celestial body's escape velocity. Let's quickly estimate this effect. For the Sun

  $$v_{\mathrm{esc}}^{\odot} \simeq 1156\ \frac{\mathrm{km}}{\mathrm{s}} \sim 3 \times 10^{03}\ c$$

  while the Sun's core temperature (the dark matter particles sink to the center after multiple scattering inside the Sun) is

  $$T_{\mathrm{core}}^{\odot} \sim 10^7\ K \sim 1\ \mathrm{keV} \sim m_\chi \cdot v_\chi^2.$$

  This gives, for the typical dark matter thermal velocities in the core of the Sun

  $$v_\chi \sim c \cdot \left(\frac{1\ \mathrm{keV}}{m_\chi}\right)^{1/2} \gtrsim v_{\mathrm{esc}}^{\odot} \to m_\chi \lesssim 0.1\ \mathrm{GeV}.$$

  Bottom line: for dark matter particles in the "preferred" WIMP mass range we can safely neglect evaporation.

  Exercise #8: Estimate evaporation in the case of the Earth: what is the relevant dark matter particle mass range for which evaporation matters in the Earth?

- The annihilation rate

  $$A^{\odot} \simeq \frac{\langle \sigma v \rangle}{V_{\mathrm{eff}}},$$

  where V_{eff} is an effective volume which depends on where WIMPs live inside the Sun; let us use the following (reasonable) guess for the density profile of the sunk WIMPs in the Sun:

  $$n(r) = n_0\ \exp\left(-\frac{m_\chi \phi_{\mathrm{grav}}(r)}{T^{\odot}}\right).$$

  We can choose to estimate V_{eff} by identifying an effective radius R_{eff} corresponding to the condition

  $$\frac{m_\chi\ \phi_{\mathrm{grav}}(R_{\mathrm{eff}})}{T^{\odot}} \simeq 1 \quad \to \quad T^{\odot} \simeq \frac{G_N \rho^{\odot} \frac{4\pi}{3} R_{\mathrm{eff}}^3 m_\chi}{R_{\mathrm{eff}}}$$

$$\rightarrow R_{\rm eff} \sim 10^9 \text{ cm} \left(\frac{m_\chi}{100 \text{ GeV}}\right)^{1/2} \quad \rightarrow \quad V_{\rm eff} \sim 10^{28} \text{ cm}^3 \left(\frac{m_\chi}{100 \text{ GeV}}\right)^{3/2}$$

Remember that the Sun's radius is approximately $R^\odot \sim 7\times 10^{10}$ cm, so this radius is smaller than the Sun's radius for reasonably light WIMPs.

Neglecting evaporation, even I can solve the differential equation above, and calculate the quantity we are really interested in, the annihilation rate in the Sun:

$$\Gamma_A = \frac{1}{2} A^\odot [N(t^\odot)]^2 = \frac{C^\odot}{2}\left[\tanh(\sqrt{C^\odot A^\odot}\; t^\odot)\right]^2,$$

with

$$t^\odot \sim 4.5 \text{ Byr} \sim 10^{17} \text{ s}$$

the Sun's age (not to be confused with the Sun's core temperature $T^\odot$!). One thing we learn from the solution above is that equilibrium between capture and annihilation is reached if

$$t^{\rm eq} \equiv \frac{1}{\sqrt{C^\odot A^\odot}} \ll t^\odot.$$

Do we expect equilibrium or not, for nominal WIMP parameters? Yes, we do! Let's plug in the numbers and convince ourselves of this fact: first, let's find the required annihilation rate for equilibrium

$$C^\odot \sim 10^{23} \text{ s}^{-1} \left(\frac{\sigma_{\chi-p}}{10^{-39} \text{ cm}^2}\right),$$

$$A^\odot_{\rm eq} \gg \frac{1}{(t^\odot)^2\, C^\odot} = \frac{1}{10^{34}\cdot 10^{23} \text{ s}} \sim 10^{-57} \text{ s}^{-1}.$$

Now, for vanilla WIMP dark matter

$$A^\odot = 3\times 10^{-54} \text{ s}^{-1} \left(\frac{\langle \sigma v\rangle}{3\times 10^{-26} \text{ cm}^3/\text{s}}\right)$$

so equilibrium is reached for $\sigma_{\chi-p}$ as small as about 10^{-41} cm^2.

Exercise #9: Re-do this calculation for the case of the Earth and find the critical dark matter-nucleon scattering cross section for equilibrium; note that the relevant scattering cross section in the Earth is spin-independent (as the Earth is mostly made of spin-0 Iron nuclei): do you then expect the equilibrium condition to hold for the flux of neutrinos from the center of the Earth?

If equilibrium is achieved, then

$$\Gamma_A \simeq \frac{C^{\odot}}{2}$$

and we don't care about the pair annihilation cross section (a unique case in the business of indirect dark matter detection!), and we only care about the cross section for dark matter capture. The resulting flux of neutrinos of flavor f will then be

$$\frac{\mathrm{d}N_{\nu_f}}{\mathrm{d}E_{\nu_f}} = \frac{C^{\odot}}{8\pi(D^{\odot})^2}\left(\frac{\mathrm{d}N_{\nu_f}}{\mathrm{d}E_{\nu_f}}\right)_{\mathrm{inj}}$$

where the last factor with the subscript "inj" is the "injection" spectrum of neutrinos per annihilation. Effects that complicate this discussion include neutrino oscillation, absorption of neutrinos in the Sun and many others that a few smart people out there have already kindly worked out for you.

The final step is to count the number of events we expect at IceCube or at other mega-neutrino-detectors (these are fundamentally arrays of photomultipliers reading Cherenkov light from muons produced by ν_μ charged current interactions):

$$N_{\mathrm{events}} = \int \mathrm{d}E_{\nu_\mu} \int \mathrm{d}y \left(A_{\mathrm{eff}} \cdot \frac{\mathrm{d}N_{\nu_\mu}}{\mathrm{d}E_{\nu_\mu}} \cdot \frac{\mathrm{d}\sigma}{\mathrm{dy}}(E_{\nu_\mu}, y) \cdot \left(R_\mu(E_{\nu_\mu})\right)\right),$$

where y indicates the ν_μ energy fraction transferred to the μ in the charged current interaction, $\mathrm{d}\sigma/\mathrm{dy}$ is the relevant cross section for charged current interactions, and the last factor R_μ indicates the muon range in the relevant material the detector lives in (for example, Antarctica ice for IceCube).

The most promising dark matter pair annihilation final states in this business are those producing a "hard" spectrum of muon neutrinos, i.e. energetic neutrinos. These by all means include W^+W^- and ZZ pairs, that dump out prompt muon neutrinos from the leptonic decay modes of the gauge bosons; luckily, for example in supersymmetry, these are exactly the preferred final states for wino- and higgsino-like lightest neutralinos.

The typical flux sensitivity threshold we want to hit to get an interesting signal is about hundreds of muons per km-squared per year, and the typical energy thresholds are 100 GeV for IceCube, which is improved down to 10 GeV for DeepCore and that could go down to the order of a GeV for the further thickly instrumented portion of the detector to be named PINGU.

Light from Dark Matter

There are two key ways to get light out of dark matter:

(i) *Prompt* photons from the annihilation or decay event, and
(ii) *Secondary* photons from radiative processes associated with stable, charged particles produced by the dark matter annihilation or decay event (in practice, the most important ones are electrons and positrons)

Prompt photons are produced either by the two-photon decay of neutral pions $\pi^0 \to \gamma\gamma$ dumped by the hadronization chain of strongly interacting annihilation products, or by internal bremsstrahlung off of charged particles in the intermediate or final state; this second contribution is typically "harder", i.e. more energetic, than the first one. Gamma rays from neutral pion decay have the nice spectral feature that I ask you to find in the next exercise.[d]

> **Exercise #10:** Show that independent of the π^0 spectrum, the differential spectrum of the resulting gamma rays, $\mathrm{d}N_\gamma^{\pi^0}/\mathrm{d}E_\gamma$ is symmetric around $E_\gamma = m_\pi/2$ on a log scale in energy.

Secondary photons originate as the counterpart of the key energy loss processes for electrons and positrons we discussed in the previous lecture: inverse Compton and Synchrotron. To qualitatively understand the features of inverse Compton emission, it is useful to commit to memory the average energy $\langle E_0' \rangle$ of the up-scattered photon (with an original initial energy E_0) as a function of the Lorentz factor $\gamma_e = E_e/m_e$ of the impinging high-energy electron:

$$\langle E_0' \rangle \sim \frac{4}{3}\gamma_e^2 \, E_0.$$

The relevant numbers for E_0 are as follows:

$$\begin{aligned} &\text{CMB} : E_0 \sim 2 \times 10^{-4} \text{ eV} \\ &\text{starlight} : E_0 \sim 1 \text{ eV} \\ &\text{dust} : E_0 \sim 0.01 \text{ eV} \end{aligned}$$

so for a typical electron-positron injection energy from dark matter

$$E_e \sim \frac{m_\chi}{10} \to \gamma_e \sim 2 \times 10^4 \left(\frac{m_\chi}{100 \text{ GeV}}\right)$$

[d]I remember this problem well, as it was asked to me during my PhD entrance exam by my advisor-to-be!

and

$$E'_{\rm CMB} \sim 10^5 \text{ eV} \left(\frac{m_\chi}{100 \text{ GeV}}\right)^2 .$$

Inverse Compton emission from dark matter therefore produces hard X-ray photons in the hundreds of keV range. This is great news, as a brand new NASA telescope, Nustar, is looking at the sky exactly at that energy range [33]! The inverse Compton light from starlight and dust falls, instead, in the low-energy gamma-ray regime.

In the monochromatic approximation, synchrotron emission peaks at

$$\frac{\nu_{\rm sync}}{\rm MHz} \simeq 2 \cdot \left(\frac{E_e}{\rm GeV}\right)\left(\frac{B}{\mu \rm G}\right)^{1/2}$$

and the synchrotron power scales like B^2. Dark matter annihilation thus produces a rich, multi-wavelength emission spectrum that goes well beyond the gamma-ray band. An example of the spectrum expected e.g. from the nearby Coma cluster of galaxies is shown in Fig. 7, left, from Ref. [34]. Note that the various secondary emission peaks appear exactly where the formulae above would predict them to be!

While secondary emission is always present, it involves the additional steps of accounting for the diffusion and energy losses of the $e^\pm$ produced by dark matter annihilation. Prompt gamma-ray emission, on the other hand, is simpler, and it only involves identifying a dark matter structure and a particle dark matter model; we will thus here make a few estimates for this prompt emission only, which for nominal WIMPs produces photons in the gamma-ray energy range. Also, for definiteness we will talk about dark matter annihilation - dark matter decay is even simpler!

What are the optimal targets and the expected detection rates for gamma ray searches for dark matter? The flux of photons produced by dark matter annihilation from a given direction ψ in the sky and within a solid angle $\Delta\Omega$ is

$$\phi_\gamma = \frac{\Delta\Omega}{4\pi}\left\{\frac{1}{\Delta\Omega}\int {\rm d}\Omega \int {\rm d}l(\psi)\ (\rho_{\rm DM})^2\right\}\frac{\langle\sigma v\rangle}{2m_\chi^2}\frac{{\rm d}N_\gamma}{{\rm d}E_\gamma},$$

where the last factor is, in fact, a sum of the prompt gamma-ray spectrum from every annihilation final state f:

$$\frac{{\rm d}N_\gamma}{{\rm d}E_\gamma} = \sum_f \frac{{\rm d}N_\gamma^f}{{\rm d}E_\gamma},$$

and where the term in curly brackets is often referred to as the "J-factor", and is a function of the solid angle $\Delta\Omega$ and of course of the direction in

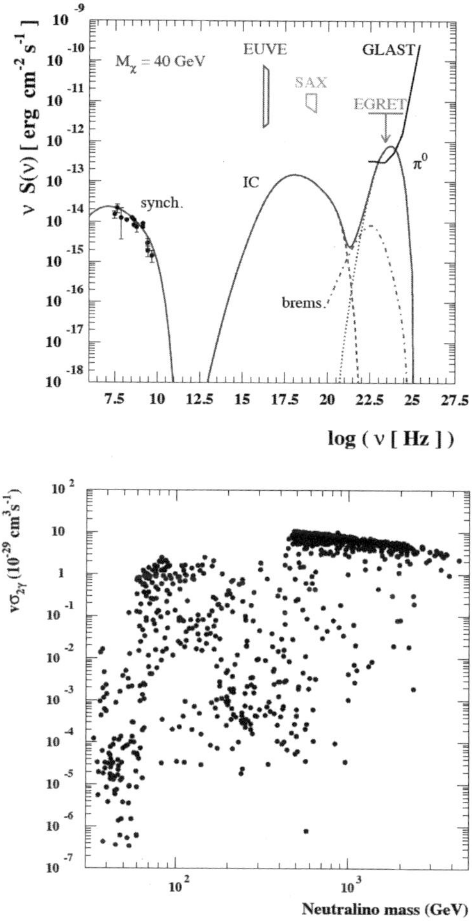

Fig. 7. Top: The multi-wavelength emission spectrum from the pair-annihilation of a dark matter particle with $m_\chi = 40$ GeV in the Coma cluster of galaxies, from Ref. [34]. Bottom: the pair annihilation cross section into two photons for MSSM neutralinos, from Ref. [35].

the sky, $J = J(\Delta\Omega, \psi)$, and carries units of $\mathrm{GeV}^2/\mathrm{cm}^5$. The solid angle $\Delta\Omega$ should be optimized for a given gamma-ray detector and for a given target, field of view and angular resolution to maximize (typically) the signal to noise. It turns out that the relevant solid angles correspond to an angular extent of about one degree, or $\Delta\Omega \sim 10^{-3}$ sr for the Fermi-LAT at an energy of about a GeV, down to an angular extent of 0.1 degrees, or $\Delta\Omega \sim 10^{-5}$ sr for ACT, or for Fermi in the high-energy regime.

Let me give you a "laundry list" of potential interesting targets to search for a gamma-ray signal from dark matter; for most of these targets the "J factor" is approximately the same for a solid angle corresponding to 1 deg or 0.1 deg:

(1) Dwarf Spheroidal Galaxies
- Draco, $J \sim 10^{19}$ GeV^2/cm^5, $\pm$ a factor 1.5;
- Ursa Minor, $J \sim 10^{19}$ GeV^2/cm^5, $\pm$ a factor 1.5;
- Segue, $J \sim 10^{20}$ GeV^2/cm^5, $\pm$ a factor 3

(2) Local Milky-Way-like galaxies
- M31, $J \sim 10^{20}$ GeV^2/cm^5

(3) Local clusters of galaxies
- Fornax, $J \sim 10^{18}$ GeV^2/cm^5
- Coma, $J \sim 10^{17}$ GeV^2/cm^5
- Bullet, $J \sim 10^{14}$ GeV^2/cm^5

(4) Galactic center
- 0.1°: $J \sim 10^{22} \ldots\ 10^{25}$ GeV^2/cm^5
- 1°: $J \sim 10^{22} \ldots\ 10^{24}$ GeV^2/cm^5

To have a detection, we need to have enough photon counts, possibly a lot:

$$N_\gamma \sim \int_{E_\gamma \text{ range}} \mathrm{d}E_\gamma\ \phi_\gamma \cdot A_{\mathrm{eff}}(E_\gamma) \cdot T_{\mathrm{obs}}$$

The following table gives a rule of thumb for the relevant energy ranges, effective areas and observing time for current and future gamma-ray observatories:

	Fermi-LAT	H.E.S.S.	CTA
E_γ range	0.1 to 300 GeV	0.1 to 10 TeV	10 GeV to 10 TeV
A_{eff}	~ 1 m^2	$\sim 10^5$ m^2	$\sim 10^6$ m^2
T_{obs}	$\sim 10^8$ s	$\sim 10^6$ s	$\sim 10^6$ s

It is instructive to calculate the minimal J factor needed to get at least *some* gamma-ray signal from dark matter. Consider for example Fermi-LAT: over the LAT energy range, typically

$$\int \mathrm{d}E_\gamma \frac{\mathrm{d}N_\gamma}{\mathrm{d}E_\gamma} \sim \frac{m_\chi}{\mathrm{GeV}},$$

so that

$$\phi_\gamma = (\Delta\Omega \cdot J)\,\frac{1}{8\pi}\frac{\langle\sigma v\rangle}{m_\chi^2}\cdot m_\chi \sim\ 10^{-32}\left(\frac{1}{\mathrm{cm}^2\ \mathrm{s}}(\frac{J}{\mathrm{GeV}^2/\mathrm{cm}^5}\right)$$

and

$$N_\gamma \sim A_{\rm eff}\cdot T_{\rm obs}\cdot\phi_\gamma \sim\ 10^{-20}\ \frac{J}{\mathrm{GeV}^2/\mathrm{cm}^5},$$

where we put in the nominal values for the effective area and observing time as in the Table above, so, we want

$$J \gtrsim 10^{20}\ \mathrm{GeV}^2/\mathrm{cm}^5.$$

This is a bit bigger than the individual dwarf spheroidal galaxies' J factors quoted above. In fact, combining observations of all (non-detected) dwarf galaxies gives one of the tightest (in my personal opinion *the* tightest) constraints to date on the dark matter pair-annihilation rate as a function of mass: dwarf galaxies are a virtually background free target, with

$$J^{\rm tot} \sim \mathrm{few}\times 10^{20}\ \mathrm{GeV}^2/\mathrm{cm}^5,$$

therefore the resulting limits are

$$\langle\sigma v\rangle_{\rm lim}\ \sim\ 3\times 10^{-26}\ \frac{\mathrm{cm}^3}{\mathrm{s}}\ \left(\frac{30\ \mathrm{GeV}}{m_\chi}\right).$$

A clear-cut signal of dark matter annihilation that could be detected with gamma-ray detectors is the close-to-monochromatic line from the direct annihilation

$$\chi\chi\ \to\ \gamma\gamma.$$

Since $E_\chi\ \sim\ m_\chi$, $E_\gamma\ \simeq\ m_\chi$ and the line is almost monochromatic. Dark matter particles, being dark, ought better not directly couple to photons, and the naive expectation is that the $\gamma\gamma$ amplitude be loop-suppressed, i.e.

$$\frac{\langle\sigma v\rangle_{\gamma\gamma}}{\langle\sigma v\rangle_{\rm tot}} \sim \frac{\alpha^2}{16\pi^2}.$$

This naive estimate gives the correct ballpark over a wide range of parameters in the MSSM, as Fig. 7, right (taken from Ref. [35]), illustrates (but things can get a lot worse, and not much better, than the naive estimate!). As you might have heard, a monochromatic line at 130 (or perhaps 135) GeV might indeed be present in the Fermi LAT data [36, 37], although a solid confirmation (possibly only with the new generation LAT analysis software Pass 8) that this is not an instrumental effect, and that the signal is indeed as statistically robust as it (frankly very much) looks, has yet to come. Stay tuned. Clearly, very exciting days lie ahead!

Acknowledgments

I would like to thank Elena Pierpaoli for inviting me to lecture at TASI 2012. It was fun, and it was great to meet so many, promising young fellas. Many thanks to the local organizing committee and to the administrative staff as well. My research is partly supported by the Department of Energy, under grant DE-FG02-04ER41286.

References

[1] E. W. Kolb, (Ed.) and M. S. Turner, (Ed.), *"The Early Universe. Reprints,"* REDWOOD CITY, USA: ADDISON-WESLEY (1988) 719 P. (FRONTIERS IN PHYSICS, 70)
[2] S. Dodelson, *"Modern cosmology,"* Amsterdam, Netherlands: Academic Pr. (2003) 440 p
[3] J. L. Feng and J. Kumar, Phys. Rev. Lett. **101**, 231301 (2008) [arXiv:0803.4196 [hep-ph]].
[4] K. Griest and M. Kamionkowski, Phys. Rev. Lett. **64**, 615 (1990).
[5] B. W. Lee and S. Weinberg, Phys. Rev. Lett. **39**, 165 (1977).
[6] E. A. Baltz, eConf C **040802**, L002 (2004) [astro-ph/0412170].
[7] S. Profumo, Phys. Rev. D **78**, 023507 (2008) [arXiv:0806.2150 [hep-ph]].
[8] C. Wainwright and S. Profumo, Phys. Rev. D **80**, 103517 (2009) [arXiv:0909.1317 [hep-ph]].
[9] P. Gondolo and G. Gelmini, Nucl. Phys. B **360**, 145 (1991).
[10] K. Griest and D. Seckel, Phys. Rev. D **43**, 3191 (1991).
[11] J. Edsjo and P. Gondolo, Phys. Rev. D **56**, 1879 (1997) [hep-ph/9704361].
[12] D. Hooper and S. Profumo, Phys. Rept. **453**, 29 (2007) [hep-ph/0701197].
[13] M. Citron, J. Ellis, F. Luo, J. Marrouche, K. A. Olive and K. J. de Vries, arXiv:1212.2886 [hep-ph].
[14] P. Salati, Phys. Lett. B **571**, 121 (2003) [astro-ph/0207396].
[15] S. Profumo and P. Ullio, JCAP **0311**, 006 (2003) [hep-ph/0309220].
[16] M. R. Buckley and S. Profumo, Phys. Rev. Lett. **108**, 011301 (2012) [arXiv:1109.2164 [hep-ph]].
[17] T. Bringmann, New J. Phys. **11**, 105027 (2009) [arXiv:0903.0189 [astro-ph.CO]].
[18] S. Profumo, K. Sigurdson and M. Kamionkowski, Phys. Rev. Lett. **97**, 031301 (2006) [astro-ph/0603373].
[19] J. M. Cornell and S. Profumo, JCAP **1206**, 011 (2012) [arXiv:1203.1100 [hep-ph]].
[20] N. Arkani-Hamed, A. Delgado and G. F. Giudice, Nucl. Phys. B **741**, 108 (2006) [hep-ph/0601041].
[21] S. D. McDermott, H. -B. Yu and K. M. Zurek, Phys. Rev. D **85**, 023519 (2012) [arXiv:1103.5472 [hep-ph]].
[22] G. G. Raffelt, *"Stars as laboratories for fundamental physics: The as-*

trophysics of neutrinos, axions, and other weakly interacting particles," Chicago, USA: Univ. Pr. (1996) 664 p

[23] D. Spolyar, K. Freese, P. Gondolo, A. Aguirre, P. Bodenheimer, J. A. Sellwood and N. Yoshida, PoS IDM **2008**, 077 (2008) [arXiv:0901.4574 [astro-ph.CO]].

[24] G. D. Mack, J. F. Beacom and G. Bertone, Phys. Rev. D **76**, 043523 (2007) [arXiv:0705.4298 [astro-ph]].

[25] M. Cirelli, G. Corcella, A. Hektor, G. Hutsi, M. Kadastik, P. Panci, M. Raidal and F. Sala *et al.*, JCAP **1103**, 051 (2011) [Erratum-ibid. **1210**, E01 (2012)] [arXiv:1012.4515 [hep-ph]].

[26] O. Adriani *et al.* [PAMELA Collaboration], Phys. Rev. Lett. **105**, 121101 (2010) [arXiv:1007.0821 [astro-ph.HE]].

[27] O. Adriani *et al.* [PAMELA Collaboration], Nature **458**, 607 (2009) [arXiv:0810.4995 [astro-ph]].

[28] T. Aramaki, S. E. Boggs, W. W. Craig, H. Fuke, F. Gahbauer, C. J. Hailey, J. E. Koglin and N. Madden *et al.*, Adv. Space Res. **46**, 1349 (2010).

[29] M. Ackermann *et al.* [Fermi LAT Collaboration], Phys. Rev. Lett. **108**, 011103 (2012) [arXiv:1109.0521 [astro-ph.HE]].

[30] S. Profumo, Central Eur. J. Phys. **10**, 1 (2011) [arXiv:0812.4457 [astro-ph]].

[31] L. Gendelev, S. Profumo and M. Dormody, JCAP **1002**, 016 (2010) [arXiv:1001.4540 [astro-ph.HE]].

[32] M. S. Longair, (ed.), *"High-energy astrophysics. Vol. 1: Particles, photons and their detection,"* Cambridge, UK: Univ. Pr. (1992) 418 p

[33] T. E. Jeltema and S. Profumo, arXiv:1108.1407 [astro-ph.HE].

[34] S. Colafrancesco, S. Profumo and P. Ullio, Astron. Astrophys. **455**, 21 (2006) [astro-ph/0507575].

[35] L. Bergstrom and P. Ullio, Nucl. Phys. B **504**, 27 (1997) [hep-ph/9706232].

[36] C. Weniger, JCAP **1208**, 007 (2012) [arXiv:1204.2797 [hep-ph]].

[37] M. Su and D. P. Finkbeiner, arXiv:1206.1616 [astro-ph.HE].

Chapter 5

Modified Gravity

Fabian Schmidt

Theoretical Astrophysics, California Institute of Technology M/C 350-17, Pasadena, California 91125-0001, USA;
Department of Astrophysical Sciences, Princeton University, Princeton, NJ 08544
E-mail: fabians@caltech.edu

These lectures provide a brief introduction into modified gravity theories, i.e. theories that deviate from General Relativity. We focus on theories that satisfy the Einstein Equivalence Principle, which are characterized by a single metric governing the motions of bodies and dynamics of fields. Further, we emphasize models that have received interest in the cosmological context, and which modify gravity on large scales.

Contents

1. Motivation

To our current knowledge, gravity is described successfully by Einstein's theory of General Relativity (GR) when including a small, positive cosmological constant Λ. Moreover, GR is a beautiful theory in that it is the simplest metric theory of gravity that satisfies the equivalence principle. So a natural question to ask at the beginning of these lectures is: why modify gravity?

One of the reasons is the observation of a recent accelerated expansion of the Universe. If we apply GR to a homogeneous and isotropic Universe (for which there are good observational grounds), we obtain an equation for the scale factor $a(t)$,

$$\frac{\ddot{a}}{a} = -\frac{4\pi G_N}{3}(\rho + 3p) + \frac{1}{3}\Lambda. \tag{1}$$

Thus, unless $\Lambda > 0$, acceleration implies that the pressure $p < -\rho/3$. Thus, we have two options: accept a positive Λ, or add an exotic component to the stress-energy budget of the Universe with negative pressure (now commonly referred to as Dark Energy). The main issue with Λ is its value: the effective energy density in Λ is

$$\rho_\Lambda = \frac{\Lambda}{8\pi G} = M_{\rm Pl}^2\Lambda \sim 10^{-47}\text{GeV}^4 \sim (10^{-3}\,\text{eV})^4. \tag{2}$$

On the other hand, quantum field theory leads us to estimate a vacuum energy of

$$\rho_{\rm vac} \sim M_{\rm Pl}^4 \sim 10^{76}\text{GeV}^4. \tag{3}$$

Thus, the various contributions to $\rho_{\rm vac}$ need to cancel to some 120 significant digits to yield an extremely small, but non-zero value. This "fine-tuning" problem has led many researchers to look for more natural explanations. One of the possibilities is to modify GR, and correspondingly Eq. (1) derived from it.

Another reason is that, like any other theory, we would like to test General Relativity. In order to do that, we need some guidance as to what modifications are (reasonably) possible, and what observational consequences they have. This is especially important for gravity since it describes an extremely wide range of phenomena, from the Solar System to Black Holes and the Big Bang/early Universe. Ideally, we would like to test as many aspects of gravitational phenomenology as possible. On the other hand, the wide range of applicability of GR implies that there are many constraints a viable modified gravity model has to satisfy; of course it is not sufficient to just pass some of the constraints.

General Relativity (GR), is remarkably weakly constrained at the present day on scales larger than a few AU. Research in modified gravity over the past decades has shown that there is still room for order unity deviations in the cosmos today, if they happen at scales much larger than the Solar System. Thus, testing gravity on cosmological scales is an interesting frontier, and the focus of much current research [1–11]. Over the past two decades, cosmology has developed from making "order-of-magnitude" statements to a precision science, mainly driven by the increasing flux of observational data. These data come from the cosmic microwave background as well as galaxy surveys from various redshifts. They thus cover scales ranging from ~ 0.1 to 10^4 Mpc (a Mpc is $\approx 2 \times 10^{11}$ AU $\approx 3 \times 10^{22}$ m $\approx 1.5 \times 10^{38}$ GeV^{-1}). Moreover, above ~ 1 Mpc, gravity is the only relevant force in cosmology. It is then natural to try to best use the large cosmological data sets to test our theory of gravity.

These lectures attempt to provide an introduction to the field of modified gravity, i.e. of theories that depart from GR. In most sections, we follow the notation of Clifton et al. [12] who provide an excellent comprehensive review on the subject. The main goals are to first, introduce some popular representative models, and second, present the observational constraints from local tests in the Solar system and cosmology. A large number of modified gravity models have been proposed in the literature (see [12]), and we thus have to make a very subjective selection. Here, I will focus on models that leave testable observational signatures in large-scale measurements, i.e. lab tests, Solar System tests, and cosmology. Thus, these constitute "IR" modifications to gravity, as opposed to the "UV" modifications discussed in the context of string theory, supergravity, etc., which only potentially lead to signatures in particle colliders or in the very early Universe.

We use the natural units of high energy physics throughout, with $c = \hbar = 1$. However, we keep $G_N = 1/(8\pi M_{\rm Pl}^2)$ explicit (unlike [12]).

The outline is as follows. §2 summarizes some basic results on metric theories of gravity, which provide theoretical constraints on modified gravity, as well as on local tests of gravity which provide observational constraints. §3 presents a subjective selection of some important examples of modified gravity models. Finally, in §4 we discuss observational tests from cosmology.

2. Foundations of Gravity

Einstein's motivation for conceiving GR was the *equivalence principle* (EP): the laws of physics should not depend on the observer's state of motion. In other words, there should not be a preferred frame that the laws of physics make reference to. If we exclude gravity, applying this to electrodynamics immediately leads to special relativity. When including gravity, we in fact have several choices as to how strictly we enforce the EP. In order of increasing restriction,

- *Weak equivalence principle* (WEP): any uncharged test body moves ("free-falls") on the same spacetime trajectory, given initial position and velocity, independently of composition. Here, test body means a body whose self-gravity and backreaction on the gravitational field can be neglected.
- *Einstein equivalence principle* (EEP): in addition to the WEP, the laws of physics are the same for any observer following such a test body ("free-falling observer").
- *Strong equivalence principle* (SEP): WEP also holds for any non-test, but small body. Small here means that any finite-size correction falls off as r^{-3} or faster, where r is the distance to the body. As an example, the SEP implies that if we substitute the Earth for a black hole of the same mass, it will follow the Earth's orbit around the Sun.

Note that different authors choose different naming conventions. As throughout, we follow [12]. The WEP can be tested by measuring the acceleration of test bodies of different composition. The best current bound on violations of the WEP, as quantified by the relative difference in acceleration, is [13]

$$\eta = 2\frac{|a_1 - a_2|}{a_1 + a_2} < 4 \times 10^{-13}. \tag{4}$$

The WEP is so deeply engrained in our understanding of what "gravity" is that many researchers think of a theory as a true modification to gravity only if it satisfies the WEP (already apparent in the way the mass of a test body cancels when equating Newton's second and third law $ma = -GMm/r^2$). The validity of EEP implies the existence of a unique metric and connection which determine both the geodesics on which any test particle moves, and the behavior of other forces such as the electromagnetic force. This is what one commonly thinks of as "metric theory of gravity"; theories violating the EEP involve preferred frames induced by vector or tensor fields. Most well-known modified gravity models satisfy the EEP, but may violate the SEP. In the following, we will only consider EEP-satisfying models.

We now discuss some relevant theorems which provide further limitations on the theory space available, or supply insights on the repercussions of modified gravity.

2.1. *Theorems*

Consider a spacetime with $D = 4$ dimensions and a theory described by an action

$$S = \int d^4x\sqrt{-g}\mathcal{L}\left(g_{\mu\nu}(x), \partial_\alpha g_{\mu\nu}(x), ...\right). \tag{5}$$

Note that we pull out a factor of $\sqrt{-g}$ so that $\mathcal{L}$ transforms as a scalar under coordinate transformations (whereas [12] include $\sqrt{-g}$ in $\mathcal{L}$ so that their $\mathcal{L}$ is not a scalar). **Lovelock's theorem** states that the only possible second-order field equations that can be derived from such an action are Einstein's equations:

$$G_{\mu\nu} + \Lambda g_{\mu\nu} = 8\pi G_N T_{\mu\nu}, \tag{6}$$

by suitable definition of the constants G_N and Λ. Here the stress-energy tensor is defined as in GR through the derivative of the matter Lagrangian with respect to $g^{\mu\nu}$:

$$T_{\mu\nu} \equiv \frac{2}{\sqrt{-g}}\frac{\partial\left(\sqrt{-g}\mathcal{L}_m\right)}{\partial g^{\mu\nu}}. \tag{7}$$

Comment. Consider the case of a scalar field with potential,

$$\mathcal{L}_m = \frac{1}{2}g_{\mu\nu}\nabla^\mu\phi\nabla^\nu\phi + V(\phi). \tag{8}$$

The equation of motion is obtained by varying the action,

$$\Box\phi + \frac{\partial V}{\partial\phi} = 0. \tag{9}$$

If we perturb around a minimum of the potential, this becomes

$$\Box\phi + m^2\phi^2 = 0, \quad m^2 \equiv \left.\frac{\partial^2 V}{\partial\phi^2}\right|_{\min}. \tag{10}$$

To derive Eq. (7), we begin with

$$\begin{aligned}\frac{\partial}{\partial g_{\mu\nu}}\left(\sqrt{-g}(\nabla\phi)^2\right) &= \frac{1}{2}\sqrt{-g}\nabla^\mu\phi\nabla^\nu\phi + \left[\frac{1}{2}(\nabla\phi)^2 + V(\phi)\right]\frac{\partial}{\partial g_{\mu\nu}}\sqrt{-g} \\ &= \frac{1}{2}\sqrt{-g}\nabla^\mu\phi\nabla^\nu\phi + \left[\frac{1}{2}(\nabla\phi)^2 + V(\phi)\right]\left(-\frac{1}{2}\sqrt{-g}\, g^{\mu\nu}\right).\end{aligned} \tag{11}$$

We thus obtain for the stress-energy tensor of a scalar field:

$$T^{\mu\nu} = \nabla^\mu\phi\nabla^\nu\phi - g^{\mu\nu}\left[\frac{1}{2}(\nabla\phi)^2 + V(\phi)\right]. \tag{12}$$

As example, the energy density in flat space is obtained as

$$T^{00} = \dot{\phi}^2 + \left[-\frac{1}{2}\dot{\phi}^2 + \frac{1}{2}(\vec{\nabla}\phi)^2 + V(\phi)\right] = \frac{1}{2}\dot{\phi}^2 + \frac{1}{2}(\vec{\nabla}\phi)^2 + V(\phi), \tag{13}$$

as expected.

This result severely limits our options in modifying gravity. We can choose either of the following options:

(1) Let $\mathcal{L}$ be a function of not only the metric. For example, one can allow the connection coefficients $\Gamma^\mu_{\alpha\beta}$ as separate degrees of freedom, or consider a non-local action.
(2) Allow for field equations that involve higher than second derivatives of the metric.
(3) Increase the number of spacetime dimensions ($D > 4$).

In the following, we will see examples of the second and third kind.

Birkhoff's theorem states that the most general spherically symmetric solution to Einstein's equations in vacuum is the Schwarzschild solution. In plain words, the spacetime outside an arbitrary isolated spherically symmetric distribution of matter (without angular momentum) in vacuum is independent of the exact arrangement of the matter, and its radius. This is effectively what is used in the Newtonian derivation of the expansion of the Universe: we cut out a sphere of matter of mass $M = 4\pi/3\bar{\rho}r^3$ from the homogeneous Universe (with mean matter density $\bar{\rho}$) and collapse it to some small radius. The equation of motion of particles outside this mass is then simply

$$\ddot{r} = -\frac{G_N M}{r^2} = -\frac{4\pi G_N}{3}\bar{\rho}r, \tag{14}$$

which is the Newtonian analogue of Eq. (1) for a dust-filled Universe ($p = \Lambda = 0$). If we now suppose that a modification to gravity yields an accelerating Universe without Λ by modifying Eq. (1), while still satisfying Birkhoff's theorem, we can invert this reasoning to obtain the modified Schwarzschild solution given the scale factor $a(t)$ for a dust-filled Universe [14]. By matching the FRW solution to the Schwarschild solution at the edge of the sphere, we obtain

$$g_{rr} = \left(1 - r^2 H^2\right)^{-1}, \tag{15}$$

where r is the Schwarzschild radial coordinate and $H = \dot{a}/a$ is the Hubble parameter. In GR for a dust-filled Universe, $H^2 = (8\pi G_N/3)\bar{\rho}$ and hence $r^2H^2 = 2G_N M/r$, yielding the standard Schwarzschild metric. Now, we assume that significant modifications to the Friedmann equation appear "today", when $H \approx H_0$. Thus, we expect modifications to the Schwarzschild metric, Eq. (15), to scale as

$$\frac{(g_{rr})_{\text{mod gr.}}}{(g_{rr})_{\text{GR}}} - 1 = \frac{2G_N M/r}{r^2 H_0^2} = \frac{2G_N M}{r^3 H_0^2} = \frac{r_s r_H^2}{r^3}, \tag{16}$$

where r_s is the Schwarzschild radius of the mass, and $r_H = H_0^{-1} \approx$ 4200 Mpc is the current horizon scale. Thus, if a modified gravity model yields acceleration without Λ while satisfying Birkhoff's theorem, we expect modifications to the gravitational field of isolated bodies to appear at $r \sim r_* \equiv (r_s r_H^2)^{1/3}$. This is much smaller than r_H, which one might have naively expected as the scale where gravity is modified. Interestingly, braneworld models show precisely this scaling, even though Birkhoff's theorem is not valid in these models.

More general stationary solutions in GR are severely restricted as well, which is phrased precisely in the **no-hair theorems**. One important implication is that a black hole cannot support any scalar charge. Consider the case of a light scalar field coupling to ordinary matter (of which there is no observational evidence of course). If a star with some scalar charge made up of the sum of the charges of its constituent particles collapses to a black hole, the no-hair theorem implies that in GR it has to radiate away all scalar charge to produce a black hole uncharged with respect to this field. We will see that many models that invoke an additional scalar field actually lead to the same effect, even though they are not GR.

2.2. *PPN formalism and Solar System tests*

The parametrized post-Newtonian framework (PPN) is a general expansion of a stationary, spherically symmetric and asymptotically flat spacetime, as explained in detail in [15, 16]. It is most commonly applied to the Solar System, whose mass is dominated by the Sun. We expand the metric around Minkowski,

$$g_{\mu\nu} = \eta_{\mu\nu} + h^{(1)}_{\mu\nu} + h^{(2)}_{\mu\nu} + h^{(3)}_{\mu\nu} + ..., \tag{17}$$

where $h_{\mu\nu}^{(n)} = \mathcal{O}(\varepsilon^n)$. The expansion parameter ε should be much less than 1 in order for this expansion to make sense. By definition,

$$\begin{aligned} U &= \frac{G_N M}{r} = \mathcal{O}(\varepsilon^2) \\ v &= \mathcal{O}(\varepsilon) \\ \frac{P}{\rho} &= \mathcal{O}(\varepsilon^2). \end{aligned} \tag{18}$$

Note that this implies that U satisfies $\nabla^2 U = -4\pi G_N \rho$. The stress-energy tensor for example can be written as

$$\begin{aligned} T_{00} &= \rho\left(1 + v^2 + \Pi - h_{00}\right) + \mathcal{O}(\varepsilon^6) \\ T_{0i} &= -\rho v_i + \mathcal{O}(\varepsilon^5) \\ T_{ij} &= \rho v_i v_j + p\delta_{ij} + \mathcal{O}(\varepsilon^6). \end{aligned} \tag{19}$$

Here, ρ is the rest-mass density, v is the velocity of matter, and Π is the non-rest-mass contribution to the energy density (internal energy such as thermal energy), divided by ρ. The most interesting metric perturbation is at second order, which can be written as

$$\begin{aligned} h_{00}^{(2)} &= 2GU \\ h_{0i}^{(2)} &= 0 \\ h_{ij}^{(2)} &= 2\gamma G U \delta_{ij}. \end{aligned} \tag{20}$$

Here we have introduced two dimensionless parameters: G (not to be confused with the locally measured value of Newton's constant, G_N), and γ. γ describes the spatial curvature generated by a given matter distribution, in units of the Newtonian potential. In GR, the weak field solution is simply recovered by $G = \gamma = 1$. The higher order metric corrections $h_{\mu\nu}^{(n)}$ can be similarly written down, though they become progressively more complicated.

For a given theory, we can then solve the field equations perturbatively given the stress-energy tensor, and through a coordinate change bring the metric into the form Eq. (20), which is dictated by spherical symmetry. We can then identify the dimensionless PPN parameters such as G and γ. Note that there is no guarantee that the metric in a given theory can be brought into *exactly* PPN form; the PPN formalism is merely designed to encompass as broad a range of theories as possible, and to allow for a model-independent phrasing of observational constraints. The precise definitions of the parameters are an inspired guess to allow different observations to constrain independent parameters.

We now consider important observational constraints on the PPN parameters in the Solar System. The deflection of light is given by an integral over the geodesic equation:

$$\begin{aligned}\delta\theta^i &= \frac{1}{2}\int d\lambda \left[h_{00}{}^{,i} + (h_{jk}{}^{,i})\hat{n}^j\hat{n}^k\right] \\ &= \frac{1}{2}\int d\lambda \left[2G\partial^i U + 2\gamma G\partial^i U\right] \\ &= G(1+\gamma)\int d\lambda\, \partial^i U, \end{aligned} \tag{21}$$

where $\hat{n}$ is the spatial unit vector determining the direction of the photon momentum, and we have used that h_{ij} vanishes at spatial infinity (since the spacetime is asymptotically flat). λ is the affine parameter, for which we choose the distance along the geodesic. In GR, $1+\gamma = 2$. This is the famous factor of two increase in the deflection over the Newtonian prediction, which was confirmed in the 1919 Solar eclipse and made Einstein famous.

Similarly, we can obtain the time delay, which is given by

$$\delta t = G(1+\gamma)\int d\lambda\, U. \tag{22}$$

Assuming the mass distribution is known, we know U, and measurements can thus put constraints on $G(1+\gamma)$. However, there is a further measurement we can use: the Earth's orbit around the Sun is governed by the geodesic equation for a non-relativistic body:

$$\frac{d^2x^i}{dt^2} = \frac{1}{2}h_{00}{}^{,i} = G\partial^i U. \tag{23}$$

Since we can measure this acceleration very precisely, we have a good knowledge of G which combined with the above yields a constraint on $1+\gamma$. Observationally, the best constraint comes from the time delay measured towards the Cassini probe [17], and is given by

$$\gamma - 1 = (2.1 \pm 2.3)10^{-5}. \tag{24}$$

Note that in fact Newton's constant G_N (contained in U) also drops out, so that the constraint on γ is tighter than our absolute knowledge of G_N. This is an important constraint for many modified gravity theories, as we will see.

3. Modified Gravity Scenarios

3.1. *Scalar-tensor theories*

In GR, the gravitational force is mediated by two massless spin-2 (tensor) degrees of freedom. The simplest modification to GR is to add an additional scalar field to the theory. This class of theories is referred to as *scalar-tensor theories.* Many modified gravity models invented in the cosmological domain turn out to be effectively scalar-tensor theories.

Scalar-tensor theories are most easily described in terms of their action. We generalize the *Einstein-Hilbert action* for GR,

$$S = \int d^4x\sqrt{-g}\left[\frac{1}{16\pi G_N}(R - 2\Lambda) + \mathcal{L}_m(\chi, g_{\mu\nu})\right], \tag{25}$$

to

$$S = \int d^4x\sqrt{-g}\left[\frac{1}{16\pi G_N}\left(\phi R - \frac{\omega(\phi)}{\phi}\nabla_\mu\phi\nabla^\mu\phi - 2\Lambda(\phi)\right) + \mathcal{L}_m(\chi, g_{\mu\nu})\right]. \tag{26}$$

Here, ∇_μ denotes covariant derivatives with respect to $g_{\mu\nu}$, and $\mathcal{L}_m(\chi, g_{\mu\nu})$ denotes the Lagrangian of matter fields χ which are minimally coupled to the metric $g_{\mu\nu}$; i.e., they follow the geodesics of that metric. Note that the scalar field ϕ does not couple directly to matter. It is only through its effect on $g_{\mu\nu}$ that it indirectly influences the dynamics of matter. We refer to the theory as described by the action Eq. (26) as in the *Jordan frame.* $\omega(\phi)$ and $\Lambda(\phi)$ are free functions (we have used the freedom to redefine the field $\phi \to f(\phi)$ to bring the first term in the Lagrangian into the standard form ϕR). We can think of the action Eq. (26) as obtained by making Newton's constant G_N a dynamic quantity which can vary from point to point, $G_N \to G_N/\phi(x)$.

The equations of motions in this scalar-tensor theory can be obtained by varying the action with respect to ϕ and $g_{\mu\nu}$. This is left as exercise. We obtain two coupled non-linear (or quasi-linear, to be precise) partial differential equations for $g_{\mu\nu}$ and ϕ,

$$\begin{aligned}\phi G_{\mu\nu} + \left[\Box\phi + \frac{1}{2}\frac{\omega}{\phi}(\nabla\phi)^2 + \Lambda\right]g_{\mu\nu} - \nabla_\mu\nabla_\nu\phi - \frac{\omega}{\phi}\nabla_\mu\phi\nabla_\nu\phi &= 8\pi G_N T_{\mu\nu}\\ 2\omega\Box\phi + \left(\omega' - \frac{\omega}{\phi}\right)(\nabla\phi)^2 - 2\phi\Lambda' &= -\,\phi R,\end{aligned}$$

where we have multiplied the second line with ϕ. The first line corresponds to the modified Einstein equation. Replacing the r.h.s. of the second line with the trace of the modified Einstein equation, we obtain the standard form of the scalar field equation:

$$(2\omega + 3)\Box\phi + \omega'(\nabla\phi)^2 + 4\Lambda - 2\phi\Lambda' = 8\pi G_N T. \tag{27}$$

We see that ϕ is sourced by the trace of the stress-energy tensor,

$$T \equiv g_{\mu\nu}T^{\mu\nu}. \tag{28}$$

The GR limit of the theory is obtained by letting $\omega \to \infty$ while $\omega'/\omega^2 \to 0$, and $\Lambda \to$ const. In that case, the scalar field equation reduces to $\Box\phi = 0$, leading to $\phi =$ const. This constant field does not have any dynamical effects and can be absorbed into a redefined G_N.

The fact that ϕ is sourced by T is a generic property of scalar-tensor theories. For a non-relativistic body, $T = -\rho$, so that the overall scalar charge is proportional to

$$\left|\int d^3x\, T\right| = M. \tag{29}$$

On the other hand, for a relativistic body, i.e. a body with Kepler velocities approaching the speed of light, $|T| \ll \rho$ so that

$$\left|\int d^3x\, T\right| \ll M. \tag{30}$$

In a non-rigorous way we can think of a black hole as an extreme example where $T = 0$ everywhere. A black hole will thus not source ϕ, in keeping with the no-hair theorems.

The physics of scalar-tensor theories becomes more apparent when performing a *conformal transformation.* That is, we write the action in terms of a fictitious metric $\bar{g}_{\mu\nu}$ defined through

$$g_{\mu\nu} = e^{2\Gamma(x)}\bar{g}_{\mu\nu}. \tag{31}$$

Note that the since the line element transforms as $ds^2 = e^{2\Gamma}d\bar{s}^2$, a null geodesic (defined through $ds^2 = 0$) in the metric $g_{\mu\nu}$ remains a null geodesic in $\bar{g}_{\mu\nu}$. Hence the "causal structure" of the spacetime remains invariant under Eq. (31). We can work out how the various terms in Eq. (26) can be written in terms of $\bar{g}_{\mu\nu}$. Using that $g^{\mu\nu} = e^{-2\Gamma}\bar{g}^{\mu\nu}$, we have

$$\begin{aligned}(\nabla\phi)^2 &= e^{-2\Gamma}\bar{g}^{\mu\nu}\partial_\mu\phi\partial_\nu\phi = e^{-2\Gamma}(\bar{\nabla}\phi)^2\\ \Box\phi &= g^{\mu\nu}\nabla_\mu\nabla\nu\phi = e^{-2\Gamma}\left(\bar{\Box}\phi + 2(\bar{\nabla}\phi)\cdot(\bar{\nabla}\Gamma)\right),\end{aligned} \tag{32}$$

where

$$\bar{\Box}\phi = \bar{g}^{\mu\nu}\bar{\nabla}_\mu\bar{\nabla}_\nu\phi; \quad (\bar{\nabla}\phi)\cdot(\bar{\nabla}\Gamma) = \bar{g}^{\mu\nu}\bar{\nabla}_\mu\phi\bar{\nabla}_\nu\Gamma \tag{33}$$

and $\bar{\nabla}_\mu$ denotes the covariant derivative with respect to $\bar{g}_{\mu\nu}$. Further, $\sqrt{-g} = e^{4\Gamma}\sqrt{-\bar{g}}$, and

$$\sqrt{-g}\phi R = e^{2\Gamma}\phi\sqrt{-\bar{g}}\left[\bar{R} - 6\bar{\Box}\Gamma - 6(\bar{\nabla}\Gamma)^2\right], \tag{34}$$

where $\bar{R}$ is the Ricci scalar with respect to the metric $\bar{g}_{\mu\nu}$. Thus, if we choose $e^{2\Gamma} = \phi^{-1}$, then we obtain a term $\sqrt{-\bar{g}}\bar{R}$ in the action, which looks like the Einstein-Hilbert action for GR. The action Eq. (26) then becomes

$$S = \int d^4x\sqrt{-\bar{g}}\left[\frac{1}{16\pi G_N}\left(\bar{R} - \frac{3+2\omega(\phi)}{2\phi^2}(\bar{\nabla}\phi)^2 - 2e^{4\Gamma}\Lambda\right)\right.$$
$$\left. + e^{4\Gamma(\phi)}\mathcal{L}_m(\chi; g_{\mu\nu} = e^{2\Gamma(\psi)}\bar{g}_{\mu\nu})\right], \tag{35}$$

where we have performed an integration by parts on the term $(3/\phi)\bar{\Box}\phi$. We now implicitly define a new scalar field $\psi(\phi)$ through

$$\frac{\partial\Gamma(\phi)}{\partial\psi} = -\frac{1}{2}\frac{\partial\ln\phi}{\partial\psi} = \left[2(3+2\omega(\phi))\right]^{-1/2}, \tag{36}$$

so that

$$\frac{1}{\phi^2}(\bar{\nabla}\phi)^2 = \frac{2}{3+2\omega(\phi(\psi))}(\bar{\nabla}\psi)^2, \tag{37}$$

and introduce a new potential

$$V(\psi) = e^{4\Gamma}\Lambda(\phi(\psi)). \tag{38}$$

Note that if $\omega(\phi) = \text{const}$, corresponding to the simplest scalar-tensor theories, we have $\psi \propto -\ln\phi$. With this Eq. (35) becomes

$$S = \int d^4x\sqrt{-\bar{g}}\left[\frac{1}{16\pi G_N}\bar{R} - \frac{1}{8\pi G_N}\left(\frac{1}{2}\bar{\nabla}_\mu\psi\bar{\nabla}^\mu\psi + V(\psi)\right)\right.$$
$$\left. + e^{4\Gamma(\psi)}\mathcal{L}_m(\chi; g_{\mu\nu} = e^{2\Gamma(\psi)}\bar{g}_{\mu\nu})\right]. \tag{39}$$

We refer to this as the *Einstein frame* action, since the gravitational part looks like GR, in terms of the metric $\bar{g}_{\mu\nu}$, with a minimally coupled scalar field ψ with self-interaction described by the potential $V(\psi)$.[a] On the other hand, matter still couples to the *physical* metric $g_{\mu\nu}$, given by Eq. (31). Written in the Einstein frame, the scalar field ψ thus exhibits a peculiar universal coupling to matter through the $e^{2\Gamma(\psi)}$ factor. The fact that $\bar{g}_{\mu\nu}$ is not the physical metric according to which matter moves can sometimes

[a]Note that in our convention the field ψ is dimensionless. In order to obtain a normalized field, one can define $\tilde{\psi} = \psi/\sqrt{8\pi G_N} = M_{\rm Pl}\psi$.

cause confusion when calculating observables. It is useful to cross-check any calculation with the equivalent Jordan frame formulation.

We can also define a fictitious conformal stress-energy tensor,

$$\bar{T}^{\mu\nu} \equiv \frac{2}{\sqrt{-\bar{g}}} \frac{\partial\left(\sqrt{-\bar{g}}\, e^{4\Gamma}\mathcal{L}_m(\chi; g_{\mu\nu} = e^{2\Gamma}\bar{g}_{\mu\nu})\right)}{\partial \bar{g}_{\mu\nu}} = e^{6\Gamma}T^{\mu\nu}. \tag{40}$$

Note that $\bar{T}^{\mu\nu}$ is not conserved. The field equations then become

$$\bar{G}_{\mu\nu} = 8\pi G_N \left[\bar{T}_{\mu\nu} + \frac{1}{8\pi G_N}\left\{\bar{\nabla}_\mu\psi\bar{\nabla}_\nu\psi - \left(\frac{1}{2}\bar{\nabla}_\alpha\psi\bar{\nabla}^\alpha\psi + V(\psi)\right)\bar{g}_{\mu\nu}\right\}\right] \tag{41}$$

$$\bar{\Box}\psi - \frac{\partial V}{\partial\psi} = -8\pi G_N \alpha \bar{T}, \tag{42}$$

where $\alpha = \partial\Gamma/\partial\psi = [2(3+2\omega)]^{-1/2}$, and $\bar{T} = \bar{g}_{\mu\nu}\bar{T}^{\mu\nu}$. These equations are straightforwardly interpreted as the Einstein equations for $\bar{g}_{\mu\nu}$, sourced by matter through $\bar{T}_{\mu\nu}$, and by the kinetic and potential energy of the scalar field ψ; and the scalar field equation with self-interaction, again sourced by $\bar{T}$. In many cases, it is easier to deal with these Einstein-frame equations, since physical approximations can be made (for example, in many quasi-Newtonian contexts we can neglect the stress-energy contributed by the scalar field in Eq. (41)).

The Einstein frame also allows us to gain some intuition on the phenomenological effects of scalar-tensor theories. Since $\bar{g}_{\mu\nu}$ is governed by Einstein's equations (assuming the stress-energy of ψ can be neglected, which is the case in non-relativistic systems), we can easily derive the PPN metric to second order, Eq. (20), in this theory:

$$\begin{aligned} -1 + h_{00}^{(2)} &= e^{2\Gamma}(-1+2U) \quad \Rightarrow \quad h_{00}^{(2)} = 2\,(U - \bar{\alpha}\,\delta\psi) + \mathcal{O}(\varepsilon^4) \\ \delta_{ij} + h_{ij}^{(2)} &= e^{2\Gamma}(1+2U)\delta_{ij} \quad \Rightarrow \quad h_{ij}^{(2)} = 2\,(U + \bar{\alpha}\,\delta\psi)\,\delta_{ij} + \mathcal{O}(\varepsilon^4), \end{aligned} \tag{43}$$

where $\delta\psi = \psi - \bar{\psi}$, $\bar{\psi}$ is the background value of the field (which can be chosen to be zero in the case of a Minkowski background but not for cosmological solutions), and $\bar{\alpha} = \partial\Gamma/\partial\psi|_{\bar{\psi}}$. Thus, scalar-tensor theories affect the motion of non-relativistic bodies, which depend on g_{00}, differently than the propagation of light, which probes $g_{00} + g_{ij}\hat{n}^i\hat{n}^j$. In fact, photons are not affected by ψ at all at order ε^2. This goes back to the invariance of null geodesics under conformal transformations. We can then derive the PPN γ parameter through the second order PPN metric, Eq. (20):

$$\gamma = \frac{h_{ij}^{(2)}\hat{n}^i\hat{n}^j}{h_{00}^{(2)}} = \frac{1+\alpha\psi/U}{1-\alpha\psi/U} \tag{44}$$

3.2. Brans-Dicke theory with and without potential

The simplest scalar-tensor theory is obtained by setting ω = const and Λ = const. This model is referred to as *Brans-Dicke theory* after the initial authors. In this case, the field ψ does not have a potential term and is thus a massless scalar field with infinite range, i.e. Eq. (42) yields (in the static case)

$$\psi = -2\alpha U. \tag{45}$$

The PPN parameter γ in this theory is then

$$\gamma = \frac{1-2\alpha^2}{1+2\alpha^2} = \frac{1+\omega}{2+\omega}. \tag{46}$$

We see that as $\omega \to \infty$, $\gamma \to 1$ and we recover GR. We can then use the Solar System constraint Eq. (24) to put a lower limit on ω:

$$\omega \gtrsim 40,000. \tag{47}$$

There are also cosmological constraints from the abundance of light elements synthesized during Big Bang nucleosynthesis (BBN). This is because, as we have noted, $G = G_N/\phi$ in scalar-tensor theories, leading to a different expansion rate at a fixed temperature than in GR. However, these constraints are weaker than the Solar System constraints.

Equation (47) constrains this theory to modify gravity by less than $\sim 10^{-5}$ throughout the Universe. It is thus not very interesting to investigate its phenomenological consequences. In the next section we will see how to get around this bound.

A slightly more general scalar-tensor theory adopts ω = const, but allows for a potential $V(\psi)$. Consider the case where $V(\psi)$ has a minimum at ψ_0. In the static case with no sources, ψ will assume this equilibrium value everywhere. Now let us assume that we do have sources, but they are weak in the sense that they only slightly perturb ψ away from ψ_0, $\delta\psi = \psi - \psi_0 \ll 1$. Then, Eq. (42) becomes

$$\bar{\Box}\delta\psi - m^2\delta\psi = -8\pi G_N \alpha \bar{T}, \quad m^2 \equiv \left.\frac{\partial^2 V}{\partial \psi^2}\right|_{\psi_0}. \tag{48}$$

In other words, $\delta\psi$ behaves like a massive scalar field. We can then derive the modified acceleration $\mathbf{a} = \nabla g_{00}$ experienced by a non-relativistic object through Eq. (43). For a static source with mass M that is much smaller than λ_C, we have

$$\delta\psi(r) = -2\alpha U(r) e^{-r/\lambda_C}, \tag{49}$$

where $\lambda_C = 1/m$ is the Compton length[b] of the field ψ. The modified gravitational acceleration thus becomes

$$\mathbf{a} = \nabla\left[U(r)\left(1 + 2\alpha^2 e^{-r/\lambda_C}\right)\right]$$
$$2\alpha^2 = \frac{1}{3+2\omega}. \tag{50}$$

If λ_C is much greater than an astronomical unit (1.5×10^8 km), the standard Brans-Dicke constraint Eq. (47) still applies.

3.3. *Chameleon theories*

Chameleon theories are another subset of scalar-tensor theories, where $V(\psi)$ is monotonically decreasing ("run-away potential"). For example, $V(\psi) = V_0\psi^{-n}$ or $V(\psi) = V_0\exp(-\psi)$. Further, we choose for simplicity $\omega = \text{const}$, and thus $\alpha = 1/\sqrt{2(3+2\omega)} = \text{const}$. Consider a stationary non-relativistic matter distribution. Then, the equation of motion for ψ, Eq. (42) becomes (recall $\bar{T} \approx T = -\rho$)

$$\nabla^2\psi = \frac{\partial V_{\text{eff}}}{\partial\psi} \tag{51}$$
$$V_{\text{eff}} = V(\psi) + 8\pi G_N \alpha\,\rho\,\psi. \tag{52}$$

Thus, the scalar field effectively sees a density-dependent potential. As long as $\partial V/\partial\psi < 0$ and $\alpha > 0$, the minimum of the potential shifts to smaller ψ as ρ increases. Furthermore, the curvature around the potential minimum, i.e. the mass $m^2 \equiv \partial^2 V/\partial\psi^2$ increases with ρ (for a run-away potential).

In cosmology, the background value of the field $\bar{\psi}$, and its mass $\bar{m}^2 = 1/\bar{\lambda}_C^2 = \partial^2 V/\partial^2\psi|_{\min}$ are then set by the background density $\bar{\rho}$. Note that since $\bar{\rho}$ decreases as a^{-3}, $\bar{\psi}$ will generically *increase* towards late times in cosmology, whereas the mass $\bar{m}^2$ will decrease, increasing the reach λ_C of the field. On the other hand, as structure forms in the Universe, the density becomes much larger than the background density in some regions, suppressing the field. This is known as the *chameleon mechanism* [18], as the field adapts to its environment like a chameleon.

This can be seen by going back to Eq. (48), which applies when perturbations to the field $\delta\psi = \psi - \bar{\psi}$ are much smaller than 1. More precisely, we subtract the field equation for $\bar{\psi}$ (Eq. (42) with $\rho = \bar{\rho}$) from the full field

[b] Recall that we choose natural units $c = \hbar = 1$. Otherwise, we should think of $V(\psi)$ as having dimension $[\text{length}]^{-2}$, and correspondingly m as dimension $[\text{length}]^{-1}$.

equation, and restrict to the static case to obtain:

$$\nabla^2 \delta\psi - \bar{m}^2 \delta\psi = 8\pi G_N \alpha \delta\rho. \tag{53}$$

The solution for a spherically symmetric source (with boundary condition $\delta\psi \to 0$ as $r \to \infty$) is then given by Eq. (49). For scales $r \ll \lambda_C$, it reduces to $\delta\psi = -2\alpha U$. where we have assumed that the source is much smaller than $\bar{\lambda}_C$ and $U(r)$ is the Newtonian potential of the source as before. If the source is sufficiently heavy, the condition $\delta\psi \ll \bar{\psi}$ will be violated at some $r_{\rm ch}$. At that radius, the field will move appreciably from its background equilibrium value, and will instead satify the simplified equation

$$\frac{\partial V_{\rm eff}}{\partial \psi} = 0. \tag{54}$$

In other words, ψ and λ_C will be suppressed (we also call this *screened*) from their background value by some power of $\rho/\bar{\rho}$. In the Solar System, $\rho/\bar{\rho} \gtrsim 10^6$. This is sufficient to suppress λ_C to values less than a millimeter, evading both Solar System and laboratory tests for *any* value of the coupling constant α. In other words, chameleon models are able to evade the stringent constraint Eq. (47).

However, this will only happen for sufficiently heavy objects. In order to obtain a rough estimate when this happens, we equate the approximate solution $\delta\psi = -2\alpha U$ to $-\bar{\psi}$, corresponding to the case where the field is completely screened:

$$\delta\psi(r_{\rm ch}) = -\bar{\psi} \quad \Rightarrow U(r_{\rm ch}) = \left|\frac{\bar{\psi}}{2\alpha}\right|. \tag{55}$$

In other words, if there is a finite radius at which the Newtonian potential well of a body exceeds $|\bar{\psi}/2\alpha|$, then this body will be *chameleon-screened* inside of $r_{\rm ch}$. The external ψ field of such a body is then only sourced by the mass outside $r_{\rm ch}$, which for a dense body is merely a *thin shell.* Thus, Eq. (55) is also referred to as "thin-shell condition". Clearly, we need this to happen in the Solar System in order to pass Solar System and laboratory constraints. Initially, [19] thought of the Earth as fulfilling the thin-shell condition, which would allow interesting effects to happen for satellites orbiting the Earth at some distance. However, as discussed by [20], in fact the dominant contribution to the local potential well $U(r)$ is provided by the Milky Way's dark matter halo, whose mass can be inferred from the galactic rotation curve. Thus, the Solar System is screened *if and only if* the galactic halo is screened at some $r_{\rm ch} \sim 10$ kpc or more, where

$U(r) \sim$ few $\times 10^{-6}$. This implies an upper bound to the background field value for any chameleon theory:

$$\left|\frac{\bar{\psi}}{\alpha}\right| \lesssim \text{few} \times 10^{-6}. \tag{56}$$

As we will see below, this still allows for interesting effects to happen in other regions of the cosmos, unlike Brans-Dicke models.

3.4. $f(R)$ gravity

$f(R)$ gravity is the first modified gravity model proposed as an alternative explanation for the accelerated expansion of the Universe [21]. We write the gravitational action as

$$S = \frac{1}{16\pi G_N} \int d^4x\sqrt{-g}\,[R + f(R)]\,. \tag{57}$$

GR with cosmological constant is clearly recovered through $f(R) = -2\Lambda$. Choosing $f(R) = aR + b$ just corresponds to a rescaling of G_N. Only if f is a non-linear function of R do we obtain an actual modification of GR. The field equations can be obtained through variation of the action Eq. (57),

$$(1 + f_R)G_{\mu\nu} - \left(\frac{f}{2} - \frac{1}{2}f_R R - \Box f_R\right) g_{\mu\nu} - \nabla_\mu\nabla_\nu f_R = 8\pi G_N\, T_{\mu\nu}. \tag{58}$$

Here, $f_R \equiv df/dR$. Since R contains second derivative of the metric, these equations are fourth order in derivatives if f_R is not simply a constant, i.e. if f is a non-linear function. Usually, higher order field equations are a bad sign, since they often entail instabilities. We will see however that there is a wide class of well-behaved $f(R)$ models.

Equation (57) can be transformed into a somewhat more transparent form. We use a trick, defining a new field ϕ through

$$\phi = 1 + f_R(R). \tag{59}$$

As long as we assume a non-trivial $f(R)$ model with $f_{RR} \equiv d^2f/dR^2 \neq 0$, this relation is invertible to $R(\phi)$. This transformation can be seen as a Legendre transformation of variables. We further define a potential

$$\Lambda(\phi) = \frac{1}{2}\left[R(\phi)\phi - f(R(\phi)) - R(\phi)\right]. \tag{60}$$

Then, Eq. (57) becomes

$$S = \frac{1}{16\pi G_N} \int d^4x\sqrt{-g}\,[\phi R - 2\Lambda(\phi)]\,. \tag{61}$$

This is just a scalar-tensor theory of the form Eq. (26), although a peculiar one with $\omega = 0$, where f_R plays the role of scalar field (this can already be noticed by comparing Eq. (58) with Eq. (27)). While in GR there is an algebraic relation,

$$R = 8\pi G_N T, \tag{62}$$

in $f(R)$ this relation is replaced by a field equation for f_R, i.e. the trace of Eq. (58):

$$3\Box f_R - R + f_R R - 2f = 8\pi G_N T. \tag{63}$$

This corresponds to a scalar-tensor theory with a potential $\partial V/\partial f_R = R - f_R R + 2f$. The curvature of the potential around the potential minimum is

$$m^2(f_R) = \lambda_C^{-2} = \frac{1}{3}\left(\frac{1+f_R}{f_{RR}} - R\right) \approx \frac{1}{3} f_{RR}^{-1}, \tag{64}$$

where the approximate equality holds in the weak field limit, $f_R \ll 1$. $f(R)$ gravity has attracted significant interest in the realm of cosmology. We can divide models into two classes: (i) f is an increasing function of R; in this case the modifications become important at high spacetime curvature, e.g. in the early Universe. (ii) f is a decreasing function of R; the modifications become important at low spacetime curvature, in particular in the late Universe. The first kind has attracted interest in the context of inflation, while the second kind has been proposed as an alternative to Dark Energy. We will discuss the second kind here.

Since we have a free function to choose, $f(R)$ gravity can in fact reproduce essentially any expansion history. One possible choice is

$$f(R) = -2\Lambda \frac{R/R_c}{R/R_c + 1}, \tag{65}$$

which when $R \gg R_c$ becomes

$$f(R) \approx -2\Lambda \left(1 - \frac{R_c}{R}\right) + \mathcal{O}((R_c/R)^2), \tag{66}$$

and thus effectively provides a cosmological constant, with corrections becoming important in regions of *small* spacetime curvature. On the other hand, as $R/R_c \to 0$, $f(R) \to 0$ and thus the model does not employ a true cosmological constant. The field f_R is given in the large R limit by

$$f_R = -2\Lambda \frac{R_c}{R^2} = f_{R0} \frac{R_0^2}{R^2} \tag{67}$$

where we have defined a new dimensionless parameter $f_{R0} < 0$, which gives the field value corresponding to the background curvature of the Universe today, R_0. Thus, $R(f_R) \propto f_R^{-1/2}$, and we see that Eq. (63) includes a "run-away" potential just like chameleon models. Matching Eq. (63) in the quasi-static case neglecting the potential ($\nabla^2 f_R = -8\pi G_N/3\,\rho$) immediately leads to the necessary requirement for the chameleon mechanism

$$U(r_{\rm ch}) = \frac{2}{3}|\bar{f}_R|. \tag{68}$$

Since $\bar{f}_R = f_{R0}$ today, this implies a constraint of $|f_{R0}| \lesssim 10^{-6}$ today from Solar System constraints, a significant constraint on $f(R)$ models designed to produce an accelerated expansion. On the other hand, the Compton wavelength of such models (Eq. (64)) is still in the range of several Mpc, and thus can leave imprints in cosmological structure.

3.5. *Braneworld models*

Braneworld models are a class of higher-dimensional models of gravity. We know from lab tests and particle experiments that matter fields and the three forces apart from gravity follow a four-dimensional spacetime. There are thus two options for introducing extra dimensions: first, one can compactify them to a scale $R \lesssim 1\text{TeV}^{-1}$, thus hiding them from current measurements. This is the standard approach adopted in string theory. Models with warped extra dimensions (such as the Randall-Sundrum model) can also be included in this class, as they do not modify the large-distance behavior of gravity.

Another option is to think of our four spacetime dimensions as a submanifold (*brane*) embedded in a higher-dimensional spacetime. If we confine all matter and interactions to this brane, apart from gravity, we can reproduce the observable Universe. The key point is to let gravity propagate into the higher dimension(s) (*bulk*), allowing for interesting signatures in the gravity sector. One motivation for this behavior comes, again, from string theory: fermions and vector bosons are modeled as open strings, which can be confined to a brane by attaching their endpoints to it (Dirichlet boundary condition, hence D-branes). On the other hand, gravitons are modeled as closed strings which cannot be attached to branes, and hence propagate through the entire bulk. Here, I will only consider this second class, since compact extra dimensions do not have any large-scale observable signatures.

3.5.1. *The Dvali-Gabadadze-Porrati (DGP) Model*

The Dvali-Gabadadze-Porrati model [22] consists of a spatially three-dimensional brane in a 4+1 dimensional Minkowski bulk. Matter and all interactions except gravity are confined to the brane. The gravitational action consists of the five-dimensional Einstein-Hilbert action, plus a 4D Einstein-Hilbert term which ensures that gravity becomes four-dimensional on small scales:

$$S_{\text{grav}} = \frac{1}{16\pi\, G^{(5)}} \int d^5X \sqrt{-g^{(5)}} R^{(5)} + S_{\text{boundary}} + \frac{1}{16\pi\, G^{(4)}} \int_{\text{brane}} d^4x \sqrt{-g^{(4)}} \left(R^{(4)} + \mathcal{L}_m \right). \quad (69)$$

Here, X, $g^{(5)}$ stand for the bulk coordinates and metric, while x, $g^{(4)}$ are the induced coordinates and metric on the brane, and $R^{(5)}$, $R^{(4)}$ denote the corresponding Ricci scalars. Note that $g^{(4)}$ is not an independent degree of freedom but fixed by the 5D metric $g^{(5)}$. In the following, we will drop the $^{(4)}$ notation where no confusion can arise. The boundary term is added to the action in order to ensure that variation with respect to $g^{(5)}$ leads to the correct five-dimensional Einstein equations (e.g., [23]).

The two gravitational constants, or Planck masses $M_{\text{Pl}\,(5)}$, $M_{\text{Pl}\,(4)}$ appearing in Eq. (69) can be related via a length scale, the *crossover scale*:

$$r_c \equiv \frac{1}{2}\frac{G^{(5)}}{G^{(4)}} = \frac{1}{2}\frac{M^2_{\text{Pl}\,(4)}}{M^3_{\text{Pl}\,(5)}}. \quad (70)$$

On scales above r_c, gravity becomes five-dimensional, with forces falling off as $1/r^3$. Below r_c, gravity is four-dimensional, but not Einsteinian gravity, a point to which we return below. This behavior can be summarized as a modified graviton propagator given by

$$\frac{1}{k^2 + k/r_c}. \quad (71)$$

Since all matter is thought as confined to the brane, the five-dimensional metric has to obey non-trivial junction conditions over the brane [24]. Assuming an empty Minkowski bulk, a spatially flat brane, and a homogeneous and isotropic matter distribution on the brane, the junction conditions lead to the following analogue of the Friedmann equation for the scale factor of the induced metric on the brane:

$$H^2 \pm \frac{H}{r_c} = \frac{8\pi\, G}{3}\rho, \quad H \equiv \frac{\dot{a}}{a}. \quad (72)$$

The junction conditions constrain only the magnitude of the jump in the derivative of the metric, leaving two possible branches of the theory which are determined by the sign on the left hand side of Eq. (72). The branch with the '$-$' sign asymptotes to a late-time de Sitter-Universe, $H = 1/r_c =$ const., and is correspondingly called the *self-accelerating branch.* On the other hand, the $+$ sign leads to the so-called *normal branch*, which does not lead to any acceleration. Even in the self-accelerating branch, the expansion history is different from ΛCDM, and corresponds to an effective dark energy with $w_{\rm eff} \to -1/2$ in the matter-dominated era at high redshifts.

3.5.2. *Cosmological Perturbations and Brane Bending Mode*

The propagation of light and particles on the DGP brane is completely determined by the perturbed 4D Friedmann-Robertson-Walker metric. However, in order to determine the evolution of the metric potentials on the brane, it is necessary to solve the full 5D Einstein equations [9, 25]. An additional scalar degree of freedom appears, the so-called*brane-bending mode* φ which couples to matter. φ is directly proportional to the displacement of the brane in the extra dimension, so that the second derivative of φ corresponds to a bending of the brane. In our convention, φ is dimensionless, instead of being scaled to the Planck mass $M_{\rm Pl\,(4)}$.

In the decoupling limit of DGP [26], when setting gravitational interactions to 0, the self-interactions of the φ field remain constant. Hence, while perturbations of the metric higher than linear order can be neglected for cosmological studies, it is crucial to consider the self-interactions in the brane-bending mode. The field equation reads

$$\Box\varphi + \frac{r_c^2}{3\beta}[(\Box\varphi)^2 - (\nabla_\mu\nabla_\nu\varphi)(\nabla^\mu\nabla^\nu\varphi)] = \frac{8\pi G_N}{3\beta}\delta\rho, \tag{73}$$

where $\delta\rho = \rho_m - \overline{\rho_m}$ is the matter density perturbation, and the function $\beta(a)$ is given by:

$$\beta(a) = 1 \pm 2H(a)\,r_c\left(1 + \frac{\dot{H}(a)}{3H^2(a)}\right). \tag{74}$$

Note that β is always negative in the self-accelerating branch, signaling that gravity is weakened by a repulsive brane-bending mode. This is not expected for a scalar field, and in fact it has been shown that perturbations around the de Sitter limit of the self-accelerating branch are ghosts (i.e. the kinetic term has the wrong sign). The normal branch does not suffer from this issue.

The brane-bending mode influences the dynamics of particles and light in the same way as in standard scalar-tensor theories. In particular,

$$\begin{aligned} g_{00} &= g_{00}^{\mathrm{GR}} + \frac{1}{2}\varphi \\ g_{ij} &= g_{ij}^{\mathrm{GR}} - \frac{1}{2}\varphi\, \delta_{ij}. \end{aligned} \tag{75}$$

Thus, we again see that dynamics of non-relativistic objects couple to ϕ, whereas light does not.

3.5.3. *Weak Brane Regime*

In the limit where the interaction terms (quadratic terms in Eq. (73)) are small, Eq. (73) can be linearized, yielding a standard Poisson equation:

$$\nabla^2 \varphi_L = \frac{8\pi G_N}{3\beta}\delta\rho \quad \Rightarrow \quad \varphi_L = -\frac{2}{3\beta}U. \tag{76}$$

In this regime, also called weak-brane phase, $\varphi = \varphi_L$ becomes proportional to the Newtonian potential U, corresponding to a constant rescaling of the gravitational constant through Eq. (20) and Eq. (75):

$$G_N \to G_{\mathrm{eff}} = \left(1 + \frac{1}{3\beta}\right) G_N. \tag{77}$$

Substituting the linear solution into Eq. (73), we obtain a rough estimate for the overdensity $\delta \equiv \delta\rho/\bar{\rho}_m$ at which the non-linear interactions become important:

$$1 \sim \frac{r_c^2}{3\beta}(\nabla^2\varphi_L)^2 \,/\, \nabla^2\varphi_L = \frac{8\pi G\, r_c^2}{9\beta^2}\delta\rho = \frac{3H_0^2 r_c^2}{9\beta^2}\Omega_m\, \delta. \tag{78}$$

For a self-accelerating DGP model that leads to cosmic acceleration today, $r_c \sim H_0^{-1}$, $\beta(a=1) \sim -1$, and the prefactor in Eq. (78) is of order unity today. Hence, the self-coupling of the brane bending mode φ becomes important as soon as the matter density field becomes non-linear, $\delta \sim 1$.

3.5.4. *Vainshtein Effect*

In general, Eq. (73) is difficult to solve in full generality due to the non-linearity in the derivative terms. However, an instructive test case, a spherically symmetric matter distribution, can be solved analytically. In the spherically symmetric case, Eq. (73) becomes [25]:

$$\left(\frac{d^2}{dr^2}+\frac{2}{r}\frac{d}{dr}\right)(\varphi+\frac{1}{3\beta}\Xi)=\frac{8\pi G_N}{3\beta}\delta\rho, \tag{79}$$
$$\Xi\equiv 2r_c^2\int_0^r\frac{dr'}{r'}\left(\frac{d\varphi}{dr'}\right)^2.$$

For simplicity, we assume a spherical mass M of radius R with uniform density, and set $a=1$. Then, we can integrate Eq. (79) once and obtain the gravitational acceleration in DGP:

$$g=g_N+\frac{1}{2}\frac{d\varphi}{dr}=g_N[1+\Delta(r)], \tag{80}$$

where g_N is the Newtonian acceleration of the spherical mass, and:

$$\Delta(r)=\frac{2}{3\beta}\begin{cases} r^3/r_*^3\left(\sqrt{1+\left(\frac{r_*}{r}\right)^3}-1\right), r\geq R\\ \\ R^3/r_*^3\left(\sqrt{1+\left(\frac{r_*}{R}\right)^3}-1\right), r<R.\end{cases} \tag{81}$$

Here, r_* denotes a characteristic scale of the solution, the *Vainshtein radius*:

$$r_*^3=\frac{8\,r_c^2 r_s}{9\beta^2}, \tag{82}$$

and $r_s=2GM$ is the gravitational radius of the mass. For very large distances, $r\gg r_*$, $\Delta(r)$ approaches the constant $1/(3\beta)$, which exactly matches the linear solution, Eq. (77). Note also that by substituting $\delta\rho\sim M/r^3$ in Eq. (78), we see that the non-linearity criterion is simply proportional to $(r_*/r)^3$.

In the opposite limit, $r\ll r_*$, φ becomes suppressed with respect to the linear solution, and $\Delta(r)$ approaches the small constant $2/(3\beta)(R/r_*)^{3/2}$ inside the mass (assuming $R\ll r_*$). This *Vainshtein effect* [27, 28] amounts to restoring GR in deep potential wells, while far away from the mass, gravity is in the scalar-tensor regime. Interestingly, outside the mass but well within the Vainshtein radius, $\Delta(r)$ scales as $(r/r_*)^{3/2}$, and thus $g_N\Delta\propto r^{-1/2}$. This departure from a pure $1/r^2$ force leads to anomalous precession of objects such as the moon. Constraints from precise measurements of the moon's orbit through laser ranging could become competitive with future measurements.

Another exact solution to Eq. (73) exists: a density field with planar symmetry, $\delta\rho(\mathbf{x})=\delta\rho(z)$. By symmetry, $\varphi=\varphi(z)$ and we see that the two non-linear terms in Eq. (73) exactly cancel, so that we are left with the linear solution Eq. (76). The planar and spherically symmetric solutions can be considered as two limiting cases for understanding the φ field behavior in the cosmological context.

4. Testing Gravity with Cosmology

We now turn to the different ways with which cosmological observations can constrain modified gravity models. We have already seen that Solar System tests provide stringent constraints on AU scales, but that modified gravity models can pass these tests through non-linear screening mechanisms while still yielding significant modifications on larger scales (above ~ 1 Mpc $\approx 2 \times 10^{11}$ AU).

We can separate cosmological probes into those that measure the expansion history and geometry of the background Universe, and those that probe the evolution of structure (departures from the smooth background). The treatment of perturbations can further be separated into the regime where density perturbations are linear, on large scales, and the regime in which they become non-linear, on smaller scales, to form dark matter halos, galaxies, clusters, etc. The latter regime can only rigorously be treated using simulations. The scale at which structure becomes non-linear today (redshift $z = 0$) is roughly 30 Mpc (depending on the exact definition).

4.1. *Expansion history and early Universe*

The expansion history of the Universe can be described through the Hubble parameter $H(t)$. Qualitatively, the Universe transitioned from radiation-dominated ($H \propto a^{-2}(t)$) to matter-dominated ($H \propto a^{-3/2}(t)$) at redshift $z_{\rm eq} \approx 10^4$. Many observations now indicate that the Universe transitioned to Dark Energy-dominated at $z \approx 1$.

In order of decreasing redshift, we have the following probes of $H(t)$:

- **Big Bang Nucleosynthesis:** Measurements of the abundances of light elements (in particular D and He) probe the expansion rate at $t_{\rm BBN} \approx 3$ min after the Big Bang, through its influence on the nuclear reaction network. Since the Universe was radiation-dominated at that time, the absolute measurement of $H(t_{\rm BBN})$ can be used together with the measurements on late-time cosmology to constrain any deviation of G_N at $t_{\rm BBN}$ from its value today, to the level of $\sim$5%. This puts constraints on scalar-tensor theories where G_N evolves as $\bar{\phi}^{-1}$.

- **Recombination and Cosmic Microwave Background (CMB):** The CMB puts stringent constraints on the density of matter (both dark and baryonic) and radiation in the Universe at the time of recombination $t_{\rm rec}$. Together with measurements of the Hubble constant, it also provides a stringent constraint on the angular diameter distance to redshift $z = 1100$, and the curvature. So far, measurements are consistent with a flat Universe $\Omega_K = 0$.
- **Supernova Ia:** The measurement of the apparent magnitude of Supernovae as function of observed redshift provides a measurement of the luminosity distance for redshifts $z \approx 0.1 - 1.7$.
- **Baryon Acoustic Oscillations:** The CMB provides an accurate measurement of the acoustic sound horizon at recombination. This scale is also imprinted in the matter power spectrum and thus the power spectrum of large-scale structure tracers. A measurement of the angular scale of this feature at as given redshift provides a measurement of the angular diameter distance to that redshift. This technique has so far been applied to $z \sim 0.2 - 0.8$.

The last three measurements are used to infer the accelerated expansion. Using the Friedmann equation in GR, we can write (assuming a flat Universe)

$$H^2 = \frac{8\pi G_N}{3}\left[\bar{\rho}_m + \rho_{\rm DE}\right] = H_0^2\left[\Omega_m a^{-3} + \Omega_{\rm DE} a^{-3(1+w)}\right], \qquad (83)$$

parametrizing the "Dark Energy" through its density parameter and its equation of state w (assumed to be constant here). A cosmological constant corresponds to $w = -1$. Current measurements are consistent with $w = -1$, with error bars at redshifts $z < 0.5$ of $\sigma(w) \approx 0.1$. Theories such as $f(R)$ and DGP of course modify Eq. (83); however, assuming we have only matter in the Universe, we can phrase the modification as an *effective* Dark Energy density $\rho_{\rm DE,eff}$. For example, in the case of self-accelerating DGP, this Dark Energy density has $w_{\rm eff} \approx -0.8$ today, reducing to -0.5 at higher redshifts. CMB, Supernovae, and BAO are sufficient to rule out this expansion history at the $4 - 5\sigma$ level. In the case of $f(R)$ gravity, deviations in the Friedmann equation are of order f_R, and expansion history observations rule out amplitudes of $|f_{R0}| \sim 1$.

One of the main motivations for studying these models was as working examples to see what phenomenological signatures to look for when testing gravity. The expansion history of the Universe has one major drawback as test of gravity: we can only test gravity under the assumption of a cosmological constant or Dark Energy component. If we allow for unspecified Dark Energy, then *any* expansion history is possible in the GR+DE framework, rendering the expansion history a blunt probe of gravity. The growth of structure on the other hand is not directly influenced by a cosmological constant or Dark Energy,[c] and thus can serve as a test of GR.

4.2. *Growth of structure*

We have no way of predicting the initial seed perturbations of the Universe as a function of position; they are stochastic. However, we can obtain information on the stress-energy content of the Universe and gravity through statistical measures of the perturbations.

In the commonly chosen conformal-Newtonian gauge, the perturbed FRW metric can be written as (see Julien Lesgourgues' lectures)

$$ds^2 = a^2(\eta)\left[-(1+2\Psi)d\eta^2 + (1-2\Phi)d\mathbf{x}^2\right], \tag{84}$$

where $a(\eta)$ is the scale factor, η denotes conformal time, and Φ, Ψ are the cosmological potentials. In General Relativity (GR), $\Phi = \Psi$ in the absence of anisotropic stress, which is the case at late times in ΛCDM and most dark energy models. In addition, the potentials are conventionally re-expressed in terms of the matter overdensity $\delta_m = (\rho - \bar{\rho})/\bar{\rho}$ using the Poisson equation:

$$\nabla^2\Psi \overset{\mathrm{GR}}{=} 4\pi G a^2 \rho_m\, \delta_m = -\frac{3}{2}\Omega_m H_0^2 a^{-1}\delta_m, \tag{85}$$

where the derivatives on the l.h.s. are in terms of comoving coordinates $\mathbf{x}$ (hence the factor a^2 on the r.h.s.). In case of modified gravity, both $\Phi = \Psi$ and Eq. (85) do not necessarily hold. In particular, we have seen that the scalar degree of freedom in scalar-tensor theories contributes differently to Φ and Ψ:

$$\begin{aligned}\Psi &= \Psi_{\mathrm{GR}} + \alpha\psi\\ \Phi &= \Phi_{\mathrm{GR}} - \alpha\psi.\end{aligned} \tag{86}$$

The dynamics of matter is governed by Ψ, while gravitational lensing measures $\Phi - \Psi$ and thus is not directly influenced by the scalar field.

[c]Unless we couple DE to matter, in which case the distrinction between DE and modified gravity becomes increasingly blurred.

4.2.1. *Linear Scales*

On large scales, the matter density contrast δ_m is much less than one. This is the simplest case to consider since we can linearize all equations in terms of perturbations. In that case, Fourier modes of δ_m evolve independently, and can be written in ΛCDM as

$$\delta_m(k,z) = D(z)\delta_m(k,z_i) \tag{87}$$

where z_i is some large initial redshift and D is called *growth factor*. The amplitude of $\delta_m(k,z_i)$ is measured well (on a certain range of scales) by the CMB at $z_i \approx 1100$. If we restrict ourselves to scales much smaller than the horizon (where most measurements of perturbations are done), then in GR the growth factor obeys

$$\ddot{D}(t) + 2H(t)\dot{D}(t) \overset{\text{GR}}{=} 4\pi G_N \bar{\rho}_m(t) D(t). \tag{88}$$

Phrased in terms of conformal time η, where $dt = ad\eta$, this becomes

$$\frac{d^2D(\eta)}{d\eta^2} + aH\frac{dD}{d\eta} \overset{\text{GR}}{=} 4\pi G_N \bar{\rho}_m(\eta) a^2 D(\eta) = \frac{3}{2}\Omega_m H_0^2 a^{-1} D(\eta). \tag{89}$$

The r.h.s. is just $\nabla^2\Psi$ divided by $\delta_m(k,z_i)$, and we have used Eq. (85). We have seen that scalar-tensor theories change the potential Ψ, i.e. the 00-component of the metric, by adding a contribution from the scalar field. In DGP, the brane-bending mode modifies $G_N \to G_N(1+1/3\beta)$ (on linear scales), thus

$$\ddot{D} + 2H\dot{D} \overset{\text{DGP}}{=} 4\pi G_N \left(1 + \frac{1}{3\beta}\right) \bar{\rho}_m(t)\, D. \tag{90}$$

For a given expansion history and initial perturbations, DGP thus predicts a different amplitude of density perturbations, for perturbations of all (sub-horizon) scales. In $f(R)$, forces are enhanced by 4/3 within the reach $\lambda_C = 1/m_{f_R}$ of the field (which in $f(R)$ is a function of scale factor due to the evolution of the background R), so that

$$\ddot{D} + 2H\dot{D} \overset{f(R)}{=} 4\pi G_N \left(\frac{4}{3} - \frac{1}{3}\frac{m_{f_R}(a)^2}{k^2/a^2 + m_{f_R}(a)^2}\right) \bar{\rho}_m(t)\, D. \tag{91}$$

Here, k/a is the physical wavenumber of the perturbation, k being comoving. This tells us that the growth factor D is now a function of k, in contrast to GR, and grows towards larger k (smaller scales).

The growth factor can in principle be measured through the clustering of large-scale structure tracers. However, those typically have an unknown bias parameter relating the amplitude of matter clustering to that of the tracer, which is degenerate with the amplitude of the growth factor. For this reason, other observables are more interesting:

- **Velocities:** the velocities of large-scale structure tracers are measurable statistically through the Doppler effect: the observed redshift is a sum of the cosmological redshift, which is a function of distance, and the Doppler shift, which is given by the line-of-sight component of the velocity. This can be measured through the distortion (*redshift distortion*) of the correlation function of tracers. Velocities of large-scale structure are determined by the gradient of the 00-component of the metric $\nabla\Psi$ (since these motions are non-relativistic). In the linear regime, they probe the derivative of the growth factor, $dD/d\ln a$. Combining this with the clustering of tracers allows us to measure $d\ln D/d\ln a$, which is directly influenced by modifications to gravity.
- **Gravitational lensing:** lensing denotes the deflection of light rays by spacetime perturbations; it is an inherently relativistic effect. Lensing is determined by the combination $\Phi+\Psi$ of the metric (i.e., $h_{00}+h_{ij}\hat{n}^i\hat{n}^j$ in our notation above). On large scales (where linear perturbation theory applies), gravitational lensing is measured through the weak statistical distortion of galaxy images.

In testing gravity, our goal is to disentangle gravity from other non-standard effects which can modify the growth of structure, such as modified initial conditions, massive neutrinos, etc.. The most promising approach to disentangling these effects is to compare the two potentials Ψ and $\Phi+\Psi$, by employing both velocity and lensing measurements. This is essentially an extension of the Solar System constraint on the PPN parameter γ, which also compares lensing (time delay and deflection) with non-relativistic dynamics (Earth's orbit).

4.2.2. *Non-Linear Probes*

The bulk of cosmological information at redshifts probed by large-scale structure is in the non-linear regime (i.e., δ_m is not much less than 1). In order to properly calculate predictions of interesting modified gravity models, we need to perform N-body simulations which solve the non-linear field equations together with the growth of structure. This is currently the forefront of research, but several interesting results have been obtained. For example, the abundance of massive galaxy clusters is very sensitive to changes in the growth factor D. Eq. (91) has been used in this way to place constraints on $f(R)$ gravity from observations [29].

One possibility to directly test gravity is the comparison of dynamical mass estimates of objects such as clusters and galaxies with lensing mass estimates. For this, we typically average over many objects to reduce the scatter in the individual dynamics and environment of objects. We can also look for signatures of the non-linear screening mechanisms. For example, the chameleon mechanism predicts a transition in the modified gravity effects as a function of the potential well (mass) of the objects considered [30].

References

[1] P. Zhang, M. Liguori, R. Bean and S. Dodelson (2007).
[2] B. Jain and P. Zhang, *ArXiv e-prints* **709** (September 2007).
[3] Y.-S. Song and K. Koyama, *ArXiv e-prints* **802** (February 2008).
[4] L. Knox, Y.-S. Song and J. A. Tyson, *Phys. Rev. D* **74**, 023512 (July 2006).
[5] F. Schmidt, *Phys. Rev. D* **78**, 043002 (August 2008).
[6] Y.-S. Song, *ArXiv Astrophysics e-prints* (February 2006).
[7] S. Tsujikawa and T. Tatekawa, *ArXiv e-prints* **804** (April 2008).
[8] P. Zhang, *Phys. Rev. D* **73**, 123504 (June 2006).
[9] Y.-S. Song, I. Sawicki and W. Hu, *Phys. Rev. D* **75**, 064003 (March 2007).
[10] F. Schmidt, M. Liguori and S. Dodelson, *Phys. Rev. D* **76**, 083518 (October 2007).
[11] J. Uzan, *ArXiv e-prints* (August 2009).
[12] T. Clifton, P. G. Ferreira, A. Padilla and C. Skordis, *Phys. Rep.* **513**, 1 (March 2012).
[13] S. Schlamminger, K.-Y. Choi, T. A. Wagner, J. H. Gundlach and E. G. Adelberger, *Physical Review Letters* **100**, p. 041101 (February 2008).
[14] A. Lue, R. Scoccimarro and G. Starkman, *Phys. Rev. D* **69**, p. 044005 (February 2004).
[15] C. M. Will, *Living Reviews in Relativity* **9**, 3 (March 2006).
[16] C. M. Will, *Theory and Experiment in Gravitational Physics* (UK: Cambridge University Press, March 1993).
[17] B. Bertotti, L. Iess and P. Tortora, *Nature* **425**, 374 (September 2003).
[18] J. Khoury and A. Weltman, *Phys. Rev. D* **69**, 044026 (February 2004).
[19] J. Khoury and A. Weltman, *Physical Review Letters* **93**, 171104 (October 2004).
[20] W. Hu and I. Sawicki, *Phys. Rev. D* **76**, 064004 (September 2007).
[21] S. M. Carroll, V. Duvvuri, M. Trodden and M. S. Turner, *Phys. Rev. D* **70**, 043528 (August 2004).
[22] G. Dvali, G. Gabadadze and M. Porrati, *Physics Letters B* **485**, 208 (July 2000).
[23] G. W. Gibbons and S. W. Hawking, *Phys. Rev. D* **15**, 2752 (May 1977).
[24] C. Deffayet, *Physics Letters B* **502**, 199 (March 2001).

[25] K. Koyama and F. P. Silva, *Phys. Rev. D* **75**, 084040 (April 2007).

[26] A. Nicolis and R. Rattazzi, *Journal of High Energy Physics* **6**, 59 (June 2004).

[27] A. I. Vainshtein, *Physics Letters B* **39**, 393 (1972).

[28] C. Deffayet, G. Dvali, G. Gabadadze and A. Vainshtein, *Phys. Rev. D* **65**, 044026 (February 2002).

[29] F. Schmidt, A. Vikhlinin and W. Hu, *Phys. Rev. D* **80**, 083505 (October 2009).

[30] F. Schmidt, *Phys. Rev. D* **81**, p. 103002 (May 2010).

Chapter 6

TASI 2012 Lectures on Inflation

Leonardo Senatore

Stanford Institute for Theoretical Physics
Department of Physics, Stanford University, Stanford, CA 94306,
Kavli Institute for Particle Astrophysics and Cosmology,
Stanford University and SLAC, Menlo Park, CA 94025,
CERN, Theory Division, 1211 Geneva 23, Switzerland
E-mail: leonardo@stanford.edu

Planning to explore the beginning of the Universe? A lightweight *guide du routard* for you.

Contents

1. Introduction

The purpose of these TASI lectures on Inflation is to introduce you to the currently preferred theory of the beginning of the universe: the theory of Inflation. This is one of the most fascinating theories in physics. Starting from the shortcomings of the standard big bang theory, we will see how a period of accelerated expansion solves these issues. We will then move on to explain how inflation can give such an accelerated expansion (**lecture 1**). We will then move on to what is the most striking prediction of inflation, which is the possibility that quantum fluctuations during this epoch are the source of the cosmological perturbations that seed galaxies and all structures in the universe (**lecture 2**). We will then try to generalize the concept of inflation to develop a more modern description of this theory. We will introduce the Effective Field Theory of Inflation. We will learn how to compute precisely the various cosmological observables, and how to simply get the physics out of the Lagrangians (**lecture 3**). Finally, in the last lecture (**lecture 4**), we will discuss one of the most important observational signatures of inflation: the possible non-Gaussianity of the primordial density perturbation. We will see how a detection of a deviation from gaussianity would let us learn about the inflationary Lagrangian and make the sky a huge particle detector. Time permitting (unfortunately there will be no time in reality), we will discuss one of the conceptually most beautiful regimes of inflation, the regime of eternal inflation, during which quantum effects become so large to change the asymptotics of the whole space-time.

A video of these lecture is available at [1]. These notes are written as a natural complement to those lectures.

Notation

$$c = \hbar = 1 \ , \qquad M_{\rm Pl}^2 = \frac{1}{8\pi G} \ . \tag{1}$$

2. Lecture 1

Intro on Inflation: how brave it was.

Notice that we will perform calculations more explicitly when they are less simple. So in this first lecture we will skip some passages. General homework of this class: fill in the gaps.

2.1. *FRW cosmology*

We begin by setting up the stage with some basic concepts in cosmology to highlight the shortcoming of the standard big bang picture. Some of this lecture has overlap with Prof. Bertshinder's lectures.

The region of universe that we see today seems to be well described by an homogenous and isotropic metric. The most general metric satisfying these symmetries can be put in the following form

$$ds^2 = -dt^2 + a(t)^2\left(\frac{dr^2}{1-kr^2} + r^2\left(d\theta^2 + \sin^2\theta d\phi^2\right)\right) \tag{2}$$

We see that this metric represents a slicing of space-time with spatial slices Σ that are simply rescaled by the scale factor a as time goes on. If $k=0$, we have a flat space, if $k=+1$, the space Σ describes a sphere, while if $k=-1$ we have an hyperbolic space. A fundamental quantity is of course the Hubble rate

$$H = \frac{\dot{a}}{a} \tag{3}$$

which has units of inverse time. It is useful for us to put the metric (2) into the following form

$$ds^2 = -dt^2 + a(t)^2\left(d\chi^2 + S_k(\chi^2)\left(d\theta^2 + \sinh^2\theta d\phi^2\right)\right) \tag{4}$$

where

$$r^2 = S_k(\chi^2) = \begin{cases} \sin^2\chi & \text{if } k=-1, \\ \chi^2 & \text{if } k=0, \\ \sin^2\chi & \text{if } k=+1. \end{cases} \tag{5}$$

χ plays the role of a radius. Let us now change coordinates in time (it is General Relativity at the end of the day!) to something called conformal time

$$\tau = \int^\tau \frac{dt}{a(t)}\,. \tag{6}$$

Now the FRW metric becomes

$$ds^2 = a(\tau)^2\left[-d\tau^2 + d\chi^2 + S_k(\chi^2)\left(d\theta^2 + \sin^2\theta d\phi^2\right)\right] \tag{7}$$

In these coordinates it is particularly easy to see the casual structure of space-time. This is determined by how light propagates on null geodesic $ds^2 = 0$. Since the space is isotropic, geodesic solutions have constant θ and ϕ. In this case we have

$$\chi(\tau) = \pm\tau + \text{const.} \tag{8}$$

These geodesics move at 45 degrees in the $\tau - \chi$ plane, as they would in Minkowski space. This is so because apart for the angular part, the metric in (7) is conformally flat: light propagates as in Minwkoski space in the coordinates $\tau - \chi$. Notice that this is not so if we had used t, the proper time for comoving (i.e. fixed FRW-slicing spatial coordinates) observers.

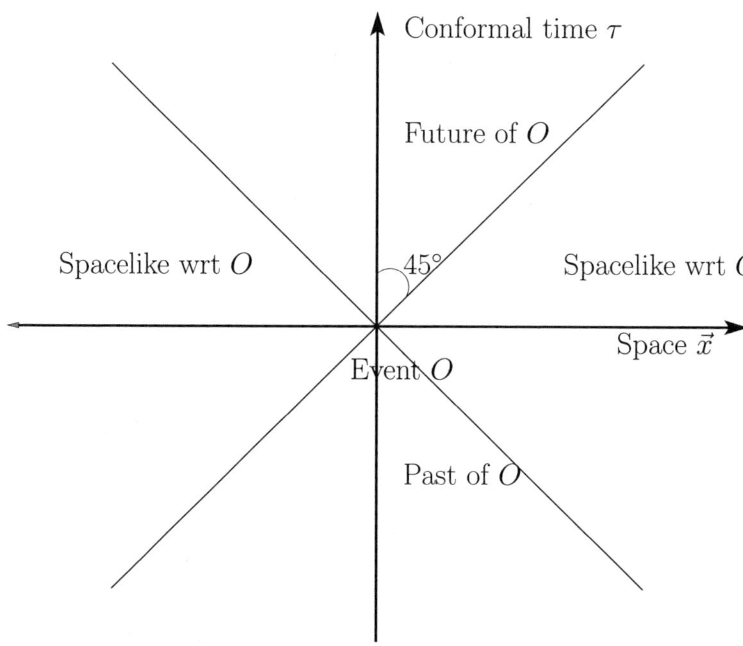

Fig. 1. Propagation of signals in the $\tau - \chi$ plane.

It is interesting to notice that is we declare that the universe started at some time t_i, then there is a maximum amount of time for light to have travelled. A point sitting at the origin of space (remember that we are in a space-translation invariant space), by the time t could have sent a signal at most to a point at coordinate χ_p given by

$$\chi_p(\tau) = \tau - \tau_i = \int_{t_i}^{t} \frac{dt}{a(t)} \ . \tag{9}$$

The difference is conformal time is equal to the maximum coordinate-separation a particle could have travelled. Notice that the geodesic distance on the spacial slice between two point one particle-horizon apart is obtained by multiplying the coordinate distance with the scale factor:

$$d_p(t) = a(\tau)\chi_p(\tau) \ . \tag{10}$$

The presence of an horizon for cosmologies that begin at some definite time will be crucial for the motivation of inflation.

It will be interesting for us to notice that there is a different kind of horizon, called event horizon. If we suppose that time ends at some point t_{end} (sometimes this t_{end} can be taken to ∞), then there is a maximum coordinate separation between two points beyond which no signal can be sent from the first point to reach the second point by the time t_{end}. This is called event horizon, and it is the kind of horizon associated to a Schwartshild black hole. From the same geodesic equation, we derive

$$\chi_e(\tau) = \tau_{end} - \tau = \int_{\tau}^{\tau_{\rm end}} \frac{dt}{a(t)} \ . \tag{11}$$

Clearly, as $\tau \to \tau_{\rm end}$, $\chi_e \to 0$.

We have seen that the casual structure of space-time depends on when space-time started and ended, and also on the value of $a(t)$ at the various times, as we have to do an integral. In order to understand how $a(t)$ evolves with time, we need to use the equations that control the *dynamics* of the metric. These are the Einstein equations

$$G_{\mu\nu} = \frac{T_{\mu\nu}}{M_{\rm Pl}^2} \ . \tag{12}$$

These in principle 10 equations reduce for an FRW metric to just two. Indeed, by the symmetries of space-time, in FRW slicing, we must have

$$T^{\mu}{}_{\nu} = \begin{pmatrix} \rho & 0 & 0 & 0 \\ 0 & -p & 0 & 0 \\ 0 & 0 & -p & 0 \\ 0 & 0 & 0 & -p \end{pmatrix} \tag{13}$$

and the Einstein equations reduce to

$$H^2 = \left(\frac{\dot{a}}{a}\right)^2 = \frac{1}{3M_{\rm Pl}^2}\rho - \frac{k}{a^2} \tag{14}$$

$$\dot{H} + H^2 = \frac{\ddot{a}}{a} = -\frac{1}{6}\left(\rho + 3p\right) \ . \tag{15}$$

The first equation is known as Friedamnn equation. These two equations can be combined to give the energy conservation equation (this follows from the Bianchi identity $0 = \nabla_\mu G^\mu_\nu = \nabla_\mu T^\mu_\nu$):

$$\frac{d\rho}{dt} + 3H(\rho + p) = 0 \ . \tag{16}$$

This is a general-relativistic generalization of energy conservation. (Homework: make sense of it by considering dilution of energy and work done by pressure.) By defining a constant equation of state w

$$p = w\rho \ , \tag{17}$$

energy conservation gives

$$\rho \propto a^{-3(1+w)} \tag{18}$$

and

$$a(t) \propto \begin{cases} t^{\frac{2}{3(1+w)}} & w \neq 1 \\ e^{Ht} & w = -1 \ . \end{cases} \tag{19}$$

Notice that indeed $\rho_{matter} \propto a^{-3}$, $\rho_{radiation} \propto a^{-4}$. Notice also that is a is power low with t to an order one power, than $H \sim 1/t$. That is, the proper time sets the scale of H at each time.

The standard big bang picture is the one in which it is hypothesized that the universe was always dominated by "normal" matter, with $w > 0$. In order to see the shortcomings of this picture, it is useful to define the present energy fractions of the various constituents of the universe. If we have various components in the universe

$$\rho = \sum_i \rho_i \ , \qquad p = \sum_i p_i \ , \quad w_i = \frac{p_i}{\rho_i}. \tag{20}$$

We can define the present energy fraction of the various components by dividing each density by the "critical density" ρ_{cr} (the density that would be required to make the universe expand with rate H_0 without the help of anything else)

$$\Omega_{i,0} = \frac{\rho_0^i}{\rho_{cr,0}} \ . \tag{21}$$

We also define

$$\Omega_{k,0} = -\frac{k}{a(t_0)^2 H_0^2} \tag{22}$$

as a measure of the relative curvature contribution. By setting as it is usually done $a(t_0) = a_0 = 1$, we can recast the Friedmann equation in the following form

$$\left(\frac{H^2}{H_0^2}\right) = \sum_i \Omega_{i,0} a^{-3(1+w_i)} + \Omega_{k,0} a^{-2} \ . \tag{23}$$

At present time we have $\sum_i \Omega_{i,0} + \Omega_{k,0} = 1$.

One can define also time dependent energy fractions

$$\Omega_i(a) = \frac{\rho_i(a)}{\rho_{cr}(a)} \ , \qquad \Omega_k(a) = -\frac{k}{a^2 H^2(a)} \ . \tag{24}$$

Notice that $\rho_{cr} = 3M_{\rm Pl}^2 H^2$ is indeed time dependent. The Friedmann equation becomes

$$\Omega_k(a) = 1 - \sum_i \Omega_i(a) \ . \tag{25}$$

2.2. *Big Bang shortcomings*

We are now going to highlight some of the shortcoming of the big bang picture that appear if we assume that its history has always been dominated by some form of matter with $w \geq 0$. We will see that upon this assumptions, we are led to very unusual initial conditions. Now, this leads us to a somewhat dangerous slope, which catches current physicists somewhat unprepared. Apart for Cosmology, Physics is usually the science that predicts the evolution of a certain given initial state. No theory is generally given for the initial state. Physicists claim that if you tell them on which state you are, they will tell you what will be your evolution (with some uncertainties). The big bang puzzles we are going to discover are about the very peculiar initial state the universe should have been at the beginning of the universe if 'normal' matter was always to dominate it. Of course, it would be nice to see that the state in which the universe happens to begin in is a natural state, in some not-well defined sense. Inflation was indeed motivated by providing an attractor towards those peculiar looking initial conditions.[a] We should keep in mind that there could be other reasons for selecting a peculiar initial state for the universe.

[a]Luckly, we will see that inflation does not do just this, but it is also a predictive theory.

2.2.1. *Flatness Problem*

Let us look back at

$$\Omega_k(a) = -\frac{k}{a^2 H^2(a)} \ , \tag{26}$$

and let us assume for simplicity that the expansion is dominated by some form of matter with equation of state equal to w. We have then $a \sim t^{\frac{2}{3(1+w)}}$ and we have

$$\dot{\Omega}_k = H\Omega_k(1+3w) \ , \qquad \frac{\partial \Omega_k}{\partial \log a} = \Omega_k(1+3w) \ . \tag{27}$$

If we assume that $w > -1/3$, then this shows that the solution $\Omega_k = 0$ is un unstable point. If $\Omega_k > 0$ at some point, Ω_k keeps growing. Viceversa, if $\Omega_k < 0$ at some point, it keeps decreasing. Of corse at most $\Omega_k = \pm 1$, in which case $w \to -1/3$ if $k < 0$, or otherwise the universe collapses if $k > 0$.

The surprising fact is that Ω_k is now observed to be smaller than about 10^{-2}: very close to zero. Given the content of matter of current universe, this mean that in the past it was even closer to zero. For example, at the BBN epoch, it has to be $|\Omega_k| \lesssim 10^{-18}$, at the Planck scale $|\Omega_k| \lesssim 10^{-63}$. In other words, since curvature redshifts as a^{-2}, it tends to dominate in the future with respect to other forms of matter (non relativistic matter redshifts as a^{-3}, radiation as a^{-4}). So, if today curvature is not already dominating, it means that it was very very very negligible in the past. The value of Ω_k at those early times represents a remarkable small number. Why at that epoch Ω_k was so small?

Of course one solution could be that $k = 0$ in the initial state of the universe. It is unknown why the universe should choose such a precise state initially, but it is nevertheless a possibility. A second alternative would be to change at some time the matter content of the universe, so that we are dominated by some matter content with $w < -1/3$. We will see that inflation provides this possibility in a very simple way.[b]

2.2.2. *Horizon Problem*

An even more dramatic shortcoming of the standard big bang picture is the horizon problem. Let us assume again that the universe is dominated by

[b]Another possibility would be to imagine the universe underwent a period of contraction, like in the bouncing cosmologies. Curvature becomes subdominant in a contracting universe.

some form of matter with equation of state w. Let us compute the particle horizon:

$$\chi_P(\tau) = \tau - \tau_i = \int_{\tau_i(t_i)}^{\tau(t)} \frac{dt'}{a(t')} = \int_{a_i}^{a} \frac{da}{Ha^2} \sim a^{(1+3w)/2} - a_i^{(1+3w)/2} \,. \quad (28)$$

We notice that if $w > -1/3$ (notice, the same $-1/3$ as in the flatness problem), then in an expanding universe the horizon grows with time. This is very bad. It means that at every instant of time, new regions that had never been in causal contact before come into contact for the first time. This means that they should look like very different from one another (unless the universe did not decide to start in a homogenous state). But if we look around us, the universe seems to be homogenous on scales that came into causal contact only very recently. Well, maybe they simply equilibrate very fast? Even if this unlikely possibility were to be true, we can make the problem even sharper when we look at the CMB. In this case we can take a snapshot of casually disconnected regions (at the time at which they were still disconnected), and we see that they look like the same. This is the horizon problem.

Notice that if $w > -1/3$ the particle horizon is dominated by late times, and so we can take $a_i \simeq 0$ in its expression. In this way we have that the current physical horizon is

$$d_p \sim a\tau \sim t \sim \frac{1}{H} \,. \quad (29)$$

For this kind of cosmologies where $w > -1/3$ at all times, the Hubble length is of order of the horizon. This is what has led the community to often use the ill-fated name "horizon" for "Hubble". "Hubble is the horizon" is parametrically true only for standard cosmologies, it is not true in general. We will try to avoid calling Hubble as the horizon in all of these lectures, even though sometimes habit will take a toll.

Notice however that the horizon problem goes away if we assume the universe sit there for a while at the singularity.

Let us look again at the CMB. Naive Horizon scale is one degree ($l \sim 200$), and fluctuations are very small on larger scales. How was that possible?

Apart for postulating an ad hoc initial state, we would need also to include those perturbations in the initial state... This is getting crazy! (though in principle possible) We will see that inflation will provide an attractive solution.

The problem of the CMB large scale fluctuations is a problem as hard as the horizon one.

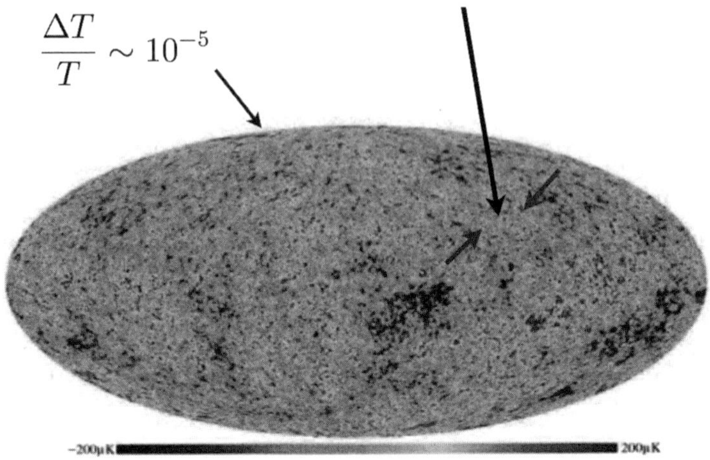

Fig. 2. The naive horizon H^{-1} at the time of recombination (among the two short arrows), is much smaller than the scale over which we see statistical homogeneity.

2.2.3. *Solving these Problems: Conditions*

In order to solve these two problems, we need to have some form of energy with $w < -1/3$. We can say it somewhat differently, by noticing that in order for Ω_k to decrease with time, since

$$\Omega_k = -\frac{k}{(aH)^2} \tag{30}$$

we want an epoch of the universe in which aH increases with time. Equivalently, $1/(aH)$ decreases with time. $1/(aH)$ is sometimes called "comoving Horizon", ... a really bad name in my humble opinion. You can notice that since $1/H$ is the particle horizon in standard cosmologies, $1/(aH)$ identifies the comoving coordinate distance between two points one naive-Horizon apart. If this decreases with time, then one creates a separation between the true particle horizon, and the naive particle horizon. Two points that naively are separated by a $1/(aH)$ comoving distantce are no more separated by a particle horizon. Even more simply, the formula for the particle horizon reads

$$\tau = \int_{t_i}^{t} \frac{da}{(aH)^2} \; . \tag{31}$$

If $(aH)^{-1}$ is large in the past, then the integral is dominated by the past, and the actual size of the horizon has nothing to do with present time quantities such as the Hubble scale at present. In standard cosmologies the opposite was happening: the integral was dominated by late times.

Let us formulate the condition for $(aH)^{-1}$ to decrease with time in equivalent forms.

- Accelerate expansion: it looks like that this condition implies that the universe must be accelerating in that epoch:

$$\frac{\partial \frac{1}{(aH)}}{\partial t} < 0 \quad \Rightarrow \quad \ddot{a} > 0 \; . \tag{32}$$

  This implies that $k/(aH)$ decreases: physical wavelengths become longer than H^{-1}.
- As we stressed, this should imply $w < -1/3$. Let us verify it. From Friedman equation

$$0 < \ddot{a} = -\frac{a}{6}(\rho + 3p) = \frac{a\rho}{6}(1 + 3w) \quad \Rightarrow \quad w < -1/3 \quad \text{if } \rho > 0 \; . \tag{33}$$

Inflation, in its most essential definition, is the postulation of a phase with $w < -1/3$ in the past of our universe.[c]

Is it possible to see more physically what is going on? In a standard cosmology, the scale factor goes to zero at finite conformal time. For $w > -1/3$, we have that

$$a \sim \tau^{2/(1+3w)} \tag{34}$$

implying the existence of a singularity $a \to 0, H \to \infty$ as $\tau \to 0$. This is why we had to stop there. This is the big bang moment in standard cosmology. This however implies that there is a beginning of time, and that the particle horizon is order τ. This is the source of the problems we discussed about.

However, if we have a phase in which $w < -1/3$, then the singularity in the past is pushed way further back, and the actual universe is much

[c]If there is only one field involved, than scale invariance of the perturbations and the requirement that the solution is an attractor forces $w \simeq -1$. This is a theorem [8].

longer than what τ indicates. For example, for inflation $H \sim$ const. and $a(\tau) = -\frac{1}{H\tau}$, with $\tau \in [-\infty, \tau_{end}]$, $\tau_{end} \leq 0$. In general τ can be extended to negative times, in this way making the horizon much larger than $1/H$.

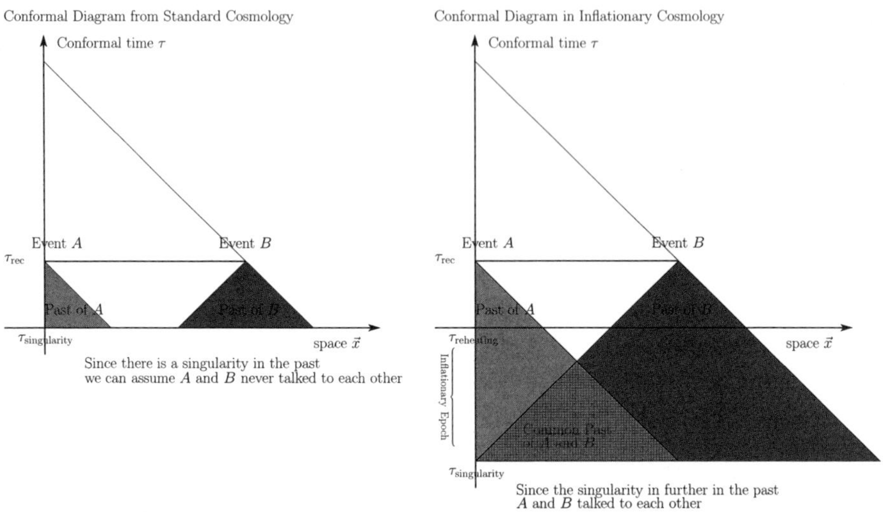

Fig. 3. How inflation solve the horizon problem: in the past, there is much more time than what there would have naively been without inflation.

2.3. *The theory of inflation*

Inflation is indeed a period of the history of the universe that is postulated to have happened before the standard big bang history. Direct observation of BBN products tell us that the universe was radiation dominated at $t \sim 1 - 100$ sec, which strongly suggests that inflation had to happen at least earlier than this. More specifically, inflation is supposed to be a period dominated by a form of energy with $w \simeq -1$, or equivalently $H \simeq$ const. How can this be achieved by some physical means? (See Fig. 3)

2.3.1. *Simplest Example*

The simplest example of a system capable of driving a period of inflation is a scalar field on top a rather flat potential. These kinds of models are called "slow roll inflation" and were the ones initially discovered to drive

inflation. Let us look at this

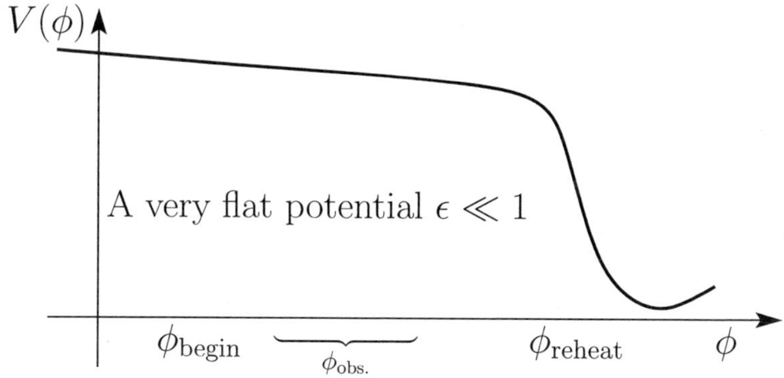

Fig. 4. A simple inflationary model.

The scalar field plus gravity has the following action

$$S = \int d^4x\sqrt{-g}\left[\frac{M_{\text{Pl}}^2}{2}R + \frac{1}{2}g^{\mu\nu}\partial_\mu\phi\partial_\nu\phi - V(\phi)\right] . \tag{35}$$

The first term is the Einstein Hilbert term of General Relativity (GR). The second and third terms represent the action of a scalar field S_ϕ. The idea of inflation is to fill a small region of the initial universe with an homogeneously distributed scalar field sitting on top of its potential $V(\phi)$. Let us see what happens, by looking at the evolution of the space-time. We need the scalar field stress tensor:

$$T_{\mu\nu}^{(\phi)} = -\frac{2}{\sqrt{-g}}\frac{\delta S_\phi}{\delta g^{\mu\nu}} = \partial_\mu\phi\partial_\nu\phi - g_{\mu\nu}\left(\frac{1}{2}\partial_\rho\phi\partial^\rho\phi + V(\phi)\right) . \tag{36}$$

For an homogenous field configuration, this leads to the following energy density and pressure

$$\rho_\phi = \frac{1}{2}\dot{\phi}^2 + V(\phi) \quad \text{obviously} \tag{37}$$

$$p_\phi = \frac{1}{2}\dot{\phi}^2 - V(\phi) \quad \text{notice the sign of } V . \tag{38}$$

Therefore the equation of state is

$$w_\phi = \frac{p_\phi}{\rho_\phi} = \frac{\frac{1}{2}\dot{\phi}^2 - V(\phi)}{\frac{1}{2}\dot{\phi}^2 + V(\phi)} . \tag{39}$$

We see that if the potential energy dominates over the kinetic energy, we have

$$\dot{\phi}^2 \ll V(\phi) \quad \Rightarrow \quad w_\phi \simeq -1 < -\frac{1}{3} \tag{40}$$

as we wished. Notice that this means that

$$\epsilon = -\frac{\dot{H}}{H^2} \sim \frac{\dot{\phi}^2}{V} \ll 1 \, . \tag{41}$$

The equation of motion for the scalar field is

$$\frac{\delta S}{\delta \phi} = \frac{1}{\sqrt{-g}} \partial_\mu (\sqrt{-g} \partial^\mu \phi) + V_{,\phi} = 0 \quad \Rightarrow \quad \ddot{\phi} + 3H\dot{\phi} + V_{,\phi} = 0 \, . \tag{42}$$

This equation of motion is the same as the one of a particle rolling down its potential. This particle is subject to friction though the $H\dot{\phi}$ term. Like for a particle trajectory, this means that the solution where $\dot{\phi} \simeq V_\phi/(3H)$ is an attractor 'slow-roll' solution if friction is large enough. Being on this trajectory requires

$$\eta = -\frac{\ddot{\phi}}{H\dot{\phi}} \ll 1 \, . \tag{43}$$

We have therefore found two "slow roll parameters":

$$\epsilon = -\frac{\dot{H}}{H^2} \ll 1 \, , \qquad \eta = -\frac{\ddot{\phi}}{H\dot{\phi}} \ll 1 \, . \tag{44}$$

The first parameters being much smaller than one means that we are on a background solution where the Hubble rate changes very slowly with time. The second parameter means that we are on an attractor solution (so that the actual solution does not depend much from the initial conditions), and also that this phase of accelerated expansion ($w \simeq -1$, $a \sim$ Exp(Ht)) will last for a long time. Indeed, one can check that

$$\frac{\dot{\epsilon}}{H\epsilon} \sim \mathcal{O}(\epsilon, \eta) \, . \tag{45}$$

We will see that the smallness of η is really forced on us by the scale invariance of the cosmological perturbations.

Once we assume we are on the slow roll solution, then we can express them in terms of the potential terms. We have

$$\epsilon \simeq \frac{M_{\rm Pl}^2}{2} \left(\frac{V_{,\phi}}{V} \right)^2 \, , \qquad \eta \simeq M_{\rm Pl}^2 \frac{V_{,\phi\phi}}{V} - \frac{M_{\rm Pl}^2}{2} \left(\frac{V_{,\phi}}{V} \right)^2 \, . \tag{46}$$

On this solution we also have

$$\dot{\phi} \simeq \frac{V_{,\phi}}{3H} , \qquad H \simeq \frac{V(\phi)}{3M_{\rm Pl}^2} \simeq {\rm const} , \qquad a \sim e^{3Ht} . \tag{47}$$

When does inflation end? By definition, inflation ends when w ceases to be close to -1. This means that

$$\epsilon \sim \eta \sim 1 . \tag{48}$$

More concretely, we see that the field that starts on top of his potential will slowly roll down until two things will happen: Hubble will decrease, providing less friction, and the potential will become too steep to guaranteed that the kinetic energy is negligible with respect to potential energy. We call the point in field space where this happens ϕ_{end}. At that point, a period dominated by a form of energy with $w > -1/3$ is expected to begin. We will come back in a second on it.

Duration of inflation: For the moment, let us see how long inflation needs to last. The number of e-foldings of inflation is defined as the logarithm of the ratio of the scale factor at the end of inflation and at the beginning of inflation. For a generic initial point ϕ, we have

$$N^{to\ end}(\phi) = \log\left(\frac{a_{end}}{a}\right) \simeq \int_t^{t_{end}} H dt = \int_\phi^{\phi_{end}} \frac{H}{\dot{\phi}} d\phi \simeq \int_{\phi_{end}}^{\phi} \frac{V}{V_{,\phi}} d\phi , \tag{49}$$

where in the third passage we have used that $a \sim e^{Ht}$, and in the last passage we have used the slow roll solutions.

The horizon and flatness problems are solved in inflation very simply. During inflation

$$\Omega_k = -\frac{k}{a^2H^2} \propto \frac{1}{a^2} \to 0 . \tag{50}$$

So, if we start with $\Omega_k \sim 1$ at the onset of inflation, and we wish to explain why $\Omega_k(a_{BBN}) \sim 10^{-18}$, we need about 20 e-foldings of inflation. This is so because at the end of inflation we have

$$\Omega_k(a_{end}) \simeq \Omega_k(a_{in})\frac{a_{in}^2}{a_{end}^2} \sim \frac{a_{in}^2}{a_{end}^2} = 2\log N \tag{51}$$

and this must be equal to the curvature we expect at the beginning of the FRW phase (that we can assume to be equal to the end of inflation)

$$\Omega_k(a_{end}) = \Omega_k(a_0)\frac{a_{end}^2H_I^2}{a_0^2H_0^2} \sim 10^{-2}\frac{a_{end}^2H_I^2}{a_0^2H_0^2} \quad \Rightarrow \quad N = \log\left(\frac{a_{end}H_I}{a_0H_0}\right) . \tag{52}$$

In this case however we would need the hot-big-bang period to be start after inflation directly with BBN-like temperatures. If the universe started at higher temperatures, say the GUT scale, we would need about 60 e-foldings of inflation. So, you see that the required number of e-foldings depends on the starting temperature of the universe, but we are in the realm of several tens.

The horizon problem is solved by asking that the region we see in the CMB was well inside the horizon. Since the contribution to the particle horizon from the radiation and the matter dominated eras is too small to account for the isotropy of the CMB, we can can assume that the integral that defines the particle horizon is dominated by the period of inflation. If t_L is the time of the last scattering surface, we have

$$d_p = a(t_L)\int_{t_in}^{t_{end}} \frac{dt}{a(t)} \simeq \frac{a(t_L)}{a_{end}H_I}e^N \ , \tag{53}$$

where we have used that $a(t) = a(t_{end})e^{H_I(t_{end}-t)}$. The particle horizon has to be bigger than the region that we can see now of the CMB. This is given by the angular diameter distance of the CMB last scattering surface. It is simply the physical distance between two points that now are one Hubble radius far apart, at the time t_L:

$$d_L = \frac{a(t_L)}{H_0 a_0} \ . \tag{54}$$

To solve the horizon problem we need

$$d_p \gtrsim d_L \quad \Rightarrow \quad N \gtrsim \log\left(\frac{a_I H_I}{a_0 H_0}\right) \ . \tag{55}$$

This is the same number as we need to solve the flatness problem, so we find the same number of e-foldings is needed to solve the horizon problem as are necessary to solve the flatness problems.

2.4. *Reheating*

But we still miss a piece of the story. How inflation ends? So far, we have simply seen that as $\epsilon \sim 1$ the accelerated phase stops. At this point, typically the inflaton begins to oscillate around the bottom of the potential. In this regime it drives the universe as if it were dominated by non-relativistic matter. The equation for the inflation indeed reads

$$\frac{\partial \rho_\phi}{\partial t} + (3H + \Gamma)\,\rho_\phi = 0 \tag{56}$$

(Homework: derive this expression). For $\Gamma = 0$, this is the dilution equation for non-relativistic matter. Γ represents the inflation decay rate. Indeed, in this period of time the inflation is supposed to decay into other particles. These thermalize and, once the inflation has decayed enough, start dominating the universe. This is the start of the standard big-bang universe.

2.5. *Simplest models of inflation*

2.5.1. *Large Field Inflation*

The simplest versions of inflation are based on scalar fields slowly rolling down their potential. These typically fall into two categories: large fields and small fields. Large field models are those characterized by a potential of the form

$$V(\phi) = \frac{\phi^\alpha}{M^{\alpha-4}}. \tag{57}$$

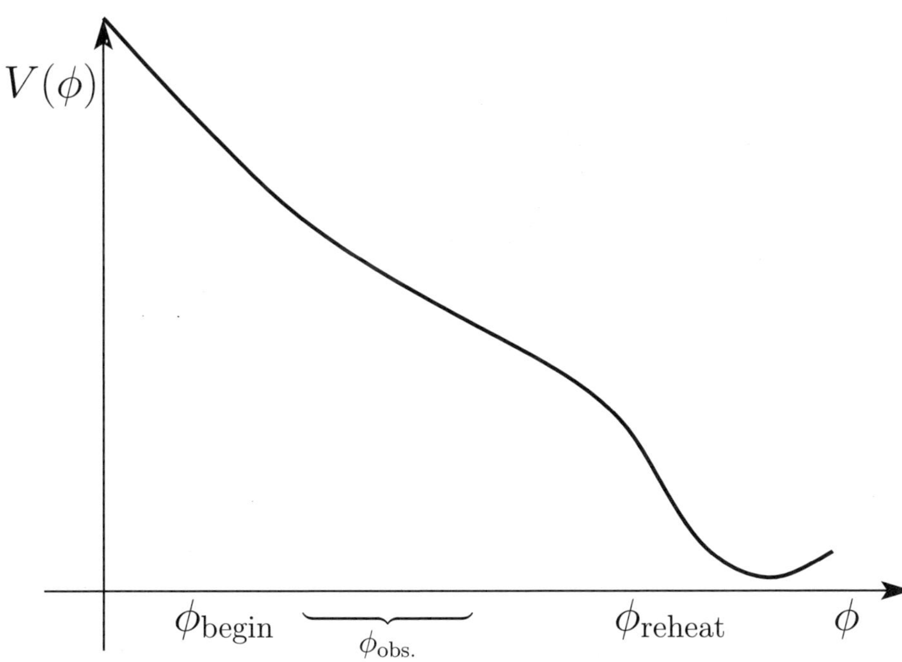

Fig. 5. A "large-field" inflationary model.

For any M and α, if we put the scalar field high enough, we can have an inflationary solution. Let us see how this happens by imposing the slow roll conditions.

$$\epsilon \sim M_{\rm Pl}^2 \left(\frac{V_{,\phi}}{V}\right)^2 \sim \alpha^2 \frac{M_{\rm Pl}^2}{\phi^2} \; . \tag{58}$$

For $\alpha \sim 1$, we have

$$\epsilon \ll 1 \qquad \Rightarrow \qquad \phi \gg M_{\rm Pl} \; . \tag{59}$$

The field vev has to be super planckian. Further, notice that the field travels an amount of order

$$\Delta\phi = \int_{\phi_{in}}^{\phi_{end}} d\phi = \int_{t_{in}}^{t_{end}} \dot{\phi} dt \simeq \frac{\dot{\phi}}{H} \int_{Ht_{in}}^{Ht_{end}} d(Ht) = \frac{\dot{\phi}}{H} N_e \sim \epsilon^{1/2} N_e \, M_{\rm Pl} \tag{60}$$

For $\epsilon \sim 1/N_e$ and not too small, the field excursion is of order $M_{\rm Pl}$. This is a pretty large field excursion (this explains the name large field models). But notice that in principle there is absolutely nothing bad about this. The energy density of the field is of order $\phi^\alpha/M^{\alpha-4} \sim \left(\frac{M_{\rm Pl}}{M}\right)^\alpha M^4$ and needs to be smaller than $M_{\rm Pl}^4$ for us to be able to trust general relativity and the semiclassical description of space-time. This is realized once $M \gg M_{\rm Pl}$ (for $\alpha = 4$ we have $V = \lambda\phi^4$ and we simply require $\lambda \ll 1$). So far so good from the field theory point of view. Now, ideally some of us would like to embed inflationary theories in UV complete theories of gravity such as string theory. In this case the UV complete model need to be able to control all $M_{\rm Pl}$ suppressed operators. This is possible, though sometimes challenging, depending on the scenario considered. This is a lively line of research.

2.5.2. *Small Field Inflation*

From (60) we see that if we wish to have a $\Delta\phi \ll M_{\rm Pl}$, we need to have ϵ very very small. This is possible to achieve in models of the form

$$V(\phi) = V_0 \left(1 - \left(\frac{\phi}{M}\right)^2\right) \; . \tag{61}$$

In this case, we have

$$\epsilon \simeq \frac{M_{\rm Pl}^2 \phi^2}{M^4} \tag{62}$$

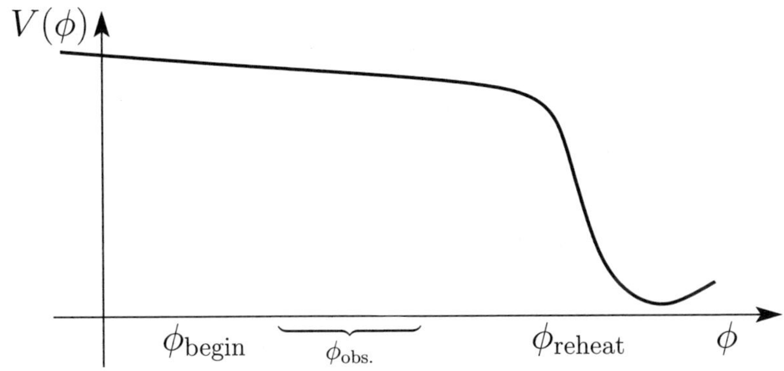

Fig. 6. A "small-field" inflationary model.

that becomes smaller and smaller as we send $\phi \to 0$. Of course, we need to guarantee a long enough duration of inflation, which means that $\phi \sim \Delta\phi \sim \epsilon^{1/2} M_{\rm Pl} N_e$. Both conditions are satisfied by taking $M \gtrsim M_{\rm Pl} N_e$.

2.5.3. *Generalizations*

Over the thirty years since the discovery of the first inflationary models, there have been a very large number of generalizations. From fields with a non-trivial kinetic terms, such as DBI inflation and Ghost Inflation, to theories with multiple fields or with dissipative effects. We will come back to these models later, when we will offer a unified description.

2.6. *Summary of Lecture 1*

- Standard Big Bang Cosmology has an horizon and a flatness problem. Plus, who created the density fluctuations in the CMB?
- A period of early acceleration solves the horizon and flatness problems.
- Inflation, here for the moment presented in the simplest form of a scalar field rolling downs a flat potential, solves them.

3. Lecture 2: Generation of Density Perturbations

This is the most **exciting, fascinating and predicting part**. It is the most predicting part, because we will see that this is what makes inflation predictive. While the former cosmological shortcomings that we saw so far were what motivated scientists such as Guth to look for inflation, cosmological perturbations became part of the story well after inflation was formulated. The fact that inflation could source primordial perturbations was indeed realized only shortly after the formulation of inflation. At that time, CMB perturbations were not yet observed, but the fact that we observed galaxies today, and the fact that matter grows as $\delta \propto a$ in a matter dominated universe predicted that some perturbations had to exist on the CMB. The way inflation produces these perturbations is both exciting and beautiful. It is simply beautiful because it shows that quantum effects, that are usually relegated to the hardly experiencable world of the small distances, can be exponentiated in the peculiar inflationary space-time to become actually the source of all the cosmological perturbations, and ultimately of the galaxies and of all the structures that are present in our universe. With inflation, quantum effects are at the basis of the formation of the largest structures in the universe. This part is also when inflation becomes more intellectually exciting. We will see that there is a very interesting quantum field theory that happens when we put some field theory in a accelerating space-time. And this is not just for fun, it makes predictions that we are actually testing right now in the universe!

The calculation of the primordial density perturbations can be quite complicated. Historically, it has taken some time to outstrip the description of all the irrelevant parts and make the story simple. This is typical of all parts of science and of all discoveries. Therefore, I will give you what I consider the simplest and most elegant derivation. Even with this, the calculation is quite complicated. Therefore we will first see how we can estimate the most important characteristics of the perturbations without doing any calculations. Only later, we will do the rigorous, and now simple, calculation.[d]

3.1. *Simple Derivation: Real space*

In this simple derivation we will drop all numerical factors. We will concentrate on the physics.

Let us expand the field around the background solution. Since the world is quantum mechanical, if the lowest energy state is not an eigenstate of the field operator $\hat{\phi}|0\rangle \neq \phi|0\rangle$, then

$$\phi = \phi_0(t) + \delta\phi(\vec{x}, t) \tag{63}$$

[d]General lesson I think I have learned from my teachers: always know the answer you have to get before starting a difficult calculation.

Notice that if we change coordinates

$$x^\mu \to x'^\mu = x^\mu + \xi^\mu \tag{64}$$

then

$$\delta\phi(\vec{x},t) \to \widetilde{\delta\phi}(\vec{x}'^\mu) - \dot{\phi}_0(t)\xi^0 \tag{65}$$

$\delta\phi$ does not transform as a scalar, it shifts under time diffeomorphisms (diffs.). The actual definition of $\delta\phi$ *depends* on the coordinates chosen. This has been the problem that has terrified the community for a long time, and made the treatment of perturbations in inflation very complicated.[e] Instead, we will simply ignore this subtlety, as it is highly irrelevant. Indeed, we are talking about a scalar field, very much like the Higgs field. When we study the Higgs field we do not bother about specifying the coordinates.

So why we should do it now? We do not even bother of writing down the metric perturbations, so why we should do it now? We will later justify why this is actually possible in more rigorous terms. Let us therefore proceed, and expand the action for the scalar field at quadratic order in an unperturbed FRW metric:

$$\begin{aligned} S &= \int d^4x a^3 \left[\mathcal{L}_0 + \left.\frac{\delta\mathcal{L}}{\delta\phi}\right|_0 \delta\phi + \left.\frac{\delta^2\mathcal{L}}{\delta\phi^2}\right|_0 \delta\phi^2\right] \\ &= S_0 + \int d^4x e^{3Ht}\left[-g^{\mu\nu}\partial_\mu\delta\phi\partial_\nu\phi\right] \ . \end{aligned} \tag{66}$$

Notice that the term linear in $\delta\phi$ is called the tadpole term, and if we expand around the solution of the background equations $\delta S/\delta\phi|_0 = 0$ it vanishes. We have used that $\sqrt{-g} = a^3 = e^{3Ht}$. The action contains simply a kinetic term for the inflation. The potential terms are very small, because the potential is very flat, so that we can neglect it.

- Let us concentrate on very small wavelengths (high-frequencies). $\omega \gg H$. $\Delta\vec{x} \ll H^{-1}$ (see Fig. 7). In that regime, we can clearly neglect the expansion of the universe, as we do when we do LHC physics (this is nothing but the equivalence principle at work: at distances much shorter than the curvature of the universe we live in flat space). We are like in Minkowski space, and therefore

$$\langle\delta\phi(\vec{x},t)\delta\phi(\vec{x}',t)\rangle_{vac.} \sim \text{ something} \sim [\text{length}]^{-2} \ , \tag{67}$$

[e]Of course, at the beginning things were new, and it was very justified not to get things immediately in the simplest way.

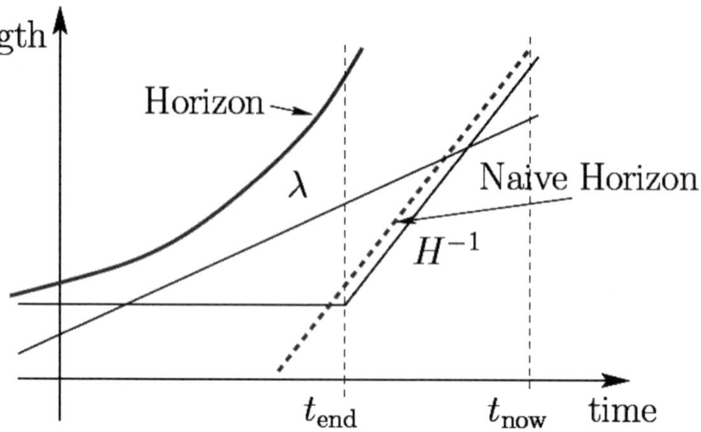

Fig. 7. Relative ratios of important length scales as a function of time in the inflationary universe. Modes start shorter than H^{-1} during inflation and become longer than H^{-1} during inflation.

just by dimensional analysis. Since there is no length scale or mass scale in the Lagrangian (remember that H is negligible), then the only length in the system is $\Delta\vec{x}$. We have

$$\langle \delta\phi(\vec{x},t)\delta\phi(\vec{x}',t)\rangle_{vac.} \sim \frac{1}{|\Delta\vec{x}|^2} \,. \tag{68}$$

Notice that the two point function decreases as we increase the distance between the two points: this is why usually quantum mechanics is segregated to small distances.

- But the universe is slowly expanding wrt $1/|\Delta\vec{x}|$, so the physical distance between to comoving points grows (slowly) with a:

$$|\Delta\vec{x}| \to |\Delta\vec{x}(t)| \propto a(t) \quad \Rightarrow \quad \langle \delta\phi(\vec{x},t)\delta\phi(\vec{x}',t)\rangle_{vac.} \sim \frac{1}{|\Delta\vec{x}|^2(t)} \tag{69}$$

- Since H is constant (it would be enough for the universe to be accelerating), at some point we will have

$$|\Delta\vec{x}|(t) \sim H^{-1} \tag{70}$$

and keeps growing. At this point, the Hubble expansion is clearly not a slow time scale for the system, it is actually very important. In particular, if two points are one Hubble far apart, then we have[f]

$$v_{\rm relative} \gtrsim v_{\rm light} \ . \tag{71}$$

Notice that this is not in contradiction with the principle of relativity: the two points simply stop communicating. But then gradients are irrelevant, and the value of ϕ and $\vec{x}$ should be unaffected by the value of ϕ at $\vec{x}'$. Since any value of $\delta\phi$ is as good as the others (if you look at the action, there is no potential term that gives difference in energy to different values of $\delta\phi$). The two point function stops decreasing and becomes constant

$$\langle\delta\phi(\vec{x},t)\delta\phi(\vec{x}',t)\rangle_{vac.} \sim \frac{1}{|\Delta\vec{x}|^2 = H^{-2}} \sim H^2 \quad \text{as} \quad \Delta\vec{x}\to\infty \ . \tag{72}$$

So, we see that the two point function stops decreasing and as $\Delta\vec{x}$ becomes larger than H^{-1}, it remains basically constant of order H^2. This means that there is no scale in the two point functions, once the distance is larger than H^{-1}. An example of a scale dependent two point function that we could have found could have been: $\langle\delta\phi(\vec{x},t)\delta\phi(\vec{x}',t)\rangle \sim H^4|\vec{x}|^2$. This does not happen here, and we have a scale invariant spectrum.

3.2. *Simple Derivation: Fourier space*

Let us see at the same derivation, working this time in Fourier space. The action reads

$$S = \int d^4x e^{3Ht}\left[-g^{\mu\nu}\partial_\mu\delta\phi\partial_\nu\delta\phi\right] = \int dtd^3k\; a^3\left(\dot{\delta\phi}_{\vec{k}}\dot{\delta\phi}_{-\vec{k}} - \frac{k^2}{a^2}\delta\phi_{\vec{k}}\delta\phi_{-\vec{k}}\right) \ . \tag{73}$$

- Each Fourier mode evolves independently. This is a quadratic Lagrangian!
- Each Fourier mode represents a quantum mechanical harmonic oscillator (apart for the overall factor of a^3), with a time-dependent frequency

$$\omega(t) \sim \frac{k}{a(t)} \ . \tag{74}$$

The canonically normalized harmonic oscillator is $\delta\phi_{can} \sim a^{-3/2}\delta\phi$

[f]Very roughly speaking. In more rigorous terms, one point is beyond the apparent event horizon of the other.

- Let us focus on one Fourier mode. At sufficiently early times, we have

$$\omega(t) \simeq \frac{k}{a} \gg H \ . \tag{75}$$

In this regime, as before, we can neglect the expansion of the universe and therefore any time dependence. Then we are as if we were in Minkowski space, and therefore we must have, for a canonically normalized scalar field (i.e. harmonic oscillator)

$$\langle \delta\phi^2_{can,k} \rangle \sim \frac{1}{\omega(t)} \qquad \Rightarrow \qquad \langle \delta\phi^2_k \rangle \sim \frac{1}{a^3} \cdot \frac{1}{\omega(t)} \ . \tag{76}$$

- While $\omega \gg H$, ω slowly decreases with time $\dot{\omega}/\omega \sim H \ll \omega$, so the two point function follows adiabatially the value on the vacuum. This happens until $\omega \sim H$ and ultimately $\omega \ll H$. At this transition, called freeze-out, the adiabatic approximation breaks down. What happens is that no more evolution is possible, because the two points are further away than an Hubble scale, and so they are beyond the event horizon. Equivalently the harmonic oscillator now has an overdamping friction term $\ddot{\delta\phi}_{\vec{k}} + 3H\dot{\delta\phi}_{\vec{k}} = 0$ that now is relevant. Since this happen when

$$\omega \sim \frac{k}{a(t_{f.o.})} \sim H \qquad \Rightarrow \qquad a_{f.o.} \sim \frac{k}{H} \tag{77}$$

where $f.o.$ stray for freeze-out. By substituting in the two point function, we obtain

$$\langle \delta\phi^2_k \rangle \sim \frac{1}{a^3_{f.o.}} \cdot \frac{1}{\omega(t_{f.o.})} \sim \frac{H^2}{k^3} \tag{78}$$

This is how a scale invariant two-point function spectrum looks like in Fourier space. It is so because in Fourier space the phase space goes as $d^3k \sim k^3$, so, if the power spectrum goes as $1/k^3$, we have that each logarithmic interval in k-space contributes equally to the two-point function in real space. In formulae

$$\langle\, \delta\phi(\vec{x})^2 \big|^{E_2}_{E_1} \rangle \sim \int_{E_1}^{E_2} d^3k \langle \delta\phi^2_k \rangle \sim H^2 \log\left(\frac{E_1}{E_2}\right) \tag{79}$$

This is simply beautiful, at least in my opinion. In Minkowski space quantum mechanics is segregated to small distances because

$$\langle \delta\phi(\vec{x},t)\delta\phi(\vec{x}',t) \rangle_{vac.} \sim \frac{1}{|\Delta\vec{x}|^2} \tag{80}$$

In an inflationary space-time (it locally looks like a de Sitter space, but, contrary to de Sitter space, it ends), we have that on very large distances

$$\langle\delta\phi(\vec{x},t)\delta\phi(\vec{x}',t)\rangle_{vac.} \sim H^2 \gg \frac{1}{|\Delta\vec{x}|^2} \qquad \text{for} \qquad \Delta\vec{x} \gg H^{-1} \tag{81}$$

At a given large distance, quantum effects are much larger than what they would have naively been in Minkowski space, and this by a huge amount once we consider that in inflation scales are stretched out of the horizon by a factor of order e^{60}.

Since we are all physicists here, we can say that this is a remarkable story for the universe.

Further, it tells us that trough this mechanisms, by exploring cosmological perturbations we are studying quantum mechanics, and so fundamental physics.

But still, we need to make contact with observations.

3.3. *Contact with observation: Part 1*

In the former subsection we have seen that the scalar field develops a large scale-invariant two-point function at scales longer than Hubble during inflation. How these become the density perturbations that we see in the CMB and then grow to become the galaxies?

Let us look at what happens during inflation. Let us take a box full of inflation up in the potential, and let inflation happen. In each point in space, the inflaton will roll down the potential and inflaton will end when the inflation at each location will reach a point $\phi(\vec{x},t_{end}) = \phi_{end}$. We can therefore draw a surface of constant field $\phi = \phi_{end}$. Reheating will start, and in every point in space reheating will happen in the same way: the only thing that changes between the various points is the value of the gradient of the fields, but for the modes we are interested in, these are much much longer than the horizon, and so gradients are negligible; also the velocity of the field matters, but since we are on an attractor solution, we have the same velocity everywhere. At this point there is no difference between the various points, and so reheating will happen in the same way in every location. In the approximation in which re-heating happens instantaneously, the surfaces $\phi = \phi_{end}$ are equal temperature surfaces (if reheating is not instantaneous, then the equal temperature surface will be displaced later, but nothing will change really in the conclusions), and so equal energy density surfaces. Now, is this surface an equal time surface? In the limit in which there no quantum fluctuations for the scalar field, it would be so, but quantum fluctuations make it perturbed. How a quantum fluctuation will affect the duration of inflation at each point? Well, a jump $\delta\phi$ will move the inflaton towards or far away from the end of inflation. This means that the duration of inflation in a given location will be perturbed, and consequently the overall expansion of the universe when $\phi = \phi_{end}$ will be different. We therefore have a $\phi = \phi_{end}$ surface which locally looks like an unperturbed universe, the only difference is that the have a difference local scale factor.[g] These are the curvature perturbations that we call ζ. In formulas

$$\delta\phi \Rightarrow \delta t_{inflation} \sim \frac{\delta\phi}{\dot{\phi}} \Rightarrow \delta\text{expansion} \sim \zeta \sim \frac{\delta a}{a} \sim H\delta t_{inflation} \sim \frac{H}{\dot{\phi}}\delta\phi \,. \tag{82}$$

Here we defined in an approximate way $\zeta \sim \delta a/a$. We will define it rigorously later on. So, the power spectrum of the curvature perturbation is given by

$$\langle\zeta_{\vec{k}}\zeta_{\vec{k}'}\rangle = \frac{H^2}{\dot{\phi}^2}\langle\phi_{\vec{k}}\phi_{\vec{k}'}\rangle = (2\pi)^3\delta^3(\vec{k}+\vec{k}')\frac{H^4}{\dot{\phi}^2}\cdot\frac{1}{k^3} \equiv (2\pi)^3\delta^3(\vec{k}+\vec{k}')P_\zeta \tag{83}$$

$$P_\zeta = \frac{H^4}{\dot{\phi}^2}\cdot\frac{1}{k^3} \simeq \frac{H^2}{M_{\rm Pl}^2\epsilon} \tag{84}$$

where in the second passage we have used the slow roll expressions.

It is the time-delay, stupid!.[h] It is important to realize that the leading mechanism through which inflation generates perturbations is by the time delay induced by the inflation fluctuations, not by the fluctuations in energy during inflation. It took some time for the community to realise this. Let us be sure about this. In slow roll inflation the potential needs to be very flat, we can therefore work by expanding in the smallness of the slow roll parameters. How large are the metric perturbations? Well, the difference in energy associated to a jump of the inflation is about

$$\delta\rho \sim V'\delta\phi \sim \sqrt{\epsilon}H^3M_{\rm Pl} \quad \Rightarrow \quad \delta g^{\mu\nu} \sim \frac{\delta\rho}{\rho} \sim \sqrt{\epsilon}\frac{H}{M_{\rm Pl}} \,. \tag{85}$$

This means that the curvature perturbation due to this effects has actually

[g]Notice that since this surface has the same energy but different overall expansion: by GR, there are must a curvature for space.

[h]No offense to anybody: this is just a famous quote from Bill Clinton in his campaign to become president in 1992.

an ϵ upstairs, so, in the limit that ϵ is very small, this is a subleading contribution. Notice indeed that the time-delay effect has an ϵ *downstairs*: the flatter is the potential, the longer it takes to make-up for the loosed or gained ϕ-distance, and so the more δexpansion you get. This is ultimately the justification of why we could do the correct calculation without having to worry *at all* about metric perturbations.

3.3.1. ζ *Conservation for Modes Longer than the Horizon*

Why we cared to compute the power spectrum of $\zeta \sim \delta a/a$? Why do we care of ζ and not of something else? The reason is that this is the quantity that it is conserved during all the history of the universe from when a given mode becomes longer than H^{-1}, to when it becomes shorter the H^{-1} during the standard cosmology. This is very very important. We know virtually nothing about the history of the universe from when inflation ends to say BBN. In order to trust predictions of inflation, we need something to be constant during this epoch, so that we can connect to when we know something about the universe. Proving this constancy in a rigorous way requires some effort, and it is a current topic of research to prove that this conservation holds at quantum level. For the moment, it is easy to give an heuristic argument. The ζ fluctuation is defined as the component of the metric that represents the perturbation to the scale factor $a_{\rm eff} = a(1+\zeta)$. Let us consider the regime in which all modes are longer than the Hubble scale. The universe looks locally homogenous, with everywhere the same energy density, exactly the same universe, with the only difference that in each place the scale factor is valued $a(1+\zeta)$ instead of a. But remember that the metric, apart for tensor modes, is a constrained variable fully determined by the matter fluctuations. Since matter is locally unperturbed, how can it change in a time dependent way the evolution of the scale factor? Impossible Indeed locally ζ can be set to zero by a rescaling of coordinates. The scale factor will evolve as in an unperturbed universe, and therefore ζ will be constant in time. This will happen until gradients will become shorter than Hubble again, so that local dynamics will be able to feel that the universe is not really unperturbed, and so ζ will start evolving.

We should think that it is indeed ζ that sources directly the temperature perturbations we see in the CMB. We should think that $P_\zeta \sim 10^{-10}$.

Some proofs of the conservation of ζ outside of the horizon at tree level are given in [9].

3.4. *Scale invariance and tilt*

As we saw, the power spectrum of ζ is given by

$$P_\zeta(k) = \frac{H^4}{\dot\phi^2} \cdot \frac{1}{k^3} \simeq \frac{H^2}{M_{\rm Pl}^2 \epsilon} \frac{1}{k^3} \,. \tag{86}$$

This is a scale invariant power spectrum. The reason why it is scale invariant is because every Foureir mode sees exactly the same history: it starts shorter than H^{-1}, becomes longer than H^{-1}, and becomes constant. In the limit in which H and $\dot\phi$ are constant (we are in an attractor solution, so $\ddot\phi$ is just a function of ϕ), then every Fourier mode sees the same history and so the power in each mode is the same. In reality, this is only an approximation. Notice that the value of H and of $\dot\phi$ depend slightly on the position of the scalar field. In order to account of this, the best approximation is to evaluate for each mode H and $\dot\phi$ at the time when the mode crossed Hubble and became constant. This happens at the k-dependent $t_{f.o.}(k)$ freezing time defined by

$$\begin{aligned} \omega(t_{f.o.}) \simeq \frac{k}{a(t_{f.o.})} &= H(t_{f.o.}) \\ \Rightarrow \quad t_{f.o.}(k) &\simeq \frac{1}{H(t_{f.o.}(k))} \log\left(\frac{H(t_{f.o.}(k))}{k}\right) \end{aligned} \tag{87}$$

This leads to a deviation from scale invariance of the power spectrum. Our improved version now reads

$$P_\zeta = \frac{H(t_{f.o.}(k))^4}{\dot\phi(t_{f.o.}(k))^2} \cdot \frac{1}{k^3} \,. \tag{88}$$

A measure of the scale dependence of the power spectrum is given by the tilt, defined such that the k-dependence of the power spectrum is approximated by the form

$$P_\zeta \sim \frac{1}{k^3} \left(\frac{k}{k_0}\right)^{n_s - 1} \tag{89}$$

where k_0 is some pivot scale of reference. We therefore have

$$n_s - 1 = \frac{\partial \log(k^3 P_k)}{\partial \log k} = \left.\frac{d \log\left(\frac{H^4}{\dot\phi^2}\right)}{d \log k}\right|_{k/a\sim H} = \left.\frac{d \log\left(\frac{H^4}{\dot\phi^2}\right)}{H dt} \frac{H dt}{d \log k}\right|_{k/a\sim H} \tag{90}$$

where we have used the fact that the solution is a function of k though the ratio k/a as this is the physical wavenumber. At this point we can use that

$$d\log k = d\log(aH) \simeq H dt \tag{91}$$

to obtain

$$n_s - 1 \simeq -2\frac{\dot{H}}{H^2} + 2\frac{\ddot{\phi}}{H\dot{\phi}} = 4\epsilon - 2\eta \ . \tag{92}$$

The tilt of the power spectrum is of order of the slow roll parameters, as expected. How come we were able to compute the tilt of the power spectrum that is slow roll suppressed, though we neglected metric fluctuations, that are also slow roll suppressed? The reason is that the correction to the power spectrum due to the tilt become larger and larger as k becomes more and more different from k_0. Metric fluctuations are expected to give a finite correction of order slow roll to the power spectrum, but not one that is enhanced by the difference of wave numbers considered. This is the same approximation we do in Quantum Field Theory when we use the running of the couplings (which is log enhanced), without bothering of the finite corrections. The pivot scale k_0 is in this context analogous to the renormalization scale.

3.5. *Energy scale of inflation*

We can at this point begin to learn something about inflation. Remember that the power spectrum and its tilt are of order

$$P_\zeta \sim \frac{H^2}{M_{\rm Pl}^2 \epsilon}\frac{1}{k^3} \ , \qquad n_s - 1 = 4\epsilon - 2\eta \ , \tag{93}$$

with, for slow roll inflation

$$H^2 \simeq \frac{V(\phi)}{M_{\rm Pl}^2} \tag{94}$$

From observations of the CMB, we know that

$$P_\zeta \sim 10^{-10} \ , \qquad n_s - 1 \sim 10^{-2} \ . \tag{95}$$

Knowledge of these two numbers is not enough to reconstruct the energy scale of inflation. However, if we assume for the moment that $\eta \sim \epsilon$, a reasonable assumption that however it is sometimes violated (we could have $\epsilon \ll \eta$), then we get

$$\frac{H}{M_{\rm Pl}} \sim 10^{-6} \ , \qquad H \sim 10^{-13}{\rm GeV} \ , \qquad V \sim \left(10^{15}{\rm GeV}\right)^4 \ . \tag{96}$$

These are remarkably large energy scales. This is the energy scale of GUT, not very distance from the Plank scale. Inflation is really beautiful. Not only it has made quantum fluctuations the origin of all the structures of the universe, but it is likely that these are generated by physics at very high energy scales. These are energy scales that unfortunately we will probably never be able to explore at particle accelerators. But these are energy scales that we really would like to be able to explore. We expect very interesting new physics to lie there: new particles, possibly GUT theories, and even maybe string theory. We now can explore them with cosmological observations!

3.6. *Statistics of the fluctuations: Approximate gaussianity*

Let us go back to our action of the fluctuations of the scalar field. Let us write again the action in Fourier space, but this time it turns out to be simpler to work in a finite comoving box of volume V. We have

$$\phi(x) = \frac{1}{V}\sum_{\vec{k}} \phi_k e^{i\vec{k}\cdot\vec{x}} \ . \tag{97}$$

Notice that the mass dimensions of $\phi_{\vec{k}}$ is -2. To get the action, we need the following manipulation

$$\begin{aligned}\int d^3x\ \phi(x)^2 &= \frac{1}{V^2}\sum_{k,k'}\int d^3x\ e^{i(\vec{k}+\vec{k}')\cdot\vec{x}}\phi_k\ \phi_{k'} \\ &\simeq \frac{1}{V^2}\sum_{k,k'}\delta^3(\vec{k}+\vec{k}')\phi_k\ \phi_{k'} \simeq \frac{1}{V}\sum_{k,k'}\delta_{\vec{k},-\vec{k}'}\phi_k\ \phi_{k'} = \frac{1}{V}\sum_{k,k'}\phi_k\ \phi_{-k'}\end{aligned} \tag{98}$$

The action therefore reads

$$S_2 = \frac{1}{V}\sum_k a^3\left(\dot{\phi}_{\vec{k}}\dot{\phi}_{-\vec{k}} + \frac{k^2}{a^2}\phi_{\vec{k}}\phi_{-\vec{k}}\right) \ . \tag{99}$$

Let us find the Hamiltonian. We need the momentum conjugate to $\phi_{\vec{k}}$.

$$\Pi_{\vec{k}} = \frac{\delta S_2}{\delta\dot{\phi}_{\vec{k}}} = \frac{a^3}{V}\dot{\phi}_{-\vec{k}} \tag{100}$$

The Hamiltonian reads

$$\begin{aligned}H &= \sum_{\vec{k}}\Pi_{\vec{k}}\dot{\phi}_{\vec{k}} - \frac{1}{V}\sum_k a^3\left(\dot{\phi}_{\vec{k}}\dot{\phi}_{-\vec{k}} + \frac{k^2}{a^2}\phi_{\vec{k}}\phi_{-\vec{k}}\right) \\ &= \sum_{\vec{k}}\frac{V}{a^3}\Pi_{\vec{k}}\Pi_{-\vec{k}} + \frac{1}{V}\frac{k^2}{a^2}\phi_{\vec{k}}\phi_{-\vec{k}}\end{aligned} \tag{101}$$

If we concentrate on early times where the time dependence induced by Hubble expansion is negligible, we have, for each $\vec{k}$ mode, the same Hamiltonian as an Harmonic oscillator, which reads (again, remember that I am dropping all numerical factors)

$$H = \frac{P^2}{m} + m\omega^2 x^2 \tag{102}$$

We can therefore identify

$$m = \frac{a^3}{V} , \qquad \phi_{\vec{k}} = x , \qquad \omega = \frac{k}{a}. \tag{103}$$

The vacuum wave function for an harmonic oscillator is a Gaussian

$$|0\rangle = \int dx \; e^{-m\omega x^2} |x\rangle \tag{104}$$

which tells us that the vacuum wave function for each Fourier mode $\vec{k}$ reads

$$|0\rangle_{k/a \gg H} = \sum_{\phi_{\vec{k}}} e^{-\frac{a^3}{V}\frac{k}{a}\phi_{\vec{k}}^2} |\phi_k\rangle \ . \tag{105}$$

Since all Fourier mode evolve independently, for the set of Fourier modes that have $k/a \gg H$, we can write

$$|0\rangle_{k_i/a \gg H} = \prod_{\vec{k}_i \gg Ha} \sum_{\phi_{\vec{k}}} e^{-\frac{a^3}{V}\frac{k_i}{a}\phi_{\vec{k}_i}^2} |\phi_{\vec{k}_i}\rangle \ . \tag{106}$$

For each Fourier mode, at early time we have a Gaussian wave function with width $V^{1/2}/(k^{1/2}a)$.

Let us follow the evolution of the wave function with time. As discussed, at early times when $k/a \gg H$, the wave function follows adiabatically the wave function of the would be harmonic oscillator with those time dependent mass and frequency given by (103). However, as the frequency drops below the Hubble rate, the natural time scale of the harmonic oscillator becomes too slow to keep up with Hubble expansion. The state gets frozen on the parameters that it had when $\omega(t) \sim H$. Bu substituting $k/a \to H$, $a \to k/H$,the wave function at late times becomes

$$|0\rangle_{k_i/a \ll H} = \prod_{\vec{k}_i \ll Ha} \sum_{\phi_{\vec{k}}} e^{-\frac{1}{V}\frac{k_i^3}{H^2}\phi_{\vec{k}_i}^2} |\phi_{\vec{k}_i}\rangle \ . \tag{107}$$

This is a Gaussian in field space. Its width is given by

$$\langle \phi_{\vec{k}} \phi_{-\vec{k}} \rangle = \delta_{\vec{k},-\vec{k}} V \frac{H^2}{k^3} \simeq (2\pi)^3 \delta^3(\vec{k}+\vec{k}') \frac{H^2}{k^3} \quad \text{as} \quad V \to \infty \ . \tag{108}$$

We recover the same result of before for the power spectrum. We additionally see that the distribution of values of $\phi_{\vec{k}}$ are Gaussianly distributed. Notice that we are using a quite unusual base of the Hilbert space of a quantum field theory (more used when one talks about the path integral), which is the $|\phi\rangle$ eigenstates base instead of the usual Fock base with occupation numbers. This base is sometimes more useful, as we see here.

So, we learn that the distribution is Gaussian. This result could have been expected. At the end, (so far!), we started with a quadratic Lagrangian, the field theory is free, and so equivalent to an harmonic oscillator, which, in its vacuum, is Gaussianly distributed. We will see in the last lecture that when we consider interacting field theories the distribution will not be Gaussian anymore! Indeed, the statement that cosmological perturbations are so far Gaussian simply means that the field theory describing inflation is a weakly coupled quantum field theory in its vacuum. We will come back to this.

3.7. *Why does the universe looks classical?*

So far we have seen that the cosmological fluctuations are produced by the quantum fluctuations of the inflation in its vacuum state. But then, why does the universe looks classical? The reason is the early vacuum state for each wave number becomes a very classical looking state at late times. Let us see how this happens.

The situation is very simple. We saw in the former subsection that the vacuum state at early times is the one of an harmonic oscillator with frequency $k/a \gg H$. However the frequency is red shifting, and at some point it becomes too small to keep up with Hubble expansion. At that point, while the frequency goes to zero, the state remains trapped in the vacuum state of the would-be harmonic oscillator with frequency $k/a \sim H$. The situation is very similar to what happens to the vacuum state of an harmonic oscillator when one opens up very abruptly the width of the potential well.

This is an incredibly squeezed state with respect to the ground state of the harmonic oscillator with frequency $\omega \sim e^{-60}H$. This state is no more the vacuum state of the late time harmonic oscillator. It has a huge occupation number, and it looks classical.

Let us check that indeed that wave function is semiclassical. The typical condition to check if a wavefunction is well described by a semiclassical approximation is to check if the ϕ-length scale over which the amplitude of

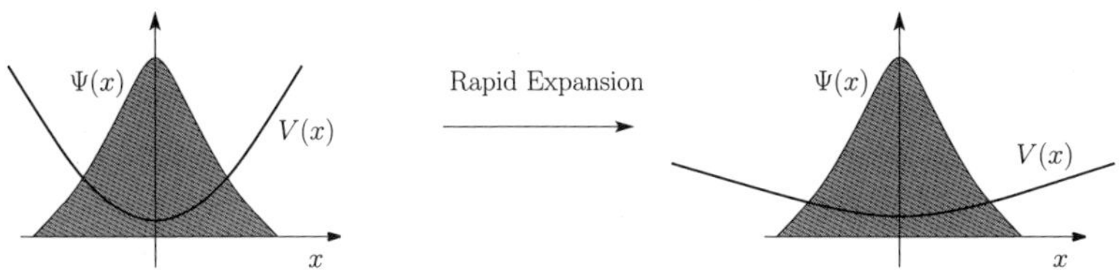

Fig. 8. Formation of a squeezed state by the rapid expansion of the universe.

the wavefunction changes is much longer than the ϕ-length scale over which the phase changes. To obtain the wavefunction at late times, we performed the sudden approximation of making the frequency instantaneously zero. This corresponds to make an expansion in $k/(aH)$. In our calculation we obtained a real wavefunction (109). This means that the phase must have been higher order in $k/(aH) \ll 1$, in the sense that it should be much more squeezed than the width of the magnitude, much more certain the outcome: the time-dependent phase has decayed away. We therefore can write approximately

$$|0\rangle_{k_i/a\ll H,guess} \sim \prod_{\vec{k}_i \ll Ha} \int d\phi_{\vec{k}_i}\, e^{-\frac{1}{V}\frac{k^3}{H^2}\phi^2_{\vec{k}_i}\left[1+i\frac{aH}{k}\right]}|\phi_{\vec{k}_i}\rangle \tag{109}$$

We obtain:

$$\begin{aligned} \Delta\phi_{\text{Amplitude Variation}} &\sim \frac{H}{k^{3/2}}\frac{1}{V^{1/2}} \\ \Delta\phi_{\text{Phase Variation}} &\sim \frac{H}{k^{3/2}}\frac{1}{V^{1/2}}\left(\frac{k}{aH}\right)^{1/2} . \end{aligned} \tag{110}$$

So

$$\frac{\Delta\phi_{\text{Phase Variation}}}{\Delta\phi_{\text{Amplitude Variation}}} \sim \left(\frac{k}{aH}\right)^{1/2} \to 0 . \tag{111}$$

So we see that the semiclassicality condition is satisfied at late times.

Notice furthermore that the state of the inflation is a very squeezed state. The variance of $\delta\phi$ is huge. Since we have just verified that the system is classical, this means that the system has approached a classical stochastical description. A nice discussion of this, stated not exactly in this language, is given in [11].

Of course, later in the universe, local environmental correlations will develop that will decorrelate the quantum state. But we stress that the system is semiclassical even before decorrelation effects are taken into account.

3.8. *Tensor*

Before moving on, let us discuss briefly the generation of tensor modes. In order to do that, we need to discuss about the metric fluctuations. (Remarkably, this is the first time we have to do that).

3.8.1. *Helicity Decomposition of Metric Perturbations*

A generically perturbed FRW metric can be put in the following form

$$ds^2 = -(1+2\Phi)dt^2 + 2a(t)B_i dx^i + a(t)^2\left[(1-2\Psi)\delta_{ij} + E_{ij}\right] . \tag{112}$$

For background space-times that have simple transformation rules under rotation (FRW for example is invariant), it is useful to decompose these perturbations according to their transformation properties under rotation under one axis. A perturbation of wavenumber $\vec{k}$ has elicit λ if under a rotation along the $\hat{k}$ of angle θ, transforms simply by multiplication by $e^{i\lambda\theta}$:

$$\delta g \to e^{i\lambda\theta}\delta g . \tag{113}$$

Scalars have helicity zero, vectors have helicity one, and tensors have helicity two. It is possible to decompose the various components of $\delta g_{\mu\nu}$ in the following way:

$$\Phi,\ \Psi \tag{114}$$

have helicity zero. We can then write

$$B_i = \partial_i B_S + \tilde{B}_{V,i} \tag{115}$$

where $\partial^i \tilde{B}_{V,i} = 0$. B_S is a scalar, B_V is a vector. Finally

$$E_{ij} = E^S_{ij} + E^V_{ij} + \gamma_{ij} \tag{116}$$

where

$$\begin{aligned} E^S_{ij} &= \frac{1}{\partial^2}\left(\partial_i\partial_j - \frac{1}{3}\delta_{ij}\partial^2\right)\tilde{E}^S \\ E^V_{ij} &= \frac{1}{2\partial^2}\left(\partial_i\tilde{E}^V_j + \partial_j\tilde{E}^V_i\right) , \quad \text{with} \quad \partial_i\tilde{E}^i = 0 \\ \partial_i\gamma_{ij} &= 0, \quad \gamma_i{}^i = 0 . \end{aligned} \tag{117}$$

with $\partial^2 = \delta^{ij}\partial_i\partial_j$. $\tilde{E}^S$ is a scalar, $\tilde{E}^V$ is a vector, and γ is a tensor.

Now, it is possible to show that at linear level, in a rotation invariant background, scalar, vector and tensor modes do not couple and evolve independently (you can try to contract the vectors together it does not work: you cannot make it).

Under a change of coordinate

$$x^\mu \to \tilde{x}^\mu = x^\mu + \xi^\mu \tag{118}$$

these perturbations change according to the transformation law of the metric

$$\tilde{g}^{\mu\nu} = \frac{\partial \tilde{x}^\mu}{\partial x^\rho}\frac{\partial \tilde{x}^\nu}{\partial x^\sigma} g^{\rho\sigma} \tag{119}$$

The change of coordinates ξ^μ can also be decomposed into a scalar and a vector component

$$\xi^0_S\,, \quad \xi^i_S = \partial^i \xi \tag{120}$$

$$\xi^0_V = 0\,, \quad \partial_i \xi^i_V = 0\,. \tag{121}$$

At linear level, different helicity metric perturbations do not get mixed and they are transformed only by the change coordinates with the same helicity (for the same reasons as before). For this reasons, we see that tensor perturbations are invariant. They are gauge invariant. This is not so for scalar and vector perturbations. For example, scalar perturbations transform as the following

$$\Phi \to \Phi - \dot{\xi}^0_S \tag{122}$$

$$B_S \to B_S + \frac{1}{a}\xi^0_S - a\dot{\xi} \tag{123}$$

$$E \to E - B_S \tag{124}$$

$$\Psi \to \Psi - H\alpha\,. \tag{125}$$

The fact that tensor modes are gauge invariant and uncoupled (at linear level!) means that we can write the metric for them as

$$g_{ij} = a^2\left(\delta_{ij} + \gamma_{ij}\right)\,, \tag{126}$$

and set to zero all other perturbations (including $\delta\phi$). By expanding the action for the scalar field plus GR at quadratic order, one obtains an action of the form (actually only the GR part contributes, and the following action could just be guessed)

$$S \sim \int d^4x\, a^3\, M_{\rm Pl}^2 \left[(\dot\gamma_{ij})^2 - \frac{1}{a^2}(\partial_l \gamma_{ij})^2 \right]$$

$$\sim \sum_{s=+,\times} \int dt d^3k\, a^3\, M_{\rm Pl}^2 \left[\dot\gamma^s_{\vec k} \dot\gamma^s_{-\vec k} - \frac{k^2}{a^2} \gamma^s_{\vec k} \gamma^s_{-\vec k} \right] \tag{127}$$

where in the last passage we have decomposed the generic tensor mode in the two possible polarization state

$$\gamma_{ij}^{+,\times} = \gamma_s(t) e_{ij}^{(+,\times)} \tag{128}$$

In matrix form, for a mode in the $\hat k = \hat z$ direction

$$\gamma = \begin{pmatrix} \gamma^\times & \gamma^+ & 0 \\ \gamma^+ & -\gamma^\times & 0 \\ 0 & 0 & 0 \end{pmatrix} \tag{129}$$

$$\gamma_{ij} = \int d^3k \sum_{s=+,\times} e^s_{ij}(k) \gamma^s_{\vec k}(t) e^{i\vec k\cdot \vec x} \tag{130}$$

$$\epsilon^s_{ii} = k^i \epsilon_{ij} = 0 \qquad \epsilon^s_{il}\epsilon^{s'}_{lj} = \delta_{ij} \tag{131}$$

We see that the action for each polarization is the same as for a normal scalar field, just with a different canonical normalization. The two polarization are also independent (of course), and therefore, without having to do any calculation, we obtain the power spectrum for gravity waves to be

$$\langle \gamma^s_{\vec k} \gamma^{s'}_{\vec k'} \rangle = (2\pi)^3 \delta^3(\vec k + \vec k') \delta_{s,s'} \frac{H^2}{M_{\rm Pl}^2} \frac{1}{k^3}\ . \tag{132}$$

Notice that the power spectrum depends only on one unknown quantity H. This means that if we detect gravitational waves from inflation, we could measure the energy scale of inflation. ... Actually, this was a "theorem" that was believed to hold until last september. At that time new mechanisms further than the vacuum fluctuations have been identified that could dominate the ones produced by vacuum fluctuations and that could be detectable [20].

By now we are expert: the tilt of gravity waves power spectrum is given by

$$n_t - 1 = -2\epsilon \tag{133}$$

as only the variation of H is involved.

The measurement of this tilt would give us a measurement of ϵ. Again, until recently this was thought to be true, and unfortunately (and luckily) things have changed now, and the above formula for the tilt holds only for the simplest models of inflation.

Notice further that if we were to measure the amplitude of the gravitational waves and their tilt, then, under the hypothesis of standard slow roll inflation, we would know H and ϵ. In this same hypothesis therefore we would therefore predict the size of the ζ power spectrum. If this would hold, we would discover that inflation happened in the slow roll inflation way. This is called consistency condition for slow roll single field inflation.

Notice that, in standard slow roll inflation (this is true only for the simple inflationary scenarios), the power in gravity waves is smaller than the one in scalars by a factor of $\epsilon \ll 1$. This means that if gravity waves are detected, ϵ cannot be too small, and therefore the field excursion during inflation is over planckian: $\Delta\phi \gtrsim M_{\rm Pl}$. This is known as the Lyth's bound [10].

Finally, notice that this signal is proportional to $\hbar$. Such a measurement would be the first direct evidence that GR is quantized. We have never seen this (frankly there are no doubt that gravity is quantized but still better to see it in experiments.)

3.9. *Summary of Lecture 2*

- the quantum fluctuations of the scalar field naturally produce a scale invariant spectrum of perturbations
- they become curvature perturbations at the end of inflation
- they look like classical and (quasi) Guassian
- Quantum mechanical effects are at the source of the largest structures in the universe
- The Energy scale of inflation could be as high as the GUT scale, opening the possiblity to explore the most fundamental laws of physics from the cosmological observations
- Tensor modes are also produced. If seen, first evidence of quantization of gravity.
- Everything is derived without hard calculations

Now we are ready to see how we check for this theory in the data.

4. Lecture 3: Contact with Observations (Part 2)

Absolutely, the best way we are testing inflation is by the observation of the cosmological perturbations.

You had already several classes on the evolution of perturbations in the universe and how they connect to observations. Here I will simply focus on the minimum amount of information that we need to establish what this observations are really telling us about Inflation. I will focus just on CMB, for brevity. The story is very similar also for large scale structures.

4.1. *CMB basics*

For a given perturbation $\delta X(k,\tau)$ at a given time τ and with Fourier mode k, we can define its transfer function for the quantity X at that time τ and for the Fourier mode k as

$$\delta X = T(k,\tau,\tau_{in})\zeta_k(\tau_{in}) \ . \tag{134}$$

This must be so in the linear approximation. We can take τ_{in} early enough so that the mode k is smaller than aH, in this way $\zeta_k(\tau_{in})$ represents the constant value ζ took at freeze out during inflation.

For the CMB temperature, we perform a spherical harmonics decomposition

$$\frac{\delta T}{T}(\tau_0,\hat{n}) = \sum_{l,m} a_{lm} Y_{lm}(\hat{n}) \tag{135}$$

and the by statistical isotropy the power spectra reads

$$\langle a_{lm} a_{l'm'}\rangle = C_l^{TT}\delta_{ll'}\delta_{mm'} \ . \tag{136}$$

Since the temperature anisotropy are dominated by scalar fluctuations, we have

$$a_{lm} = \int d^3k \Delta_l(k)\zeta_k Y_{lm}(\hat{k}) \ , \quad \Rightarrow \quad C_l = \int dk\, k^2\, \Delta_l(k)^2 \ \ P_\zeta(k) \ , \tag{137}$$

$\Delta_l(k)$ contains both the effect of the transfer functions and also of the projection on the sky.

- **large scales:** If we look at very large scales, we find modes that were still outside H^{-1} at the time of recombination (see Fig. 9. Nothing could have happened to them.

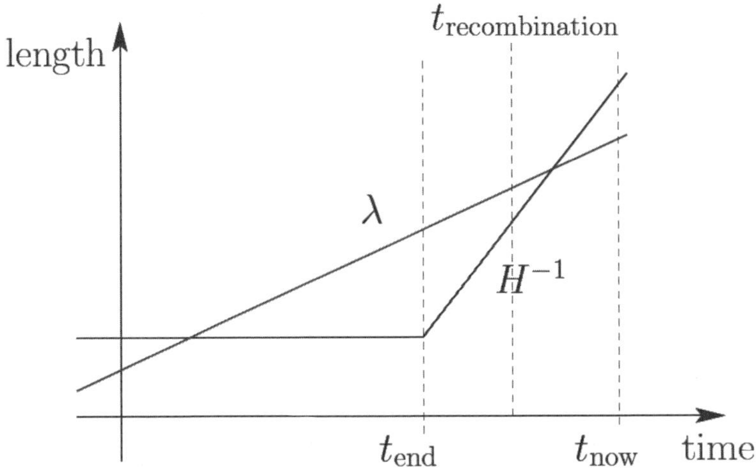

Fig. 9. Relative ratios of important length scales as a function of time in the inflationary universe. There are length scales that we can see now that were longer than H^{-1} at the time of recombination.

As in Lesgourges' class you have been told, there has been no evolution and only projection effects.

$$\Delta_l(k) \simeq j_l(k(\tau_0 - \tau_{\rm rec})) \quad \Rightarrow \quad C_l \simeq \int dk\, k^2 P_\zeta\; j_l^2(k(\tau_0 - \tau_{\rm rec})) \tag{138}$$

$j_l^2(k(\tau_0 - \tau_{rec}))$ is sharply peaked at $k(\tau_0 - \tau_{\rm rec}) \sim l$, so we can approximately perform the integral, to obtain

$$\begin{aligned} C_l &\simeq k^3 P_\zeta\big|_{k=l/(\tau_0-\tau_{\rm rec})} \times \int d\log x\; j_l^2(x) \\ &\sim k^3 P_\zeta\big|_{k=l/(\tau_0-\tau_{\rm rec})} \times \frac{1}{l(l+1)} \end{aligned} \tag{139}$$

$$\Rightarrow l(l+1)C_l \;\text{ is flat, equivalently } l\text{-independent.} \tag{140}$$

- **small scales**. On short scales, mode entered inside H^{-1} and begun to feel both the gravitational attraction of denser zones, but also their pressure repulsion. This leads to oscillatory solutions.

$$\ddot{\delta T} + c_s^2 \nabla^2 \delta T \simeq F_{\rm gravity}(\zeta) \tag{141}$$

$$\Rightarrow \delta T_k \simeq A_{\vec{k}} \cos(k\eta) + B_{\vec{k}} \sin(k\eta) = \tilde{A}_{\vec{k}} \cos(k\eta + \phi_{\vec{k}}) \tag{142}$$

Here $A_{\vec{k}}$ and $B_{\vec{k}}$ depend on the initial conditions. In inflation, we have

$$\tilde{A}_{\vec{k}} \simeq \frac{1}{k^3} , \qquad \phi_{\vec{k}} = 0 . \tag{143}$$

All the modes are in phase coherence. Notice, dynamics and wavenumber force all mode of a fixed wavenumber to have the same frequency. However, they need not have necessarily the same phase. Inflation, or superHubble fluctuations, forces $\zeta \simeq \frac{\delta T}{T}$ =const on large scales, which implies $\phi_{\vec{k}} = 0$. This is what leads to acoustic oscillations in the CMB

$$\begin{aligned} & \delta T(\vec{k}, \eta) \sim \delta T_{in}(\vec{k}) \times \cos(k\eta) \\ \Rightarrow \quad & \delta T(\vec{k}, \eta_0) \sim \delta T_{in}(\vec{k}) \times \cos(k\eta_{rec}) \\ \Rightarrow \quad & \langle \delta T(\vec{k}, \eta_0) \rangle \sim \langle \delta T^2_{\vec{k}} \rangle \cos^2(k\eta_{rec}) \end{aligned} \tag{144}$$

we get the acoustic oscillations.

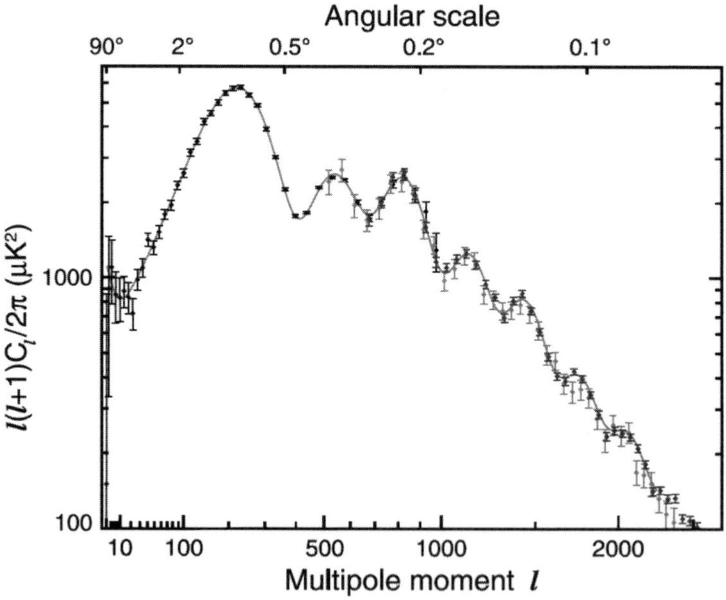

Fig. 10. Power spectrum of the CMB fluctuations. Oscillations are clearly seeable. Picture is taken from [13], which combines the result of several CMB experiments such as WMAP, SPT [14] and ACT [15].

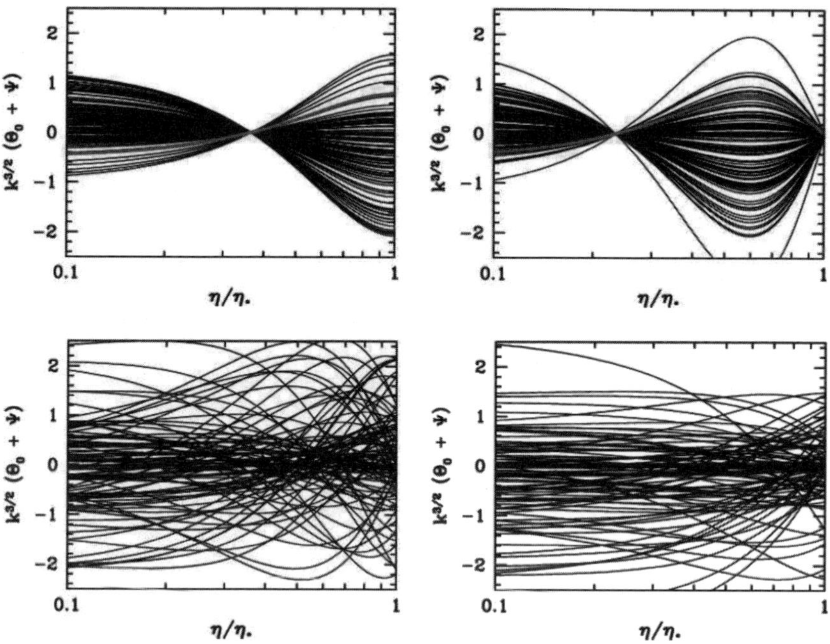

Fig. 11. On top: time evolution of two different modes that have different initial amplitude, but all the same phases. We see that the typical size of the amplitude at the time of recombination is different for different modes. We obtain oscillations in the power spectrum. On bottom: time evolution of two different modes with different amplitudes and phases. We see that the typical size of the fluctuations at the time of recombination is independent of the wavenumber. The power spectrum has no oscillations and is featureless. These pictures are taken from [16].

This is the greatest qualitative verification of inflation so far. Acoustic oscillations told us that the horizon was much larger than H^{-1} at recombination and that there were constant superHubble perturbation before recombination. This is very very non-trivial prediction of inflation. Notice that scale invariance of the fluctuations was already guessed to be in the sky (Harrison-Zeldovich spectrum) at the time of formulation of inflation, but nobody knew of the acoustic oscillations at that time. CMB experiments found them!

This is a very important *qualitative* verification of inflation that we get from the CMB. But it is not a quantitative confirmation. Information on the quantitative part is very limited.

Fig. 12. A nice picture of the CMB as measured by the WMAP experiment [12]. There is a correlation not only in the intensity of the radiation, but also in its polarization, that can be represented as a bi-dimensional vector living on the 2-sphere.

4.2. *What did we verify of inflation so far?*

Let us give a critical look at what we learnt about inflation so far form the observational point of view.

There have been three *qualitative* theoretical predictions of inflation that have been verified so far. One is the oscillations in the CMB, another is the curvature of the universe, of order $\Omega_k \sim 10^{-2}$. At the time inflation was formulated, Ω_k could have been of order one. It is a natural prediction of inflation that lasts a little more than the necessary amount to have $\Omega_k \ll 1$. The third is that the perturbations are Gaussian to a very good approximation: the signature of a weakly coupled field theory.

But what did we learn at a *quantitative* level about inflation so far? Just two numbers, not so much in my opinion unfortunately. This is so because all the beautiful structures of the peaks in the CMB (and also in Large Scale Structures) is just controlled by well known Standard Model physics at 1 eV of energy. The input from inflation are the qualitative initial conditions for each mode, and quantitatively the power spectrum and its tilt

$$P_\zeta \simeq \frac{H^2}{M_{\rm Pl}^2 \epsilon} \sim 10^{-10} \ , \quad n_s - 1 \simeq 4\epsilon - 2\eta \simeq -6 \times 10^{-2} \ , \qquad (145)$$

just two numbers fit it all.

This is a pity, because clearly cosmological data have much more information inside them. Is there something more to look for?

4.3. *CMB polarization*

One very interesting observable is the CMB polarization. The CMB has been already observed to be partially polarized. Polarization of the CMB can be represented as the set of vectors tangent to the sphere, the direction of each vector at each angular point representing the direction of the polarization coming from the point, and its length the fractional amount.

CMB polarization in induced by Thomson scattering in the presence of a quadruple perturbation. Information on cosmological perturbations is carried over by the correlation of polarization (very much the same as the correlation of temperature). It is useful to define two scalar fields that live on the sphere.

Polarization can be decomposed into the sum of the fields, E and B, that have very different angular patter.

Scalar perturbations induce E polarization, and they are being measured with greater and greater accuracy. However tensor perturbations induce both E and B polarization. See [5] for more details. This means that a discovery of B modes would be a detection of tensor modes produced during inflation (there are some B modes produced by lensing, but they are only on small angular scales). So far there is no evidence of them, but even if we saw them, what we would learn about inflation?

We will learn a great qualitative point. Producing scale invariant tensor perturbations is very hard, because tensor perturbation tend to depend only on the nature of the space-time. Scale invariant tensor modes would represent most probably that an early de Sitter phase happened and so that inflation did happen.

At a quantitative level, however, we would just learn two numbers: the amplitude and the tilt of the power spectrum. In the simplest models of inflation, the amplitude of the power spectrum gives us direct information about H, and if the signal is detectable, it would teach us about the energy scale of inflation. Its scale invariance would teach us that H is constant with time: this is the definition of inflation.

However recently new mechanism for produce large and detectable tensor modes have been found, which disentangle the measurement of B modes from a measurement of H, at least in principle [20]. While the overall size is different, the signal is still scale invariant.

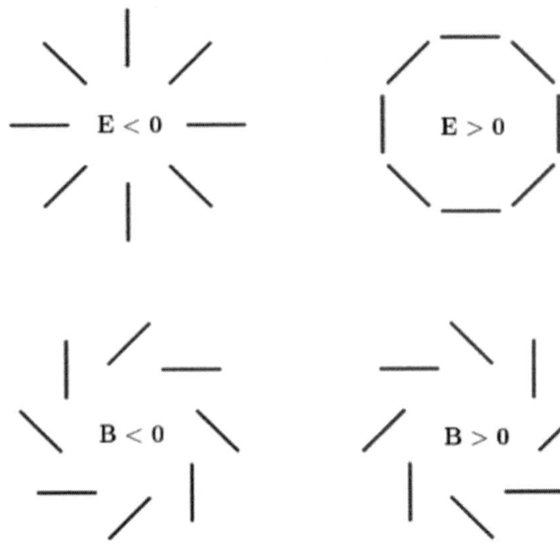

Fig. 13. We normally decompose the vector field on the sphere that represent the polarisation in terms of E and B vector fields that have the above typical behaviour.

So, the question really remains: is it there something more to look for?

4.4. *Many more models of inflation*

Indeed, there are many more models of inflation than standard slow roll that we discussed.

DBI Inflation. One remarkable example is DBI inflation [17]. This described the motion of a brane in ADS space. Since the brane has a speed limit, an inflationary solution happens when the brane is moving at the speed of light. At that point special relativistic effects slow down the brane, and you have inflation, even though the brane is moving at the speed of light. The brane fluctuations in this case play the role of the inflaton.

This model, though it happens in a totally different regime than slow roll inflation, it is totally fine with the observations we looked at so far. It turns out that the power spectrum scales in a different way that in slow roll models. We have a speed of sound $c_s \ll 1$

$$\omega^2 \sim c_s^2 k^2 \,. \tag{146}$$

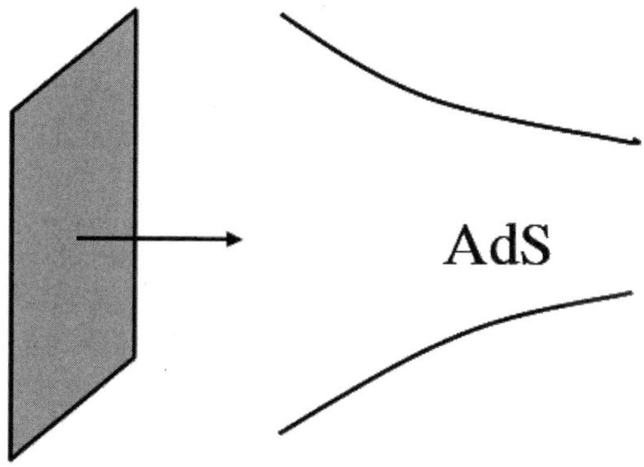

Fig. 14. Inflation can be realised by a brane moving relativistically in AdS space.

This affects the power spectrum in the following way

$$P_\zeta \sim \frac{H^2}{\epsilon M_{\rm Pl}^2 c_s} \sim 10^{-10} \,, \qquad n_s - 1 \simeq 4\epsilon - 2\eta + \frac{\dot{c}_s}{H c_s} \sim 10^{-2} \,. \qquad (147)$$

Given than to match the CMB we need just these two inputs from the inflationary model, it is pretty expectable that they can be fixed. And indeed this happens.

This inflationary model had the remarkable features that non-gaussianities were detectably large. The skewness of the distribution of the fluctuations was

$$\frac{\langle\zeta^3\rangle}{\langle\zeta^2\rangle^{3/2}} \sim \frac{1}{c_s^2}\langle\zeta^2\rangle^{1/2} \gg 10^{-5} \qquad (148)$$

where we used that $\langle\zeta^2\rangle^{1/2} \sim 10^{-5}$ For comparison, the same number is standard inflation is of order $\epsilon\langle\zeta^2\rangle^{1/2} \ll 10^{-5}$. While for standard slow roll inflation this is undetectably small, it is detectable for DBI inflation.

This opens up a a totally new possible observational signature, and the possibility to distinguish and to learn about models that would be indistinguishable at the level of the two point function.

Non-Gaussianity!!

Ghost inflation. Ghost inflation is another peculiar looking model [18]. It consists of a scalar field with the wrong sign kinetic term (a ghost).

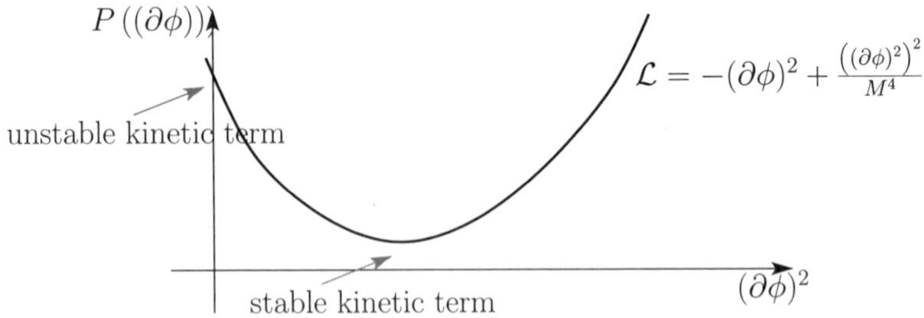

Fig. 15. The Ghost Inflation model.

This triggers an instability that condensate in a different vacuum, where $\dot{\phi} = \text{const}$ even in the absence of potential. This leads to inflation. The fluctuations have a dispersion relation of the form

$$\omega^2 \simeq \frac{k^4}{M^2} \tag{149}$$

which is extremely non-relativistic.

Again, this model is totally fine in fitting observations of the power spectrum, but it produces a large and detectable non-Gaussianity.

This are new models, some inspired by string theory. But they have new signatures. So, the question is: how generic are these signatures? What are the generic signatures of inflation?

In order to do that, we need an approach that is very general, and looks at inflation in its most essential way: we go to the Effective Field Theory approach.

4.5. *The Effective Field Theory of Inflation*

Effective Field Theories (EFTs) have played the role of the guiding principle for particle physics and even condensed matter physics. EFTs have the capacity of synthesizing the relevant physics at the energy scale of interests. Effects of higher energy, largely irrelevant, physics are encoded in the coefficients of the higher dimension operators. It is *the way* to explore the phenomenology at a given energy scale. What we are going to do next is to develop the effective field theory of inflation. In doing so, we can look at inflation as the theory of a Goldstone boson: the Goldstone boson of time translations.

Review of Goldstone bosons. Goldstone bosons are ubiquitous in particle physics (they got Nambu the well deserved 2008 nobel prize!). Let us consider the simplest theory of a $U(1)$ global symmetry $\phi \to e^{i\alpha}\phi$ that is spontaneously broken because of a mexican hat like potential $\phi \to \langle\phi\rangle$. Then there is Goldstone boson π that non-linearly realizes the symmetry $\pi \to \pi + \alpha$.

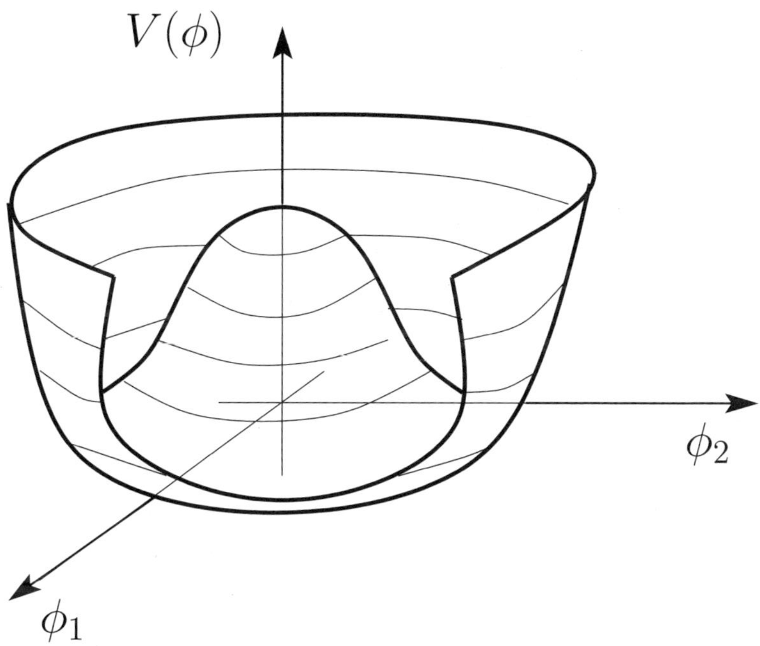

Fig. 16. Mexican-hat potential for a complex scalar field that leads to spontaneous breaking of a $U(1)$ symmetry.

$$\mathcal{L} = \partial_\mu\phi^\star\partial^\mu\phi - m^2\phi^\star\phi + \lambda\phi^{\star 2}\phi^2 \quad \to \quad \phi = \frac{m}{\lambda^{1/2}}e^{i\,\pi(\vec{x},t)}\,. \tag{150}$$

The action for the field π is therefore the one of a massless scalar field endowed with a shift symmetry

$$\mathcal{L}_\pi = (\partial\pi)^2 + \frac{1}{m/\lambda^{1/2}}(\partial\pi)^4 + \ldots \tag{151}$$

the higher derivative operators being suppressed by powers of the high energy scale $m/\lambda^{1/2}$.

A famous example of Goldstone bosons are the pions of the Chiral Lagrangian, that represent the Goldstone boson that non-linearly realise the $SU(2)$ chiral flavor symmetry, and they represent in the UV theory of QCD bound states of quark and antiquark. Notice that pions represent emergent scalar fields: there is no fundamental scalar field in QCD.

Inflation as the theory of a Goldstone boson. How do we build the EFT of Inflation. In order to do that, we need to think of inflation in its most essential way. What we really know about inflation is that it is a period of accelerated expansion, where the universe was quasi de sitter. However, it could not be exactly de Sitter, because it has to end. This means that time-translation is spontaneously broken, and we will therefore consider that there is a physical clock measuring time and forcing inflation to end.

No matter what this clock is, we can use coordinate invariance of GR to go to the frame where these physical clock is set to zero. This can be done by choosing spatial slices where the fluctuations of the clock are zero, by performing a proper time diffs from any coordinate frame. As an example, if the inflaton was a fundamental scalar field (we are not assuming that, but just to make example) and we are in a coordinate frame where $\delta\phi(\vec{x},t) \neq 0$, we can perform a time diff. $t \to \tilde{t} = t + \delta t(\vec{x},t)$, such that (at linear order, it can be generalized to arbitrary non-linear order)

$$0 = \tilde{\delta\phi}(\vec{x},t) = \delta\phi(\vec{x},t) - \dot{\phi}_0(t)\delta t(\vec{x},t) \tag{152}$$

Now, suppose we are in this frame. We follow the rules of EFT. They say we have to write the action with the degrees of freedom that are available to us. This is just the metric fluctuations. We have to expand in fluctuations, and write down all operators compatible with the symmetries of the problem. In our case we can arbitrarily change spatial coordinates within the various spacial slices, on each spatial slice in a different way. This means that the residual gauge symmetry is time-dependent spatial diff.s:

$$x^i \to \tilde{x}^i = x^i + \xi^i(t,\vec{x}) \ . \tag{153}$$

Further, still following the EFT procedure, we expand in perturbations and go to the order up to which we are interested (for example, quadratic order for 2-point functions, cubic order for 3-point functions, quartic order for 4-point functions, and so on), and then expand, at each order in the fluctuations, in derivatives, higher derivative terms being suppressed by the ratio of the energy scale E of the problem versus some high energy scale Λ.

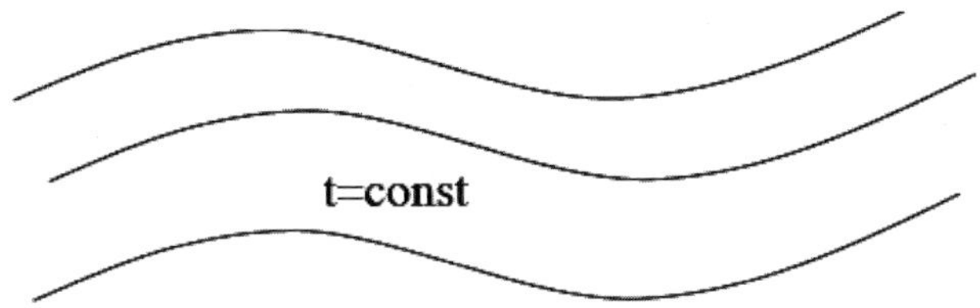

Fig. 17. If there is a clock-field driving inflation, then there is a privileged time-slicing where this clock is taken as uniform.

4.5.1. *Construction of the Action in Unitary Gauge*

What is the most general Lagrangian in this unitary gauge? Here we will follow [21] closely. One must write down operators that are functions of the metric $g_{\mu\nu}$, and that are invariant under the (linearly realized) time dependent spatial diffeomorphisms $x^i \to x^i + \xi^i(t, \vec{x})$. Spatial diffeomorphisms are in fact unbroken. Besides the usual terms with the Riemann tensor, which are invariant under all diffs, many extra terms are now allowed, because of the reduced symmetry of the system. They describe the additional degree of freedom eaten by the graviton. For example it is easy to realize that g^{00} is a scalar under spatial diffs, so that it can appear freely in the unitary gauge Lagrangian.

$$\tilde{g}^{00} = \frac{\partial \tilde{t}}{\partial x^\mu}\frac{\partial \tilde{t}}{\partial x^\nu} g^{\mu\nu} = \delta^0_\mu \delta^0_\nu g^{\mu\nu} = g^{00} \; . \tag{154}$$

Polynomials of g^{00} are the only terms without derivatives. Given that there is a preferred slicing of the spacetime, one is also allowed to write geometric objects describing this slicing. For instance the extrinsic curvature $K_{\mu\nu}$ of surfaces at constant time is a tensor under spatial diffs and it can be used in the action. If n^μ is the vector orthogonal to the equal time slices, we have

$$K_{\mu\nu} = h_\nu{}^\sigma \nabla_\sigma n_\nu \; , \tag{155}$$

with ∇ being the covariant derivative, and $h_{\mu\nu}$ the induced metric on the spatial slices

$$h_{\mu\nu} = g_{\mu\nu} + n_\mu n_\nu \; . \tag{156}$$

Notice that generic functions of time can multiply any term in the action. The most generic Lagrangian can be written as (see App. A and B of [21] for a proof)

$$S = \int d^4x\, \sqrt{-g}\Big[\frac{1}{2}M_{\rm Pl}^2 R - c(t)g^{00} - \Lambda(t) + \frac{1}{2!}M_2(t)^4(\delta g^{00})^2 + \frac{1}{3!}M_3(t)^4(\delta g^{00})^3 + -\frac{\bar{M}_1(t)^3}{2}(\delta g^{00})\delta K^{\mu}{}_{\mu} - \frac{\bar{M}_2(t)^2}{2}\delta K^{\mu}{}_{\mu}{}^2 - \frac{\bar{M}_3(t)^2}{2}\delta K^{\mu}{}_{\nu}\delta K^{\nu}{}_{\mu} + \ldots\Big] , \tag{157}$$

where the dots stand for terms which are of higher order in the fluctuations or with more derivatives. $\delta g^{00} = g^{00} + 1$. We denote by $\delta K_{\mu\nu}$ the variation of the extrinsic curvature of constant time surfaces with respect to the unperturbed FRW: $\delta K_{\mu\nu} = K_{\mu\nu} - a^2 H h_{\mu\nu}$ with $h_{\mu\nu}$ is the induced spatial metric. Notice that only the first three terms in the action above contain linear perturbations around the chosen FRW solution, all the others are explicitly quadratic or higher. Therefore the coefficients $c(t)$ and $\Lambda(t)$ will be fixed by the requirement of having a given FRW evolution $H(t)$, *i.e.* requiring that tadpole terms cancel around this solution. Before fixing these coefficients, it is important to realize that this simplification is not trivial. One would expect that there are an infinite number of operators which give a contribution at first order around the background solution. However one can write the action as a polynomial of linear terms like $\delta K_{\mu\nu}$ and $g^{00}+1$, so that it is evident whether an operator starts at linear, quadratic or higher order. All the linear terms besides the ones in Eq. (157) will contain derivatives and they can be integrated by parts to give a combination of the three linear terms we considered plus covariant terms of higher order. We conclude that *the unperturbed history fixes $c(t)$ and $\Lambda(t)$, while the difference among different models will be encoded into higher order terms.*

We can now fix the linear terms imposing that a given FRW evolution is a solution. As we discussed, the terms proportional to c and Λ are the only ones that give a stress energy tensor

$$T_{\mu\nu} = -\frac{2}{\sqrt{-g}}\frac{\delta S_{\rm matter}}{\delta g^{\mu\nu}} \tag{158}$$

which does not vanish at zeroth order in the perturbations and therefore contributes to the right hand side of the Einstein equations. During inflation we are mostly interested in a flat FRW Universe

$$ds^2 = -dt^2 + a^2(t)d\vec{x}^2 \tag{159}$$

so that Friedmann equations are given by

$$H^2 = \frac{1}{3M_{\rm Pl}^2}\left[c(t)+\Lambda(t)\right] \tag{160}$$

$$\frac{\ddot{a}}{a} = \dot{H} + H^2 = -\frac{1}{3M_{\rm Pl}^2}\left[2c(t)-\Lambda(t)\right] . \tag{161}$$

Solving for c and Λ we can rewrite the action (157) as

$$S = \int d^4x\, \sqrt{-g}\, \Big[\, \frac{1}{2}M_{\rm Pl}^2 R + M_{\rm Pl}^2 \dot{H} g^{00} - M_{\rm Pl}^2(3H^2+\dot{H}) + \frac{1}{2!}M_2(t)^4(\delta g^{00})^2 + \frac{1}{3!}M_3(t)^4(\delta g^{00})^3 - \frac{\bar{M}_1(t)^3}{2}(\delta g^{00})\delta K^\mu{}_\mu - \frac{\bar{M}_2(t)^2}{2}\delta K^\mu{}_\mu{}^2 - \frac{\bar{M}_3(t)^2}{2}\delta K^\mu{}_\nu \delta K^\nu{}_\mu + \ldots\Big] . \tag{162}$$

As we said all the coefficients of the operators in the action above may have a generic time dependence. However we are interested in solutions where H and $\dot{H}$ do not vary significantly in one Hubble time. Therefore it is natural to assume that the same holds for all the other operators. With this assumption the Lagrangian is approximately time translation invariant.[i] Therefore the time dependence generated by loop effects will be suppressed by a small breaking parameter.[j] This assumption is particularly convenient since the rapid time dependence of the coefficients can win against the friction created by the exponential expansion, so that inflation may cease to be a dynamical attractor, which is necessary to solve the homogeneity problem of standard FRW cosmology.

It is important to stress that this approach does describe the most generic Lagrangian not only for the scalar mode, but also for gravity. High energy effects will be encoded for example in operators containing the perturbations in the Riemann tensor $\delta R_{\mu\nu\rho\sigma}$. As these corrections are of higher order in derivatives, we will not explicitly talk about them below.

[i] The limit in which the time shift is an exact symmetry must be taken with care because $\dot{H} \to 0$. This implies that the spatial kinetic term for the Goldstone vanishes, as we will see in the discussion of Ghost Inflation.

[j] Notice that this symmetry has nothing to do with the breaking of time diffeomorphisms. To see how this symmetry appears in the ϕ language notice that, after a proper field redefinition, one can always assume that $\dot{\phi} =$ const. With this choice, invariance under time translation in the unitary gauge Lagrangian is implied by the shift symmetry $\phi \to \phi$ + const. This symmetry and the time translation symmetry of the ϕ Lagrangian are broken down to the diagonal subgroup by the background. This residual symmetry is the time shift in the unitary gauge Lagrangian.

Let us give some examples of how to write simple models of inflation in this language. A model with minimal kinetic term and a slow-roll potential $V(\phi)$ can be written in unitary gauge as

$$\int d^4x\,\sqrt{-g}\left[-\frac{1}{2}(\partial\phi)^2 - V(\phi)\right] \to \int d^4x\,\sqrt{-g}\left[-\frac{\dot{\phi}_0(t)^2}{2}g^{00} - V(\phi_0(t))\right]. \tag{163}$$

As the Friedmann equations give $\dot{\phi}_0(t)^2 = -2M_P^2\dot{H}$ and $V(\phi(t)) = M_{\rm Pl}^2(3H^2 + \dot{H})$ we see that the action is of the form (162) with all but the first three terms set to zero. Clearly this cannot be true exactly as all the other terms will be generated by loop corrections: they encode all the possible effects of high energy physics on this simple slow-roll model of inflation.

A more general case includes all the possible Lagrangians with at most one derivative acting on each ϕ: $L = P(X,\phi)$, with $X = g^{\mu\nu}\partial_\mu\phi\partial_\nu\phi$. Around an unperturbed solution $\phi_0(t)$ we have

$$S = \int d^4x\,\sqrt{-g}\;P(\dot{\phi}_0(t)^2 g^{00}, \phi(t)) \tag{164}$$

which is clearly of the form above with $M_n^4(t) = \dot{\phi}_0(t)^{2n}\partial^n P/\partial X^n$ evaluated at $\phi_0(t)$. Terms containing the extrinsic curvature contain more than one derivative acting on a single scalar and will be crucial in the limit of exact de Sitter, $\dot{H} \to 0$. They reproduce ghost inflation and new models that are discovered in this set up.

4.5.2. *Action for the Goldstone Boson*

The unitary gauge Lagrangian is very general, but it is clearly not very intuitive. For example, in a particular limit, it contains standard slow roll inflation. But where is the scalar degree of freedom? This is so complicated because it is the unitary gauge Lagrangian of a spontaneously broken gauge symmetry.

Goldstone boson equivalence theorem. The unitary gauge Lagrangian describes three degrees of freedom: the two graviton helicities and a scalar mode. This mode will become explicit after one performs a broken time diffeomorphism (Stückelberg trick) as the Goldstone boson which non-linearly realizes this symmetry. In analogy with the equivalence theorem for the longitudinal components of a massive gauge boson [19], we expect that the physics of the Goldstone decouples from the two graviton helicities at short distance, when the mixing can be neglected. Let us review briefly what happens in a non-Abelian gauge theory before applying the same method in our case.

The unitary gauge action for a non-Abelian gauge group A^a_μ is

$$S = \int d^4x \; -\frac{1}{4}\mathrm{Tr}\, F_{\mu\nu}F^{\mu\nu} - \frac{1}{2}m^2 \mathrm{Tr}\, A_\mu A^\mu \;, \tag{165}$$

where $A_\mu = A^a_\mu T^a$. Under a gauge transformation we have

$$A_\mu \to U A_\mu U^\dagger + \frac{i}{g} U \partial_\mu U^\dagger \equiv \frac{i}{g} U D_\mu U^\dagger \;. \tag{166}$$

The action therefore becomes

$$S = \int d^4x \; -\frac{1}{4}\mathrm{Tr}\, F_{\mu\nu}F^{\mu\nu} - \frac{1}{2}\frac{m^2}{g^2}\mathrm{Tr} D_\mu U^\dagger D_\mu U \;. \tag{167}$$

The mass term was not gauge invariant, and so we have factors of U in that term. The gauge invariance can be "restored" writing $U = \exp\left[iT^a\pi^a(t,\vec{x})\right]$, where π^a are scalars (the Goldstones) which transform non-linearly under a gauge transformation Λ as

$$e^{iT^a\tilde{\pi}^a(t,\vec{x})} = \Lambda(t,\vec{x})\, e^{iT^a\pi^a(t,\vec{x})} \;. \tag{168}$$

Notice that if for a moment we consider the case in which the gauge theory is a $U(1)$ theory, we would have

$$\Lambda = e^{i\alpha(\vec{x},t)} \;, \qquad \Rightarrow \qquad \pi \to \tilde{\pi} = \pi + \alpha \tag{169}$$

π shifts under a gauge transformation. This is a non-linear transformation because 0 is not mapped into 0. Gauge invariance has been restored by reintroducing a dynamical field that however, transforms non-linearly. Gauge invariance is non-linearly realized.

Going to canonical normalization

$$\frac{m^2}{g^2}(\partial\pi)^2 \quad \Rightarrow \quad \pi_c \equiv m/g \cdot \pi \tag{170}$$

we see that the Goldstone boson self-interactions become strongly coupled at the scale $4\pi m/g$, which is parametrically higher than the mass of the gauge bosons. The advantage of reintroducing the Goldstones is that for energies $E \gg m$ the mixing between them and the transverse components of the gauge field becomes irrelevant, so that the two sectors decouple. Mixing terms in eq. (166) are in fact of the form

$$\frac{m^2}{g} A^a_\mu \partial^\mu \pi^a = m A^a_\mu \partial^\mu \pi^a_c \tag{171}$$

which are irrelevant with respect to the canonical kinetic term $(\partial\pi_c)^2$ for $E \gg m$.

Notice that from expanding the term $D_\mu U D^\mu U$ we obtain irrelevant (i.e. non-renormalizable) terms of the form

$$\frac{m^2}{g^2}\pi^2(\partial\pi)^2 \sim \frac{1}{m^2/g^2}\pi_c^2(\partial\pi_c)^2 \,. \tag{172}$$

This is an operator that becomes strongly coupled and leads to unitarity violation at energies $E \sim 4\pi m/g$.

In the window $m \ll E \ll 4\pi m/g$ the physics of the Goldstone π is weakly coupled and it can be studied neglecting the mixing with transverse components.

Let us follow the same steps for our case of broken time diffeomorphisms. Let us concentrate for instance on the two operators:

$$\int d^4x\,\sqrt{-g}\left[A(t) + B(t)g^{00}(x)\right] \,. \tag{173}$$

Under a broken time diff. $t \to \widetilde{t} = t + \xi^0(x)$, $\vec{x} \to \vec{\widetilde{x}} = \vec{x}$, g^{00} transforms as:

$$g^{00}(x) \to \widetilde{g}^{00}(\widetilde{x}(x)) = \frac{\partial\widetilde{x}^0(x)}{\partial x^\mu}\frac{\partial\widetilde{x}^0(x)}{\partial x^\nu}g^{\mu\nu}(x)\,. \tag{174}$$

The action written in terms of the transformed fields is given by:

$$\int d^4x\,\sqrt{-\widetilde{g}(\widetilde{x}(x))}\left|\frac{\partial\widetilde{x}}{\partial x}\right|\left[A(t) + B(t)\frac{\partial x^0}{\partial\widetilde{x}^\mu}\frac{\partial x^0}{\partial\widetilde{x}^\nu}\widetilde{g}^{\mu\nu}(\widetilde{x}(x))\right] \,. \tag{175}$$

Changing integration variables to $\widetilde{x}$, we get:

$$\begin{aligned}&\int d^4\widetilde{x}\,\sqrt{-\widetilde{g}(\widetilde{x})}\,\Big[A(\widetilde{t} - \xi^0(x(\widetilde{x})))\\ &\quad + B(\widetilde{t} - \xi^0(x(\widetilde{x})))\frac{\partial(\widetilde{t} - \xi^0(x(\widetilde{x})))}{\partial\widetilde{x}^\mu}\frac{\partial(\widetilde{t} - \xi^0(x(\widetilde{x})))}{\partial\widetilde{x}^\nu}\widetilde{g}^{\mu\nu}(\widetilde{x})\Big] \,.\end{aligned} \tag{176}$$

The procedure to reintroduce the Goldstone is now similar to the gauge theory case. Whenever ξ^0 appears in the action above, we make the substitution

$$\xi^0(x(\widetilde{x})) \to -\widetilde{\pi}(\widetilde{x})\,. \tag{177}$$

This gives, dropping the tildes for simplicity:

$$\int d^4x\sqrt{-g(x)}\left[A(t+\pi(x)) + B(t+\pi(x))\frac{\partial(t+\pi(x))}{\partial x^\mu}\frac{\partial(t+\pi(x))}{\partial x^\nu}g^{\mu\nu}(x)\right] \,. \tag{178}$$

One can check that the action above is invariant under diffs at all orders (and not only for infinitesimal transformations) upon assigning to π the transformation rule

$$\pi(x) \to \widetilde{\pi}(\widetilde{x}(x)) = \pi(x) - \xi^0(x) \; . \tag{179}$$

With this definition π transforms as a scalar field plus an additional shift under time diffs. Notice that diff. invariant terms did not get a π.

Applying this procedure to the unitary gauge action (162) we obtain

$$\begin{aligned} S = \int d^4x \, \sqrt{-g} \, & \left[\frac{1}{2} M_{\rm Pl}^2 R - M_{\rm Pl}^2 \left(3H^2(t+\pi) + \dot{H}(t+\pi) \right) \right. \\ & + M_{\rm Pl}^2 \dot{H}(t+\pi) \left((\partial_\mu (t+\pi) \partial_\nu (t+\pi) g^{\mu\nu} \right) \\ & + \frac{M_2(t+\pi)^4}{2!} \left(\partial_\mu (t+\pi) \partial_\nu (t+\pi) g^{\mu\nu} + 1 \right)^2 \\ & \left. + \frac{M_3(t+\pi)^4}{3!} \left(\partial_\mu (t+\pi) \partial_\nu (t+\pi) g^{\mu\nu} + 1 \right)^3 + \ldots \right] , \end{aligned} \tag{180}$$

where for the moment we have neglected for simplicity terms that involve the extrinsic curvature.

This action is rather complicated, and at this point it is not clear what is the advantage of reintroducing the Goldstone π from the unitary gauge Lagrangian. In analogy with the gauge theory case, the simplification occurs because, at sufficiently short distances, the physics of the Goldstone can be studied neglecting metric fluctuations (this is nothing but the equivalence principle). As for the gauge theory case, the regime for which this is possible can be estimated just looking at the mixing terms in the Lagrangian above. In eq.(180) we see in fact that quadratic terms which mix π and $g_{\mu\nu}$ contain fewer derivatives than the kinetic term of π so that they can be neglected above some high energy scale. In general the answer will depend on which operators are present. Let us here just do the simplest case in which only the tadpole terms are relevant ($M_2 = M_3 = \ldots = 0$). This includes the standard slow-roll inflation case. The leading mixing with gravity will come from a term of the form

$$\sim M_{\rm Pl}^2 \dot{H} \dot{\pi} \delta g^{00} \; . \tag{181}$$

We see that

$$\begin{aligned} &\text{Kinetic term} \sim M_{\rm Pl}^2 \dot{H} \delta g^{00} \to M_{\rm Pl}^2 \dot{H} \left(\partial_\mu (t+\pi) \partial_\nu (t+\pi) g^{\mu\nu} \right) \supset M_{\rm Pl}^2 \dot{H} \dot{\pi}^2 \\ &\text{Mixing term} \sim M_{\rm Pl}^2 \dot{H} \delta g^{00} \to M_{\rm Pl}^2 \dot{H} \left(\partial_\mu (t+\pi) \partial_\nu (t+\pi) g^{\mu\nu} \right) \supset M_{\rm Pl}^2 \dot{H} \delta g^{00} \dot{\pi} \end{aligned} \tag{182}$$

δg^{00} is a constrained variable, it is the gravitational potential, and it is determined by π. At short distances, the Newtonian approximation holds:

$$M_{\rm Pl}^2 H \partial_i \delta g^{00} \sim\sim \dot{H} M_{\rm Pl}^2 \partial_i \pi \quad \Rightarrow \quad \delta g^{00} \sim \frac{\dot{H}}{H}\pi \,. \tag{183}$$

We have

$$\frac{\text{Mixing term}}{\text{Kinetic term}} \sim \frac{\delta g^{00}}{\dot{\pi}} \sim \frac{\dot{H}}{H}\frac{\pi}{\dot{\pi}} \sim \frac{\dot{H}}{EH} \ll 1 \quad \Rightarrow \quad E \gg \epsilon H \,, \tag{184}$$

where we have used that at energies of order E, $\partial_t \sim E$. The mixing term is negligible in the UV (GR equivalence principle). The actual scale $E_{\rm mix}$ at which the mixing can be neglected depends on the actual operators turned on, but it is guaranteed that at energies $E \gg E_{\rm mix}$ we can neglect the mixing terms.

In the regime $E \gg E_{\rm mix}$ the action dramatically simplifies to

$$S_\pi = \int d^4x \sqrt{-g} \left[\frac{1}{2} M_{\rm Pl}^2 R - M_{\rm Pl}^2 \dot{H} \left(\dot{\pi}^2 - \frac{(\partial_i \pi)^2}{a^2} \right) \right. \\ \left. + 2M_2^4 \left(\dot{\pi}^2 + \dot{\pi}^3 - \dot{\pi}\frac{(\partial_i \pi)^2}{a^2} \right) - \frac{4}{3} M_3^4 \dot{\pi}^3 + \ldots \right] . \tag{185}$$

Given an inflationary model, one is interested in computing predictions for present cosmological observations. From this point of view, it seems that the decoupling limit (185) is completely irrelevant for these extremely infrared scales. However, as for standard single field slow-roll inflation, one can prove that there exists a quantity, the usual ζ variable, which is constant out of the horizon at any order in perturbation theory

Therefore the problem is reduced to calculating correlation functions just after horizon crossing. We are therefore interested in studying our Lagrangian with an IR energy cutoff of order H. If the decoupling scale $E_{\rm mix}$ is smaller than H, the Lagrangian for π (185) will give the correct predictions up to terms suppressed by $E_{\rm mix}/H$. When this is not the case, nothing dramatic happens: we simply have to keep also the metric fluctuations.

This is the justification of the calculations we did in lecture 2.

As we discussed, we are assuming that the time dependence of the coefficients in the unitary gauge Lagrangian is slow compared to the Hubble time, that is, suppressed by some generalized slow roll parameters. This implies that the additional π terms coming from the Taylor expansion of the coefficients are small. In particular, the relevant operators, *i.e.* the ones which dominate moving towards the infrared, like the cubic term, are unimportant at the scale H and have therefore been neglected in the Lagrangian (185). They can be nevertheless straightforwardly included, as done in [22, 23].

In conclusion, with the Lagrangian (185) one is able to compute all the observables which are not dominated by the mixing with gravity, like for example the non-Gaussianities in standard slow-roll inflation [9, 25]. Notice however that the tilt of the spectrum can be calculated, at leading order, with the Lagrangian (185). As we saw earlier, its value can in fact be deduced simply by the power spectrum at horizon crossing computed neglecting the mixing terms. It is important to stress that our approach does not lose its validity when the mixing with gravity is important so that the Goldstone action is not sufficient for predictions. The action (162) contains all the information about the model and can be used to calculate all predictions even when the mixing with gravity is large.

Let us stress a few points

- The above Lagrangian is very simple, and it unifies all single-degree-of-freedon inflationary models.
- It describes the theory of the fluctuations, which is what we are actually testing.
- It is analogous to the Chiral Lagrangian of particle physics. Indeed, it is telling us that from the experimental point of view, inflation is the theory of a Goldstone boson
- Since it encodes all possible single-clock models on inflation, it allows to prove theorems on the possible signals.
- It also allows to explore all possible signatures.
- What is forced by symmetries, what are the allowed operators and what is possible to do is made clear. For example, the coefficient of $(\partial_i\pi)^2$ is fixed to be $\dot{H}M_{\rm Pl}^2$. This is not the case for $\dot{\pi}^2$. This tells us that at leading order in derivatives it is impossible to violate the null energy condition. $\dot{H} > 0$ implies that the spatial kinetic term for π has the negative-energy sign, and so it leads to an uncontrollable instability. The EFT also tells you how this problem can be fixed, by adding higher derivative terms. Indeed all currently known ways to violate the null Energy Condition (NEC) that are currently known have been found in this context.
- This formalism is very prone to do with it what we normally do for the beyond the standard model physics: one can add symmetries to enhance operators with respect to others, or one can try to UV complete some specific models.

- Being explicitly a theory for the fluctuations, it allows to assess the important of operators very easily. For example, in the standard treatment with scalar fields, an operators $(\partial\phi)^8$ contributes to the quadratic action with $\dot{\phi}_0^6(\partial\delta\phi)^2$. This is also very useful for studying loop corrections. At a fixed order in fluctuations and derivatives, in the EFT there is a finite number of counter terms, while this is not so with the scalar field theory. Indeed the EFT formalism was crucial to prove the constancy of ζ at quantum level [29].

4.6. *Rigorous calculation of the power spectrum in unitary gauge*

We are now ready to see the new spectacular signatures of inflation. But I really feel that it is time for us to do a rigorous calculation. Notice that we got so far without having to do one at all. Pretty good I would say. However, there is little more rewarding that seeing your simple estimates being confirmed by a somewhat tricky calculation.

We just saw that we could neglect metric perturbations for standard slow roll inflation. And indeed we did the correct calculation neglecting them. Additionally, we saw that using π makes it explicit this fact. In order to see that we did not loose anything, we will now do the calculation in an un-intuitive gauge: the so called ζ-gauge or Maldacena-gauge. This is one of the gauges that are possible in our unitary gauge. Even though it is unintuitive, it is good for something. Indeed, it is the absolutely best gauge to study the tricky infrared properties of ζ, the variable we ultimately need to compute. In this gauge, we will see that in the infrared ζ becomes constant. Unfortunately, as we discussed, unitary gauges are the worst possible gauges to see the decoupling of matter perturbations from metric perturbations. I am not aware of a gauge which is equally nice both in the UV and the IR at the same time. I will however show you later how to do the calculation using π.

We said that we want to compute the correlation function of ζ. Let us write the metric in the so-called ADM parametrization

$$ds^2 = -N^2dt^2 + h_{ij}\left(dx^i + N^i dt\right)\left(dx^j + N^j dt\right) \ . \tag{186}$$

We have to quantize a system with Gauge redundancy. In our case the gauge freedom (sometimes historically and wrongly called gauge symmetry) is time-dependent spatial diff invariance. The quantization is tricky, but it is the same as for gauge theories. Just a different symmetry group. The procedure is the following (see Weinberg's QFT I and II books).

- Expand the action. In ADM parametrization, it reads

$$S = \frac{1}{2}\int \sqrt{h}\ \Big[NR^{(3)} + \frac{1}{N}\left(E_{ij}E^{ij} - E_i^{i2}\right) \\ -2M_{\rm Pl}^2\dot{H}\cdot\frac{1}{N} - M_{\rm Pl}^2(3H^2+\dot{H})\cdot N\Big] \qquad (187)$$

where

$$E_{ij} = \frac{1}{2}\left[\partial_t h_{ij} + \nabla_i N_j + \nabla_j N_i\right] \qquad (188)$$

and ∇ is the covariant derivative with respect to h_{ij}.
Now take equations of motions with respect to all fluctuating degrees of freedom.

$$\frac{\delta S}{\delta\, \delta g^{\mu\nu}} = 0\ . \qquad (189)$$

- For semplicity, we do the calculation for $M_{2,\ldots} = 0$ (this includes slow roll inflation). In this case the equations of motion for N and N_i take the following form

$$\nabla_i\left[N^{-1}\left(E_j^i - \delta E\right)\right] = 0 \qquad (190)$$

$$M_{\rm Pl}^2\left[R^{(3)} - \frac{1}{N^2}\left(E_{ij}E^{ij} - E_i^{i2}\right)\right] - (3H^2+\dot{H}) + 2M_{\rm Pl}^2\dot{H}\cdot\frac{1}{N^2} = 0\ .$$

These two equations are extremely important. Notice that no time derivative acts on N nor on N_i. These tells us that N and N_i are *constrained* variables: they are known once you specify what the other degrees of freedom do. They are *not* independent degrees of freedom. They are very much (and not by chance) like the gravitation potential in Newtonian gravity, or the Electric potential in electrostatic.
- Let us count the degrees of freedom. We started with the metric, which has 10 components. But we have 3-independent gauge generators for the spatial diffs. This means that we can set 3 of these components to any value we want (including 0). This means that they are not degrees of freedom. For example we can set to zero 3 components of h_{ij}. Then from above, we see that N, N^i are 4 constrained variables. So they are also not degrees of freedom. We are left with

$$\text{number of degrees of freedom} = 10 - 3 - 4 = 3\ . \qquad (191)$$

Does this work? We should have the two elicities of the graviton and the matter degree of freedom (equivalent to π): 3. Ok, we are on!

- Fix a gauge. Fix the spatial diffs by fixing the spatial metric to be
$$h_{ij} = a^2 \delta_{ij} e^{2\zeta} \ . \tag{192}$$
I am neglecting tensor perturbations here, because as said at quadratic level they do not mix. This gauge is called ζ-gauge or Maldacena-gauge.
- In this gauge you can clearly see why $\zeta = \delta a/a$. Assuming that N and N_i go to their unperturbed value when $k/(aH) \to 0$,[k] then we see that we are in an perturbed FRW (as $\delta\phi = 0$), with just a δa.
- The constrained variables N and N^i are constrained, and so we can solve for them in terms of the only remaining degree of freedom: ζ. The solution reads
$$N = 1 + \frac{\dot{\zeta}}{H} \ , \qquad N_i = \partial_i \left(-\frac{1}{a^2} \frac{\zeta}{H} - \frac{\dot{H}}{H^2} \frac{1}{\partial^2} \dot{\zeta} \right) \ . \tag{193}$$
- Plug back this values for N and N^i in the action. Notice, you can do this only because they are constrained variables. The action now reads
$$S = \int d^4x \, a^3 \left(-\frac{\dot{H}}{H^2} \right) M_{\rm Pl}^2 \left[\dot{\zeta}^2 - \frac{1}{a^2} (\partial_i \zeta)^2 \right] \ . \tag{194}$$
- Let us quantize the system. Follow textbook: find
$$\Pi_\zeta = \frac{\delta \mathcal{L}}{\delta \dot{\zeta}} = -2a^3 \left(\frac{\dot{H}}{H^2} \right) \dot{\zeta} \tag{195}$$
$$[\zeta, P_\zeta] = i \ . \tag{196}$$
This is a quadratic Lagrangian, so we simply expand the fourier components of ζ in annihilation and creation operators
$$\hat{\zeta}_{\vec{k}}(t) = \zeta^{cl}_{\vec{k}}(t) a^\dagger_{\vec{k}} + \zeta^\star_{cl}(t) a_{-\vec{k}} \tag{197}$$
with ζ^{cl} satisfying the equation of motion (Heisemberg equation for $\hat{\zeta}$)
$$0 = \frac{\delta L}{\delta \zeta} = \frac{d\left(a^3 \left(-\frac{\dot{H}}{H^2} \dot{\zeta}^{cl}_k \right) \right)}{dt} + \frac{\dot{H}}{H^2} a k^2 \zeta^{cl}_k \ . \tag{198}$$
This is a second order equation, that requires two initial conditions. This condition can be found in the following way. We define the vacuum state as the state annihilated by $a_{\vec{k}}$:
$$a_{\vec{k}} |0\rangle = 0 \tag{199}$$

[k] Indeed this will be true because N, N_i are constrained variables sourced by the gradients of ζ.

but what this state is actually depends on what we choose as ζ^{cl}. How do we choose it? Well, we know that at early times, the mode $k/a \gg 1$, so we would like the solution to be the same as in Minkowski space (this is GR!). In other words, the vacuum state for modes well inside H^{-1} should be the same as in flat space. This give the following condition

$$\zeta_k^{cl}(-k\eta \gg 1) \sim \frac{-i}{(2\epsilon)^{1/2} M_{\rm Pl} a(\eta)^3} \frac{1}{(2k/a(\eta))^{1/2}} e^{ik\eta} \tag{200}$$

for $\frac{k}{aH} = -k\eta \gg 1$. Notice that the exponential reads $k\eta \simeq \frac{k}{a} a\eta \simeq k_{\rm phys} t$. The pre factors come from the canonical normalization. This is the solution that we would get for an harmonic oscillator $1/\sqrt{2\omega}$ after we take into account of the rescaling to make the field canonical.

- Now we can solve the linear equation. Since at early times the Hubble expansion is negligible, and at late times ζ goes to a constant, we can neglect the time dependence of $H, \dot{H}$, and evaluate those terms at freeze out (it is possible to solve that equation exactly at first order in slow roll parameters. You can do this yourself). Using Mathematica, the solution reads

$$\zeta_k^{cl}(\eta) = \frac{1}{(2\epsilon)^{1/2} M_{\rm Pl}} \frac{1}{(2k)^{3/2}} (1 - ik\eta) e^{ik\eta} \,. \tag{201}$$

- We can now compute the power spectrum:

$$\begin{aligned} \langle 0 | \zeta_{\vec{k}}(\eta) \zeta_{\vec{k}'}(\eta') | 0 \rangle &= (2\pi)^3 \delta^3(\vec{k} + \vec{k}') \\ &\times \frac{1}{(2\epsilon)^{1/2} M_{\rm Pl}} \frac{1}{(2k)^{3/2}} (1 - ik\eta) e^{ik\eta} \frac{1}{(2\epsilon)^{1/2} M_{\rm Pl}} \frac{1}{(2k)^{3/2}} (1 + ik\eta') e^{-ik\eta'} \end{aligned} \tag{202}$$

when $k\eta \ll 1$ and $k\eta' \ll 1$, we obtain

$$\langle \zeta_{\vec{k}} \zeta_{\vec{k}'} \rangle_{\rm late} = (2\pi)^3 \delta^3(\vec{k} + \vec{k}') \, \frac{1}{k^3} \cdot \frac{H^4}{4(-\dot{H}) M_{\rm Pl}^2} \,. \tag{203}$$

Which nicely reproduces the results we found with our estimates (but now we even got the factor of 4!).

- One can compute correlation functions not on the vacuum state. Vacuum is somewhat better justified, though generalizations have been considered (see for example [30]).
- We could have done exactly the same calculation in a gauge where π is not zero, and we fix space and time diffs so that $h_{ij} = a^2 \delta_{ij}$. The only subtlety is that at late times π is *not* constant, but ζ is the constant quantity. We therefore need a relationship between π and ζ. This is

given by performing a time-diff $\delta t = \pi$ to go from π-gauge to ζ-gauge. Quite intuitively, the relationship is

$$\zeta = -H\pi \ . \tag{204}$$

So, one computs $\langle\pi\pi\rangle$ up to freeze out, and then one switches to ζ. Very simply

$$\langle\zeta_{\vec{k}}\zeta_{\vec{k}'}\rangle_{\rm late} = H^2_{f.o.}\langle\pi_{\vec{k}}\pi_{\vec{k}'}\rangle_{f.o.} \tag{205}$$

and one has never to talk about constraint equations and metric variables.

4.6.1. *The Various Limits of Single Field Inflation*

Slow-roll Inflation and High Energy Corrections

The simplest example of the general Lagrangian (162) is obtained by keeping only the first three terms, which are fixed once we know the background Hubble parameter $H(t)$, and setting to zero all the other operators of higher order: $M_2 = M_3 = \bar{M}_1 = \bar{M}_2 \ldots = 0$. In the ϕ language, this corresponds to standard slow-roll inflation, with no higher order terms. We have already done this case, both using π or using ζ.

Notice however that not all observables can be calculated from the π Lagrangian (185): this happens when the leading result comes from the mixing with gravity or is of higher order in the slow-roll expansion. For example, as the first two terms of Eq. (185) do not contain self-interactions of π, the 3-point function $\langle\zeta(\vec{k}_1)\zeta(\vec{k}_2)\zeta(\vec{k}_3)\rangle$would be zero. One is therefore forced to look at subleading corrections, taking into account the mixing with gravity in Eq. (180).

Obviously our choice of setting to zero all the higher order terms cannot be exactly true. At the very least they will be radiatively generated even if we put them to zero at tree level. The theory is non-renormalizable and all interactions will be generated with divergent coefficients at sufficiently high order in the perturbative expansion. As additional terms are generated by graviton loops, they may be very small. For example it is straightforward to check that starting from the unitary gauge interaction $M_{\rm Pl}^2\dot{H}g^{00}$ a term of the form $(\delta g^{00})^2$ will be generated with a logarithmically divergent coefficient $M_2^4 \sim \dot{H}^2\log\Lambda$. This implies that one should assume $M_2^4 \gtrsim \dot{H}^2$.[1]

[1]The explicit calculation of logarithmic divergences in a theory of a massless scalar coupled to gravity has been carried out a long time ago in [28].

This lower limit is however very small. For example the dispersion relation of π will be changed by the additional contribution to the time kinetic term: this implies, as we will discuss thoroughly below, that the speed of π excitations deviates slightly from the speed of light, by a relative amount $1 - c_s^2 \sim M_2^4/(|\dot{H}|M_{\rm Pl}^2) \sim |\dot{H}|/M_{\rm Pl}^2$. Using the normalization of the scalar spectrum, we see that the deviation from the speed of light is $\gtrsim \epsilon^2 \cdot 10^{-10}$. A not very interesting lower limit.

The size of the additional operators will be much larger if additional physics enters below the Planck scale. In general this approach gives the correct parametrization of all possible effects of new physics. As usual in an effective field theory approach, the details of the UV completion of the model are encoded in the higher dimension operators. This is very similar to what happens in physics beyond the Standard Model. At low energy the possible effects of new physics are encoded in a series of higher dimensional operators compatible with the symmetries [26]. The detailed experimental study of the Standard model allows us to put severe limits on the size of these higher dimensional operators. The same can be done in our case, although the set of conceivable observations is unfortunately much more limited.

Small Speed of Sound and Large Non-Gaussianities

The Goldstone action (185) shows that the spatial kinetic term $(\partial_i\pi)^2$ is completely fixed by the background evolution to be $M_{\rm Pl}^2\dot{H}(\partial_i\pi)^2$. In particular only for $\dot{H} < 0$, it has the "healthy" negative sign. This is an example of the well studied relationship between violation of the null energy condition, which in a FRW Universe is equivalent to $\dot{H} < 0$, and the presence of instabilities in the system. Notice however that the wrong sign of the operator $(\partial_i\pi)^2$ is not enough to conclude that the system is pathological: higher order terms like $\delta K^{\mu}{}_{\mu}{}^2$ may become important in particular regimes, as we will discuss thoroughly below.

The coefficient of the time kinetic term $\dot{\pi}^2$ is, on the other hand, not completely fixed by the background evolution, as it receives a contribution also from the quadratic operator $(\delta g^{00})^2$. In eq. (185) we have

$$\left(-M_{\rm Pl}^2\dot{H} + 2M_2^4\right)\dot{\pi}^2 \, . \tag{206}$$

To avoid instabilities we must have $-M_{\rm Pl}^2\dot{H} + 2M_2^4 > 0$. As time and spatial kinetic terms have different coefficients, π waves will have a "speed of sound" $c_s \neq 1$. This is expected as the background spontaneously breaks

Lorentz invariance, so that $c_s = 1$ is not protected by any symmetry. As we discussed in the last section, deviation from $c_s = 1$ will be induced at the very least by graviton loops.[m] The speed of sound is given by

$$c_s^{-2} = 1 - \frac{2M_2^4}{M_{\rm Pl}^2 \dot{H}} \,. \tag{207}$$

This implies that in order to avoid superluminal propagation we must have $M_2^4 > 0$ (assuming $\dot{H} < 0$). Superluminal propagation would imply that the theory has no Lorentz invariant UV completion [27]. In the following we will concentrate on the case $c_s \leq 1$.

Using the equation above for c_s^2 the Goldstone action can be written at cubic order as

$$\begin{aligned} S_\pi = \int d^4x \, \sqrt{-g} \, \Bigg[& -\frac{M_{\rm Pl}^2 \dot{H}}{c_s^2} \left(\dot{\pi}^2 - c_s^2 \frac{(\partial_i \pi)^2}{a^2} \right) \\ & + M_{\rm Pl}^2 \dot{H} \left(1 - \frac{1}{c_s^2} \right) \left(\dot{\pi}^3 - \dot{\pi} \frac{(\partial_i \pi)^2}{a^2} \right) - \frac{4}{3} M_3^4 \dot{\pi}^3 \ldots \Bigg] \,. \end{aligned} \tag{208}$$

From the discussion in section (4.5.2) we know that the mixing with gravity can be neglected at energies $E \gg E_{\rm mix} \simeq \epsilon H$.

The calculation of the 2-point function follows closely the case $c_s = 1$ if we use a rescaled momentum $\bar{k} = c_s k$ and take into account the additional factor c_s^{-2} in front of the time kinetic term. We obtain

$$\begin{aligned} \langle \zeta(\vec{k}_1) \zeta(\vec{k}_2) \rangle &= (2\pi)^3 \delta(\vec{k}_1 + \vec{k}_2) \frac{1}{c_{s*}} \cdot \frac{H_*^4}{4 M_{\rm Pl}^2 |\dot{H}_*|} \frac{1}{k_1^3} \\ &= (2\pi)^3 \delta(\vec{k}_1 + \vec{k}_2) \frac{1}{c_{s*}} \cdot \frac{H_*^2}{4 \epsilon_* M_{\rm Pl}^2} \frac{1}{k_1^3} \,. \end{aligned} \tag{209}$$

The variation with time of the speed of sound introduces an additional contribution to the tilt

$$n_s = \frac{d}{d \log k} \log \frac{H_*^4}{|\dot{H}_*| c_{s*}} = \frac{1}{H_*} \frac{d}{dt_*} \log \frac{H_*^4}{|\dot{H}_*| c_{s*}} = 4 \frac{\dot{H}_*}{H_*^2} - \frac{\ddot{H}_*}{\dot{H}_* H_*} - \frac{\dot{c}_{s*}}{c_{s*} H_*} \,. \tag{210}$$

[m] If we neglect the coupling with gravity and the time dependence of the operators in the unitary gauge Lagrangian (so that $\pi \to \pi + \text{const}$ is a symmetry), $c_s = 1$ can be protected by a symmetry $\partial_\mu \pi \to \partial_\mu \pi + v_\mu$, where v_μ is a constant vector. Under this symmetry the Lorentz invariant kinetic term of π changes by a total derivative, while the operator proportional to M_2^4 in Eq. (185) is clearly not invariant, so that $c_s = 1$. Notice that the theory is not free as we are allowed to write interactions with more derivatives acting on π. This symmetry appears in the study of the brane bending mode of the DGP model.

From the action (208) we clearly see that the same operator giving a reduced speed of sound induces cubic couplings of the Goldstones of the form $\dot{\pi}(\nabla\pi)^2$ and $\dot{\pi}^3$. The non-linear realization of time diffeomorphisms forces a relation between a reduced speed of sound and an enhanced level of the 3-point function correlator, *i.e.* non-Gaussianities. Indeed remember that the ϕ-wavefunction was a Gaussian in the vacuum state simply because the action was quadratic in the fields. Interactions will lead to deviation from a Gaussian wavefunction: i.e. non-Gaussianities.

To estimate the size of non-Gaussianities, one has to compare the non-linear corrections with the quadratic terms around freezing, $\omega \sim H$. In the limit $c_s \ll 1$, the operator $\dot{\pi}(\nabla\pi)^2$ gives the leading contribution, as the quadratic action shows that a mode freezes with $k/a \sim H/c_s$, so that spatial derivatives are enhanced with respect to time derivatives. Notice indeed that

$$H \sim \omega \sim c_s \frac{k}{a} \,, \quad \Rightarrow \quad \frac{k}{a(t_{f.o.})} \sim \frac{H}{c_s} \gg H \,. \tag{211}$$

The level of non-Gaussianity will thus be given by the ratio:

$$\frac{\mathcal{L}_{\dot{\pi}(\nabla\pi)^2}}{\mathcal{L}_2} \sim \frac{H\pi\left(\frac{H}{c_s}\pi\right)^2}{H^2\pi^2} \sim \frac{H}{c_s^2}\pi \sim \frac{1}{c_s^2}\zeta \,, \tag{212}$$

where in the last step we have used the linear relationship between π and ζ. Taking $\zeta \sim 10^{-5}$ we have an estimate of the size of the non-linear correction. Usually the magnitude of non-Gaussianities is given in terms of the parameters $f_{\rm NL}$, which are parametrically of the form:

$$\frac{\mathcal{L}_{\dot{\pi}(\nabla\pi)^2}}{\mathcal{L}_2} \sim f_{\rm NL}\zeta \,. \tag{213}$$

The leading contribution will thus give

$$f^{\rm equil.}_{{\rm NL},\,\dot{\pi}(\nabla\pi)^2} \sim \frac{1}{c_s^2} \,. \tag{214}$$

The superscript "equil." refers to the momentum dependence of the 3-point function, which in these models is of the so called equilateral or orthogonal form. This is physically clear in the Goldstone language as the relevant π interactions contain derivatives, so that they die out quickly out of the horizon; the correlation is only among modes with comparable wavelength.

In the Goldstone Lagrangian (208) there is an additional independent operator, $-\frac{4}{3}M_3^4\dot{\pi}^3$, contributing to the 3-point function, coming from the unitary gauge operator $(\delta g^{00})^3$. We thus have two contributions of the form $\dot\pi^3$ which give

$$f^{\rm equil.}_{{\rm NL},\,\dot\pi^3} \sim 1 - \frac{4}{3}\frac{M_3^4}{M_{\rm Pl}^2|\dot H|c_s^{-2}} \,. \tag{215}$$

The size of the operator $-\frac{4}{3}M_3^4\dot\pi^3$ is not constrained by the non-linear realization of time diffeomorphisms: it is a free parameter. In DBI inflation [17] we have $M_3^4 \sim M_{\rm Pl}^2|\dot H|c_s^{-4}$, so that its contribution to non-Gaussianities is of the same order as the one of eq. (214). The same approximate size of the M_3^4 is obtained if we assume that both the unitary gauge operators $M_2^4(\delta g^{00})^2$ and $M_3^4(\delta g^{00})^3$ become strongly coupled at the same energy scale.

Cutoff and Naturalness

As discussed, for $c_s < 1$ the Goldstone action contains non-renormalizable interactions. Therefore the self-interactions among the Goldstones will become strongly coupled at a certain energy scale, which sets the cutoff of our theory. This cutoff can be estimated looking at tree level partial wave unitarity, *i.e.* finding the maximum energy at which the tree level scattering of πs is unitary. The calculation is straightforward, the only complication coming from the non-relativistic dispersion relation. The cutoff scale Λ turns out to be

$$\Lambda^4 \simeq 16\pi^2 M_2^4\frac{c_s^7}{(1-c_s^2)^2} \simeq 16\pi^2 M_{\rm Pl}^2|\dot H|\frac{c_s^5}{1-c_s^2} \,. \tag{216}$$

The same result can be obtained looking at the energy scale where loop corrections to the $\pi\pi$ scattering amplitude become relevant. As expected the theory becomes more and more strongly coupled for small c_s, so that the cutoff scale decreases. On the other hand, for $c_s \to 1$ the cutoff becomes higher and higher. This makes sense as there are no non-renormalizable interactions in this limit and the cutoff can be extended up to the Planck scale. This cutoff scale is obtained just looking at the unitary gauge operator $(\delta g^{00})^2$; depending on their size the other independent operators may give an even lower energy cutoff. Notice that the scale Λ indicates the maximum energy at which our theory is weakly coupled and make sense; below this scale new physics must come into the game. However new physics can appear even much below Λ.

If we are interested in using our Lagrangian for making predictions for cosmological correlation functions, then we need to use it at a scale of order the Hubble parameter H during inflation. We therefore need that this energy scale is below the cutoff, $H \ll \Lambda$. Using the explicit expression for the cutoff (216) in the case $c_s \ll 1$ one gets

$$H^4 \ll M_{\rm Pl}^2 |\dot{H}| c_s^5 \tag{217}$$

which can be rewritten using the spectrum normalization (209) as an inequality for the speed of sound

$$c_s \gg P_\zeta^{1/4} \simeq 0.003 \,. \tag{218}$$

A theory with a lower speed of sound is strongly coupled at $E \simeq H$. Not surprisingly this value of the speed of sound also corresponds to the value at which non-Gaussianity are of order one: the theory is strongly coupled at the energy scale H relevant for cosmological predictions.

Let us comment on the naturalness of the theory. One may wonder whether the limit of small c_s is natural or instead loop corrections will induce a larger value. The Goldstone self-interactions, $\dot{\pi}(\nabla\pi)^2$ and $(\nabla\pi)^4$ for example, will induce a radiative contribution to $(\nabla\pi)^2$. It is easy to estimate that these contributions are of order $c_s^{-5}\Lambda^4/(16\pi^2 M_2^4)$, where Λ is the UV cutoff, *i.e.* the energy scale at which new physics enters in the game. We can see that it is impossible to have large radiative contribution; even if we take Λ at the unitarity limit (216), the effect is of the same order as the tree level value. This makes sense as the unitarity cutoff is indeed the energy scale at which loop corrections become of order one.

We would like also to notice that the action (185) is natural from an effective field theory point of view [24]. The relevant operators are in fact protected from large renormalizations if we assume an approximate shift symmetry of π. In this case the coefficients of the relevant operators are sufficiently small and they will never become important for observations as cosmological correlation functions probe the theory at a fixed energy scale of order H: we never go to lower energy. Clearly here we are only looking at the period of inflation, where an approximate shift symmetry is enough to make the theory technically natural; providing a graceful exit from inflation and an efficient reheating are additional requirements for a working model which are not discussed in our formalism.

De-Sitter Limit and the Ghost Condensate

In the previous section we saw that the limit $c_s \to 0$ is pathological as the theory becomes more and more strongly coupled. However we have neglected in our discussion the higher derivative operators in the unitary gauge Lagrangian (162)

$$\int d^4x\sqrt{-g}\left(-\frac{\bar{M}_2(t)^2}{2}\delta K^\mu{}_\mu{}^2-\frac{\bar{M}_3(t)^2}{2}\delta K^\mu{}_\nu\delta K^\nu{}_\mu\right) . \tag{219}$$

These operators give rise in the Goldstone action to a spatial kinetic term of the form

$$\int d^4x\,\sqrt{-g}\left[-\frac{\bar{M}^2}{2}\frac{1}{a^4}(\partial_i^2\pi)^2\right] , \tag{220}$$

where $\bar{M}^2=\bar{M}_2^2+\bar{M}_3^2$. Notice that we obtain the very non-relativistic dispersion relation

$$\omega^2\sim\frac{k^4}{M^2} . \tag{221}$$

This models naturally leads to large non-Gaussianities.

De-Sitter Limit Without the Ghost Condensate

In this section we want to study the effect of the operator

$$\int d^4x\,\sqrt{-g}\,\left(-\frac{\bar{M}_1(t)^3}{2}\delta g^{00}\delta K^\mu{}_\mu\right) . \tag{222}$$

on the quadratic π action. We will see that, if the coefficient of this operator is sufficiently large, we obtain a different de Sitter limit, where the dispersion relation at freezing is of the form $\omega^2\propto k^2$, instead of the Ghost Condensate behavior $\omega^2\propto k^4$.

For simplicity we can take $\bar{M}_1$ to be time independent. Reintroducing the Goldstone we get a 3-derivative term of the form $-\bar{M}_1^3\dot{\pi}\nabla^2\pi/a^2$.[n] This would be a total time derivative without the time dependence of the scale factor $a(t)$ and of the metric determinant. Integrating by parts we get a standard 2-derivative spatial kinetic term

$$-\int d^4x\sqrt{-g}\,\frac{\bar{M}_1^3H}{2}\left(\frac{\partial_i}{a}\pi\right)^2 . \tag{223}$$

In the exact de Sitter limit, $\dot{H}=0$, and taking $M_2\sim\bar{M}_1\sim M$, this operator gives a dispersion relation of the form

$$c_s^2=\frac{H}{M}\ll 1 \tag{224}$$

and naturally to large non-Gaussianities.

[n]The operator gives also a contribution to $\dot{\pi}^2$ proportional to H. We will assume that this is small compared to $M_2^4\dot{\pi}^2$.

This and the Ghost condensate case are finally the only known ways to violet the Null Energy Condition in a stable way [31].

5. Lecture 4: Non-Gaussianity: Who Are You?

We have seen in the former section that we can have inflationary models with large self-interactions. We said that they produce some non-Gaussianity. Indeed we saw that in the limit of free-theory the vacuum wavefunction was a Gaussian. This was because the Lagrangian was quadratic and each fourier mode was like an harmonic oscillator. But if the action is slightly non-linear, than we can imagine some slight non-Gaussianity. Something like, just symbolically:

$$|0\rangle_{k_i/a\ll H} \sim \prod_{\{\zeta_{\vec{k}}\}} e^{-\frac{\zeta^2_{\vec{k}_i}}{\sigma_{\zeta_{\vec{k}_i}}}-\frac{\zeta_{\vec{k}_i}\zeta_{\vec{k}_j}\zeta_{\vec{k}_i+\vec{k}_j}}{C(\vec{k}_1,\vec{k}_2,\vec{k}_1+\vec{k}_2)}}|\{\zeta_{\vec{k}}\}\rangle\ . \tag{225}$$

This would mean that a signal like the three-point function

$$\langle\zeta_{\vec{k}_1}\zeta_{\vec{k}_2}\zeta_{\vec{k}_3}\rangle = (2\pi^3)\delta^{(3)}(\vec{k}_1+\vec{k}_2+\vec{k}_3)\bar{F}(\vec{k}_1,\vec{k}_2,\vec{k}_3) \tag{226}$$

would not be zero. Current limits sets

$$\frac{\langle\zeta^3\rangle}{\langle\zeta^2\rangle} \lesssim 10^{-2} \sim \frac{1}{N_{pix}^{1/2}} \tag{227}$$

which is a very small number! Look at the plot.

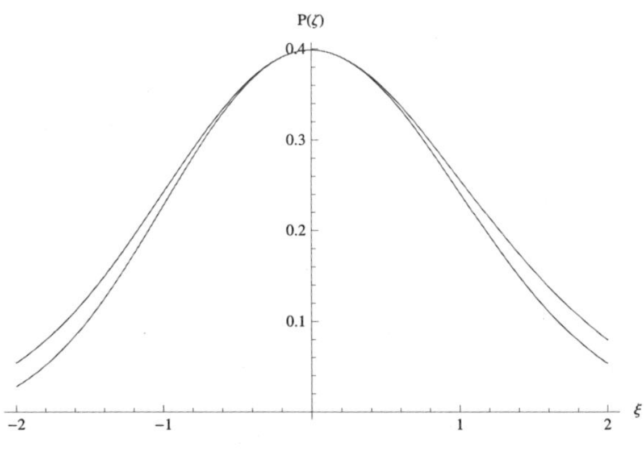

Fig. 18. Plot of a Gaussian Distribution and of a non-Gaussian distribution that has skewness approximately equal to 1. Can you tell which one is which? Cosmological observations are constraining the skewness of the distribution of the CMB radiation to be less than a percent of the one of the figure. Wow!

Being a limit on a statistics, the limit scale as $N_{pix}^{-1/2}$. For WMAP, we have indeed about 10^5 modes.

But what this tells us is that a detection of non-Gaussianities would be associated to the interacting part of the Lagrangian, which is really the interesting part of the Lagrangian! And we are talking of interactions at extremely high energies! Interactions contain so much more information that they would allow us to learn about the real dynamics that drove inflation.

Clearly, since non-Gaussianities are small, it is expectable that the leading signature will appear in the 3-point function.

Let us look at the function F. So far it depends on 9 variables. But let us use the symmetries of the problem. By the cyclic invariance of the correlation function (remember that at late times we are semiclassical), we can set $k_1 \geq k_2 \geq k_3$. Translation invariance forces the sum of the three momenta to be zero: they must form a closed triangle. We are down to 6. We can use to rotational invariance to point $\vec{k}_1$ in the $\hat{x}$ direction, and $\vec{k}_2$ in the $x - y$ plane. We are down to 3 variables. Additionally, the 3-point function should be scale invariant, because two triplet of modes, one an overall rescaling of the other, see approximately the same history. We can use this to set the modulus of $k_1 = 1$. The overall k_1 dependence has to be $1/k_1^6$, so that the real space 3-point function

$$\langle \zeta(x)^3 \rangle = \int d^3k_1 d^3k_2 d^3k_3 \, \langle \zeta_{\vec{k}_1} \zeta_{\vec{k}_2} \zeta_{\vec{k}_3} \rangle \tag{228}$$

receives the same contribution from each logarithmic interval.

So, in terms of degrees of freedom, we are down to two variables

$$\langle \zeta_{\vec{k}_1} \zeta_{\vec{k}_2} \zeta_{\vec{k}_3} \rangle = (2\pi^3)\delta^{(3)}(\vec{k}_1 + \vec{k}_2 + \vec{k}_3) \frac{1}{k_1^6} F(x_2, x_3) \tag{229}$$

$$x_2 = k_2/k_1 \ , \quad x_3 = k_3/k_1 \ , \quad x_3 \leq x_2 \leq 1 \ , \quad x_3 \leq x_2 \ .$$

This is a *huge* amount of information. Remember that because of the various symmetries, the 2-point function had to go as $1/k^3$, and so it dependent only on one *number*. Because of the slight deviation of scale invariance, we had also the tilt, which is just a second *number*. Here with non-Gaussianities, we are talking about a *function of 2 parameters*. This is ∞ numbers! this is a huge amount of information, incomparable with respect to the information contained in the 2-point function. Indeed, it has the same amount of information as a 2-2 scattering as a function of angles. And this is not little thing: we learn about spin of particles and nature of interactions from this. Let us plot F. A useful quantity to plot is a quantity the resembles the signal to noise ratio in each triangular configuration [34]. It is

$$\left.\frac{S}{N}\right|_{triangle} \sim x_2^2 x_3^2 F(1, x_2, x_3) \tag{230}$$

which is a function of the triangular shape. A typical shape is the following:

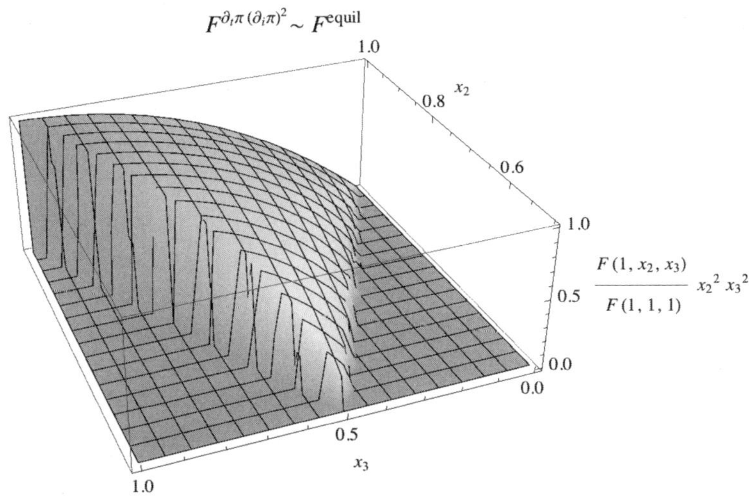

Fig. 19. A shape of non-gaussianities. We will explain the details more later, but you can see that it contains very non-trivial information.

Isn't this a beauty? It has a lot of information. Such a detection would really make us confident that something very non-trivial was going on in the sky. It would also teach a lot about the dynamics that drove inflation.

5.1. *Computation of 3-point function*

Let us see how to compute this F. In the EFT of inflation, we have seen that at leading order in derivatives we have two interaction operators: $\dot{\pi}^3$ and $\dot{\pi}(\partial_i\pi)^2$. Let us compute the shape due to the first, as an example.

This is nothing by a QFT exercise, just follow the rules:

We have an interacting theory. Very much as we do when computing scattering amplitudes or correlation functions in Minkowski, we go to the interaction picture. We split the Hamiltonian in

$$H = H_0 + H_{int} \tag{231}$$

and evolve the operators with H_0 and the state with H_{int}. Since the evolution under H_0 is completely understood, we need simply to evolve the state with the interaction picture evolutor

$$U_{int}(t, t_{in}) = Te^{-i\int_{t_{in}}^{t} dt' H_{int}(t')} \tag{232}$$

where T denotes time ordering.

What we would like to compute is the expectation value of $\zeta_{\vec{k}_1}\zeta_{\vec{k}_2}\zeta_{\vec{k}_3}$ evaluated on the initial state of the theory, which is the vacuum $|\Omega(t_{in})\rangle$, evolved to time t.

$$|\Omega(t)\rangle = U_{int}(t, t_{in})|\Omega(t_{in})\rangle \ . \tag{233}$$

We then have:

$$\begin{aligned}&\langle\Omega(t)|\zeta_{\vec{k}_1}\zeta_{\vec{k}_2}\zeta_{\vec{k}_3}|\Omega(t)\rangle \\ &\quad = \langle\Omega(t_{in})|\left(\bar{T}e^{i\int_{t_{in}}^{t} dt' H_{int}(t')}\right)\zeta_{\vec{k}_1}^{int}\zeta_{\vec{k}_2}^{int}\zeta_{\vec{k}_3}^{int}\left(Te^{-i\int_{t_{in}}^{t} dt' H_{int}(t')}\right)|\Omega(t_{in})\rangle\end{aligned} \tag{234}$$

with $\bar{T}$ representing anti-time ordering and ζ^{int} the interaction picture operator.

Notice that this expectation value is taken between two *in* states. This is why it is called in-in formalism. Notice that this is different than what one usually does in scattering amplitudes, where one computes in-out correlation functions. This is the source of a series of differences with scattering amplitude. For example, the results are not independent of field redefinitions. We wish to compute correlation functions of ζ.

How do we compute the vacuum state? We know how to express well states in the Fock base, so, it would be good to express $|\Omega(t)\rangle$ in this base. It is possible to express $|\Omega(t)\rangle$ in terms of the free theory Bunch Davies vacuum with a simple rotation in the complex plane of the contour of integration of the evaluator operator. To see this, let us expand the free theory vacuum state $|0(t_1)\rangle$ in Eigenstates of the true Hamiltonian. As we will see, we need to impose the vacuum condition at early times, so that we can neglect about the time dependence of the Hamiltonian. Let us evolve

$|0(t_1)\rangle$ from t_1 to t_2

$$\begin{aligned}|0(t_2)\rangle &= e^{H(t_2-t_1)}|0(t_1)\rangle\\ &= \sum_n e^{iE_n(t_2-t_1)}|n(t_1)\rangle\ \langle n(t_1)|0(t_1)\rangle\\ &= e^{iE_0(t_2-t_1)}|\Omega(t_1)\rangle\langle\Omega(t_1)|0(t_1)\rangle\\ &\quad+\sum_{n\neq 0} e^{iE_n(t_2-t_1)}|n(t_1)\rangle\langle n(t_1)|0(t_1)\rangle\ .\end{aligned} \tag{235}$$

We see that by sending $t_1 \to -\infty(1+i\epsilon)$, we are projecting away all the states apart for $|\Omega\rangle$. So we can write

$$|\Omega\rangle_{in} = \lim_{t_1\to-\infty}\frac{1}{\langle\Omega(t_1)|0(t_1)\rangle e^{i(t_2-t_1^+)}}U(t_2,t_1^+)|0(t_1)\rangle \tag{236}$$

where the t_1^+ reminds us how the t_1 countour has been rotated. Analogously:

$$_{in}\langle\Omega| = \lim_{t_1\to-\infty}\frac{1}{\langle 0(t_1)|\Omega(t_1)\rangle e^{-i(t_2-t_1^-)}}U^\dagger(t_2,t_1^-)\langle 0(t_1)| \tag{237}$$

We are therefore led to compute

$$\begin{aligned}&\langle\Omega(t)|\zeta_{\vec{k}_1}(t)\zeta_{\vec{k}_2}(t)\zeta_{\vec{k}_3}(t)|\Omega(t)\rangle\\ &= \lim_{t_1\to-\infty}\frac{1}{|\langle 0(t_1)|\Omega(t_1)\rangle|^2 e^{-i(t_2-t_1^-)}e^{+i(t_2-t_1^+)}}\\ &\quad\times\ _{in}\langle 0|\left(\bar{T}e^{i\int_{-\infty(1+i\epsilon)}^t dt' H_{int}(t')}\right)\zeta^{int}_{\vec{k}_1}(t)\zeta^{int}_{\vec{k}_2}(t)\zeta^{int}_{\vec{k}_3}(t)\\ &\qquad\left(Te^{-i\int_{-\infty(1-i\epsilon)}^t dt' H_{int}(t')}\right)|0\rangle_{in}\ .\end{aligned} \tag{238}$$

The prefactor is not so nice. We can divide by 1:

$$1 = \langle\Omega|\Omega\rangle = \lim_{t_1\to-\infty}\frac{\langle 0(t_1)|U^\dagger(t_2,t_1^+)U(t_2,t_1^-)|0(t_1)\rangle}{|\langle 0(t_1)|\Omega(t_1)\rangle|^2 e^{-i(t_2-t_1^-)}e^{+i(t_2-t_1^+)}}$$

We obtain

$$\begin{aligned}&\langle\Omega(t)|\zeta_{\vec{k}_1}(t)\zeta_{\vec{k}_2}(t)\zeta_{\vec{k}_3}(t)|\Omega(t)\rangle\\ &=\frac{_{in}\langle 0|\left(\bar{T}e^{i\int_{-\infty(1+i\epsilon)}^t dt' H_{int}(t')}\right)\zeta^{int}_{\vec{k}_1}(t)\zeta^{int}_{\vec{k}_2}(t)\zeta^{int}_{\vec{k}_3}(t)\left(Te^{-i\int_{-\infty(1-i\epsilon)}^t dt' H_{int}(t')}\right)|0\rangle_{in}}{\langle 0(t_1)|U^\dagger(t_2,t_1^+)U(t_2,t_1^-)|0(t_1)\rangle}\end{aligned} \tag{239}$$

The denominator represents the exponential of the sum over all bubble diagrams. In the numerator, there are disconnected components that, as it happens when we compute S-matrixes, exponentiate to give a term equal to denominator times the sum over the connected diagrams. We therefore can finally write:

$$\begin{aligned}&\langle\Omega(t)|\zeta_{\vec{k}_1}(t)\zeta_{\vec{k}_2}(t)\zeta_{\vec{k}_3}(t)|\Omega(t)\rangle\\&\quad={}_{in}\langle 0|\left(\bar{T}e^{i\int_{-\infty(1+i\epsilon)}^{t}dt' H_{int}(t')}\right)\zeta^{int}_{\vec{k}_1}(t)\zeta^{int}_{\vec{k}_2}(t)\zeta^{int}_{\vec{k}_3}(t)\\&\qquad\left(Te^{-i\int_{-\infty(1-i\epsilon)}^{t}dt' H_{int}(t')}\right)|0\rangle_{in}^{\text{connected}}\end{aligned}\tag{240}$$

At leading order we can Taylor expand the exponential to obtain

$$\langle\Omega(t)|\zeta_{\vec{k}_1}\zeta_{\vec{k}_2}\zeta_{\vec{k}_3}|\Omega(t)\rangle\simeq-2\mathrm{Re}\left[\int_{-\infty(1-i\epsilon)}^{\tau}d\tau'\langle\zeta^{int}_{\vec{k}_1}(\tau)\zeta^{int}_{\vec{k}_2}(\tau)\zeta^{int}_{\vec{k}_3}(\tau)H_{int}(\tau')\rangle\right]\tag{241}$$

At this order in perturbation theory, $H_{int}=-\int d^3x\mathcal{L}_{int}$. Pay attention, this is partially non trivial! Our $\mathcal{L}_{int}$ is given by

$$\begin{aligned}\mathcal{L}_{int}&=-\frac{4}{3}M_3^4\int d^3x\;a^4\left(\frac{1}{a(\tau)}\frac{\partial\pi(\vec{x},\tau)}{\partial\tau}\right)^3\\&=-\frac{4}{3}M_3^4\int d^3k_1\,d^3k_2\,d^3k_3\;a\;\delta^3(\vec{k}_1+\vec{k}_2+\vec{k}_3)\;\pi^{int\prime}_{\vec{k}_1}(\tau)\pi^{int\prime}_{\vec{k}_2}(\tau)\pi^{int\prime}_{\vec{k}_3}(\tau)\end{aligned}\tag{242}$$

The factor a^4 is due to the fact that we are integrating in conformal time.

Use that $\zeta=-H\pi$ and that

$$\pi^{int}_{\vec{k}}(\tau)=\pi^{cl}_{\vec{k}}(\tau)a^{\dagger}_{\vec{k}}+\pi^{cl\star}_{\vec{k}}(\tau)a_{-\vec{k}}\;.\tag{243}$$

with

$$\pi^{cl}_{k}(\tau)=-\frac{1}{H}\frac{c_s}{(2\epsilon)^{1/2}M_{\rm Pl}}\frac{1}{(2c_sk)^{3/2}}(1-ic_sk\tau)e^{ic_sk\tau}\;.\tag{244}$$

Perform the Wick contraction, and then perform the integral. The integral reads:

$$\langle\Omega(t)|\zeta_{\vec{k}_1}\zeta_{\vec{k}_2}\zeta_{\vec{k}_3}|\Omega(t)\rangle=(-H^3)(-6)\times 2\times\frac{4}{3}M_3^4\tag{245}$$

$$\times\,\mathrm{Re}\left[\pi^{cl}_{\vec{k}_1}(\tau)^{\star}\pi^{cl}_{\vec{k}_2}(\tau)^{\star}\pi^{cl}_{\vec{k}_3}(\tau)^{\star}\int_{-\infty(1-i\epsilon)}^{\tau}d\tau'\pi^{cl}_{\vec{k}_1}{}'(\tau')\pi^{cl}_{\vec{k}_2}{}'(\tau')\pi^{cl}_{\vec{k}_3}{}'(\tau')a(\tau')\right]\tag{246}$$

The results gives

$$\langle\Phi_{\vec{k}_1}\Phi_{\vec{k}_2}\Phi_{\vec{k}_3}\rangle=(2\pi)^3\delta^{(3)}\left(\sum_i\vec{k}_i\right)F(k_1,k_2,k_3)\;.\tag{247}$$

$$F_{\dot{\pi}^3}(k_1, k_2, k_3) = \frac{20}{3}\left(1 - \frac{1}{c_s^2}\right)\tilde{c}_3 \cdot \Delta_\Phi^2 \cdot \frac{1}{k_1 k_2 k_3 (k_1 + k_2 + k_3)^3} \, . \tag{248}$$

where

$$\Phi = \frac{3}{5}\zeta \, , \tag{249}$$

$$\Delta_\Phi = \frac{9}{25}\frac{H^2}{4\epsilon \, c_s M_{\rm Pl}^2} \, , \qquad M_3^4 = \frac{\dot{H} M_{\rm Pl}^2}{c_s^4}\tilde{c}_3 \, . \tag{250}$$

For $\tilde{c}_3 \sim 1$, we have that the unitarity bound associated to the operator in M_3 is the same as the one from the operator in M_2.

The standard definition of f_{NL} is

$$F(k, k, k) = f_{NL} \cdot \frac{6\Delta_\Phi^2}{k^6} \, , \tag{251}$$

This allows us to define

$$f_{NL}^{\dot{\pi}(\partial_i \pi)^2} = \frac{85}{324}\left(1 - \frac{1}{c_s^2}\right) \, , \tag{252}$$

$$f_{NL}^{\dot{\pi}^3} = \frac{10}{243}\left(1 - \frac{1}{c_s^2}\right)\left(\tilde{c}_3 + \frac{3}{2}c_s^2\right) \, .$$

5.1.1. *Shape of Non-Gaussianities*

- **Huge information**
  We see that at leading order in derivatives we have two operators $\dot{\pi}^3$ and $\dot{\pi}(\partial_i \pi)^2$. Let us see the plots. We clearly see that there is a huge amount of information contained in the 3-point function. These are functions, not just numbers: they have maxima, minima, asymptotic behaviours, etc. For example, since there are two operators at leading order in derivatives, we get any linear combination of two different shapes.
- **Local shape**
  As we can see, the non-Gaussian signal from these models is always very small in the squeezed limit $k_3 \ll k_1, k_2$. This is indeed a theorem due to Maldacena [9, 32, 33]. In reality, in some humble sense we are now beyond that theorem, because we have the Lagrangian for any single-degree-of-freedom inflationary model. We have therefore access to all the shapes that single-clock inflation can do: if we see something different, we exclude single-degree-of-freedom inflation. But still it is a remarkable feature of single degree of freedom inflation that in the squeezed limit

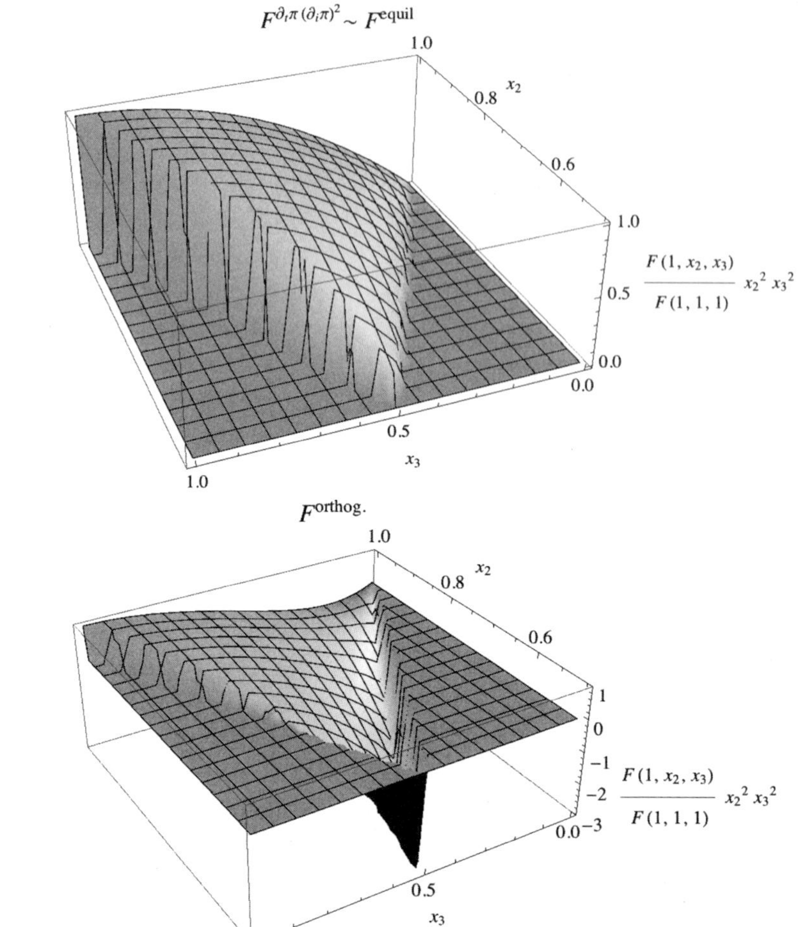

Fig. 20. Different shapes of the three-point function are obtained as we change the relative size of the operators $\dot{\pi}^3$ and $\dot{\pi}(\partial_i\pi)^2$. The shape can peak on equilateral triangles, on flattened triangles [36], or on both, as in the case of the orthogonal shape [37].

the signal is so small. Can there be inflationary models that give large 3-point function in that limit? Yes, multi filed inflation can do that.

A shape with a lot of signal there is a shape where the fluctuation ζ is defined in real space with the help of an auxiliary gaussian field:

$$\zeta(\vec{x}) = \zeta_{gaussian}(\vec{x}) + \frac{6}{5} f_{NL}^{local} \left(\zeta_{gaussian}(\vec{x})^2 - \langle\zeta_{gaussian}(\vec{x})^2\rangle\right) . \quad (253)$$

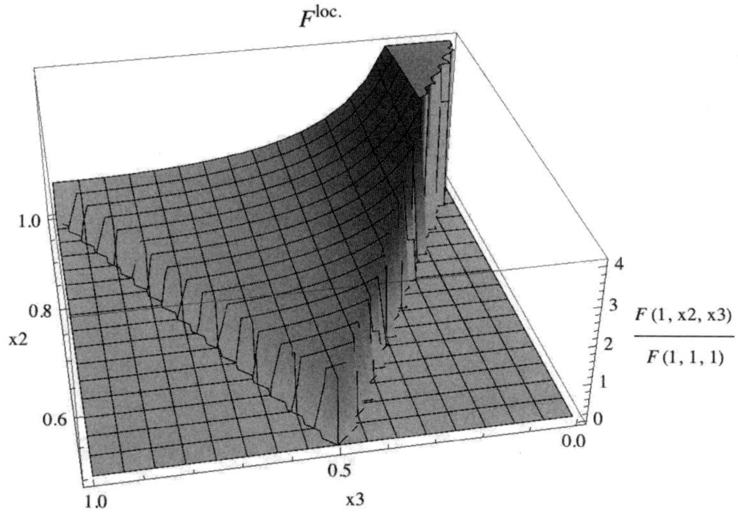

Fig. 21. The local shape has a signal peaked on the squeezed triangles. It can be produced only in multi field inflationary models. See for example [35].

Its F reads something like

$$F_{local}(k_1, k_2, k_3) = \frac{1}{k_1^3 k_2^3} + \frac{1}{k_2^3 k_3^3} + \frac{1}{k_1^3 k_3^3} \ . \tag{254}$$

Such a non-Guassianity is generated for example when the duration of inflation depends on a second field which fluctuates during inflation. For example. this could happen if the decay rate γ of the inflation is determined by a coupling that depends in turns from a light field σ.

In this way:

$$\frac{\delta a}{a} = \zeta(\vec{x}) = f(\Gamma(\{\sigma\})) \ . \tag{255}$$

Since the conversion of the σ fluctuations into $\delta a/a$ happens when all the interesting modes are outside of H^{-1}, the relation above must be local in real space:

$$\zeta(\vec{x}) = f(\Gamma(\sigma(\vec{x}))) \ . \tag{256}$$

Since the non-gaussianities are quite small, the linear term must domi-

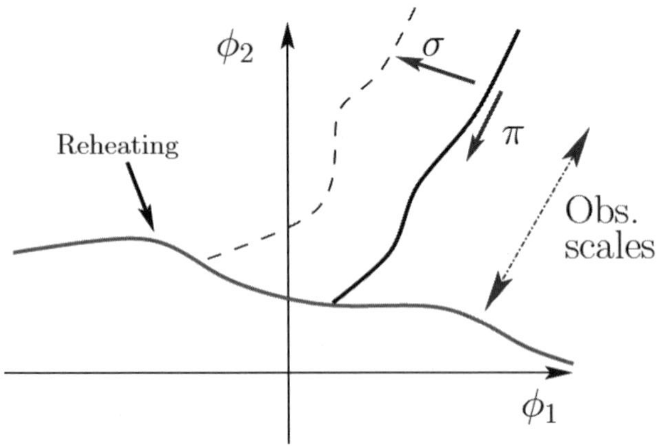

Fig. 22. Plot of typical multi field inflationary potential. Fluctuations of the second σ field affect the duration of inflation and therefore the curvature perturbation of the universe at reheating. If the relationship between σ and ζ is non-linear, then non-Gaussianities of the local kind are produced.

nate. We can taylor expand f:

$$\begin{aligned}\zeta(\vec{x}) &\simeq a_0 + a_1\sigma(\vec{x}) + a_2\sigma(\vec{x})^2 \\ &\equiv \zeta_{gaussian}(\vec{x}) + \frac{3}{5} f_{NL}^{local}\left(\zeta_{gaussian}(\vec{x})^2 - \langle\zeta_{gaussian}(\vec{x})^2\rangle\right)\end{aligned} \tag{257}$$

- **Particle physics knowledge**
  Limits on non-Gaussian signatures get translated into limits onto limits of the parameters of the inflationary Lagrangian. Cosmological observations are mapped directly into parameters of a fundamental physics Lagrangian... the sky is like a particle accelerator! This approach was developed in [37].

The is really a lot more to say about non-Guassianities and the EFT of Inflation. Non-Gaussianities have really become a large field in inflationary cosmology, and maybe this is happening also for the EFT of inflation, as this is the ideal set up to study interactions. Indeed, many additional developments have been made in this field, that I have no time to mention: EFT of multi field inflation, impose additional symmetries on π. such as Supersymmetry, discrete shift symmetry, parity, etc. ... roughly, all what we have been doing in Beyond the Standard Model physics has now motivation to be applied to inflation and the EFT of inflation offers the simple connection.

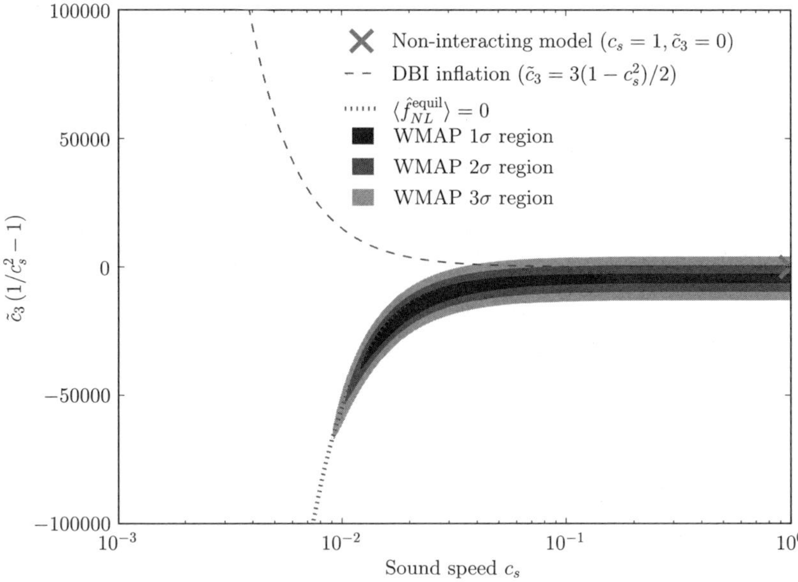

Fig. 23. WMAP 9yr constraints on models of inflation.

I leave you with the current WMAP 9yr constraints at 2σ [12]:

$$-221 < f_{NL}^{equil.} < 323 \tag{258}$$

$$-445 < f_{NL}^{orthog.} < -45 \tag{259}$$

$$-3 < f_{NL}^{loc.} < 77 \tag{260}$$

We have a 2.5σ evidence that the orthogonal shape is non-zero. Further data, and in particular Planck, will tell us.

6. Summary

This is all Guys.

In these lectures we have started from the shortcomings of Big Bang Cosmology that motivated inflation. We have seen how a period of accelerated expansion fixes all this problem. With simple estimates that are helpful to develop intuition, we have seen how inflation produces a quasi scale-invariant, quasi-Gaussian, stochastic but classical, spectrum of density perturbations, and how some qualitative predictions of inflation have been confirmed in the data. We have also seen that it would be great to have something more to look for. For this reason, we have introduced the Effective Field Theory of Inflation, which shows that Inflation is essentially a theory of a Goldstone boson. We have seen that there are new spectacular signatures in inflation: the non-Gaussianity of the density perturbation. They contain a huge amount of information, and they represent the interactions, and therefore the non-trivial dynamics, of the inflationary Lagrangian.

Inflationary physics is very ample, and there are many aspects that we could not touch. For example we did not discuss how some inflationary models are embedded in string theory, or what the beautiful phase called eternal inflation, according to which quantum effects change the asymptotic of the space-time, arise.

In any event, for all what concerns the phenomenology of Inflation and its connection to the data, you should be good to go.

Thank you very much for your attention and your interactions. Teaching at TASI has been a wonderful experience for me, and it has been a pleasure to have you around and discuss with you. I hope you'll find these lecture useful for your future research in Physics and Cosmology. It is a great moment for our field.

My best wishes.

References

[1] http://physicslearning.colorado.edu/tasi/tasi_2012/tasi_2012.htm
Disclaimer: Here it follows an incomplete list of references. I will not try to be complete and to follow the historical development. I will simply quote references that I think a student might find useful in relation to the lecture notes.

Some References related to Lecture 1

[2] A. H. Guth, "The Inflationary Universe: A Possible Solution to the Horizon and Flatness Problems," Phys. Rev. D **23** (1981) 347. *Even nowadays, an awesome paper to read.*

[3] A. D. Linde, "A New Inflationary Universe Scenario: A Possible Solution of the Horizon, Flatness, Homogeneity, Isotropy and Primordial Monopole Problems," Phys. Lett. B **108** (1982) 389. *The first working model.*

[4] A. Albrecht and P. J. Steinhardt, "Cosmology for Grand Unified Theories with Radiatively Induced Symmetry Breaking," Phys. Rev. Lett. **48** (1982) 1220. *The first working model.*

[5] S. Dodelson, "Modern cosmology," Amsterdam, Netherlands: Academic Pr. (2003) 440 p *A fantastic book for what concerns cosmological perturbations. But also the first past is good for the cosmological problems.*

[6] S. Weinberg, "Cosmology," Oxford, UK: Oxford Univ. Pr. (2008) 593 p*Thanks Weinberg for another masterpiece.*

[7] D. Baumann, "TASI Lectures on Inflation," arXiv:0907.5424 [hep-th]. *I found the first part very useful.*

[8] D. Baumann, L. Senatore and M. Zaldarriaga, "Scale-Invariance and the Strong Coupling Problem," JCAP **1105** (2011) 004 [arXiv:1101.3320 [hep-th]].

Some references related to lecture 2

[9] J. M. Maldacena, "Non-Gaussian features of primordial fluctuations in single field inflationary models," JHEP **0305** (2003) 013 [astro-ph/0210603].

[10] D. H. Lyth, "What would we learn by detecting a gravitational wave signal in the cosmic microwave background anisotropy?," Phys. Rev. Lett. **78** (1997) 1861 [hep-ph/9606387]. *The bound on field excursion.*

[11] A. H. Guth and S. -Y. Pi, "The Quantum Mechanics of the Scalar Field in the New Inflationary Universe," Phys. Rev. D **32** (1985) 1899. *About the classical description of inflationary perturbations.*

Some references related to lecture 3

[12] C. L. Bennett, D. Larson, J. L. Weiland, N. Jarosik, G. Hinshaw, N. Odegard, K. M. Smith and R. S. Hill *et al.*, "Nine-Year Wilkinson Microwave Anisotropy Probe (WMAP) Observations: Final Maps and Results," arXiv:1212.5225 [astro-ph.CO].

[13] G. Hinshaw, D. Larson, E. Komatsu, D. N. Spergel, C. L. Bennett, J. Dunkley, M. R. Nolta and M. Halpern *et al.*, "Nine-Year Wilkinson Microwave Anisotropy Probe (WMAP) Observations: Cosmological Parameter Results," arXiv:1212.5226 [astro-ph.CO].

[14] R. Keisler, C. L. Reichardt, K. A. Aird, B. A. Benson, L. E. Bleem, J. E. Carlstrom, C. L. Chang and H. M. Cho *et al.*, "A Measurement of the Damping Tail of the Cosmic Microwave Background Power Spectrum with the South Pole Telescope," Astrophys. J. **743** (2011) 28 [arXiv:1105.3182 [astro-ph.CO]].

[15] S. Das, B. D. Sherwin, P. Aguirre, J. W. Appel, J. R. Bond, C. S. Carvalho, M. J. Devlin and J. Dunkley *et al.*, "Detection of the Power Spectrum of Cosmic Microwave Background Lensing by the Atacama Cosmology Telescope," Phys. Rev. Lett. **107** (2011) 021301 [arXiv:1103.2124 [astro-ph.CO]].

[16] S. Dodelson, "Coherent phase argument for inflation," AIP Conf. Proc. **689** (2003) 184 [hep-ph/0309057].

[17] M. Alishahiha, E. Silverstein and D. Tong, "DBI in the sky," Phys. Rev. D **70** (2004) 123505 [hep-th/0404084].

[18] N. Arkani-Hamed, P. Creminelli, S. Mukohyama and M. Zaldarriaga, "Ghost inflation," JCAP **0404** (2004) 001 [hep-th/0312100].

[19] J. M. Cornwall, D. N. Levin and G. Tiktopoulos, "Derivation of Gauge Invariance from High-Energy Unitarity Bounds on the s Matrix," Phys. Rev. D **10** (1974) 1145 [Erratum-ibid. D **11** (1975) 972].

[20] L. Senatore, E. Silverstein and M. Zaldarriaga, "New Sources of Gravitational Waves during Inflation," arXiv:1109.0542 [hep-th].

[21] C. Cheung, P. Creminelli, A. L. Fitzpatrick, J. Kaplan and L. Senatore, "The

Effective Field Theory of Inflation," JHEP **0803** (2008) 014 [arXiv:0709.0293 [hep-th]].

[22] S. R. Behbahani, A. Dymarsky, M. Mirbabayi and L. Senatore, "(Small) Resonant non-Gaussianities: Signatures of a Discrete Shift Symmetry in the Effective Field Theory of Inflation," JCAP **1212** (2012) 036 [arXiv:1111.3373 [hep-th]].

[23] A. Achucarro, J. -O. Gong, S. Hardeman, G. A. Palma and S. P. Patil, "Effective theories of single field inflation when heavy fields matter," JHEP **1205** (2012) 066 [arXiv:1201.6342 [hep-th]].

[24] J. Polchinski, "Effective field theory and the Fermi surface," In *Boulder 1992, Proceedings, Recent directions in particle theory* 235-274, and Calif. Univ. Santa Barbara - NSF-ITP-92-132 (92,rec.Nov.) 39 p. (220633) Texas Univ. Austin - UTTG-92-20 (92,rec.Nov.) 39 p [hep-th/9210046].

[25] D. Seery, J. E. Lidsey and M. S. Sloth, "The inflationary trispectrum," JCAP **0701** (2007) 027 [astro-ph/0610210].

[26] R. Barbieri, A. Pomarol, R. Rattazzi and A. Strumia, "Electroweak symmetry breaking after LEP-1 and LEP-2," Nucl. Phys. B **703** (2004) 127 [hep-ph/0405040].

[27] A. Adams, N. Arkani-Hamed, S. Dubovsky, A. Nicolis and R. Rattazzi, JHEP **0610**, 014 (2006) [hep-th/0602178].

[28] G. 't Hooft and M. J. G. Veltman, "One loop divergencies in the theory of gravitation," Annales Poincare Phys. Theor. A **20** (1974) 69.

[29] G. L. Pimentel, L. Senatore and M. Zaldarriaga, "On Loops in Inflation III: Time Independence of zeta in Single Clock Inflation," JHEP **1207** (2012) 166 [arXiv:1203.6651 [hep-th]].

[30] N. Agarwal, R. Holman, A. J. Tolley and J. Lin, "Effective field theory and non-Gaussianity from general inflationary states," arXiv:1212.1172 [hep-th].

[31] P. Creminelli, M. A. Luty, A. Nicolis and L. Senatore, "Starting the Universe: Stable Violation of the Null Energy Condition and Non-standard Cosmologies," JHEP **0612** (2006) 080 [hep-th/0606090].

[32] P. Creminelli and M. Zaldarriaga, "Single field consistency relation for the 3-point function," JCAP **0410** (2004) 006 [astro-ph/0407059].

[33] C. Cheung, A. L. Fitzpatrick, J. Kaplan and L. Senatore, "On the consistency relation of the 3-point function in single field inflation," JCAP **0802** (2008) 021 [arXiv:0709.0295 [hep-th]].

Some references related to lecture 4

[34] D. Babich, P. Creminelli and M. Zaldarriaga, "The Shape of non-Gaussianities," JCAP **0408** (2004) 009 [astro-ph/0405356].

[35] L. Senatore and M. Zaldarriaga, "The Effective Field Theory of Multifield Inflation," JHEP **1204** (2012) 024 [arXiv:1009.2093 [hep-th]].

[36] P. Creminelli, A. Nicolis, L. Senatore, M. Tegmark and M. Zaldarriaga, "Limits on non-gaussianities from wmap data," JCAP **0605** (2006) 004 [astro-ph/0509029].

[37] L. Senatore, K. M. Smith and M. Zaldarriaga, "Non-Gaussianities in Single Field Inflation and their Optimal Limits from the WMAP 5-year Data," JCAP **1001** (2010) 028 [arXiv:0905.3746 [astro-ph.CO]].

Chapter 7

Jet Substructure

J. Shelton

Yale University Physics Department and
Harvard University Physics Department
High Energy Theory Group
17 Oxford Street
Cambridge, MA 02138, USA
E-mail: jshelton@physics.harvard.edu

Jet physics is a rich and rapidly evolving field, with many applications to physics in and beyond the Standard Model. These notes, based on lectures delivered at the June 2012 Theoretical Advanced Study Institute, provide an introduction to jets at the Large Hadron Collider. Topics covered include sequential jet algorithms, jet shapes, jet grooming, and boosted Higgs and top tagging.

Contents

1. Introduction

These notes are writeups of three lectures delivered at the Theoretical Advanced Study Institute in Boulder, Colorado, in June 2012. The aim of the lectures is to provide students who have little or no experience with jets with the basic concepts and tools needed to engage with the rapidly developing ideas concerning the use of jets in new physics searches at the LHC. A certain amount of familiarity with the structure of QCD, and in particular with QCD showers, is assumed.

Lecture one introduces sequential jet algorithms, and develops several main tools in substructure analyses using the boosted Higgs as an example. Lecture two delves further into jet grooming and jet shapes, and in lecture three we conclude with an overview of top tagging and BSM searches.

2. Lecture I: Jets, Subjets, and Sequential Jet Algorithms

To understand jet substructure and its applications, we must first begin by understanding jets. Jets, together with parton distribution functions and factorization theorems, are the phenomenological tool that allow us to separate out the perturbatively describable hard interactions in proton-proton collisions, and thereby enable us to make quantitative predictions for events involving strongly interacting particles. Jet cross-sections necessarily depend on the algorithm used to define a jet. There are many jet algorithms, each one with its own strengths and weaknesses.

The first jet algorithm was developed for $e^+e^- \to$ hadron events by Sterman and Weinberg in 1977 [1]. In this algorithm events are declared to have two jets if all but a fraction ϵ of the total energy in the event can be contained within two cones of half-angle δ. That is, radiation off of one of the initial partons must be sufficiently hard,

$$E_{rad} > \epsilon \tag{1}$$

and at sufficiently wide angles from either of the other jets,

$$\theta_{min} > \delta \tag{2}$$

for the radiation to be resolved as a separate jet. How many events have two jets and how many contain three or more obviously depends on the exact values chosen for ϵ and δ. For all sufficiently large ϵ/E_{tot} and δ, the partonic cross-section for radiation of an extra parton into the region of phase space defined by Eqs. (1) and (2) is sufficiently isolated from the soft and collinear singular regions of phase space that rates and distributions can be calculated reliably in perturbation theory. Of course, this is a partonic calculation, and to fully match the partonic picture onto reconstructed sprays of hadrons requires some additional theoretical machinery to describe such effects as (for example) hadronization. For our purposes, however, a parton shower picture will suffice.

The Sterman-Weinberg algorithm is the ur-example of a *cone* algorithm. While cone algorithms present a very intuitive picture of parton radiation, they can be somewhat clumsy in practice, particularly as the number of jets increases, and they are not in active use in most experiments today. Other algorithms can deal much more flexibly with high jet multiplicity. One such flexible algorithm is the JADE algorithm, developed by the JADE collaboration in the late 1980s, also for $e^+e^- \to$ hadrons [2, 3]. Here, jets are constructed by iteratively recombining final state particles. Define a metric to measure the separation between final state particles i and j,

$$y_{ij} \equiv \frac{m_{ij}^2}{Q^2} \approx \frac{2E_iE_j(1-\cos^2\theta_{ij})}{Q^2}, \tag{3}$$

where Q is the total energy of the event. Note that y_{ij} vanishes if either i or j is soft ($E_i \to 0$ or $E_j \to 0$), or if i and j are collinear ($\cos\theta_{ij} \to 1$). We can now construct jets using the following recipe:

- Compute the interparticle distances y_{ij} for all particles in the final state, and find the pair $\{i,j\}$ with the minimum y_{ij}.
- If this minimum $y_{ij} < y_0$ for some fixed parameter y_0, combine i and j into a new particle, and go back to the previous step.
- If $y_{ij} > y_0$, declare all remaining particles to be jets.

Since clustering of particles proceeds from smaller values of y_{ij} to larger values, this recipe preferentially clusters particles that are probing the regions of phase space dominated by the soft and collinear singularities. In a sense, the algorithm is trying to combine hadrons into partons by making its best guess for the reconstructed parton shower. The JADE algorithm has only one parameter, the separation cutoff y_0, and clearly can handle different jet multiplicities in an efficient way by varying y_0. It is the ur-example of a *sequential recombination algorithm*, and the ancestor of all jet algorithms in wide use at the LHC.

The most direct descendent of the JADE algorithm is the k_T *algorithm* [4], which replaces the particle energy factor E_iE_j in the Jade metric, Eq. (3), with the factor $\min(E_i^2, E_j^2)$:

$$y_{ij} = \frac{2\min(E_i^2, E_j^2)(1-\cos^2\theta_{ij})}{Q^2}. \tag{4}$$

This still ensures that the metric goes to zero when either $E_i \to 0$ or $E_j \to 0$ are soft, but has the advantage that the relative softness of a particle depends only on its own energy, and not that of the other particle in the pair. This fixes up a technical drawback to the JADE algorithm, where $y_{ij}^{(JADE)} \propto E_i E_j$ allows two very soft particles to be combined even if they are at very wide angles from each other. Using $y_{ij} \propto \min(E_i^2, E_j^2)$ means soft particles will get preferentially clustered with nearby harder particles instead.

For small θ_{ij}, the numerator of Eq. (4) can be written as simply $k_\perp^2$, the transverse momentum of the softer particle relative to the harder particle—hence the name of the algorithm. In this form the metric is directly related to QCD splitting functions.

To create a version of the k_T algorithm that can be used at hadron colliders, where the total energy Q^2 is unknown, both the algorithm and the metric have to be adapted [5]. In the metric, we simply use longitudinally boost-invariant quantities p_T and ΔR instead of E and $\cos\theta_{ij}$, and let the metric become dimensionful,

$$d_{ij} = \frac{\min(p_{T,i}^2, p_{T,j}^2)\Delta R_{ij}^2}{R^2}. \tag{5}$$

The angular parameter R introduced here will replace y_0 as determining the cutoff for combining particles, as we will see. We need in addition to define the quantities

$$d_{iB} = p_{T,i}^2 \tag{6}$$

for each particle i, since we need to also consider splittings from the beam.

The recombination algorithm now works as follows:

- Compute d_{ij} and d_{iB} for all particles in the final state, and find the minimum value.
- If the minimum is a d_{iB}, declare particle i a jet, remove it from the list, and go back to step one.
- If the minimum is a d_{ij}, combine particles i and j, and go back to step one.
- Iterate until all particles have been declared jets.

This algorithm is usually what is meant by when the k_T algorithm is referred to, but you may occasionally see it referred to as the *inclusive* k_T algorithm, as there is a related ("*exclusive*") variant [6]. Note that the parameter R functions as an angular cut-off: two particles separated by a distance $R_{ij} > R$ will never be combined, regardless of the p_T's of the particles (this does not necessarily preclude both particles being clustered into the same jet later). In fact, with this jet algorithm, arbitrarily soft particles can become jets. Therefore jets are customarily returned down to some finite p_T cutoff, typically tens of GeV.

Because the k_T algorithm clusters particles beginning with soft particles and working its way up to harder particles, the algorithm tends to construct irregular jets which depend on the detailed distribution of soft particles in an event. For this reason, k_T jets are not especially practical for hadron colliders: irregular jets are hard to calibrate, and the jets are quite sensitive to unrelated radiation in the event.

Other sequential algorithms are obtained by using different metrics. The *Cambridge-Aachen* or C-A algorithm is obtained by taking [7]

$$d_{ij} = \frac{\Delta R_{ij}^2}{R^2}, \qquad d_{iB} = 1. \tag{7}$$

This metric clusters particles based only on their angular separation, giving a nicely geometric interpretation of jets. The C-A algorithm still reflects aspects of the QCD parton shower, in particular the *angular ordering* of emissions. However, it is less directly related to the structure of QCD parton splitting functions than the k_T algorithm is, and represents a compromise between reflecting the structure of the parton shower and maintaining some insensitivity to soft radiation.

The *anti-k_T* algorithm entirely abandons the idea of mimicking the parton shower [8]. Here, the metric is

$$d_{ij} = \min\left(\frac{1}{p_{T,i}^2}, \frac{1}{p_{T,j}^2}\right)\frac{\Delta R_{ij}^2}{R^2}, \qquad d_{iB} = \frac{1}{p_{T,i}^2}. \tag{8}$$

With this metric, particles are clustered beginning with the *hardest* particles. This means that the most energetic cores of jets are found first. As soft particles clustered later have a minimal impact on the larger four-momentum of the jet core, the anti-k_T algorithm tends to cluster particles out to distances R from the core of a jet, yielding very regular jets. Anti-k_T jets are therefore much easier to calibrate at experiments, and the anti-k_T algorithm has become the default used at the LHC.

Let us conclude this section by emphasizing that all sequential jet algorithms return not only a list of jets but a *clustering sequence* for the event. Varying the radial parameter R simply acts to move the resolution scale up and down the clustering sequence, making it very easy to study how jet distributions and multiplicities depend on the angular resolution R. In particular, for the C-A algorithm, the cluster sequence regarded as a function of R has a purely geometric interpretation as resolving the event on different angular scales.

All three sequential jet algorithms discussed here also share the same *reach*, that is, regardless of the chosen metric, a splitting $P \to ij$ will not be combined if the angular distance between the daughters exceeds the chosen jet radius, $\Delta R_{ij} > R$. This means that, to leading order, perturbative computations of quantities such as jet rates are identical between all three algorithms.

Finally, the infrared and collinear safety of all three sequential jet algorithms can be easily checked by asking how the cluster sequence would change with the addition of a soft or collinear emission. For the shower-sensitive k_T and C-A metrics, infrared and collinear safety follows automatically. The anti-k_T metric is also manifestly IR- and collinear-safe, as can be seen with a little more thought: anti-k_T recombinations are clearly collinear-safe, since collinear splittings are combined near the beginning of the sequence. IR safety also follows, as soft radiation has negligible impact on the jet built out from the hard core.

2.1. *Jets at the LHC*

The main subject of these lectures are the possibilities and uses of jets to discover physics at and beyond the electroweak scale, which means, for practical purposes, at the LHC.

It is important to remember that events at LHC are a busy hadronic environment. In addition to the showering and hadronizing hard partons which we want to study, there are large amounts of soft, unassociated radiation from (1) the *underlying event*, that is, the remnants of the scattering protons; (2) possible *multiple interactions*, that is, additional collisions of partons arising from the same *p-p* collision as the hard interaction; and (3) *pile-up*, additional *p-p* collisions from other protons in the colliding bunches. These additional sources of radiation contribute a potentially sizable and largely uniform backdrop of hadronic activity that, when clustered into jets, will partially obscure the features of the hard interaction that we would like to reconstruct.

The default jets used at the LHC are formed using the anti-k_T algorithm, with cone sizes $R = 0.4, 0.6$ (at ATLAS) and $R = 0.5, 0.7$ (at CMS). These specific choices of R come from a compromise between (1) the desire to collect all the radiation from a single parton, and (2) the desire *not* to sweep up an excessive amount of unrelated radiation.

Many advances have combined to make jets at the LHC a particularly fertile field.

- *advances in experiment*: the calorimeters at ATLAS and CMS have much finer resolution than in previous experiments, allowing a much more finely grained picture of events. Moreover, *local* calibration of jets allows jets to be considered on multiple scales.
- *advances in computation*: the development of fast algorithms [9] allows broad implementation of sequential recombination.
- *advances in energy*: the LHC center of mass energy is large enough that particles with weak scale masses (i.e., Z, W, t, and H) will for the first time have an appreciable cross-section to be produced with enough of a boost to collimate the daughter partons. The simple picture that one parton corresponds to one jet breaks down badly in this case, and new tools are needed to separate out collimated perturbative decays from QCD showers.

There are several reasons to be interested in boosted particles. Very often, there is theoretical motivation to focus on a particular slice of phase space where the daughter particles are necessarily boosted. High mass resonances are the simplest such examples. For instance, a resonance ρ_C with mass $m_\rho \gtrsim 1.5$ TeV which decays to pairs of gauge bosons would yield highly boosted VV pairs.

Even in the absence of a resonance or other mechanism to preferentially populate boosted regions of phase space, looking for boosted signals can also be useful for improving the signal to background ratio. Changing the reconstruction method changes what the experimental definition of the signal is, and therefore necessarily the backgrounds change as well. This can sometimes—but not always!—be enough of an advantage to make up for the reduction in signal rate that comes from selecting only the boosted region of phase space. Background reduction comes in two forms. In high multiplicity final states, combinatoric background is often prohibitive. When some or all of the final state particles are boosted, the combinatoric background is greatly reduced. But it is also possible to use boosted selection techniques to identify regions where the background from other physics processes is intrinsically reduced.

To appreciate the need for new reconstruction techniques at the LHC, consider the production of top quarks at fixed center of mass energy $\sqrt{\hat{s}}$. Choosing some angular scale R_0, we can ask, what fraction of top quarks have all three, only two, or none of their partonic daughters isolated from

Table 1. Resolved parton multiplicities in $t\bar{t}$ events

$\sqrt{\hat{s}}$	R_0	3	2	1
1.5 TeV	0.4	0.55	0.45	—
1.5 TeV	0.6	0.2	0.6	0.2
2.0 TeV	0.6	0.1	0.45	0.45

the others at the scale R_0? This gives a zeroth order estimate of how well a jet algorithm with $R = R_0$ will be able to reconstruct the three partonic top daughters as separate jets. The answer we get depends sensitively on both R_0 and $\sqrt{\hat{s}}$:

Clearly, tops produced in the very interesting super-TeV regime $\sqrt{\hat{s}} \gtrsim$ TeV straddle the borderlines between several different topologies. It would be much more desirable to have a flexible reconstruction method that could handle semi-collimated tops in a unified way.

To see how we can go about building such reconstruction techniques, let's start by considering one of the landmark jet substructure analyses: the case of a boosted Higgs decaying into $b\bar{b}$.

2.2. *Boosted Higgs*

This analysis will introduce us to several ideas that will be important tools in our boosted analysis toolbox: fat jets, jet mass, jet grooming, and sequential de-clustering.

Searching for the Higgs in its decay to $b\bar{b}$ is very difficult at the LHC, due to overwhelming QCD backgrounds. Even in associated production, $pp \to HZ, HW$, the background processes $Z + b\bar{b}$, $W + b\bar{b}$, and even $t\bar{t}$ are overwhelming. Nonetheless, thanks to Ref. 10, $pp \to HV$, $H \to b\bar{b}$ is now an active search channel at the LHC.

To be specific, let's consider the process $pp \to HZ$, followed by $H \to b\bar{b}$, and $Z \to \ell^+\ell^-$. The traditional approach to this signal would be to look for final states with a leptonic Z and 2 b-tagged jets, construct the invariant mass of the jets, and look for a peak in the distribution of $m_{b\bar{b}}$. The new approach is instead to focus on events where the Higgs is produced with substantial p_T, $p_{T,H} > 200$ GeV, and cluster these events with a large ($R = 1.2$) jet radius, such that all of the Higgs decay products are swept up in a single fat jet. The signal is now a leptonic Z + a fat "Higgs-like" jet, and the background to this signal is now $Z+$ one fat jet rather than $Z+b\bar{b}$. What we'll see is that jet substructure offers us enough quantitative precision in what we mean by a "Higgs-like" jet to reduce the background by an extent that makes up for the acceptance price demanded by the high p_T cut.

For an unboosted search, the ultimate discriminator between signal and background is the b-$\bar{b}$ invariant mass: to find a resonance, look for a bump in the b-$\bar{b}$ mass spectrum. Now that we have boosted the Higgs and collected it into a single fat jet, the Higgs mass should be reflected in the invariant mass of the fat jet itself. To understand jet masses for the background, let's take a quick look at how jet masses are generated in QCD.

Jet mass. Partons are generally massless (we will neglect the b quark mass), but jets are not. Jet mass in QCD arises from emission during the parton shower, and as such we can calculate the leading contribution. Jet mass, like most perturbative jet properties in QCD, is dominated by the first emission. Let's consider for concreteness a quark emitting a gluon, and work in the collinear regime (small R). In this approximation, we can consider the jet in isolation from the rest of the event, neglecting interference and splash-in, and we can approximate the QCD splitting functions with the singular portions. Doing so, the amplitude to radiate an extra parton can be written as

$$d\sigma_{n+1} \approx d\sigma_n \, dz \, \frac{dt}{t} \frac{\alpha_s}{2\pi} \, \mathcal{P}(z), \tag{9}$$

where t is the virtuality of the parent P, $z = E_q/E_P$ is the fraction of the parent energy retained by the daughter quark, and the splitting function $\mathcal{P}(z)$ for $q \to qg$ is given by

$$\mathcal{P}(z) = C_F \frac{1+z^2}{1-z}. \tag{10}$$

The parent virtuality t is of course the jet mass-squared. In the collinear limit,

$$t = E_P^2 z(1-z)\theta^2 = (p_{T,P} \cosh \eta)^2 z(1-z)\theta^2. \tag{11}$$

Integrating over rapidity, we can approximate the average jet mass-squared as:

$$\langle m^2 \rangle \approx p_{T,P}^2 \int_0^{R^2} \frac{d\theta^2}{\theta^2} \int dz \, z(1-z)\theta^2 \frac{\alpha_s}{2\pi} \mathcal{P}(z). \tag{12}$$

Note the limits on the θ integral: this is where the choice of jet algorithm enters. As established above, for all sequential jet algorithms, only radiation at angles smaller than R will be clustered into the jet. Strictly, we should use a running α_s evaluated at a scale set by the relative transverse momentum of the splitting, but to get a quick estimate, let's perform the integral in the approximation that α_s is constant. We then obtain

$$\langle m^2 \rangle \approx \frac{\alpha_s}{\pi}\frac{3}{8}C_F\, p_T^2 R^2. \tag{13}$$

The jet mass scales like p_T, as it had to, and is suppressed by $(\alpha_s/\pi)^{1/2}$. To this order the mass increases linearly with R. The exact value of the numerical coefficient will in general depend on the quark versus gluon content of the jet sample. For instance, the major QCD background for a doubly b-tagged boosted Higgs comes from the splittings $g \to b\bar{b}$, where the splitting function is

$$\mathcal{P}(z) = C_A(z^2 + (1-z)^2), \tag{14}$$

giving, in the constant-α_s approximation,

$$\langle m^2 \rangle \approx \frac{\alpha_s}{\pi}\frac{1}{20}C_A\, p_T^2 R^2. \tag{15}$$

Coming back to the Higgs, consider now a splitting $P \to ij$. We have $m^2 \approx 2p_i \cdot p_j \sim p_{T,i}p_{T,j}\Delta R_{ij}^2 = z(1-z)p_{T,P}^2\Delta R_{ij}^2$. In other words, just from kinematics we can express the opening angle in terms of the parent mass and p_T:

$$\Delta R_{ij} \sim \frac{m}{p_T}\frac{1}{\sqrt{z(1-z)}} \sim \frac{2m}{p_T}. \tag{16}$$

Now consider the k_T metric evaluated on this splitting $P \to ij$:

$$y_{ij} = \min(E_{T,i}^2, E_{T,j}^2)\Delta R_{ij}^2 = p_T^2 z^2 \Delta R_{ij}^2 \approx \frac{z}{1-z}m^2. \tag{17}$$

For jets with a fixed mass m, cutting on the splitting scale y_{ij} then can separate QCD jets, which have a soft singularity $\propto 1/z$, from boosted Higgses, which have a flat distribution in z.[a]

Moreover, a boosted Higgs will go from a mass m_H to massless daughters in one step, while QCD splittings prefer to shed virtuality gradually. To see this, consider the Sudakov form factor, which exponentiates the splitting

[a]This is a little quick: not all QCD splitting functions have a soft singularity, and in particular $g \to q\bar{q}$ does not. However, $P_{g\to q\bar{q}}(z)$ is not flat in z, and in particular is minimized at the symmetric value $z = 1/2$, so cutting on y_{ij} can still help suppress this background.

functions to obtain the probability of evolving from an initial virtuality t_0 to a final virtuality t without branching:

$$\Delta(t) = \exp\left[-\int_{t_0}^{t} \frac{dt'}{t'}\, dz \frac{\alpha_s}{2\pi} \mathcal{P}(z)\right]. \tag{18}$$

Evaluating $\alpha_s = \alpha_s(t')$ and using an IR cut-off to regulate the splitting functions, at large t, one can work out that [11]

$$\Delta(t) \propto \left(\frac{t_0}{t}\right)^p \tag{19}$$

for an exponent $p > 0$, in other words, $\Delta(t) \to 0$ for large t. In other words, the probability of a QCD jet making a large jump in mass at a branching falls off as m^{-2p}.[b]

We have now identified two ways in which a Higgs boson H decaying perturbatively to $b\bar{b}$ will behave very differently from a QCD parton branching: the splitting will be symmetric, and show a sudden drop in parton mass. The search algorithm for finding a boosted Higgs looks for a splitting inside the Higgs jet that behaves like a perturbative decay, and works as follows:

- Cluster the event on a large angular scale (Ref. 10 uses $R = 1.2$), using the C-A algorithm. Large angular scales are necessary in order to get good acceptance for collecting both Higgs decay products into a single fat jet: from Eq. (16), we can see that the b-$\bar{b}$ separation for a 125 GeV Higgs boson is $R_{b\bar{b}} \lesssim 1$ for $p_T \gtrsim 200$ GeV. We choose the C-A algorithm because it is a good compromise between accurately reflecting the shower structure of QCD, and minimizing sensitivity to soft radiation in the event.
- Now, given a hard fat jet, successively *unwind* the jet by undoing the cluster sequence one branching at a time. At each branching $P \to ij$, check to see whether the splitting looks sufficiently non-QCD-like, by asking that the branching be both *hard*,

$$\max(m_i, m_j) < \mu m_P \tag{20}$$

for some parameter μ, and *symmetric*,

$$y_{ij} > y_{cut} \tag{21}$$

for some choice of y_{cut}.

[b]In fact, taking higher order corrections into account, one finds that the Sudakov form factor goes to zero even faster than polynomially for large t.

- If the splitting fails to be sufficiently hard and symmetric, discard the softer of i and j, and continue to unwind the harder.
- Continue until either an interesting splitting has been found or you run out of jet.

This procedure, often referred to as the "splitting" or "mass-drop" procedure, identifies an interesting Higgs-like splitting $H \to b\bar{b}$, which determines a characteristic angular scale $R_{b\bar{b}}$ for a particular event. Once this scale $R_{b\bar{b}}$ has been identified, we benefit greatly by using smaller scales to resolve the event, rather than the large $R = 1.2$ scale we started with.

The reason is the following: starting with such a large jet, we are guaranteed to sweep up a large amount of unassociated radiation along with the Higgs decay products. The effect of this unassociated radiation is to smear out the mass resolution. The invariant mass is especially vulnerable to distortion from even soft unassociated radiation, because evaluating $m^2 = E^2 - \vec{p}^2$ depends on large cancellations. The amount of distortion scales like

$$\frac{d\langle m^2 \rangle}{dR} \approx \Lambda_{soft} p_{T,J} R^3, \tag{22}$$

in the approximation that unassociated radiation contributes a constant energy Λ_{soft} per unit rapidity: the jet area scales like R^2, while the incremental contribution to the invariant mass from a soft particle at distance $R/2$ from the jet core contributes as $\Lambda p_{T,J} R/2$ [12]. So to recover mass resolution, it is vital to whittle down our initial fat jet to jets only as big as necessary to capture the radiation from the Higgs decay products.

In fact, we have already started whittling. The splitting procedure discards soft, wide-angle radiation clustered into the jet on its way towards finding the Higgs-like splitting. This by itself helps to clean up the mass resolution. But we can do better: given the scale $R_{b\bar{b}}$ which is our best guess at the angular separation of the Higgs' daughter particles, we can resolve the fat jet at the *filtering* scale $R_{filt} = \min(R_{b\bar{b}}/2, 0.3)$, and keep only the three hardest subjets. We keep three, rather than two, subjets in order to capture final-state radiation off of one of the b quarks.

Finally, demanding that the two hardest filtered subjets be b-tagged, Ref. 10 finds that the Higgs can be seen in this channel with 5σ significance in 30 fb^{-1} (at 14 TeV, combining $Z \to \ell^+\ell^-$, $Z \to \nu\bar{\nu}$, and $W \to \ell\nu$), and signal-to-background of $\mathcal{O}(1)$. However, we emphasize that *this and all other LHC phenomenological studies are based on expectations from Monte Carlo.* Even very sophisticated Monte Carlos necessarily capture only an

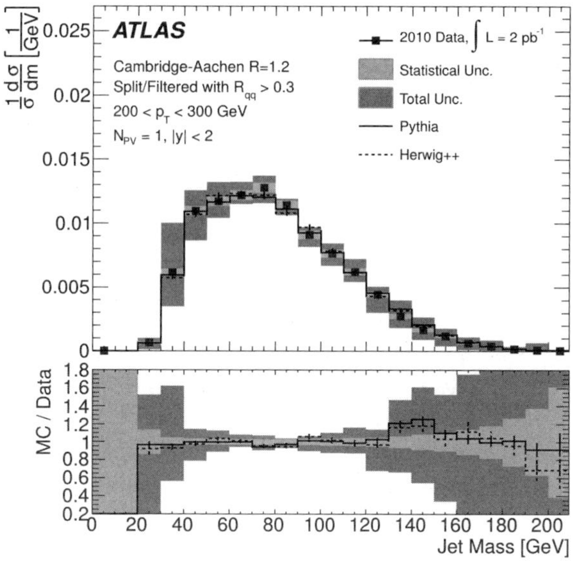

(a)

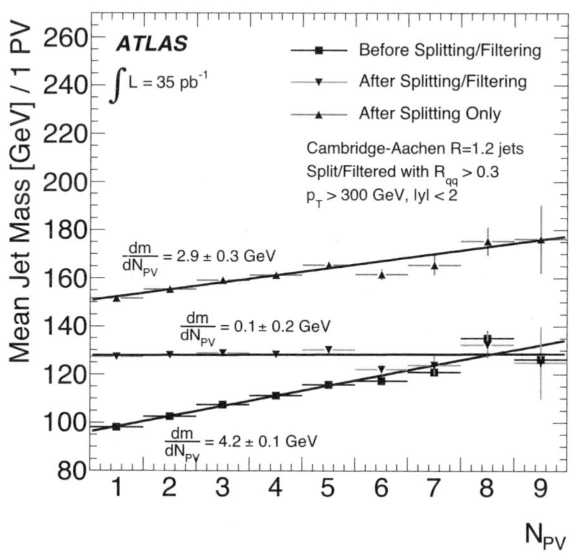

(b)

Fig. 1. (a) The distribution of jet mass for fat C-A jets (after splitting and filtering). Note the reasonable agreement between data and predictions from two different shower MCs. (b) Average jet mass as a function of the number of primary vertices N_{PV}. Note that after filtering, the jet mass has little to no dependence on N_{PV}. From Ref. 13.

approximation to the full physics of QCD. For this reason, both validation in data on one hand and formal theoretical study on the other are critical. Let us then end this section by showing a couple of the most important early experimental results. In Fig. 1, we show two plots from Ref. 13. In Fig. 1(a), we see that shower Monte Carlos do a reasonable job of predicting the spectrum of jet masses for the QCD background. In Fig. 1(b), the jet mass is plotted as a function of the number of primary vertices N_{PV} in an event, or in other words, the amount of pileup. Note that after filtering, the jet mass has little to no dependence on N_{PV}, indicating that filtering is successfully isolating the hard process. Note also that filtering is *necessary*: prior to filtering, the dependence of jet mass on N_{PV} is significant, and in the 2012 operating environment average pileup multiplicity is $N_{PV} \gtrsim 30$.

Heartened by this evidence that our theoretical techniques have a reasonable relationship with reality, we will proceed in the next section to discuss more ideas for cleaning up pileup, and more jet properties which can discriminate signals from QCD backgrounds.

3. Lecture II: Jet Grooming and Jet Shapes

Our last section concluded with a walk-through of the pioneering boosted Higgs study, where we saw examples of two topics we will be discussing in this lecture, namely jet grooming and jet shapes.

3.1. *Jet grooming*

In the boosted Higgs analysis discussed in the previous lecture, we saw that jet mass resolution was badly degraded by the presence of unassociated radiation in the jet, and introduced the process of *filtering* to mitigate these contributions. Filtering is one of several *jet grooming* algorithms, all of which are designed to "clean up" jets by subtracting the contributions of unassociated radiation.

Trimming [14], similarly to filtering, reclusters the constituents of a fat jet and retains a subset of the subjets, but has a different criterion for keeping subjets. For each jet of interest, the algorithm is:

- Recluster the constituents using some jet algorithm (the original reference specifies k_T), and resolve on a fixed small angular scale R_0.

- Keep each subjet i that passes a p_T threshhold,

$$p_{T,i} > f\Lambda_{hard} \tag{23}$$

for a cutoff parameter f_{cut} and a hard momentum scale Λ.
- The final trimmed jet is the sum of the retained subjets.

The essential idea is that radiation we want to keep tends to be distributed in clusters, reflecting a parent parton emission, while unassociated radiation we don't want to keep is more uniformly distributed. Asking that radiation cluster sufficiently on small scales then preferentially picks out the radiation which ultimately originated from a parent hard parton. The k_T algorithm was originally proposed here because it increases the chances that soft FSR will be kept: since clustering in the k_T metric works from soft up, using k_T increases the chance that a relatively soft parton emitted in the parton shower will be reconstructed and pass above the p_T threshold. But it is possible to imagine using other algorithms for the small-scale reclustering, and indeed implementations using C-A [15] or even anti-k_T [16] have been seen to be effective.

The trimming algorithm is simple to state; the detailed questions arise when we ask how the parameters should be chosen, and in any particular application parameter choices should be optimized for the specific process under consideration. Typical values for the small angular scale range between $0.2 \le R_0 \le 0.35$; for R_0 much smaller than $R_{min} = 0.2$, the finite angular resolution of the calorimeter starts to introduce irregularities. Good choices for Λ are either the total jet p_T, for dijet events or other such events where all jets have similar p_Ts, or the scalar sum transverse energy of the event, H_T, if jets have some spread over a broader range of p_Ts. Typical values for the cutoff parameter f_{cut} range between 10^{-2} (more typically for jet p_T) and 10^{-3} (for event H_T): this tends to work out to keeping subjets down to a 5 to 10 GeV threshold.

Pruning [17, 18] builds on the observation that the mass-drop algorithm improves mass resolution on boosted hard decays even before the filtering step, by discarding soft wide-angle radiation clustered into the fat jet at the final stages. In the C-A algorithm, the typical last clusterings in the fat jet are of stray soft radiation, usually unassociated with the parent particle, at wide angles to the jet core. These late, wide-angle clusterings have a disproportionate effect on jet mass.

Pruning adapts the splitting algorithm to specifically check for soft, wide-angle splittings, and throw them away. The algorithm is:

- Given a jet J, recluster its constituents with C-A, and then sequentially unwind the cluster sequence.
- At each splitting $P \to ij$, check whether the splitting is both *soft*,

$$z = \frac{\min(p_{T,i}, p_{T,j})}{p_{T,P}} < z_{cut}, \tag{24}$$

  and at *wide angle*,

$$\Delta R_{ij} > D_{cut}. \tag{25}$$

  If so, then drop the softer of i, j, and continue unwinding the harder.
- Stop when you find a sufficiently hard (or collinear) splitting.

Again, this algorithm has parameters that must be optimized specifically for each process under consideration. Typical values of z_{cut} are $z_{cut} \approx 0.1$, while the radial separation should be tuned to the expected opening angle for a hard process, $D_{cut} \approx 2m/p_T \times 1/2$.

Grooming in action. All three grooming techniques (filtering, trimming, and pruning) improve signal to background by both improving mass resolution for signal *and* suppressing QCD background. QCD jets, whose jet masses are generated by relatively softer and less symmetric emissions, are more likely to have their masses shifted substantially downward by jet grooming than collimated perturbatively decaying particles are, thus depleting the background to high-mass searches. Both the sharp gain in signal mass resolution and the depletion of the high mass background can be seen in Fig. 2.

We can also see in Fig. 2 that the different grooming techniques all act slightly differently on background massive QCD jets [15]. QCD jets with high masses dominantly have this mass generated by a relatively hard perturbative emission, which all algorithms are designed to retain, so performance between the different algorithms is similar. However, the effects of the different grooming algorithms on QCD backgrounds are still sufficiently distinct that some benefit can be obtained in applying multiple grooming algorithms [19].

At low masses, the differences between the grooming algorithms become more pronounced. QCD jets at low masses are dominated by a hard core. Filtering keeps a fixed number $N = 3$ of subjets, and therefore retains

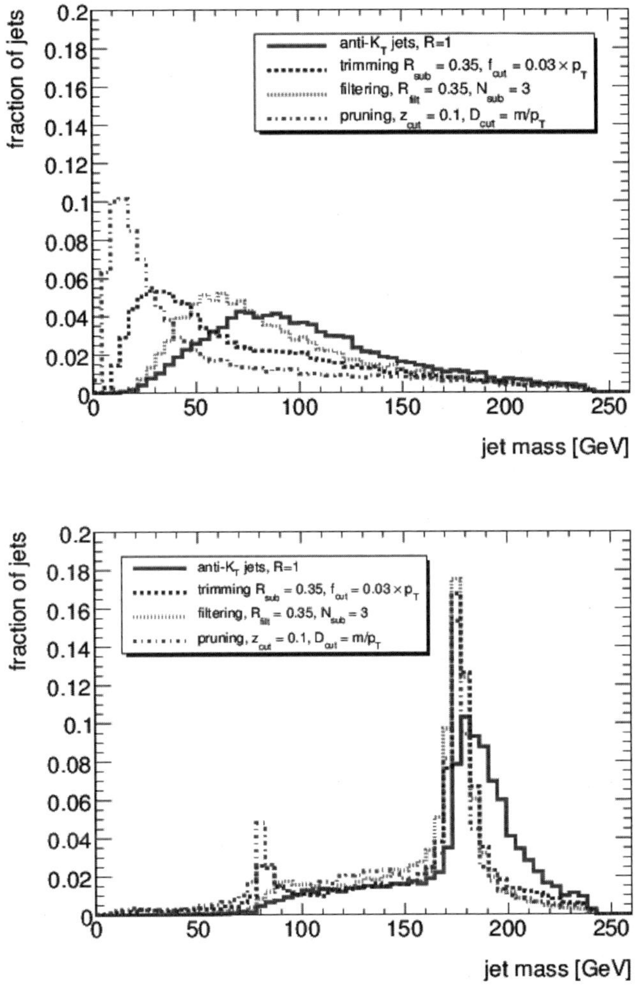

Fig. 2. The operation of filtering (green, dotted), trimming (blue, dashed), and pruning (purple, dash-dotted) on background QCD jets (top) and boosted top jets (bottom). From Ref. 15.

relatively soft radiation. Trimming, by contrast, will typically drop all radiation except that within R_{sub} of the jet core. Pruning will also typically drop all but the radiation in the core, but the resolution radius D is set to scale like m/p_T, and therefore $D \to 0$ as $m \to 0$. Thus at small masses typically $R_{prune} < R_{sub}$, so pruning acts more aggressively than trimming.

3.2. *Jet shapes*

Another feature of the boosted Higgs analysis we saw in the previous lecture was the importance of *jet mass*, which allowed us to concentrate signal in a sharp peak on top of a falling background [20]. Jet mass is an example of a *jet shape*: a function f defined on a jet J that quantifies the properties of the jet without the (explicit) use of any jet algorithm. The approach is conceptually akin to *event shapes*, which allow quantitative study of QCD without requiring specific characterization of an event in terms of jets, and indeed many jet shapes are descendants of event shapes.

Before discussing individual jet shapes, let us make two general comments. First, as we saw for jet mass, jet shapes are vulnerable to the inclusion of unassociated radiation, particularly pile-up, into jets, to a greater or lesser extent depending on the particular jet shape, and the sensitivity of the jet shape to unassociated radiation can be important. Second, one should bear in mind that any reasonable jet shape needs to be both infrared- and collinear-safe. Any linear function of particles' p_T is automatically safe; factorization theorems for other jet shapes can be proven [21].

3.2.1. *Radial Distribution of Particles Within a Jet*

The probability of a showering parton to emit a daughter parton depends on the running coupling α_s evaluated at the $k_\perp$ scale of the splitting. Jet shapes which measure the angular distribution of particles in an event are therefore measuring both the strength and the running of the strong coupling constant, and are classic probes of QCD. These jet shapes are also sensitive to the color charge of the parent parton: since $C_F < C_A$, an initial gluon will radiate more, and at wider angles, than an initial quark.

Jet broadening is a classic e^+e^- observable. Given a thrust axis $\hat{n}$, we can partition the particles i in an event into hemispheres according to $\text{sign}(\vec{p}_i \cdot \hat{n})$, which for dijet-like events is equivalent to associating each particle to a jet. Hemisphere broadening is then defined as the momentum-weighted transverse spread of the particles,

$$B_H = \frac{1}{\sum_{i\in H} |\vec{p}_i|} \sum_{i\in H} |\vec{p}_i \times \hat{n}| \tag{26}$$

where the sum runs over all particles i in a hemisphere H.

Differential and integrated jet shapes are, thanks to a historic quirk of nomenclature, names for two specific jet shapes: the so-called *differential jet shape* $\rho(r)$ and the *integrated jet shape* $\Psi(r)$, which characterize the radial distribution of radiation inside a jet. These jet shapes are also sometimes called the *jet profile*. Both of these shapes are defined on an ensemble of N jets formed with radius R. Then for $r < R$, the integrated jet shape $\Psi(r)$ is the ensemble average of the fraction of a jet's p_T which is contained within a radius r from the jet axis. Defining r_i as the distance of a constituent i from the jet axis,

$$\Psi(r) = \frac{1}{N} \sum_J \sum_{i \in J} \frac{p_T(0 < r_i < r)}{p_{T,J}}. \tag{27}$$

Here the second sum runs over all constituents i of a jet J. The differential jet shape $\rho(r)$ is then given by

$$\rho(r) = \frac{1}{\delta r} \frac{1}{N} \sum_J \sum_{i \in J} \frac{p_T(r < r_i < r + \delta r)}{p_{T,J}}. \tag{28}$$

These variables are often included in the suite of QCD precision measurements performed by experimental collaborations, as for instance in the ATLAS study [22], and are useful for validating parton shower models.

Girth is another jet shape which probes the radial distribution of radiation inside a jet. Let r_i again be the distance between a constituent i and the jet axis. Then the girth of a jet g_J is the linear radial moment of the jet,

$$g_J = \sum_{i \in J} \frac{p_{T,i} r_i}{p_{T,J}}. \tag{29}$$

In the collinear limit $\theta \to 0$, girth becomes equivalent to jet broadening (where the thrust axis is replaced by the jet axis). Girth has been shown to be particularly useful for distinguishing quark-initiated jets from gluon-initiated jets [23].

Angularities [24] are a related family of jet shapes, defined as a function of the parameter a:

$$\tau_a = \frac{1}{2E_J} \sum_{i \in J} p_{\perp,i} e^{(a-1)\eta_i}. \tag{30}$$

Here η_i is the separation in rapidity only between particle i and the jet axis, and $p_{\perp,i}$ the momentum transverse to the jet axis.

3.2.2. *Discriminating Boosted Decay Kinematics*

The radial distribution jet shapes discussed in the previous section are geared toward probing the characteristic shower structure of QCD. Here we will discuss several examples of jet shapes which target evidence of non-QCD-like substructure in jets.

Planar flow [24] considers the spread of the jet's radiation in the plane transverse to the jet axis (see also the closely related *jet transverse sphericity* shape [25]). Since QCD showers are angular-ordered, radiation subsequent to the first emission $P \to ij$ tends to be concentrated between the clusters of energy defined by i and j, leading to a roughly linear distribution of energy in the jet. By contrast, boosted three-body decays, such as boosted tops, have a more planar distribution of energy.

Define the tensor

$$I^{ab} = \frac{1}{m_J} \sum_{i \in J} \frac{p^a_{i,\perp} p^b_{i,\perp}}{E_i}, \tag{31}$$

where the indices a, b span the plane perpendicular to the jet axis, and $\vec{p}_{i,\perp}$ denotes the projection of particle i's momentum into this plane. Letting λ_1, λ_2 be the eigenvalues of I^{ab}, the planar flow of a jet is given by

$$Pf_J = \frac{4\lambda_1\lambda_2}{(\lambda_1+\lambda_2)^2} = \frac{\det I}{(\mathrm{Tr} I)^2}. \tag{32}$$

With this normalization, $Pf_J \in (0,1)$. Monte Carlo studies have demonstrated that QCD events do indeed peak at low values of Pf, while boosted top decays show a relatively flat distribution in Pf, but preliminary results show some sensitivity to shower modeling [25] and the utility of this shape in data is so far unclear.

Note that neither I^{ab} nor its eigenvalues are invariant under longitudinal boosts. For fully reconstructible events this is not a worry in theory, as all events can be considered in the reconstructed CM frame, but finite experimental resolution can become an issue in transforming from the lab frame into the CM frame.

Template overlaps define jet shapes based on (aspects of) the matrix elements for boosted object decays [26]. For example, consider the three body top quark decay with intermediate on-shell W. The phase space for this decay is (in the narrow-width approximation) determined by four parameters, which can be parameterized as the solid angle governing the two-body decays of both the t and its daughter W. Note that (1) the azimuthal angle ϕ_t is meaningful, as the detector geometry is not invariant under rotations around the top direction of motion, and (2) this phase space has both m_t and m_W built in. A series of templates describing this phase space can be generated by discretizing the four-dimensional space. To use these templates on a jet, the method of template overlaps finds the template which has best overlap with the kinematic configuration of the jet constituents according to a chosen metric. The ultimate variable is the numerical value of the best overlap, which distinguishes between QCD jets and boosted tops.

***N*-subjettiness** [27] takes a different and more general approach to probing jet substructure via jet shapes. Given N axes $\hat{n}_k$, we define N-subjettiness as

$$\tau_N = \frac{\sum_{i \in J} p_{T,i} \min(\Delta R_{ik})}{\sum_{i \in J} p_{T,i} R_0} \tag{33}$$

where R_0 is the jet radius, and ΔR_{ik} is the distance between particle i and axis $\hat{n}_k$. The smaller τ_N is, the more radiation is clustered around the chosen axes, or in other words, smaller values of τ_N indicate a better characterization of the jet J as having N (or fewer) subjets. Conversely, if τ_N is large, then a description in terms of $> N$ subjets is better.

However, as QCD alone will happily make jets with subjets, to differentiate boosted objects we need to probe not just the possible existence of subjets, but their structure. The real distinguishing power of N-subjettiness occurs when looking at *ratios*. For instance, a two-prong boosted particle such as a Higgs or W will have large τ_1 and small τ_2. QCD jets which have small τ_2 will generically have smaller τ_1 than for signal, as the QCD jets are more hierarchical; conversely, QCD jets which have large τ_1 are generally diffuse, and will have larger τ_2 as well than for signal. Thus the best single discriminating variable is τ_2/τ_1, or, more generally

$$r_N = \frac{\tau_N}{\tau_{N-1}} \tag{34}$$

for a boosted N-prong particle.

The question of how to determine the input subjet axes $\hat{n}_k$ is an interesting one. One approach, which is fast and perfectly serviceable for most applications, is to use a jet algorithm, such as exclusive k_T, to determine subjet axes. Naturally, the results then retain some dependence on the choice of jet algorithm used to find the axes. Another approach is to marginalize over all possible choices of $\hat{n}_k$, and choose the set which minimizes τ_N [28]. While this choice is computationally more intensive, it removes the dependence on the jet algorithm choice, and additionally guarantees the nice property that

$$\tau_{N-1} > \tau_N, \tag{35}$$

which holds only approximately if fixed subjet axes are used.

N-subjettiness is a conceptual descendent of the event shape N-jettiness [29], which classifies events as being N-jet-like without reference to jet algorithms.

3.2.3. *Color Flow Variables*

Beyond kinematics, boosted perturbative decays can also differ from QCD backgrounds in their color structure. Consider a color singlet such as a H or W boson decaying to a quark-antiquark pair. The daugher quark jets form a color dipole: they are color-connected to each other, but not to the rest of the event. Meanwhile, the backgrounds to these processes come from QCD dijets, which necessarily have different color connections, as we show in Fig. 3, where the radiation patterns for a color-singlet signal are plotted on the left and for a typical background on the right, as computed in the eikonal (soft) approximation. This observation has motivated work on variables which can add color flow to the suite of features which can discriminate signal from background.

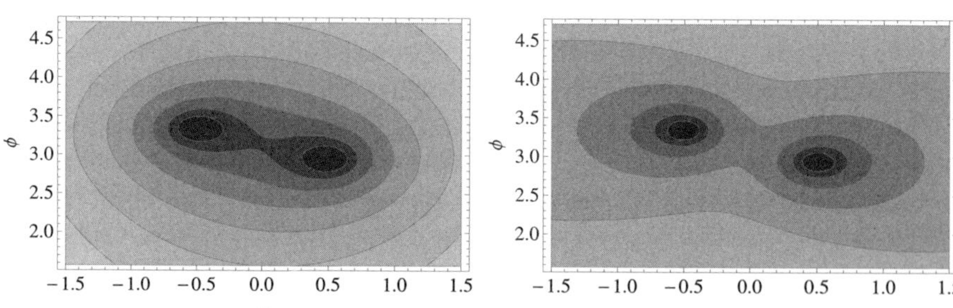

Fig. 3. Radiation patterns in the eikonal approximation for two triplet color sources color-connected to each other (left) and to the beam (right). Contours are logarithmic, and the scales in the two figures are not the same.

Jet pull [30] defines for each jet a transverse vector $\vec{t}_J$ characterizing the net directional distribution of the soft radiation surrounding the jet core. Defining $\vec{r}_i$ as the (transverse) direction of particle i from the jet axis, the pull vector is

$$\vec{t}_J = \sum_{i \in J} \frac{p_{T,i} |r_i| \vec{r}_i}{p_{T,J}}. \tag{36}$$

The direction of $\vec{t}_J$ relative to other jets in the event then is sensitive to the color connection of the jet J. Two jets which are color-connected to each other will have pull vectors pointing toward each other. Jets which are color-connected to the beam will have pull vectors pointing toward the beam. Once two interesting (sub)jets have been identified, the discriminating variable is then $\cos\theta_t$, the angle between the pull vector and the line connecting the two (sub)jet cores. An initial experimental study of pull has been carried out at D0, using the W in top events [31].

Dipolarity [32] is a jet shape which is designed to test for color dipole-like structure when the apparent particle is boosted and the two (sub)jets of interest are geometrically nearby. Since pull scales like r_i^2, it can be unduly sensitive to the detailed assignment of particles between the two (sub)jet cores to one or the other of the two (sub)jets. Dipolarity therefore uses as the relevant distance measure R_i, the transverse distance of particle i to the line segment connecting the (sub)jet cores,

$$D_J = \frac{1}{p_{T,J} R_{12}^2} \sum_{i \in J} p_{T,i} R_i^2. \tag{37}$$

Note that dipolarity requires input (sub)jet axes. The major application studied to date has been in boosted top tagging, where dipolarity can improve the identification of the boosted daughter W.

Keeping the right soft radiation. We have emphasized the need for jet grooming tools in the busy, high luminosity environment of the LHC. However, that grooming will groom away most if not all of the information about color flow. To use the information contained in an event's color flow, it is necessary to retain at least some of the soft radiation. Exactly which soft radiation is included, and at which stage in the analysis, is a question which has to be addressed case-by-case. As an example, we will discuss how the dipolarity shape can be incorporated into a boosted top tagger [32]. Top tagging will be discussed at length in the next section;

for the moment, it suffices to think of a top tagger as an algorithmic black box which acts on a fat jet to return candidate b, j_1, and j_2 subjets, and discards some radiation in the process.

The returned subjet axes define the characteristic opening scale, R_{12}, and provide the input axes for the dipolarity jet shape. As the top-tagger has discarded some of the radiation associated with the top quark in identifying the candidate subjets, to evaluate dipolarity we will need to go back to the original fat jet and include a larger subset of particles. Clearly, the radiation we'd like to include when evaluating the dipolarity of the candidate W daughters is only that associated with the two light quark jets; including radiation originating from the b would just skew the results. Let us consider only moderately boosted tops, such that the b jet is not overlapping with the other two. From the angular-ordered property of QCD showers, we know that in top events, all radiation associated with either light quark must be at angular separations less than the opening angle of the dipole, $\Delta R < R_{12}$. Thus, all radiation from the W is contained in cones of radius R_{12} around each light quark jet. The authors of Ref. 32 find that keeping all radiation within these two cones is casting too wide a net, however, and a smaller cone size of $R_{12}/\sqrt{2}$ is a better tradeoff between keeping all the radiation from the W and avoiding pollution from pileup, underlying event, and splash-in from the nearby b.

Color flow variables capture a genuine physical difference between signal and background. They have been shown, in theoretical work, to make a sizeable impact in signal significance [16, 32, 33, 45], and show great promise as tools to expand our understanding of SM and BSM physics. It is important to bear in mind, however, that these "proof of principle" analyses have all been performed using shower Monte Carlos, which capture only leading approximations to the full QCD dynamics. Just as the jet shapes discussed in Section 3.2.1 above have been and are still important tools for assessing the validity of the approximations made in the Monte Carlo generators, measuring and calibrating color flow variables in data is critical to understand the validity of the shower models and the performance of any color flow variable. This experimental program is, as of yet, in its infancy. In the meantime, theoretical studies should bear this uncertainty in mind. To estimate the uncertainties, it is useful (as it is for any novel substructure variable) to check results using more than one shower model.

4. Lecture III: Top Tagging and Searches for Physics BSM

In this section we will assemble the tools and techniques developed in the previous two sections and apply them to searches for physics beyond the standard model. By far the most universally motivated application of jet substructure techniques to BSM physics is in the hunt for TeV-scale new states which decay to electroweak-scale SM particles. The best reason for new physics to live anywhere near the weak scale is that it is partially responsible for the generation of the electroweak scale. New physics that is related to EWSB will naturally couple most strongly to those particles in the SM which feel EWSB most strongly, in particular the top quark and the EW bosons (H, W, and Z), and thus will decay preferentially to these heavy particles rather than to the light quarks and leptons which yield simpler final states. Moreover, we have compelling reasons to believe new physics will naturally decay to *boosted* SM particles. Even before the LHC turned on, the lack of deviations from SM predictions for flavor or precision electroweak observables already hinted that the likely scale for new physics was not v_{EW} as naturalness might have suggested, but rather $\Lambda \gtrsim$ few TeV. Evidence for this "little hierarchy" problem has of course only gotten stronger as the LHC has directly explored physics at TeV scales. Thus many models which address the stabilization of the EW scale will naturally give rise to final states rich in boosted tops, Higgses, W's and Z's.

In this section we will provide an introduction to top tagging at the LHC, followed by a few brief concluding comments on searching for more general BSM physics with jets.

4.1. *Top tagging*

As we established in Section 1, top pair production at the LHC covers a broad range of kinematic regimes interpolating between threshold ($\sqrt{s} = 2m_t$), where tops are well described as a six-object final state, up to TeV-scale energies, where the tops are highly collimated and are best described as a two-object final state. Top reconstruction must thus be able to flexibly cover a wide range of kinematic scenarios. In the interest of time, we will restrict our attention here to top taggers which target the hadronic decay of the top quark, although the semi-leptonic decay mode also requires interesting techniques for identification and reconstruction [34, 35].

As for jet algorithms, the "best" top tagger depends on the question being asked. In particular, different strategies are required at high p_T

($\gg m_t$) versus moderate p_T ($\gtrsim m_t$). Another question is: what signal efficiency is necessary? Every tagging technique trades off signal efficiency against background mistag rate. Depending on the search in question, the composition of the backgrounds will change, and therefore the necessary mistag rate will shift as well. For example, consider a top pair event with at least one boosted hadronic top. If the other top is also hadronic, then QCD dijets are by far the dominant background, and small QCD mistag rates are required. But if the other top is leptonic, then $W+$ jets becomes an important background, and if the top is produced in association with some new physics objects, such as $\not{E}_T$, then the backgrounds may be substantially smaller, and mistag rates may be entirely unimportant.

The aim of this section is to provide an introduction to top tagging by discussing a representative variety of top taggers. Specifically, we will consider the top taggers currently used by both LHC experiments, which work best in the highly boosted regime; the "HEP top tagger", which targets moderate p_T; and top tagging with N-subjettiness.

4.1.1. *CMS Top Tagger*

The hadronic top tagger used by CMS [36] is largely based on the "Hopkins" top tagger [37]. It builds on the techniques of the boosted Higgs "splitting/filtering" or "mass drop" analysis, which we discussed in Section 1. Thus, we again begin by clustering the event using the C-A algorithm, on large angular scales, capturing all of the top decay products in a single fat jet, which we will then unwind until we find interesting substructure. Compared to the Higgs analysis, there are two important differences. First, we are looking for at least three hard subjets, instead of two. Second, we take the fat jet radius to be noticeably smaller than we did for the Higgs case: $R = 0.8$. Using our rule of thumb, $R \sim 2p_T/m$, this means we are targeting tops with $p_T \gtrsim 500$ GeV: appropriate for production from a TeV-scale resonance. Contrast this with the boosted Higgs, which was targeting the high-p_T tails of SM associated production, where requiring large p_T imposed a significant price in signal acceptance.

Iteratively declustering the fat jet, we encounter splittings $P \to ij$. Our criterion for an interesting splitting is simply that both daughter subjets must carry a sufficiently large fraction of the *total* fat jet momentum,

$$p_{T,j} > \delta_P \, p_{T,J} \tag{38}$$

for some parameter δ_P. If a splitting fails to meet this criterion, discard the softer of i, j, and continue to unwind the harder. The splitting is rejected if it is too collinear, $|\Delta\eta_{ij}| + |\Delta\phi_{ij}| > \delta_R$, for another parameter δ_R. This procedure stops when either both i, j are softer than $\delta_P\, p_{T,J}$, or only one particle is left.

If an interesting hard, non-collinear splitting $P \to j_1 j_2$ is found, then the next step is to successively unwind both j_1 and j_2 according to the same algorithm, in search of further interesting splittings. This procedure returns a set of 2, 3, or 4 subjets. Fat jets returning only 2 subjets don't have enough substructure to be good top candidates, and are rejected. Jets which return 3 or 4 subjets do show enough substructure to be interesting, and the next step is to test whether or not they also have top-like kinematics.

As for the Higgs, the single most important discriminator is the jet mass. CMS requires that the jet mass, as computed from the sum of the returned subjets, lie within a top mass window, $m_t - 75\,\text{GeV} < m_J < m_t + 75$ GeV.

The onshell decay of the W inside the jet will also help us separate signal from background, but rather than trying to explicitly identify a pair of subjets which reconstruct a W—a procedure highly vulnerable to the misassignment of particles in overlapping jets—we will exploit the presence of the W mass scale in a less direct way.

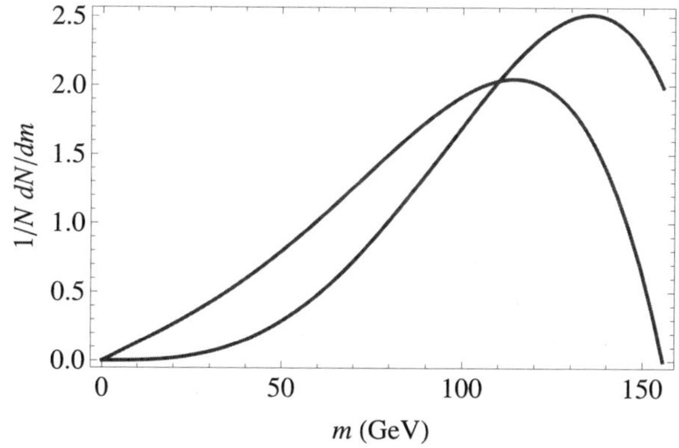

Fig. 4. Leading order distributions of $m_{b\bar{d}}$ (peaked at 110 GeV) and m_{bu} (peaked at 140 GeV) in unpolarized top decay.

The pairwise invariant masses of all possible combinations of the three daughter quarks are all governed by the mass scales in the top matrix element, m_W and m_t. The distribution of the invariant mass of the b and the $\bar{d}$-type quark (equivalent to the charged lepton in leptonic top decay) is shown in Fig. 4. The most likely value of $m_{b\bar{d}}$ is approximately 115 GeV. The invariant mass of the b and the u-type quark (equivalent to the neutrino) is peaked at even larger values. By contrast, subjet masses from QCD background processes are hierarchically smaller than the total parent jet mass. Thus instead of trying to reconstruct the W, we simply require that the *minimum* of the invariant masses formed from pairs of the three hardest subjets be sufficiently large to reject backgrounds,

$$\min(m_{12}, m_{13}, m_{23}) > 50\,\text{GeV}. \tag{39}$$

These cuts on masses, together with the substructure requirement, constitute the tagger. Note that no b-tagging information is used. Tagging b-jets is very difficult in this environment for two reasons. First, the b is embedded in a highly collimated top, so disentangling the tracks that are associated with the b from the other tracks in the jet is challenging. Second, the b itself is at very high p_T, so the opening angles of its daughter products are small, and it is difficult to get sufficient resolution from the reconstructed tracks to reconstruct the displaced vertex. Note also that the tagger doesn't require jet grooming. This is partly because the iterative decomposition procedure is performing some of that function in its own right, as it discards soft wide-angle radiation in the process of finding hard subjets (compare pruning). The smaller geometric size of the fat jets also means that pollution is not as large an effect.

4.1.2. *ATLAS Top Tagger*

We turn next to the ATLAS top tagger. Like CMS' tagger, it is optimized for high p_T, and like CMS' tagger, it is based on iterative declustering of a sequential algorithm. However, the ATLAS tagger draws on a very different set of ideas, largely based on work by Ref. 38 and the "Y-splitter" of Ref. 39.

The ATLAS top tagger begins by clustering events using the anti-k_T algorithm with $R = 1.0$. (The slightly larger jet radius means that this tagger works best at slightly lower p_T than does the CMS tagger.) Since the anti-k_T algorithm knows nothing about the singularity structure of QCD, its use is simply to identify a nicely regular initial set of particles. The next step is to take this set of particles and recluster them using the k_T algorithm.

Recall that the k_T algorithm preferentially clusters soft splittings. This means that the hardest splittings in the jet are the very last ones. Thus, there is no need to do any preliminary unwinding, and the existence of hard substructure is directly reflected in the hardness of the scales given by the k_T metric evaluated on the last few splittings in the jet:

$$d_{ij} = \min(p_{T,i}^2, p_{T,j}^2)\Delta R_{ij}^2. \tag{40}$$

Large splitting scales mean the emissions are both hard and at wide angles. The ATLAS tagger uses as inputs the splitting scales of the last three recombinations, d_{12}, d_{23}, and d_{34}. The first two splittings correspond (usually) to the identification of the three daughter partons, and the third to possible FSR from one of the partons. Since for tops the splitting d_{34} is the first which comes from the QCD shower, its scale can still be relatively large; on the other hand, for background QCD jets, the hierarchical nature of the shower means that generally $d_{34} \ll d_{23} \ll d_{12}$. Thus cuts on d_{34} maintain some discriminating power.

However, instead of cutting directly on the massive splittings d_{ij}, it is advantageous to change variables to a set which are less correlated with the jet and subjet invariant masses [25]. We define the energy sharing variables

$$z_{ij} = \frac{d_{ij}}{d_{ij} + m_{ij}^2} \approx \frac{E_j}{E_i + E_j} \tag{41}$$

where in the last step we have taken the collinear limit (and $p_{T,i} > p_{T,j}$). Notice that by performing this change of variables we have removed sensitivity to the collinear singularity, so that z_{ij} is only capturing information about the soft singularity. Meanwhile, jet invariant masses still retain information about the relative angles between the jets, so the correlation between the variables has been reduced.

The final set of variables that make up the ATLAS top tagger is then:

- The total jet mass, m_J. The tagger requires $m_J > 140$ GeV, and no upper bound: no grooming procedure is used, so the mass spectrum is distorted upwards.
- The variable Q_W, defined as the minimum pair invariant mass of the three subjets identified at the splitting scale d_{23}. This is the equivalent to cutting on the minimum pair invariant mass in the CMS tagger; only the method of finding the subjets is different. We require $Q_W > 50$ GeV.
- All three energy sharing variables, z_{12}, z_{13}, and z_{23}, which are subject to numerical cuts.

4.1.3. *HEP Top Tagger*

We turn now to the Heidelberg-Eugene-Paris top tagger, which functions on tops with $p_T \gtrsim 200$ GeV [40, 41]. In some sense this algorithm is more of an event reconstruction strategy than a top tagger. The algorithm begins by clustering the event using C-A on the extremely large angular scale $R = 1.5$, and requiring the fat jets thus formed to have $p_T > 200$ GeV. The p_T cut of 200 GeV puts us in the regime where the top is sufficiently boosted that its decay products will frequently lie in a single hemisphere. Looking at extremely fat jets is effectively identifying hemispheres in an event while avoiding the need to set any fixed angular scales for resolution within those hemispheres. This is an effective strategy for tops in this intermediate kinematic regime, where events will straddle any fixed angular scale; by unwinding C-A hemispheres, we allow the angular scales to be flexibly identified event by event.

The next step is to unwind the fat jet looking for interesting hard structure. This is done by employing a (loose) mass-drop criterion. For a splitting $P \to ij$, with $m_j < m_i$, the splitting is deemed sufficiently interesting if

$$m_j > 0.2\, m_P. \tag{42}$$

If the splitting passes this criterion, retain both i and j in the list of jets to unwind; otherwise, discard j and keep unwinding i until $m_i < 30$ GeV, at which point the unwinding stops. This unwinding procedure is performed on all subjets identified as interesting via Eq. (42). The output of this step is a list of subjets $\{j_i\}$ resulting from this iterative declustering; if there are at least 3 such subjets, then we have found enough substructure to continue.

At the next stage, we *filter* the substructures to shrink the geometric area associated with the top daughters and thereby reduce sensitivity to pileup, etc. Unlike in the Higgs case, where the mass drop criterion identified a unique angular scale $R_{b\bar{b}}$ associated with the sole hard splitting, we have a more complicated set of jets with more than one interesting splitting, and it is not immediately obvious which angular scale should be used to filter the event. The HEP top tagger determines the filter radius R_{filt} by brute force, as follows. For each possible set of three subjets that can be drawn from the $\{j_i\}$, filter them by resolving the constituents of those subjets with radius $R_{filt} = \min(0.3, \Delta R_{ij})$, and retain up to five subjets. Let m_{filt} be the invariant mass of these up-to-five filtered subjets, and select the set with m_J closest to m_t as the top candidate. These up-to-five filtered subjets are then (yet again) reclustered into three subjets, which are the candidates for the partonic top daughters.

The next step is to test whether or not the reconstructed top daughters have top-like kinematics. Again, we will exploit the presence of both the top and W mass scales. We have already used m_t to identify the best set of subjets. Unlike in the previous taggers, we will now demand evidence of the on-shell W in a more complex way. Label the three subjets returned by the previous step as $\{j_1, j_2, j_3\}$ in descending order of p_T. Of the three invariant masses m_{12}, m_{13}, and m_{23}, only two are independent. This means that the top kinematics is characterized by a specific distribution in the two-dimensional space determined by the pair invariant masses. Top jets are focused into a thin triangular annulus in this space, as two subjets reconstruct an on-shell W (the annulus is triangular since any of the m_{ij} may correspond to the W). Background, by contrast, is concentrated in regions of small pairwise invariant masses. The kinematic cuts imposed in the HEP top tagger pick out this top-like triangular annulus by asking that events lie on one of the three branches of the annulus.

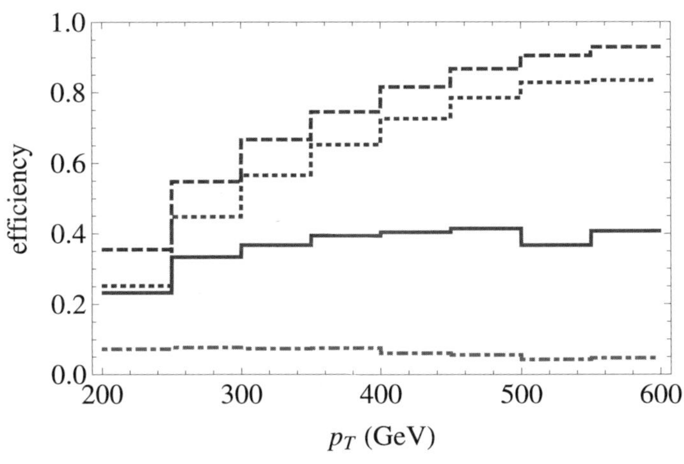

Fig. 5. HEP top tagger efficiencies on top quarks for: all top decay products within $\Delta R = 1.5$ of each other (blue, dashed); all top decay products clustered into $R = 1.5$ C-A fat jets (purple, dotted); tagged by the HEP top tagger (red, solid); tagged, but with reconstructed subjects not matching original partons (green, dot-dashed). Data from Ref. 41.

To understand how well this procedure covers the interpolating kinematic region, we plot efficiencies for tops to pass through these steps in

Fig. 5. As is evident from the blue (dashed) curve, simply demanding that all decay products of the top lie in a single hemisphere imposes a non-negligible acceptance price for tops at the low end of the p_T range, which drops quickly as the tops become more energetic. Further demanding that the top daughters all be clustered into the same fat jet results in an additional mild efficiency loss, seen in the purple (dotted) curve. The purple curve is the fraction of tops giving rise to taggable jets (neglecting the possibility of mistagged signal). The red line denotes the final efficiency of the full HEP top tagger, after the filtering and kinematic cuts. At low p_T, the fraction of taggable jets which are in fact tagged is near unity, but as the tops become more collimated, the probability of a taggable jet passing the kinematic cuts falls off, in large part because collimation and jet-particle misassignment make the W mass reconstruction less precise. At the upper end of the p_T range shown in the figure, the high-p_T top taggers are useful, and would take over. Let us also comment that there is a possibility for tops to pass the top tagger by accident, when the algorithm picks up the wrong set of jets; this is shown in the green (dot-dashed) curve.

4.1.4. *N-Subjettiness*

As we saw in section 3.2.2, N-subjettiness offers an entirely complementary test of the existence of hard substructure. A simple and highly effective top tagger can be constructed using as input variables just the jet mass and the ratio τ_3/τ_2. Further refinement is possible with a multivariate analysis which uses in addition τ_2/τ_1 as well as τ_1, τ_2, and τ_3 individually [28].

From our experience with the previous taggers, we can guess that even further improvement would be possible if some information about the W were also incorporated; since the N-subjettiness jet shape also provides a method of determining subjet axes, it naturally suggests methods for defining three subjets and computing the analog of Q_W. To the best of the author's knowledge, no such study has been publicly performed.

4.1.5. *Top Tagging Performance*

Let us now consider the performance of the top taggers which we have discussed. This task is made easier by the work performed in the BOOST 2010 [15] and BOOST 2011 [42] workshops, which compared the performance of different top taggers on the same reference sets of event samples. These event samples are publicly available online, so should you develop your own brilliant ideas about top tagging, you can cross-check the performance of your novel technique with the techniques already in the literature. In Fig. 6 we show performance curves for the high-p_T top taggers which we discussed above. Overall, these high-p_T top taggers have efficiencies on the

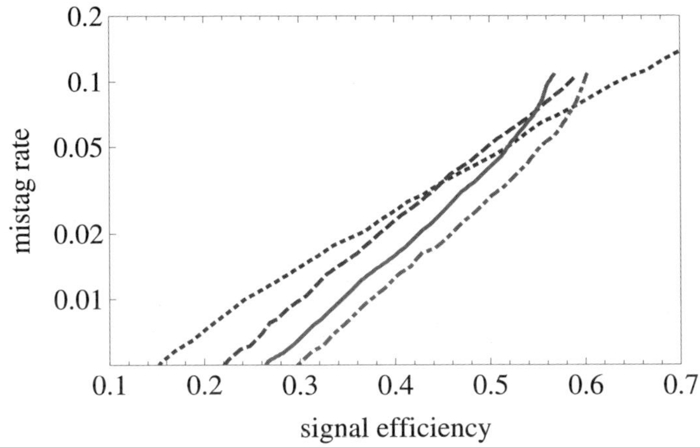

Fig. 6. Top tagging performance curves, for tops with $200\,\text{GeV} < p_T < 800\,\text{GeV}$ in the BOOST 2010 reference samples for: ATLAS (blue, dotted); CMS (red, dashed); and N-subjettiness in both the simple (green, solid) and multivariate (green, dot-dashed) versions. Data from Refs. 15, 28.

order of $\epsilon \sim 50\%$, at a (QCD) background mistag rate of $\epsilon_{fake} \sim 5\%$. (For comparison, LHC b-tagging algorithms achieve $\epsilon \sim 70\%$, with a fake rate $\epsilon_{fake} \sim 1\%$.) It is evident from the performance curves that the ATLAS tagger outperforms the CMS tagger when high signal efficiency is required, while CMS does better at lower signal efficiency. Even the simple two-variable N-subjettiness tagger outperforms both CMS and ATLAS taggers by a notable margin, except at high signal efficiency, while adding the additional multivariate discrimination to the N-subjettiness tagger provides a significant improvement. Further updates in the BOOST 2011 workshop show that (1) being more precise about modeling QCD radiation at wide angles and (2) including the effects of finite detector resolution reduce typical efficiencies to $\epsilon \sim 40\%$, at a (QCD) background mistag rate in the range $\epsilon_{fake} \sim 2 - 8\%$ depending on the tagger [42]. Incorporating finite detector resolution also tends to reduce (but not erase!) the relative advantage of N-subjettiness over the sequential decomposition-based taggers.

4.2. *BSM searches with jet substructure*

New physics produces jets with substructure when the kinematics are governed by a nontrivial hierarchy of scales. For the top examples we've been discussing, this hierarchy arises from the separation between the scale characterizing new physics and the electroweak scale:

$$\Lambda_{NP} \gg \Lambda_{EW} \gg \Lambda_{QCD}. \tag{43}$$

The little hierarchy problem results in a very strong motivation for developing tagging techniques for boosted SM objects. Besides the top tagging discussed in the previous subsection, much effort has also gone into tagging boosted W, Z, and H bosons arising from the decay of new TeV-scale particles [43–46]. This is fortunate for theorists, as, once these techniques are put into use at experiments in one context, the barrier is much lower for their adaptation in other contexts where the theoretical motivation may not be so universal.

What other kinds of BSM physics are amenable to substructure analyses? To engender events with interesting substructure, some multi-tiered hierarchy of scales is required. We will enumerate an illustrative but far from exhaustive set of examples.

Supersymmetry is one example of a new physics sector which naturally can generate multiple scales. For example, if supersymmetry is broken at very high scales, RG effects will drive the colored superpartners much heavier than those superpartners with only EW charges. Thus, at the weak scale one could naturally expect $M_{\tilde{g}} \gg M_{\chi^0}$. In the presence of a large hierarchy between gluino and neutralino, the decay products of the neutralino would be collimated. Let us further suppose that the neutralino decays via the R-parity violating udd superpotential operator, so that $\chi^0 \to qqq$. Then gluino pair production would appear as a six-jet final state, where two of the jets are actually boosted neutralinos, containing interesting substructure [47]. The very large particle content of the MSSM can easily accommodate many possible hierarchies, with different theoretical origins; see for instance Ref. 48 for another of the many possibilities.

Another way to generate a hierarchy in a BSM sector is if the new physics sector contains a broken global symmetry, so that the scale $\Lambda^{(1)}$ characterizing the lightest states is set by the magnitude of the global symmetry breaking, rather than by the overall scale of the new sector, $\Lambda^{(2)} \gg \Lambda^{(1)}$. Thus consider, for example, a composite rho ρ_C, decaying into two pseudo-Nambu-Goldstone bosons π_C, which are stable within their own sector, and therefore must subsequently decay into SM objects [49].

Hidden valley models also have this kind of multi-scale structure. Here the hierarchy is between the mass of the mediator which connects the visible and hidden sectors and the mass scale of the light states in the hidden sector,

$$\Lambda_{med} \gg \Lambda_{NP} > \Lambda_{SM}. \tag{44}$$

The mediating particle might be a SM particle, in particular the H or Z, or a novel field such as a Z' [50] or the SM LSP [51]. Exotic Higgs decays to light particles also fall under this umbrella [16, 52–56].

More generally, thinking more broadly and flexibly about jets leads to new approaches to combinatorics and event reconstruction [57], and provides novel methods to distinguish QCD events from new physics. As challenging high-multiplicity and all-hadronic final states become a larger component of the LHC program, flexible and creative jet techniques will be critical to our ability to discover and interpret the physics. Jet algorithms themselves are still an evolving field! The anti-k_T algorithm was introduced only a few years ago. As the nature of the questions that we ask about jets evolves, so do the best jet algorithms to address these questions. There is still a lot of room for new ideas!

5. Further Reading

References which were invaluable in the preparation of these lectures are the text *QCD and Collider Physics*, by Ellis, Stirling, and Webber [11], and the lecture notes "Toward Jetography" by Salam [58]. The proceedings of the BOOST 2010 and 2011 workshops [15, 42] are valuable resources for those looking for a quantitative survey of both theoretical and experimental progress in jet physics at the Tevatron and the LHC.

Acknowledgments

It is a pleasure to thank M. Schmaltz and the organizers for the opportunity to be part of such an excellent program. I thank D. Krohn for introducing me to jet substructure, and for many useful conversations during the course of our collaborations. Thanks to C. Vermilion for providing Figure 2, and M. Freytsis and D. Krohn for comments on the manuscript. Finally, thanks to my collaborators, Y. Bai, A. Falkowski, A. Thallapillal, and L.-T. Wang. I am supported by DOE grant DE-FG02-92ER40704, NSF grant PHY-1067976, and the LHC Theory Initiative under grant NSF-PHY-0969510.

References

[1] G. F. Sterman and S. Weinberg, Phys. Rev. Lett. **39**, 1436 (1977).
[2] W. Bartel *et al.* [JADE Collaboration], Z. Phys. C **33**, 23 (1986).
[3] S. Bethke *et al.* [JADE Collaboration], Phys. Lett. B **213**, 235 (1988).
[4] S. Catani, Y. L. Dokshitzer, M. Olsson, G. Turnock and B. R. Webber, Phys. Lett. B **269**, 432 (1991).
[5] S. D. Ellis and D. E. Soper, Phys. Rev. D **48**, 3160 (1993) [hep-ph/9305266].
[6] S. Catani, Y. L. Dokshitzer, M. H. Seymour and B. R. Webber, Nucl. Phys. B **406**, 187 (1993).
[7] Y. L. Dokshitzer, G. D. Leder, S. Moretti and B. R. Webber, JHEP **9708**, 001 (1997) [hep-ph/9707323].
[8] M. Cacciari, G. P. Salam and G. Soyez, JHEP **0804**, 063 (2008) [arXiv:0802.1189 [hep-ph]].
[9] M. Cacciari and G. P. Salam, Phys. Lett. B **641**, 57 (2006) [hep-ph/0512210].
[10] J. M. Butterworth, A. R. Davison, M. Rubin and G. P. Salam, Phys. Rev. Lett. **100**, 242001 (2008) [arXiv:0802.2470 [hep-ph]].
[11] R. K. Ellis, W. J. Stirling, and B. R. Webber, *QCD and Collider Physics* (Cambridge Monographs on Particle Physics, Nuclear Physics and Cosmology, Cambridge University Press, Cambridge, UK, 2003).
[12] M. Dasgupta, L. Magnea and G. P. Salam, JHEP **0802**, 055 (2008) [arXiv:0712.3014 [hep-ph]].
[13] G. Aad *et al.* [ATLAS Collaboration], JHEP **1205**, 128 (2012) [arXiv:1203.4606 [hep-ex]].
[14] D. Krohn, J. Thaler and L.-T. Wang, JHEP **1002**, 084 (2010) [arXiv:0912.1342 [hep-ph]].
[15] A. Abdesselam, E. B. Kuutmann, U. Bitenc, G. Brooijmans, J. Butterworth, P. Bruckman de Renstrom, D. Buarque Franzosi and R. Buckingham *et al.*, Eur. Phys. J. C **71**, 1661 (2011) [arXiv:1012.5412 [hep-ph]].
[16] A. Falkowski, D. Krohn, L.-T. Wang, J. Shelton and A. Thalapillil, Phys. Rev. D **84**, 074022 (2011) [arXiv:1006.1650 [hep-ph]].
[17] S. D. Ellis, C. K. Vermilion and J. R. Walsh, Phys. Rev. D **80**, 051501 (2009) [arXiv:0903.5081 [hep-ph]].
[18] S. D. Ellis, C. K. Vermilion and J. R. Walsh, Phys. Rev. D **81**, 094023 (2010) [arXiv:0912.0033 [hep-ph]].
[19] D. E. Soper and M. Spannowsky, JHEP **1008**, 029 (2010) [arXiv:1005.0417 [hep-ph]].
[20] W. Skiba and D. Tucker-Smith, Phys. Rev. D **75**, 115010 (2007) [hep-ph/0701247].
[21] S. D. Ellis, C. K. Vermilion, J. R. Walsh, A. Hornig and C. Lee, JHEP **1011**, 101 (2010) [arXiv:1001.0014 [hep-ph]].
[22] G. Aad *et al.* [Atlas Collaboration], Phys. Rev. D **83**, 052003 (2011) [arXiv:1101.0070 [hep-ex]].
[23] J. Gallicchio and M. D. Schwartz, Phys. Rev. Lett. **107**, 172001 (2011) [arXiv:1106.3076 [hep-ph]].

[24] L. G. Almeida, S. J. Lee, G. Perez, G. F. Sterman, I. Sung and J. Virzi, Phys. Rev. D **79**, 074017 (2009) [arXiv:0807.0234 [hep-ph]].
[25] J. Thaler and L.-T. Wang, JHEP **0807**, 092 (2008) [arXiv:0806.0023 [hep-ph]].
[26] L. G. Almeida, S. J. Lee, G. Perez, G. Sterman and I. Sung, Phys. Rev. D **82**, 054034 (2010) [arXiv:1006.2035 [hep-ph]].
[27] J. Thaler and K. Van Tilburg, JHEP **1103**, 015 (2011) [arXiv:1011.2268 [hep-ph]].
[28] J. Thaler and K. Van Tilburg, JHEP **1202**, 093 (2012) [arXiv:1108.2701 [hep-ph]].
[29] I. W. Stewart, F. J. Tackmann and W. J. Waalewijn, Phys. Rev. Lett. **105**, 092002 (2010) [arXiv:1004.2489 [hep-ph]].
[30] J. Gallicchio and M. D. Schwartz, Phys. Rev. Lett. **105**, 022001 (2010) [arXiv:1001.5027 [hep-ph]].
[31] V. M. Abazov *et al.* [D0 Collaboration], Phys. Rev. D **83**, 092002 (2011) [arXiv:1101.0648 [hep-ex]].
[32] A. Hook, M. Jankowiak and J. G. Wacker, JHEP **1204**, 007 (2012) [arXiv:1102.1012 [hep-ph]].
[33] D. Curtin, R. Essig and B. Shuve, arXiv:1210.5523 [hep-ph].
[34] K. Rehermann and B. Tweedie, JHEP **1103**, 059 (2011) [arXiv:1007.2221 [hep-ph]].
[35] [ATLAS Collaboration], ATL-PHYS-PUB-2009-081, ATL-PHYS-PUB-2010-008.
[36] [CMS Collaboration], CMS-PAS-JME-09-001.
[37] D. E. Kaplan, K. Rehermann, M. D. Schwartz and B. Tweedie, Phys. Rev. Lett. **101**, 142001 (2008) [arXiv:0806.0848 [hep-ph]].
[38] G. Brooijmans, ATL-PHYS-CONF-2008-008, ATL-COM-PHYS-2008-001, Jan 2008.
[39] J. M. Butterworth, B. E. Cox and J. R. Forshaw, Phys. Rev. D **65**, 096014 (2002) [hep-ph/0201098].
[40] T. Plehn, G. P. Salam and M. Spannowsky, Phys. Rev. Lett. **104**, 111801 (2010) [arXiv:0910.5472 [hep-ph]].
[41] T. Plehn, M. Spannowsky, M. Takeuchi and D. Zerwas, JHEP **1010**, 078 (2010) [arXiv:1006.2833 [hep-ph]].
[42] A. Altheimer, S. Arora, L. Asquith, G. Brooijmans, J. Butterworth, M. Campanelli, B. Chapleau and A. E. Cholakian *et al.*, J. Phys. G **39**, 063001 (2012) [arXiv:1201.0008 [hep-ph]].
[43] G. D. Kribs, A. Martin, T. S. Roy and M. Spannowsky, Phys. Rev. D **81**, 111501 (2010) [arXiv:0912.4731 [hep-ph]].
[44] A. Katz, M. Son and B. Tweedie, JHEP **1103**, 011 (2011) [arXiv:1010.5253 [hep-ph]].
[45] Y. Cui, Z. Han and M. D. Schwartz, Phys. Rev. D **83**, 074023 (2011) [arXiv:1012.2077 [hep-ph]].
[46] M. Son, C. Spethmann and B. Tweedie, JHEP **1208**, 160 (2012) [arXiv:1204.0525 [hep-ph]].

[47] J. M. Butterworth, J. R. Ellis, A. R. Raklev and G. P. Salam, Phys. Rev. Lett. **103**, 241803 (2009) [arXiv:0906.0728 [hep-ph]].
[48] J. Fan, D. Krohn, P. Mosteiro, A. M. Thalapillil and L.-T. Wang, JHEP **1103**, 077 (2011) [arXiv:1102.0302 [hep-ph]].
[49] Y. Bai and J. Shelton, JHEP **1207**, 067 (2012) [arXiv:1107.3563 [hep-ph]].
[50] M. Baumgart, C. Cheung, J. T. Ruderman, L.-T. Wang and I. Yavin, JHEP **0904**, 014 (2009) [arXiv:0901.0283 [hep-ph]].
[51] C. Cheung, J. T. Ruderman, L.-T. Wang and I. Yavin, JHEP **1004**, 116 (2010) [arXiv:0909.0290 [hep-ph]].
[52] C.-R. Chen, M. M. Nojiri and W. Sreethawong, JHEP **1011**, 012 (2010) [arXiv:1006.1151 [hep-ph]].
[53] D. E. Kaplan and M. McEvoy, Phys. Rev. D **83**, 115004 (2011) [arXiv:1102.0704 [hep-ph]].
[54] C. Englert, T. S. Roy and M. Spannowsky, Phys. Rev. D **84**, 075026 (2011) [arXiv:1106.4545 [hep-ph]].
[55] I. Lewis and J. Schmitthenner, JHEP **1206**, 072 (2012) [arXiv:1203.5174 [hep-ph]].
[56] P. Draper and D. McKeen, Phys. Rev. D **85**, 115023 (2012) [arXiv:1204.1061 [hep-ph]].
[57] A. Hook, E. Izaguirre, M. Lisanti and J. G. Wacker, Phys. Rev. D **85**, 055029 (2012) [arXiv:1202.0558 [hep-ph]].
[58] G. P. Salam, Eur. Phys. J. C **67**, 637 (2010) [arXiv:0906.1833 [hep-ph]].
[59] T. Plehn and M. Spannowsky, J. Phys. G **39**, 083001 (2012) [arXiv:1112.4441 [hep-ph]].

Chapter 8

Introduction to QCD

P. Skands

CERN TH, 1211 Geneva 23, Switzerland
E-mail: peter.skands@cern.ch

These lectures were given at TASI 2012 and are directed at a level suitable for graduate students in High Energy Physics. They are intended to give an introduction to the theory and phenomenology of quantum chromodynamics (QCD), focusing on collider physics applications. The aim is to bring the reader to a level where informed decisions can be made concerning different approaches and their uncertainties. The material is divided into five main areas: (1) fundamentals, (2) fixed-order perturbative QCD, (3) Monte Carlo event generators and parton showers, (4) Matching at Leading and Next-to-Leading Order, and (5) Soft QCD physics.

Contents

1. Introduction

When probed at very short wavelengths, QCD is essentially a theory of free "partons" — quarks and gluons — which only scatter off one another through relatively small quantum corrections, that can be systematically calculated. At longer wavelengths, of order the size of the proton $\sim 1\text{fm} = 10^{-15}\text{m}$, however, we see strongly bound towers of hadron resonances emerge, with string-like potentials building up if we try to separate their partonic constituents. Due to our inability to perform analytic calculations in strongly coupled field theories, QCD is therefore still only partially solved. Nonetheless, all its features, across all distance scales, are believed to be encoded in a single one-line formula of alluring simplicity; the Lagrangian of QCD.

The consequence for collider physics is that some parts of QCD can be calculated in terms of the fundamental parameters of the Lagrangian, whereas others must be expressed through models or functions whose effective parameters are not a priori calculable but which can be constrained by fits to data. However, even in the absence of a perturbative expansion, there are still several strong theorems which hold, and which can be used to give relations between seemingly different processes. (This is, e.g., the reason it makes sense to constrain parton distribution functions in *ep* collisions and then re-use the same ones for *pp* collisions.) Thus, in the chapters dealing with phenomenological models we shall emphasize that the loss of a factorized perturbative expansion is not equivalent to a total loss of predictivity.

An alternative approach would be to give up on calculating QCD altogether and use leptons instead. Formally, this amounts to summing inclusively over strong-interaction phenomena, when such are present. While such a strategy might succeed in replacing what we do know about QCD by "unity", however, even the most adamant chromophobe must acknowledge a few basic facts of collider physics for the next decade(s): (1) At the Tevatron and LHC, the initial states are hadrons, and hence, at the very least, well-understood and precise parton distribution functions (PDFs) will be required; (2) high precision will mandate calculations to higher orders in perturbation theory, which in turn will involve more QCD; (3) the requirement of lepton *isolation* makes the very definition of a lepton depend implicitly on QCD, and (4) the rate of jets that are misreconstructed as leptons in the experiment depends explicitly on it. Finally, (5) though many new-physics signals *do* give observable signals in the lepton sector, this is far from guaranteed. It would therefore be unwise not to attempt to solve QCD to the best of our ability, the better to prepare ourselves for both the largest possible discovery reach and the highest attainable subsequent precision.

In addition, or perhaps as a consequence, the field of QCD is currently experiencing something of a revolution. On the perturbative side, new methods to compute scattering amplitudes with very high particle multiplicities are being developed, together with advanced techniques for combining such amplitudes with all-orders resummation frameworks. On the non-perturbative side, the wealth of data on soft-physics processes from the LHC is forcing us to reconsider the reliability of the standard fragmentation models, and heavy-ion collisions are providing new insights into the collective behavior of hadronic matter. The study of cosmic rays impinging on the Earth's atmosphere challenges our ability to extrapolate fragmentation models from collider energy scales to the region of ultra-high energy cosmic rays. And finally, dark-matter annihilation processes in space can produce colored particles either directly or via decays of unstable resonances, making predictions of their spectra sensitive to the accuracy of the fragmentation modeling.

In the following, we shall focus squarely on QCD for mainstream collider physics. This includes factorization, hard processes, infrared safety, parton showers and matching, event generators, hadronization, and the so-called underlying event. While not covering everything, hopefully these topics can also serve at least as stepping stones to more specialized issues that have been left out, such as twistor-inspired techniques, heavy flavors, or forward physics, or to topics more tangential to other fields, such as lattice QCD or heavy-ion physics.

1.1. *The Lagrangian of QCD*

Quantum Chromodynamics is based on the gauge group SU(3), the Special Unitary group in 3 (complex) dimensions. In the context of QCD, we represent this group as a set of unitary 3×3 matrices with determinant one. This is called the *adjoint representation* and can be used to represent gluons in color space. Since there are 9 linearly independent unitary complex matrices, one of which has determinant -1, there are a total of 8 independent directions in the adjoint color space, i.e., the gluons are *octets*. In QCD, these matrices can operate both on each other (gluon self-interactions) and on a set of complex 3-vectors (the *fundamental* representation), the latter of which represent quarks in color space. The fundamental representation has one linearly independent basis vector per degree of SU(3), and hence the quarks are *triplets*.

The Lagrangian of QCD is

$$\mathcal{L} = \bar{\psi}_q^i (i\gamma^\mu)(D_\mu)_{ij}\psi_q^j - m_q \bar{\psi}_q^i \psi_{qi} - \frac{1}{4} F^a_{\mu\nu} F^{a\mu\nu} , \tag{1}$$

where ψ_q^i denotes a quark field with color index i, $\psi_q = (\psi_{qR}, \psi_{qG}, \psi_{qB})^T$, γ^μ is a Dirac matrix that expresses the vector nature of the strong interaction, with μ being a Lorentz vector index, m_q allows for the possibility of non-zero quark masses (induced by the standard Higgs mechanism or similar), $F^a_{\mu\nu}$ is the gluon field strength tensor for a gluon with color index a (in the adjoint representation, i.e., $a \in [1, \ldots, 8]$), and D_μ is the covariant derivative in QCD,

$$(D_\mu)_{ij} = \delta_{ij}\partial_\mu - i g_s t^a_{ij} A^a_\mu , \tag{2}$$

with g_s the strong coupling (related to α_s by $g_s^2 = 4\pi\alpha_s$; we return to the strong coupling in more detail below), A^a_μ the gluon field with (adjoint-representation) color index a, and t^a_{ij} proportional to the hermitean and traceless Gell-Mann matrices of SU(3),

$$\lambda^1 = \begin{pmatrix} 0 & 1 & 0 \\ 1 & 0 & 0 \\ 0 & 0 & 0 \end{pmatrix}, \lambda^2 = \begin{pmatrix} 0 & -i & 0 \\ i & 0 & 0 \\ 0 & 0 & 0 \end{pmatrix}, \lambda^3 = \begin{pmatrix} 1 & 0 & 0 \\ 0 & -1 & 0 \\ 0 & 0 & 0 \end{pmatrix}, \lambda^4 = \begin{pmatrix} 0 & 0 & 1 \\ 0 & 0 & 0 \\ 1 & 0 & 0 \end{pmatrix}$$

$$\lambda^5 = \begin{pmatrix} 0 & 0 & -i \\ 0 & 0 & 0 \\ i & 0 & 0 \end{pmatrix}, \lambda^6 = \begin{pmatrix} 0 & 0 & 0 \\ 0 & 0 & 1 \\ 0 & 1 & 0 \end{pmatrix}, \lambda^7 = \begin{pmatrix} 0 & 0 & 0 \\ 0 & 0 & -i \\ 0 & i & 0 \end{pmatrix}, \lambda^8 = \begin{pmatrix} \frac{1}{\sqrt{3}} & 0 & 0 \\ 0 & \frac{1}{\sqrt{3}} & 0 \\ 0 & 0 & \frac{-2}{\sqrt{3}} \end{pmatrix} . \tag{3}$$

These generators are just the SU(3) analogs of the Pauli matrices in SU(2). By convention, the constant of proportionality is normally taken to be

$$t^a_{ij} = \frac{1}{2}\lambda^a_{ij} . \tag{4}$$

This choice in turn determines the normalization of the coupling g_s, via Eq. (2), and fixes the values of the SU(3) Casimirs and structure constants, to which we return below.

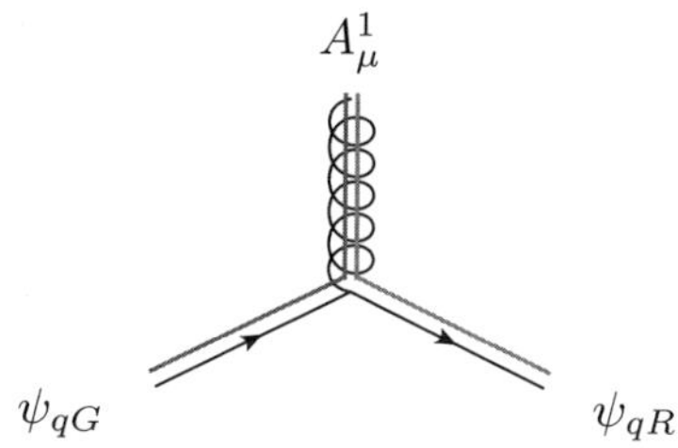

$$\propto -\tfrac{i}{2} g_s \quad \bar{\psi}_{qR} \qquad \lambda^1 \qquad \psi_{qG}$$

$$= -\tfrac{i}{2} g_s \,(1\,0\,0) \begin{pmatrix} 0 & 1 & 0 \\ 1 & 0 & 0 \\ 0 & 0 & 0 \end{pmatrix} \begin{pmatrix} 0 \\ 1 \\ 0 \end{pmatrix}$$

Fig. 1. Illustration of a qqg vertex in QCD, before summing/averaging over colors: a gluon in a state represented by λ^1 interacts with quarks in the states ψ_{qR} and ψ_{qG}.

An example of the color flow for a quark-gluon interaction in color space is given in Fig. 1.

1.2. *Color factors*

Typically, we do not measure color in the final state — instead we average over all possible incoming colors and sum over all possible outgoing ones, wherefore QCD scattering amplitudes (squared) in practice always contain sums over quark fields contracted with Gell-Mann matrices. These contractions in turn produce traces which yield the *color factors* that are associated to each QCD process, and which basically count the number of "paths through color space" that the process at hand can take.[a]

A very simple example of a color factor is given by the decay process $Z \to q\bar{q}$. This vertex contains a simple δ_{ij} in color space; the outgoing quark and antiquark must have identical (anti-)colours. Squaring the corresponding matrix element and summing over final-state colors yields a color factor of

$$e^+e^- \to Z \to q\bar{q} \quad : \quad \sum_{\text{colors}} |M|^2 \propto \delta_{ij}\delta^*_{ji} = \text{Tr}\{\delta\} = N_C = 3 \ , \qquad (5)$$

since i and j are quark (i.e., 3-dimensional fundamental-representation) indices.

A next-to-simplest example is given by $q\bar{q} \to \gamma^*/Z \to \ell^+\ell^-$ (usually referred to as the Drell-Yan process [1]), which is just a crossing of the

[a]The convention choice represented by Eq. (4) introduces a "spurious" factor of 2 for each power of the coupling α_s. Although one could in principle absorb that factor into a redefinition of the coupling, effectively redefining the normalization of "unit color charge", the standard definition of α_s is now so entrenched that alternative choices would be counter-productive, at least in the context of a supposedly pedagogical review.

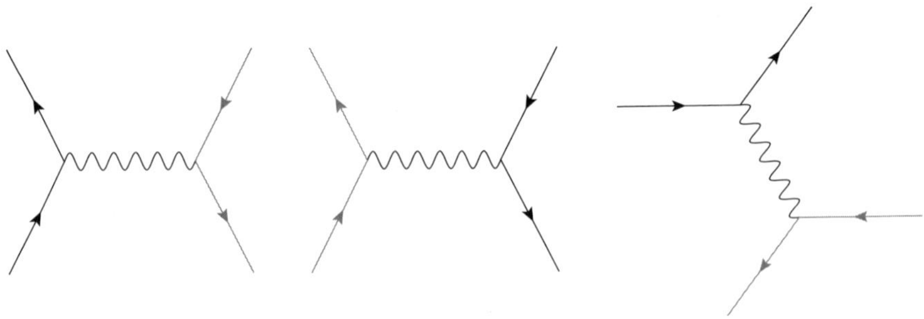

Hadronic Z decay

$e^-e^+ \to \gamma^*/Z^0 \to q\bar{q}$

$\propto N_C$

Drell-Yan

$q\bar{q} \to \gamma^*/Z^0 \to \ell^+\ell^-$

$\propto 1/N_C$

DIS

$\ell\bar{q} \overset{\gamma^*/Z^*}{\to} \ell\bar{q}$

$\propto 1$

Fig. 2. Illustration of the three crossings of the interaction of a lepton current (black) with a quark current (red) via an intermediate photon or Z boson, with corresponding color factors.

previous one. By crossing symmetry, the squared matrix element, including the color factor, is exactly the same as before, but since the quarks are here incoming, we must *average* rather than sum over their colors, leading to

$$q\bar{q} \to Z \to e^+e^- \quad : \quad \frac{1}{9}\sum_{\text{colors}} |M|^2 \propto \frac{1}{9}\delta_{ij}\delta^*_{ji} = \frac{1}{9}\text{Tr}\{\delta\} = \frac{1}{3} \,, \tag{6}$$

where the color factor now expresses a *suppression* which can be interpreted as due to the fact that only quarks of matching colors are able to collide and produce a Z boson. The chance that a quark and an antiquark picked at random from the colliding hadrons have matching colors is $1/N_C$.

Similarly, $\ell q \to \ell q$ via t-channel photon exchange (usually called Deep Inelastic Scattering — DIS — with "deep" referring to a large virtuality of the exchanged photon), constitutes yet another crossing of the same basic process, see Fig. 2. The color factor in this case comes out as unity.

To illustrate what happens when we insert (and sum over) quark-gluon vertices, such as the one depicted in Fig. 1, we take the process $Z \to 3\,\text{jets}$. The color factor for this process can be computed as follows, with the accompanying illustration showing a corresponding diagram (squared) with explicit color-space indices on each vertex:

Table 1. Trace relations for t matrices (convention-independent). More relations can be found in [2, Section 1.2] and in [3, Appendix A.3].

Trace Relation	Indices	Occurs in Diagram Squared
$\mathrm{Tr}\{t^a t^b\} = T_R\, \delta^{ab}$	$a, b \in [1, \ldots, 8]$	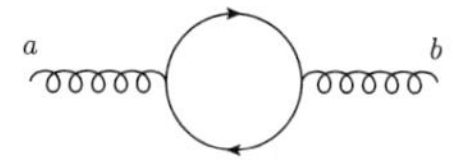
$\sum_a t^a_{ij} t^a_{jk} = C_F\, \delta_{ik}$	$a \in [1, \ldots, 8]$ $i, j, k \in [1, \ldots, 3]$	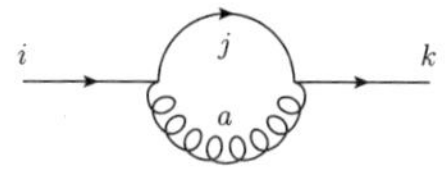
$\sum_{c,d} f^{acd} f^{bcd} = C_A\, \delta^{ab}$	$a, b, c, d \in [1, \ldots, 8]$	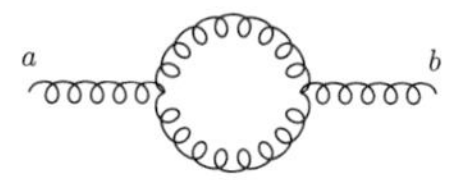
$t^a_{ij} t^a_{k\ell} = T_R \left(\delta_{jk}\delta_{i\ell} - \frac{1}{N_C}\delta_{ij}\delta_{k\ell} \right)$	$i, j, k, \ell \in [1, \ldots, 3]$	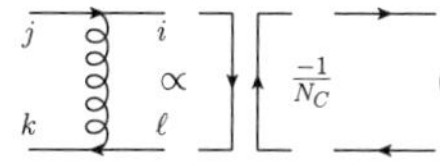 (Fierz)

$Z \to qg\bar{q}$:

$$\sum_{\text{colors}} |M|^2 \propto \delta_{ij} t^a_{jk} \left(t^a_{\ell k} \delta^*_{i\ell} \right)^* \\ = \mathrm{Tr}\{t^a t^a\} \\ = \frac{1}{2}\mathrm{Tr}\{\delta\} = 4 \ , \tag{7}$$

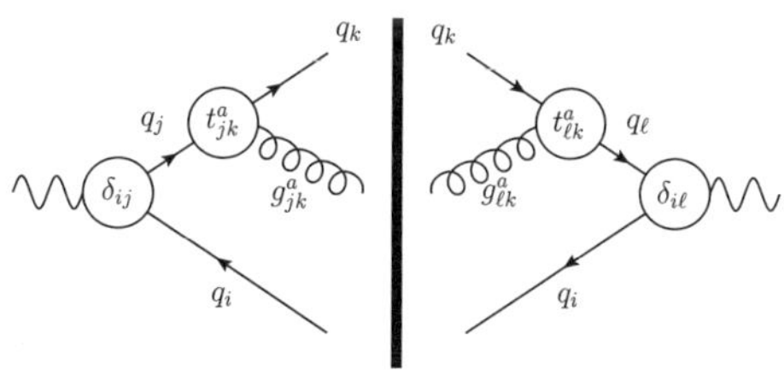

where the last $\mathrm{Tr}\{\delta\} = 8$, since the trace runs over indices in the 8-dimensional adjoint representation.

The tedious task of taking traces over SU(3) matrices can be greatly alleviated by use of the relations given in Table 1. In the standard normalization convention for the SU(3) generators, Eq. (4), the Casimirs of SU(3) appearing in Table 1 are[b]

$$T_R = \frac{1}{2} \qquad C_F = \frac{4}{3} \qquad C_A = N_C = 3 \ . \tag{8}$$

In addition, the gluon self-coupling on the third line in Table 1 involves factors of f^{abc}. These are called the *structure constants* of QCD and they enter due to the non-Abelian term in the gluon field strength tensor appearing

[b]See, e.g., [3, Appendix A.3] for how to obtain the Casimirs in other normalization conventions.

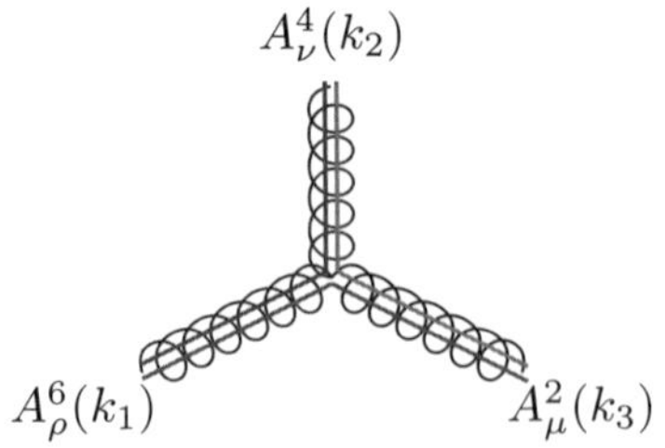

$$\propto -g_s\ f^{246}\,[(k_3-k_2)^\rho g^{\mu\nu} + (k_2-k_1)^\mu g^{\nu\rho} + (k_1-k_3)^\nu g^{\rho\mu}]$$

Fig. 3. Illustration of a ggg vertex in QCD, before summing/averaging over colors: interaction between gluons in the states λ^2, λ^4, and λ^6 is represented by the structure constant f^{246}.

in Eq. (1),

$$F^a_{\mu\nu} = \partial_\mu A^a_\nu - \partial_\nu A^a_\mu + g_s f^{abc} A^b_\mu A^c_\nu \,. \tag{9}$$

The structure constants of SU(3) are listed in the table to the right. Expanding the $F_{\mu\nu}F^{\mu\nu}$ term of the Lagrangian using Eq. (9), we see that there is a 3-gluon and a 4-gluon vertex that involve f^{abc}, the latter of which has two powers of f and two powers of the coupling. Finally, the last line of Table 1 is not really a trace relation but instead a useful so-called Fierz transformation. It is often used, for instance, in shower Monte Carlo applications, to assist in mapping between color flows in $N_C = 3$, in which cross sections and splitting probabilities are calculated, and those in $N_C \to \infty$, used to represent color flow in the MC "event record".

Structure Constants of SU(3)

$$f_{123} = 1$$

$$f_{147} = f_{246} = f_{257} = f_{345} = \frac{1}{2}$$

$$f_{156} = f_{367} = -\frac{1}{2}$$

$$f_{458} = f_{678} = \frac{\sqrt{3}}{2}$$

Antisymmetric in all indices

All other $f_{ijk} = 0$

A gluon self-interaction vertex is illustrated in Fig. 3, to be compared with the quark-gluon one in Fig. 1. We remind the reader that gauge boson self-interactions are a hallmark of non-Abelian theories and that their presence leads to some of the main differences between QED and

QCD. One should also keep in mind that the color factor for the vertex in Fig. 3, C_A, is roughly twice as large as that for a quark, C_F.

1.3. *The strong coupling*

To first approximation, QCD is *scale invariant.* That is, if one "zooms in" on a QCD jet, one will find a repeated self-similar pattern of jets within jets within jets, reminiscent of fractals such as the famous Mandelbrot set in mathematics, or the formation of frost crystals in physics. In the context of QCD, this property was originally called light-cone scaling, or Bjorken scaling after the physicist James D. Bjorken. This type of scaling is closely related to the class of angle-preserving symmetries, called *conformal* symmetries. In physics today, the terms "conformal" and "scale invariant" are used interchangeably.[c] Conformal invariance is a mathematical property of several QCD-"like" theories which are now being studied (such as $N = 4$ supersymmetric relatives of QCD). It is also closely related to the physics of so-called "unparticles", though that is a relation that goes beyond the scope of these lectures.

Regardless of the labeling, if the strong coupling did not run (we shall return to the running of the coupling below), Bjorken scaling would be absolutely true. QCD would be a theory with a fixed coupling, the same at all scales. This simplified picture already captures some of the most important properties of QCD, as we shall discuss presently.

In the limit of exact Bjorken scaling — QCD at fixed coupling — properties of high-energy interactions are determined only by *dimensionless* kinematic quantities, such as scattering angles (pseudorapidities) and ratios of energy scales.[d] For applications of QCD to high-energy collider physics, an important consequence of Bjorken scaling is thus that the rate of bremsstrahlung jets, with a given transverse momentum, scales in direct proportion to the hardness of the fundamental partonic scattering process they are produced in association with. For instance, in the limit of exact scaling, a measurement of the rate of 10-GeV jets produced in association with an ordinary Z boson could be used as a direct prediction of the rate

[c]Strictly speaking, conformal symmetry is more restrictive than just scale invariance, but examples of scale-invariant field theories that are not conformal are rare.

[d]Originally, the observed approximate agreement with this was used as a powerful argument for pointlike substructure in hadrons; since measurements at different energies are sensitive to different resolution scales, independence of the absolute energy scale is indicative of the absence of other fundamental scales in the problem and hence of pointlike constituents.

of 100-GeV jets that would be produced in association with a 900-GeV Z' boson, and so forth. Our intuition about how many bremsstrahlung jets a given type of process is likely to have should therefore be governed first and foremost by the *ratios* of scales that appear in that particular process, as has been highlighted in a number of studies focusing on the mass and $p_\perp$ scales appearing, e.g., in Beyond-the-Standard-Model (BSM) physics processes [4–7]. Bjorken scaling is also fundamental to the understanding of jet substructure in QCD, see, e.g., [8, 9].

In real QCD, the coupling runs logarithmically with the energy,

$$Q^2 \frac{\partial \alpha_s}{\partial Q^2} = \frac{\partial \alpha_s}{\partial \ln Q^2} = \beta(\alpha_s) \ , \tag{10}$$

where the function driving the energy dependence, the *beta function*, is defined as

$$\beta(\alpha_s) = -\alpha_s^2 (b_0 + b_1 \alpha_s + b_2 \alpha_s^2 + \ldots) \ , \tag{11}$$

with LO (1-loop) and NLO (2-loop) coefficients

$$b_0 = \frac{11 C_A - 4 T_R n_f}{12\pi} \ , \tag{12}$$

$$b_1 = \frac{17 C_A^2 - 10 T_R C_A n_f - 6 T_R C_F n_f}{24\pi^2} = \frac{153 - 19 n_f}{24\pi^2} \ . \tag{13}$$

In the b_0 coefficient, the first term is due to gluon loops while the second is due to quark ones. Similarly, the first term of the b_1 coefficient arises from double gluon loops, while the second and third represent mixed quark-gluon ones. At higher loop orders, the b_i coefficients depend explicitly on the renormalization scheme that is used. A brief discussion can be found in the PDG review on QCD [10], with more elaborate ones contained in [2, 11]. Note that, if there are additional colored particles beyond the Standard-Model ones, loops involving those particles enter at energy scales above the masses of the new particles, thus modifying the running of the coupling at high scales. This is discussed, e.g., for supersymmetric models in [12].

Numerically, the value of the strong coupling is usually specified by giving its value at the specific reference scale $Q^2 = M_Z^2$, from which we can obtain its value at any other scale by solving Eq. (10),

$$\alpha_s(Q^2) = \alpha_s(M_Z^2) \frac{1}{1 + b_0 \alpha_s(M_Z^2) \ln \dfrac{Q^2}{M_Z^2} + \mathcal{O}(\alpha_s^2)} \ , \tag{14}$$

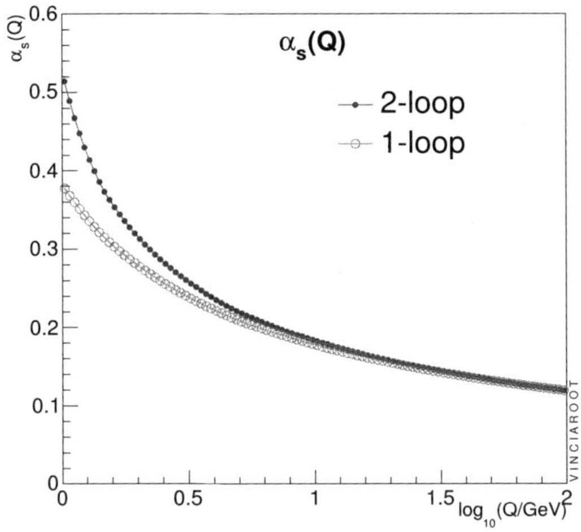

Fig. 4. Illustration of the running of α_s at 1- (open circles) and 2-loop order (filled circles), starting from the same value of $\alpha_s(M_Z) = 0.12$.

with relations including the $\mathcal{O}(\alpha_s^2)$ terms available, e.g., in [2]. Relations between scales not involving M_Z^2 can obviously be obtained by just replacing M_Z^2 by some other scale Q'^2 everywhere in Eq. (14). A comparison of running at one- and two-loop order, in both cases starting from $\alpha_s(M_Z) = 0.12$, is given in Fig. 4. As is evident from the figure, the 2-loop running is somewhat faster than the 1-loop one.

As an application, let us prove that the logarithmic running of the coupling implies that an intrinsically multi-scale problem can be converted to a single-scale one, up to corrections suppressed by two powers of α_s, by taking the geometric mean of the scales involved. This follows from expanding an arbitrary product of individualα_s factors around an arbitrary scale μ, using Eq. (14),

$$\alpha_s(\mu_1)\alpha_s(\mu_2)\cdots\alpha_s(\mu_n) = \prod_{i=1}^{n}\alpha_s(\mu)\left(1 + b_0\,\alpha_s \ln\left(\frac{\mu^2}{\mu_i^2}\right) + \mathcal{O}(\alpha_s^2)\right)$$

$$= \alpha_s^n(\mu)\left(1 + b_0\,\alpha_s \ln\left(\frac{\mu^{2n}}{\mu_1^2\mu_2^2\cdots\mu_n^2}\right) + \mathcal{O}(\alpha_s^2)\right) , \tag{15}$$

whereby the specific single-scale choice $\mu^n = \mu_1\mu_2\cdots\mu_n$ (the geometric mean) can be seen to push the difference between the two sides of the equation one order higher than would be the case for any other combination of scales.[e]

The appearance of the number of flavors, n_f, in b_0 implies that the slope of the running depends on the number of contributing flavors. Since full QCD is best approximated by $n_f = 3$ below the charm threshold, by $n_f = 4$ from there to the b threshold, and by $n_f = 5$ above that, it is therefore important to be aware that the running changes slope across quark flavor thresholds. Likewise, it would change across the threshold for top or for any colored new-physics particles that might exist, with a magnitude depending on the particles' color and spin quantum numbers.

The negative overall sign of Eq. (11), combined with the fact that $b_0 > 0$ (for $n_f \leq 16$), leads to the famous result[f] that the QCD coupling effectively *decreases* with energy, called asymptotic freedom, for the discovery of which the Nobel prize in physics was awarded to D. Gross, H. Politzer, and F. Wilczek in 2004. An extract of the prize announcement runs as follows:

> *What this year's Laureates discovered was something that, at first sight, seemed completely contradictory. The interpretation of their mathematical result was that the closer the quarks are to each other, the* weaker *is the "color charge". When the quarks are really close to each other, the force is so weak that they behave almost as free particles.* [More correctly, it is the coupling rather than the force which becomes weak as the distance decreases. The $1/r^2$ Coulomb singularity of the force is only dampened, not removed, by the diminishing coupling.] *This phenomenon is called "asymptotic freedom". The converse is true when the quarks move apart: the force becomes stronger when the distance increases.* [More correctly, it is the potential which grows, linearly, while the force becomes constant.]

Among the consequences of asymptotic freedom is that perturbation theory becomes better behaved at higher absolute energies, due to the effectively decreasing coupling. Perturbative calculations for our 900-GeV Z' boson from before should therefore be slightly faster converging than equiv-

---

[e]In a fixed-order calculation, the individual scales μ_i, would correspond, e.g., to the n hardest scales appearing in an infrared safe sequential clustering algorithm applied to the given momentum configuration.

[f]Perhaps the highest pinnacle of fame for Eq. (11) was reached when the sign of it featured in an episode of the TV series "Big Bang Theory".

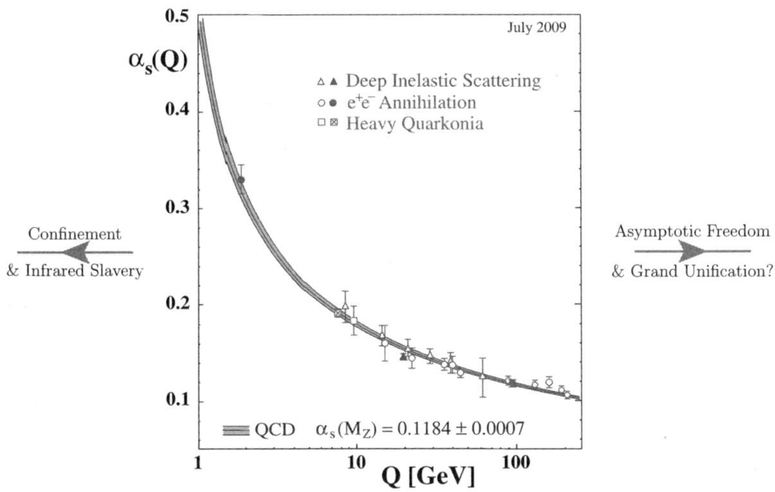

Fig. 5. Illustration of the running of α_s in a theoretical calculation (yellow shaded band) and in physical processes at different characteristic scales, from [13].

alent calculations for the 90-GeV one. Furthermore, since the running of α_s explicitly breaks Bjorken scaling, we also expect to see small changes in jet shapes and in jet production ratios as we vary the energy. For instance, since high-$p_\perp$ jets start out with a smaller effective coupling, their intrinsic shape (irrespective of boost effects) is somewhat narrower than for low-$p_\perp$ jets, an issue which can be important for jet calibration. Our current understanding of the running of the QCD coupling is summarized by the plot in Fig. 5, taken from a recent comprehensive review by S. Bethke [13].

As a final remark on asymptotic freedom, note that the decreasing value of the strong coupling with energy must eventually cause it to become comparable to the electromagnetic and weak ones, at some energy scale. Beyond that point, which may lie at energies of order $10^{15} - 10^{17}$ GeV (though it may be lower if as yet undiscovered particles generate large corrections to the running), we do not know what the further evolution of the combined theory will actually look like, or whether it will continue to exhibit asymptotic freedom.

Now consider what happens when we run the coupling in the other direction, towards smaller energies. Taken at face value, the numerical value of the coupling diverges rapidly at scales below 1 GeV, as illustrated by the curves disappearing off the left-hand edge of the plot in Fig. 5. To make this divergence explicit, one can rewrite Eq. (14) in the following form,

$$\alpha_s(Q^2) = \frac{1}{b_0 \ln \frac{Q^2}{\Lambda^2}} , \tag{16}$$

where

$$\Lambda \sim 200\,\text{MeV} \tag{17}$$

specifies the energy scale at which the perturbative coupling would nominally become infinite, called the Landau pole. (Note, however, that this only parametrizes the purely *perturbative* result, which is not reliable at strong coupling, so Eq. (16) should not be taken to imply that the physical behavior of full QCD should exhibit a divergence for $Q \to \Lambda$.)

Finally, one should be aware that there is a multitude of different ways of defining both Λ and $\alpha_s(M_Z)$. At the very least, the numerical value one obtains depends both on the renormalization scheme used (with the dimensional-regularization-based "modified minimal subtraction" scheme, $\overline{\text{MS}}$, being the most common one) and on the perturbative order of the calculations used to extract them. As a rule of thumb, fits to experimental data typically yield smaller values for $\alpha_s(M_Z)$ the higher the order of the calculation used to extract it (see, e.g., [13, 14]), with $\alpha_s(M_Z)|_{\text{LO}} \gtrsim \alpha_s(M_Z)|_{\text{NLO}} \gtrsim \alpha_s(M_Z)|_{\text{NNLO}}$. Further, since the number of flavors changes the slope of the running, the location of the Landau pole for fixed $\alpha_s(M_Z)$ depends explicitly on the number of flavors used in the running. Thus each value of n_f is associated with its own value of Λ, with the following matching relations across thresholds guaranteeing continuity of the coupling at one loop,

$$n_f = 4 \leftrightarrow 5 \quad : \quad \Lambda_5 = \Lambda_4 \left(\frac{\Lambda_4}{m_b}\right)^{\frac{2}{23}} \quad \Lambda_4 = \Lambda_5 \left(\frac{m_b}{\Lambda_5}\right)^{\frac{2}{25}} , \tag{18}$$

$$n_f = 3 \leftrightarrow 4 \quad : \quad \Lambda_4 = \Lambda_3 \left(\frac{\Lambda_3}{m_c}\right)^{\frac{2}{25}} \quad \Lambda_3 = \Lambda_4 \left(\frac{m_c}{\Lambda_4}\right)^{\frac{2}{27}} . \tag{19}$$

It is sometimes stated that QCD only has a single free parameter, the strong coupling. However, even in the perturbative region, the beta function depends explicitly on the number of quark flavors, as we have seen, and thereby also on the quark masses. Furthermore, in the non-perturbative region around or below Λ_{QCD}, the value of the perturbative coupling, as obtained, e.g., from Eq. (16), gives little or no insight into the behavior of the full theory. Instead, universal functions (such as parton densities, form factors, fragmentation functions, etc), effective theories (such as the Operator Product Expansion, Chiral Perturbation Theory, or Heavy Quark Effective Theory), or phenomenological models (such as Regge Theory or the String and Cluster Hadronization Models) must be used, which in turn depend on additional non-perturbative parameters whose relation to, e.g., $\alpha_s(M_Z)$, is not a priori known.

For some of these questions, such as hadron masses, lattice QCD can furnish important additional insight, but for multi-scale and/or time-evolution problems, the applicability of lattice methods is still severely restricted; the lattice formulation of QCD requires a Wick rotation to Euclidean space. The time-coordinate can then be treated on an equal footing with the other dimensions, but intrinsically Minkowskian problems, such as the time evolution of a system, are inaccessible. The limited size of current lattices also severely constrain the scale hierarchies that it is possible to "fit" between the lattice spacing and the lattice size.

2. Hard Processes

Our main tool for solving QCD at high energy scales, $Q \gg \Lambda_{\text{QCD}}$, is perturbative quantum field theory, the starting point for which is Matrix Elements (MEs) which can be calculated systematically at fixed orders (FO) in the strong coupling α_s. At least at lowest order (LO), the procedure is standard textbook material [3] and it has also by now been highly automated [15–24]. Here, we require only that the reader has a basic familiarity with the methods involved from graduate-level particle physics courses

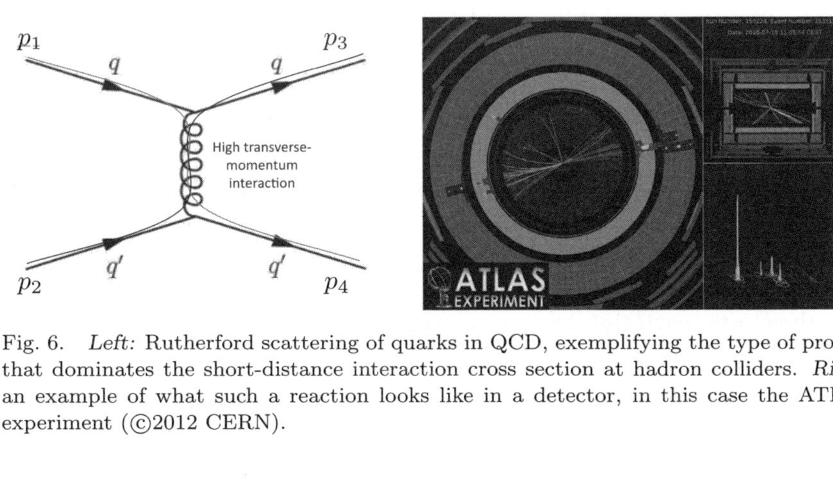

Fig. 6. *Left:* Rutherford scattering of quarks in QCD, exemplifying the type of process that dominates the short-distance interaction cross section at hadron colliders. *Right:* an example of what such a reaction looks like in a detector, in this case the ATLAS experiment (©2012 CERN).

based, e.g., on [3, 11]. Our main concern are the uses to which these calculations are put, their limitations, and ways to improve on the results obtained with them.

For illustration, take one of the most commonly occurring processes in hadron collisions: Rutherford scattering of two quarks via a t-channel gluon exchange — Fig. 6 — which at leading order has the differential cross section

$$qq' \to qq' \quad : \quad \frac{\mathrm{d}\sigma}{\mathrm{d}\hat{t}} = \frac{\pi}{\hat{s}^2}\,\frac{4}{9}\,\alpha_s^2\,\frac{\hat{s}^2+\hat{u}^2}{\hat{t}^2}\ , \tag{20}$$

with the $2 \to 2$ Mandelstam variables ("hatted" to emphasize that they refer to a partonic $2 \to 2$ scattering rather than the full $pp \to$ jets process)

$$\hat{s} = (p_1 + p_2)^2\ , \tag{21}$$

$$\hat{t} = (p_3 - p_1)^2 = -\hat{s}\frac{(1-\cos\hat{\theta})}{2}\ , \tag{22}$$

$$\hat{u} = (p_4 - p_1)^2 = -\hat{s}\frac{(1+\cos\hat{\theta})}{2}\ . \tag{23}$$

Reality, however, is more complicated; the picture on the right-hand pane of Fig. 6 shows a real dijet event, as recorded by the ATLAS experiment. The complications to be addressed when going from left to right in Fig. 6 are: firstly, additional jets, a.k.a. real-emission corrections, which can significantly change the topology of the final state, potentially shifting jets in or out of an experimentally defined acceptance region. Secondly, loop factors, a.k.a. virtual corrections, change the number of available quantum paths through phase space, and hence modify the normalization of the cross section (total *and* differential). And finally, additional corrections are generated by confinement and by the so-called underlying event. These corrections must be taken into account to complete our understanding of QCD and connect the short-distance physics with macroscopic experiments. Apart from the perturbative expansion itself, the most powerful tool we have to organize this vast calculation, is factorization.

2.1. *Factorization*

In high-energy scattering problems involving hadrons in the initial state, we immediately face the complication that hadrons are composite, with a time-dependent structure illustrated in Fig. 7; there are partons within clouds of further partons, constantly being emitted and absorbed. Thus, before we

Fig. 7. Illustration of partonic fluctuations inside a proton beam (from [25]).

can use perturbatively calculated partonic scattering matrix elements, we must first address the partonic structure of the colliding hadron(s).

For the hadron to remain intact, the fluctuations inside it must involve momentum transfers smaller than the confinement scale. Indeed, high-virtuality fluctuations are suppressed by powers of

$$\frac{\alpha_s \Lambda^2}{|k|^2} , \tag{24}$$

with Λ the confinement scale ($\sim$ 200 MeV, see section 1.3) and $|k|$ the virtuality of the fluctuation. Thus, most fluctuations occur over timescales $\sim 1/\Lambda$.

A hard perturbative probe, on the other hand, such as the exchanged photon in DIS (Fig. 2), interacts over a much shorter timescale $1/Q \ll 1/\Lambda$, during which the partonic fluctuations in the struck hadron appear almost frozen. The hard probe effectively takes an instantaneous snapshot of the hadron structure, at a characteristic resolution given by $\sim 1/Q$.

This is formalized by the *factorization theorem* [26] (see also the TASI lectures by George Sterman [27]), which expresses the independence of long-wavelength (soft) structure on the nature of the hard (short-distance) process. Originally formulated for DIS, factorization allows us to write the cross section for lepton-hadron scattering as a convolution of a non-perturbative but universal (i.e., process-independent) parton density function (PDF) and a perturbatively calculable partonic scattering cross section. Denoting the fraction of the hadron momentum carried by parton i by x_i,

$$\vec{p}_i \;=\; x_i \, \vec{p}_h \;, \tag{25}$$

we may write the lepton-hadron cross section on factorized form (see, e.g., [11, 28]),

$$\sigma_{\ell h} = \sum_i \int_0^1 \mathrm{d}x_i \int \mathrm{d}\Phi_f \, f_{i/h}(x_i, \mu_F^2) \, \frac{\mathrm{d}\hat{\sigma}_{\ell i \to f}(x_i, \Phi_f, \mu_F^2)}{\mathrm{d}x_i \, \mathrm{d}\Phi_f} \,, \tag{26}$$

with i an index running over all possible parton types[g] in the incoming hadron and f enumerating all possible (partonic) final states, with Lorentz-invariant phase space, Φ_f.

The *parton density functions* (PDFs), $f_{i/h}$, parametrize the distribution of partons inside the target hadron. They are not a priori calculable and must be constrained by fits to data. This is discussed in section 2.2.

The *partonic cross section*, $\mathrm{d}\hat{\sigma}$, knows nothing of the target hadron apart from the fact that it contained the struck parton. It is calculable within perturbation theory, as will be discussed in section 2.3.

The dividing line between the two is drawn at an arbitrary ("user-defined") scale μ_F, called the *factorization scale.* There is some arbitrariness involved in choosing a value for μ_F. Some heuristic arguments to guide in the choice of factorization scale are the following. On the long-distance side, the PDFs include a (re)summation of fluctuations inside fluctuations up to virtualities of order μ_F. It would therefore not make much sense to take μ_F significantly larger than the scales characterizing resolved particles on the short-distance side of the calculation (i.e., the particles appearing explicitly in Φ_f); otherwise the PDFs would be including sums over fluctuations that happen on timescales shorter than those probed by the physical process. Similarly, μ_F should also not be taken much lower than the scale(s) appearing in the hard process. For matrix elements characterized by a single well-defined scale, such as the Q^2 scale in DIS or the invariant-mass scale $\hat{s}$ in Drell-Yan production ($q\bar{q} \to Z/\gamma^* \to \ell^+\ell^-$), such arguments essentially fix the preferred scale choice to $\mu_F = Q$ or $\mu_F = \sqrt{\hat{s}}$, respectively, which may then be varied by a factor of 2 (or larger) around the nominal value in order to estimate uncertainties. For multi-scale problems, however, such as $pp \to Z/W + n\,$jets, there are several a priori equally good choices available, from the lowest to the highest QCD scales that can be constructed from the final-state momenta, usually with several dissenting groups of theorists arguing over which particular choice is best. Suggesting that one

[g]Typically, only quarks and gluons are included in this sum, but also photons and even leptons can in principle be included. Similarly, parton density functions are normally used to describe hadrons, but can also be defined, e.g., to describe the cloud of virtual photons (and fermion pairs) surrounding an electron.

might simply *measure* the scale would not really be an improvement, as the factorization scale is fundamentally unphysical and therefore unobservable (similarly to gauge or convention choices). One plausible strategy is to look at higher-order (NLO or NNLO) calculations, in which correction terms appear that cancel the dependence on the scale choice, stabilizing the final result. From such comparisons, a "most stable" initial scale choice can in principle be determined, which then furnishes a reasonable starting point, but we emphasize that the question *is* intrinsically ambiguous, and no golden recipe is likely to magically give all the right answers. The best we can do is to vary the value of μ_F not only by an overall factor, but also by exploring different possible forms for its functional dependence on the momenta appearing in Φ_f. A complementary useful discussion of the pros and cons of different factorization scale choices can be found in the TASI lectures by Tilman Plehn [29].

Secondly, and more technically, at NLO and beyond one also has to settle on a *factorization scheme* in which to do the calculations. For all practical purposes, students focusing on LHC physics are only likely to encounter one such scheme, the modified minimal subtraction ($\overline{\text{MS}}$) one already mentioned in the discussion of the definition of the strong coupling in section 1.3. At the level of these lectures, we shall therefore not elaborate further on this choice here.

We note that factorization can also be applied multiple times, to break up a complicated calculation into simpler pieces that can be treated as approximately independent. This will be very useful when dealing with successive emissions in a parton shower, section 3.2, or when factoring off decays of long-lived particles from a hard production process, section 3.4.

We round off the discussion of factorization by mentioning a few caveats the reader should be aware of. (See [27] for a more technical treatment.)

Firstly, the proof only applies to the first term in an operator product expansion in "twist" = mass dimension − spin. Since operators with higher mass dimensions are suppressed by the hard scale to some power, this leading twist approximation becomes exact in the limit $Q \to \infty$, while at finite Q it neglects corrections of order

$$\text{Higher Twist}: \frac{[\ln(Q^2/\Lambda^2)]^{m<2n}}{Q^{2n}} \quad (n = 2 \text{ for DIS}) \,. \tag{27}$$

In section 5, we shall discuss some corrections that go beyond this approximation, in the context of multiple parton-parton interactions.

Secondly, the proof only really applies to inclusive cross sections in DIS [26] and in Drell-Yan [30]. For all other hadron-initiated processes, factorization is an ansatz. For a general hadron-hadron process, we write the assumed factorizable cross section as:

$$\mathrm{d}\sigma_{h_1h_2} = \sum_{i,j}\int_0^1 \mathrm{d}x_i\,\mathrm{d}x_j \sum_f \int \mathrm{d}\Phi_f\, f_{i/h_1}(x_i,\mu_F^2)\, f_{j/h_2}(x_j,\mu_F^2)\, \frac{\mathrm{d}\hat{\sigma}_{ij\to f}}{\mathrm{d}x_i\,\mathrm{d}x_j\,\mathrm{d}\Phi_f}\,. \tag{28}$$

Note that, if $\mathrm{d}\hat{\sigma}$ is divergent (as, e.g., Rutherford scattering is) then the integral over $\mathrm{d}\Phi_f$ must be regulated, e.g. by imposing some explicit minimal transverse-momentum cut and/or other phase-space restrictions.

2.2. *Parton densities*

The parton density function, $f_{i/h}(x,\mu_F^2)$, represents the effective density of partons of type/flavor i, as a function of the momentum fraction[h] x_i, when a hadron of type h is probed at the factorization scale μ_F. The PDFs are non-perturbative functions which are not a priori calculable, but a perturbative differential equation governing their evolution with μ_F can be obtained by requiring that physical scattering cross sections, such as the one for DIS in Eq. (26), be independent of μ_F to the calculated orders [31]. The resulting *renormalization group equation* (RGE) is called the DGLAP[i] equation and can be used to "run" the PDFs from one perturbative resolution scale to another (its evolution kernels are the same as those used in parton showers, to which we return in section 3.2).

This means that we only need to determine the form of the PDF as a function of x a single (arbitrary) scale, μ_0. We can then get its form at any other scale μ_F by simple RGE evolution. In the context of PDF fits (constraining the form of the PDF functions by fitting cross sections to experimental data, e.g., from DIS [32, 33], Drell-Yan [34, 35], and $pp \to$ jets [36]), the reference scale μ_0 is usually taken to be relatively low, of order one or a few GeV.

The behavior of the PDFs as we evolve μ_F from a low scale, 2 GeV, to a high one, 100 GeV, is illustrated in Fig. 8, for the MSTW[j] 2008 LO[k] PDF set [38]. At low $Q = \mu_F = 2\,\mathrm{GeV}$ (left), the proton structure is

[h] Recall: the x fraction is defined in Eq. (25).

[i] DGLAP: Dokshitzer-Gribov-Lipatov-Altarelli-Parisi.

[j] MSTW: Martin-Stirling-Thorne-Watt.

[k] The "LO" means that the fit was performed using LO matrix elements in the cross section formulae.

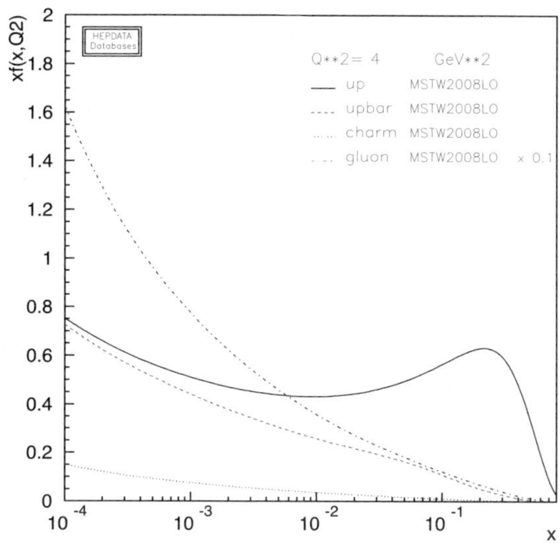

(a)

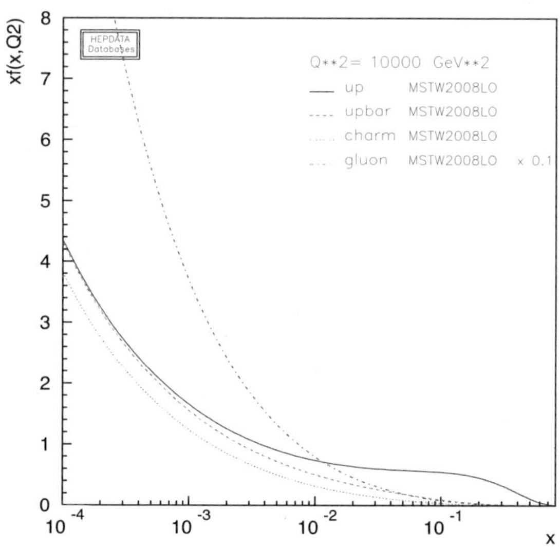

(b)

Fig. 8. Illustration of the change of the u (black), $\bar{u}$ (red, dashed), c (green, dotted), and g (blue, dot-dashed) distributions, from $Q = \mu_F = 2\,\text{GeV}$ (a) to $Q = \mu_F = 100\,\text{GeV}$ (b). Note that a factor 0.1 has been applied to the gluon distribution. Plots made using the HEPDATA online tool [37].

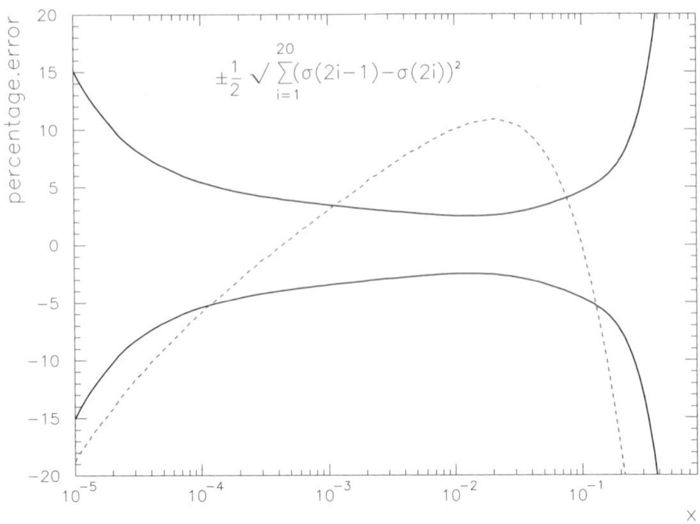

Fig. 9. Illustration of the difference between the MSTW 2008 and CTEQ6 LO gluon PDFs at $\mu_F = 10\,\text{GeV}$. All curves are normalized to the central MSTW 2008 prediction. The black solid lines show the 90% CL MSTW variations, while the dashed red line shows the CTEQ6L1 distribution.

dominated by a few hard quarks (a "valence bump" is clearly visible around $x \sim 0.2$), while at higher scales $Q = 100\,\text{GeV}$ (right) we predominantly resolve fluctuations within fluctuations, yielding increasingly large gluon- and sea-quark distributions with rather small x values, while the valence quarks play a progressively smaller role.

We note that different collaborations, like CTEQ, MSTW, NNPDF, etc., use different ansätze for the form of $f(x, \mu_0^2)$. They may also include different data in the fits, and/or treat or weight the data differently. Thus, results from different groups may not always be mutually compatible. An example is given in Fig. 9, which shows the difference between the CTEQ6L1 gluon PDF [39] (red dashed) and the MSTW 2008 LO one [38], normalized to MSTW (which would thus be a flat line at zero), at $\mu_F = 10\,\text{GeV}$. The y axis shows the relative difference between the sets, in per cent. Also shown are the 90% CL contours computed from the uncertainty variations included in the MSTW 2008 LO set (black). Using only the MSTW uncertainty band, one would arrive at an estimated $\sim 5\%$ uncertainty over most of the x range, while including the CTEQ6L1 set would increase that to $\sim 10\%$. At NLO, this discrepancy is reduced, but not removed. A significant effort is currently being undertaken within the PDF community to agree on common, and more comprehensive, ways of defining PDF uncertainty bands [36, 40]. This is complicated due to the different ways of defining $f(x, \mu_0^2)$ and due to the experimental data sets not always being fully compatible with one another. For the time being, it is recommended to try at least sets from two different groups, for a comprehensive uncertainty estimate.

Occasionally, the words *structure functions* and *parton densities* are used interchangeably. However, there is an important distinction between the two, which we find often in (quantum) physics: the former is a physical observable used to parametrize the DIS cross sections (see e.g. [11]), while the latter is a "fundamental" quantity extracted from it. In particular, since the parton densities are not, themselves, physically observable, they can only be defined within a specific factorization scheme, order by order in perturbation theory. The only exception is at leading order, at which they have the simple physical interpretation of parton number densities. When going to higher orders, we tend to keep the simple intuitive picture from LO in mind, but one should be aware that the fundamental relationship between PDFs and measured quantities is now more complicated (due to the interplay between the PDFs and the real and virtual corrections to the LO cross section), and that the parton densities no longer have a clear probabilistic interpretation starting from NLO.

The reader should also be aware that there is some ambiguity whether NLO PDFs should be used for LO calculations. In principle, the higher-order PDFs are better constrained and the difference between, e.g., an NLO and an LO set should formally be beyond LO precision, so that one might be tempted to simply use the highest-order available PDFs for any calculation. However, higher-order terms can sometimes be absorbed, at least partially, into effective lower-order coefficients. In the context of PDFs, the fit parameters of lower-order PDFs will attempt to compensate for missing higher-order contributions in the matrix elements. To the extent those higher-order contributions are *universal*, this is both desirable and self-consistent. This leads to some typical qualitative differences between LO and NLO PDFs, illustrated in Fig. 10: NLO PDFs tend to be smaller at low x and slightly larger at high x, than LO ones. Thus, it is quite possible that using an NLO PDF in conjunction with LO matrix elements can give a worse agreement with data than LO PDFs do.

Finally, another oft-raised question concerns which PDF sets to use for the parton-shower evolution in Monte Carlo generators. Importantly, the

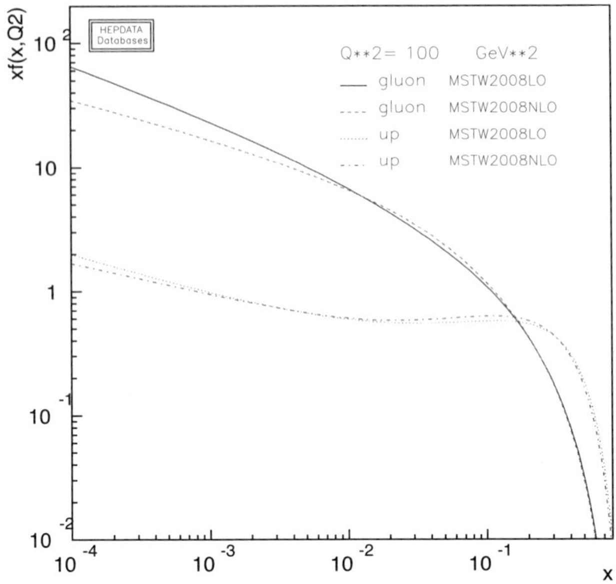

Fig. 10. Illustration of the change between PDF fits using LO and NLO matrix elements: the g distribution at LO (black) and NLO (red, dashed), and the u distribution at LO (green, dotted) and NLO (blue, dot-dashed), for the MSTW 2008 PDF sets [38], at $Q = \mu_F = 10\,\mathrm{GeV}$. Plots made using the HEPDATA online tool [37].

equations driving the initial-state showers in Monte Carlo models are only sensitive to *ratios* of PDFs [41]. Since the shower evolution typically only has leading-logarithmic (LL) precision, it should be theoretically consistent to use any (LO or better) PDF set to drive the evolution. However, similarly to above, there will be subleading differences between different choices, and one is justified in worrying about the level of physical effects that could be generated. Unfortunately, there is currently no way to ensure 100% self-consistency. Since PDF fits are not done with MC codes, but instead use analytical resummation models (see, e.g., the TASI lectures by Sterman [27]), which are not identical to their MC counterparts, the PDF fits are essentially "tuned" to a slightly different resummation than that incorporated in a given MC model. In practice, not much is known about the size and impact of this ambiguity [42]. Known differences include: the size of phase space (purely collinear massless PDF evolution vs. the finite-transverse-momentum massive MC phase space), the treatment of momentum conservation and recoil effects, additional higher-order effects explicitly or implicitly included in the MC evolution, choice of renormalization scheme and scale, and, for those MC algorithms that do not rely on collinear (DGLAP, see [11]) splitting kernels (e.g., the various kinds of dipole evolution algorithms, see [43]), differences in the effective factorization scheme.

As a baseline, we recommend simply using whatever PDF set the given MC model was originally tuned with, since this should de facto (by fitting the available data) reabsorb as much of the inconsistency as possible. Furthermore, it should be emphasized that underlying-event and minimum-bias models based on multi-parton interactions (see section 5.2) usually make the explicit assumption that the PDFs can be interpreted as physical number densities even down to very low Q and x, a property which is generally only true for LO PDFs. It must therefore be strongly discouraged to use (N)NLO PDF sets in this context.

2.3. *Fixed-order QCD*

Consider the production of an arbitrary final state, F (e.g., a Higgs boson, a $t\bar{t}$ pair, etc). Schematically, we may express the (perturbative) all-orders differential cross section for an observable $\mathcal{O}$, in the following way:

$$\left.\frac{\mathrm{d}\sigma_F}{\mathrm{d}\mathcal{O}}\right|_{\mathrm{ME}} = \underbrace{\sum_{k=0}^{\infty}\int \mathrm{d}\Phi_{F+k}}_{\Sigma\ \mathrm{legs}} \underbrace{\left|\sum_{\ell=0}^{\infty}\mathcal{M}_{F+k}^{(\ell)}\right|^2}_{\Sigma\ \mathrm{loops}} \delta\left(\mathcal{O}-\mathcal{O}(\Phi_{F+k})\right) , \qquad (29)$$

where, for compactness, we have suppressed all PDF and luminosity normalization factors. $\mathcal{M}_{F+k}^{(\ell)}$ is the amplitude for producing F in association with k additional final-state partons, "legs", and with ℓ additional loops. The sums start at $k = 0$ and $\ell = 0$, corresponding to the Leading Order for producing F, while higher terms represent real and virtual corrections, respectively.

The purpose of the δ function is to project out hypersurfaces of constant value of $\mathcal{O}$ in the full $\mathrm{d}\Phi_{F+k}$ phase space, with $\mathcal{O}(\Phi_{F+k})$ a function that defines $\mathcal{O}$ evaluated on each specific momentum configuration, Φ_{F+k}. (Without the δ function, the formula would give the total integrated cross section, instead of the cross section differentially in $\mathcal{O}$.)

We recover the various fixed-order truncations of perturbative QCD (pQCD) by limiting the nested sums in Eq. (29) to include only specific values of $k + \ell$. Thus,

$$
\begin{aligned}
k = 0,\, \ell = 0 \quad &: \quad \text{Leading Order (usually tree-level) for } F \text{ production} \\
k = n,\, \ell = 0 \quad &: \quad \text{Leading Order for } F + n\,\text{jets} \\
k + \ell \leq n, \quad &: \quad \text{N}^n\text{LO for } F \text{ (includes N}^{n-1}\text{LO for } F + 1\,\text{jet, N}^{n-2}\text{LO} \\
&\quad\;\; \text{for } F + 2\,\text{jets, and so on up to LO for } F + n\,\text{jets)}\ .
\end{aligned}
$$

For $k \geq 1$, we are not considering inclusive F production; instead, we are considering the process $F + k$ jets. If we simply integrate over all momenta, as implied by the integration over $\mathrm{d}\Phi_{F+k}$ in Eq. (29), we would be including configurations in which one or more of the k partons are collinear or soft. Such configurations are infrared divergent in QCD and hence must be regulated. Since we talk about *collinear* and *soft* divergences (the origins of which will be discussed in more detail in sections 2.4 and 3.2), cuts on *angles* and *energies* and/or cuts on combinations, like *transverse momenta*, can be used to cut away the problematic regions of phase space.

Recall, however, that pQCD is approximately scale invariant. This implies that a regularization cut on a dimensionful quantity, like energy or transverse momentum, should be formulated as a *ratio* of scales, rather than as an absolute number. For example, a jet with $p_\perp = 50\,\text{GeV}$ would be considered hard and well-separated if produced in association with an ordinary Z boson (with hard scale $M_Z = 91.2\,\text{GeV}$), while the same jet would be considered soft if produced in association with a 900-GeV Z' boson (see [4, 5] for more explicit examples).

The essence of the point is that, if the regularization scale is taken too low, logarithmic enhancements of the type

$$\alpha_s^n \ln^{m\leq 2n}\left(\frac{Q_F^2}{Q_k^2}\right) \tag{30}$$

will generate progressively larger corrections, order by order, which will spoil any fixed-order truncation of the perturbative series. Here, Q_F is the hard scale associated with the process under consideration, while Q_k is the scale associated with an additional parton, k.

A good rule of thumb is that if $\sigma_{k+1} \approx \sigma_k$ (at whatever order you are calculating), then the perturbative series is converging too slowly for a fixed-order truncation of it to be reliable. For fixed-order perturbation theory to be applicable, you must place your cuts on the hard process such that $\sigma_{k+1} \ll \sigma_k$. In the discussion of parton showers in section 3.2, we shall see how the region of applicability of perturbation theory can be extended.

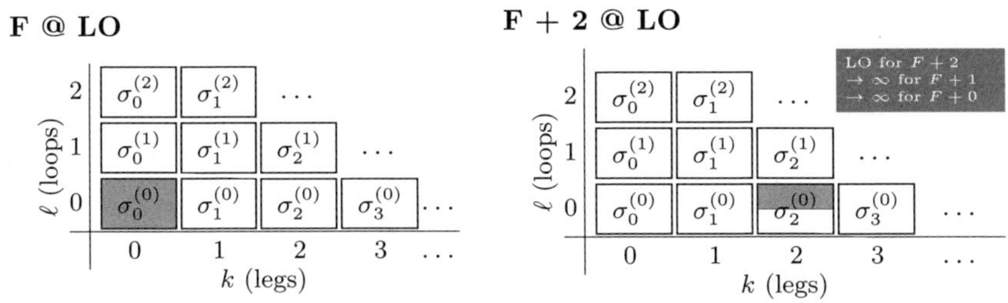

Fig. 11. Coefficients of the perturbative series covered by LO calculations. *Left:* F production at lowest order. *Right:* $F+2$ jets at LO, with the half-shaded box illustrating the restriction to the region of phase space with exactly 2 resolved jets. The total power of α_s for each coefficient is $n = k + \ell$. (Photo of Max Born from `nobelprize.org`).

The virtual amplitudes, for $\ell \geq 1$, are divergent for any point in phase space. However, as encapsulated by the famous KLN theorem [44, 45], unitarity (which essentially expresses probability conservation) puts a powerful constraint on the IR divergences,[1] forcing them to cancel exactly against those coming from the unresolved real emissions that we had to cut out above, order by order, making the complete answer for fixed $k + \ell = n$ finite. Nonetheless, since this cancellation happens between contributions that formally live in different phase spaces, a main aspect of loop-level higher-order calculations is how to arrange for this cancellation in practice, either analytically or numerically, with many different methods currently on the market. We shall discuss the idea behind subtraction approaches in section 2.4.

A convenient way of illustrating the terms of the perturbative series that a given matrix-element-based calculation includes is given in Fig. 11. In the left-hand pane, the shaded box corresponds to the lowest-order "Born-level" matrix element squared. This coefficient is non-singular and hence can be integrated over all of phase space, which we illustrate by letting the shaded area fill all of the relevant box. A different kind of leading-order calculation is illustrated in the right-hand pane of Fig. 11, where the shaded box corresponds to the lowest-order matrix element squared for $F + 2$ jets. This coefficient diverges in the part of phase space where one or both of the jets are unresolved (i.e., soft or collinear), and hence integrations can only cover the hard part of phase space, which we reflect by only shading the upper half of the relevant box.

[1]The loop integrals also exhibit UV divergences, but these are dealt with by renormalization.

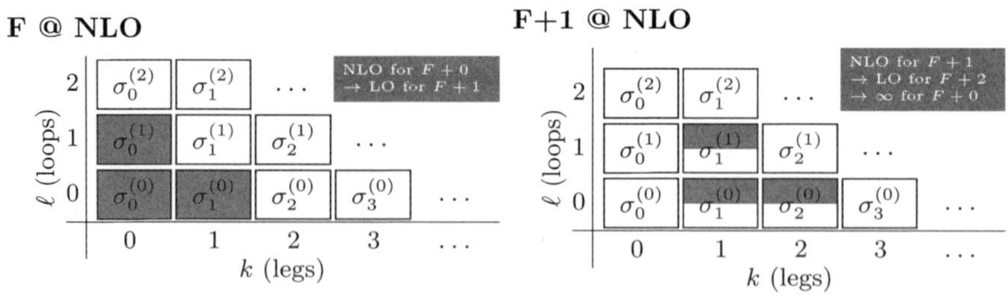

Fig. 12. Coefficients of the perturbative series covered by NLO calculations. *Left:* F production at NLO. *Right:* $F + 1\,\mathrm{jet}$ at NLO, with half-shaded boxes illustrating the restriction to the region of phase space with exactly 1 resolved jet. The total power of α_s for each coefficient is $n = k + \ell$.

Figure 12 illustrates the inclusion of NLO virtual corrections. To prevent confusion, first a point on notation: by $\sigma_0^{(1)}$, we intend

$$\sigma_0^{(1)} = \int \mathrm{d}\Phi_0 \; 2\mathrm{Re}[\mathcal{M}_0^{(1)} \mathcal{M}_0^{(0)*}] \, , \tag{31}$$

which is of order α_s relative to the Born level. Compare, e.g., with the expansion of Eq. (29) to order $k + \ell = 1$. In particular, $\sigma_0^{(1)}$ should *not* be confused with the integral over the 1-loop matrix element squared (which would be of relative order α_s^2 and hence forms part of the NNLO coefficient $\sigma_0^{(2)}$). Returning to Fig. 12, the unitary cancellations between real and virtual singularities imply that we can now extend the integration of the real correction in the left-hand pane over all of phase space, while retaining a finite total cross section,

$$\begin{aligned} \sigma_0^{\mathrm{NLO}} &= \int \mathrm{d}\Phi_0 \, |\mathcal{M}_0^{(0)}|^2 + \int \mathrm{d}\Phi_1 \, |\mathcal{M}_1^{(0)}|^2 + \int \mathrm{d}\Phi_0 \; 2\mathrm{Re}[\mathcal{M}_0^{(1)} \mathcal{M}_0^{(0)*}] \\ &= \sigma_0^{(0)} \qquad\qquad\quad + \sigma_1^{(0)} \qquad\qquad\quad + \sigma_0^{(1)} \, , \end{aligned} \tag{32}$$

with $\sigma_0^{(0)}$ the finite Born-level cross section, and the positive divergence caused by integrating the second term over all of phase space is canceled by a negative one coming from the integration over loop momenta in the third term. One method for arranging the cancellation of singularities — subtraction — is discussed in section 2.4.

However, if our starting point for the NLO calculation is a process which already has a non-zero number of hard jets, we must continue to impose that at least that number of jets must still be resolved in the final-state

F @ NNLO

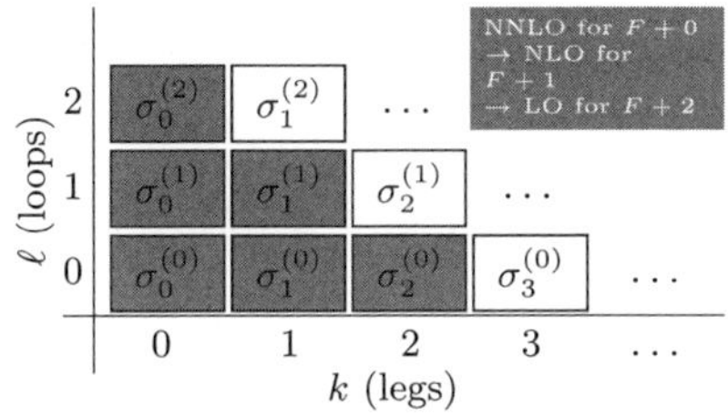

Fig. 13. Coefficients of the perturbative series covered by an NNLO calculation. The total power of α_s for each coefficient is $n = k + \ell$. Green shading represents the full perturbative coefficient at the respective k and ℓ.

integrations,

$$\begin{aligned}\sigma_1^{\rm NLO}&(p_{\perp\rm min})\\ &= \int_{p_\perp > p_{\perp\rm min}} {\rm d}\Phi_1\ |\mathcal{M}_1^{(0)}|^2 \quad + \int_{p_{\perp 1} > p_{\perp\rm min}} {\rm d}\Phi_2\ |\mathcal{M}_2^{(0)}|^2 \quad + \int_{p_\perp > p_{\perp\rm min}} {\rm d}\Phi_1\ 2{\rm Re}[\mathcal{M}_1^{(1)}\mathcal{M}_1^{(0)*}]\\ &= \sigma_1^{(0)}(p_\perp > p_{\perp\rm min}) + \sigma_2^{(0)}(p_{\perp 1} > p_{\perp\rm min}) + \sigma_1^{(1)}(p_\perp > p_{\perp\rm min})\ ,\end{aligned} \tag{33}$$

where the restriction to at least one jet having $p_\perp > p_{\perp\rm min}$ has been illustrated in the right-hand pane of Fig. 12 by shading only the upper part of the relevant boxes. In the second term in Eq. (33), the notation $p_{\perp 1}$ is used to denote that the integral runs over the phase space in which at least one "jet" (which may consist of one or two partons) must be resolved with respect to $p_{\perp\rm min}$. Here, therefore, an explicit dependence on the algorithm used to define "a jet" enters for the first time. This is discussed in more detail in the 2009 ESHEP lectures by Salam [46].

To extend the integration to cover also the case of 2 unresolved jets, we must combine the left- and right-hand parts of Fig. 12 and add the new coefficient

$$\sigma_0^{(2)} = |\mathcal{M}_0^{(1)}|^2 + 2{\rm Re}[\mathcal{M}_0^{(2)}\mathcal{M}_0^{(0)*}]\ , \tag{34}$$

as illustrated by the diagram in Fig. 13.

2.4. *The subtraction idea*

According to the KLN theorem, the IR singularities coming from integrating over collinear and soft real-emission configurations should cancel, order by order, by those coming from the IR divergent loop integrals. This implies that we should be able to rewrite e.g. the NLO cross section, Eq. (32), as

$$\sigma^{\mathrm{NLO}} = \sigma^{\mathrm{Born}} + \mathrm{Finite}\left\{\int \mathrm{d}\Phi_{F+1}\, |\mathcal{M}_{F+1}^{(0)}|^2\right\} + \mathrm{Finite}\left\{\int \mathrm{d}\Phi_F\ 2\mathrm{Re}[\mathcal{M}_F^{(1)}\mathcal{M}_F^{(0)*}]\right\}\ , \tag{35}$$

with the second and third terms having had their common (but opposite-sign) singularities canceled out and some explicitly finite quantities remaining.

The first step towards this goal is to classify all IR singularities that could appear in the amplitudes. We know that the IR limits are universal, so they can be classified using a set of process-independent functions that only has to be worked out once and for all. A widely used such set of functions are the *Catani-Seymour* (CS) dipole ones [47, 48], a method which by now has even been partially automated [49, 50]. Here, we shall instead use a formalism based on *antennae* [51–53]. The distinction between the two is basically that one antenna is made up of two dipole "ends", hence the antenna formalism tends to generate somewhat fewer terms. At NLO, however, there is no fundamental incompatibility — the antennae we use here can always be partitioned into two dipole ends, if so desired. (Note: only the antenna method has been successfully generalized to NNLO [54, 55]. Other NNLO techniques, not covered here, are *sector decomposition*, see [56, 57], and the generic formalism for hadroproduction of colorless states presented in [58].)

At NLO, the idea with subtraction is thus to rewrite the NLO cross section by adding and subtracting a simple function, $\mathrm{d}\sigma_S$, that encapsulates all the IR limits,

$$\sigma^{\mathrm{NLO}} = \sigma^{\mathrm{Born}} + \int \mathrm{d}\Phi_{F+1}\underbrace{\left(|\mathcal{M}_{F+1}^{(0)}|^2 - \mathrm{d}\sigma_S^{\mathrm{NLO}}\right)}_{\text{Finite by Universality}} + \underbrace{\int \mathrm{d}\Phi_F\ 2\mathrm{Re}[\mathcal{M}_F^{(1)}\mathcal{M}_F^{(0)*}] + \int \mathrm{d}\Phi_{F+1}\,\mathrm{d}\sigma_S^{\mathrm{NLO}}}_{\text{Finite by KLN}}\ . \tag{36}$$

The task now is to construct a suitable form for $d\sigma_S$. A main requirement is that it should be sufficiently simple that the integral in the last term can be done analytically, in dimensional regularization, so that the IR poles it generates can be canceled against those from the loop term.

To build a set of universal terms that parametrize the IR singularities of any amplitude, we start from the observation that gauge theory amplitudes factorize in the *soft limit*, as follows:

$$|\mathcal{M}_{F+1}(\ldots,i,j,k,\ldots)|^2 \stackrel{j_g\to 0}{\to}$$

$$g_s^2\, N_C \left(\frac{2s_{ik}}{s_{ij}s_{jk}} - \frac{2m_i^2}{s_{ij}^2} - \frac{2m_k^2}{s_{jk}^2} \right) |\mathcal{M}_F(\ldots,i,k,\ldots)|^2 \ , \tag{37}$$

where parton j is a soft gluon, partons i, j, and k form a chain of color-space index contractions (we say they are *color-connected*), g_s is the strong coupling, and the terms in parenthesis are called the *soft eikonal factor*. We here show it including mass corrections, which appear if i and k have non-zero rest masses, with the invariants s_{ab} then defined as

$$s_{ab} \equiv 2p_a \cdot p_b = (p_a + p_b)^2 - m_a^2 - m_b^2 \ . \tag{38}$$

The color factor, N_C, is valid for the leading-color contribution, regardless of whether the i and k partons are quarks or gluons. At subleading color, an additional soft-eikonal factor identical to the one above but with a color factor proportional to $-1/N_C$ arises for each $q\bar{q}$ pair combination. This, e.g., modifies the effective color factor for $q\bar{q} \to qg\bar{q}$ from N_C to $N_C(1 - 1/N_C) = 2C_F$, in agreement with the color factor for quarks being C_F rather than C_A.

Similarly, amplitudes also factorize in the *collinear limit* (partons i and j parallel, so $s_{ij} \to 0$), in which the eikonal factor above is replaced by the famous *Altarelli-Parisi* splitting kernels [31], which were already mentioned in section 2.2, in the context of PDF evolution. They are also the basis of conventional parton-shower models, such as those in PYTHIA [59]. We return to parton showers in section 3.2.

Essentially, what antenna functions, CS dipoles, and the like, all do, is to combine the soft (eikonal) and collinear (Altarelli-Parisi) limits into one universal set of functions that achieve the correct limiting behavior for *both* soft and collinear radiation. To give an explicit example, the *antenna function* for gluon emission from a color-connected $q\bar{q}$ pair can be derived from the matrix elements squared for the process $Z^0 \to q\bar{q} \to qg\bar{q}$ [60],

$$\frac{\left|\mathcal{M}(Z^0 \to q_i g_j \bar{q}_k)\right|^2}{\left|\mathcal{M}(Z^0 \to q_I \bar{q}_K)\right|^2} = g_s^2\, 2C_F \left[\underbrace{\frac{2s_{ik}}{s_{ij}s_{jk}}}_{\text{eikonal}} + \frac{1}{s_{IK}} \underbrace{\left(\frac{s_{jk}}{s_{ij}} + \frac{s_{ij}}{s_{jk}} \right)}_{\text{collinear}} \right] , \tag{39}$$

where we have neglected mass corrections (see [61, 62] for massive expressions) and we recognize the universal eikonal soft factor from Eq. (37) in the first term. The two additional terms are less singular, and are required to obtain the correct collinear (Altarelli-Parisi) limits as $s_{ij} \to 0$ or $s_{jk} \to 0$.

However, since the singularity structure is universal, we could equally well have used the process $H^0 \to q\bar{q} \to qg\bar{q}$ to derive the antenna function. Our antenna function would then have come out as [62],

$$\frac{\left|\mathcal{M}(H^0 \to q_i g_j \bar{q}_k)\right|^2}{\left|\mathcal{M}(H^0 \to q_I \bar{q}_K)\right|^2} = g_s^2\, 2C_F \left[\underbrace{\frac{2s_{ik}}{s_{ij}s_{jk}}}_{\text{eikonal}} + \frac{1}{s_{IK}} \underbrace{\left(\frac{s_{jk}}{s_{ij}} + \frac{s_{ij}}{s_{jk}} \right)}_{\text{collinear}} + \underbrace{\frac{2}{s_{IK}}}_{\text{finite}} \right] , \tag{40}$$

where the additional term $2/s_{IK}$ is non-singular ("finite") over all of phase space. Thus, we here see an explicit example that the singularities are process-independent while the non-singular terms are process-dependent. Since we add and subtract the same term in Eq. (36), the final answer does not depend on the choice of finite terms. We say that they correspond to different *subtraction schemes*. One standard antenna subtraction scheme, which uses the antenna function defined in Eq. (39) rather than the one in Eq. (40), is the Gehrmann-Gehrmann-de Ridder-Glover (GGG) one, given in [53].

If there is more than one color antenna in the Born-level process, the form of $\mathrm{d}\sigma_S$ is obtained as a sum over terms, each of which captures one specific soft limit and either all or "half" of a collinear one, depending on the specific scheme and the type of parton,

$$\mathrm{d}\sigma_S = \sum_j A_{IK\to ijk} \left|\mathcal{M}_F(\ldots, I, K, \ldots)\right|^2 , \tag{41}$$

with the sum running over all singular $3 \to 2$ "clusterings" of the $(F+1)$-parton state to F partons. An analysis of the different ways of partitioning the collinear singularity of gluons among neighboring antenna is beyond the scope of this introduction, but useful discussions can be found in [63–65].

2.5. *Infrared safety*

A further requirement for being able to perform calculations within perturbative QCD is that the observable be *infrared safe*. Note: by "infrared", we here mean any limit that involves a low scale (i.e., any non-UV limit), without regard to whether it is collinear or soft.

The property of infrared safety defines a special class of observables which have *minimal sensitivity* to long-distance physics, and which can be consistently computed in pQCD. An observable is infrared safe if:

(1) (*Safety against soft radiation*): Adding any number of infinitely soft particles should not change the value of the observable.
(2) (*Safety against collinear radiation*): Splitting an existing particle up into two comoving particles, with arbitrary fractions z and $1-z$, respectively, of the original momentum, should not change the value of the observable.

If both of these conditions are satisfied, any long-distance non-perturbative corrections will be suppressed by the ratio of the long-distance scale to the short-distance one to some (observable-dependent) power, typically

$$\text{IR Safe Observables: IR corrections} \quad \propto \quad \frac{Q_{\text{IR}}^2}{Q_{\text{UV}}^2} \tag{42}$$

where Q_{UV} denotes a generic hard scale in the problem, and $Q_{\text{IR}} \sim \Lambda_{\text{QCD}} \sim \mathcal{O}(1 \text{ GeV})$.

Due to this *power suppression*, IR safe observables are not so sensitive to our lack of ability to solve the strongly coupled IR physics, unless of course we go to processes for which the relevant hard scale, Q_{UV}, is small (such as minimum-bias, soft jets, or small-scale jet substructure). Even when a high scale is present, however, as in resonance decays, jet fragmentation, or underlying-event-type studies, infrared safety only guarantees us that infrared corrections are small, not that they are zero. Thus, ultimately, we run into a precision barrier even for IR safe observables, which only a reliable understanding of the long-distance physics itself can address.

To constrain models of long-distance physics, one needs infrared *sensitive* observables. Hence it is not always the case that infrared safe observables are preferable — the purpose decides the tool. Instead of the suppressed corrections above, the perturbative prediction for such observables contains logarithms of the type already encountered in Eq. (30),

$$\text{IR Sensitive Observables: IR Corrections} \quad \propto \quad \alpha_s^n \log^{m\leq 2n}\left(\frac{Q_{\mathrm{UV}}^2}{Q_{\mathrm{IR}}^2}\right) , \tag{43}$$

which grow increasingly large as $Q_{\mathrm{IR}}/Q_{\mathrm{UV}} \to 0$. As an example, consider such a fundamental quantity as particle multiplicities (= number of particles); in the absence of nontrivial infrared effects, the number of partons tends logarithmically to infinity as the IR cutoff is lowered. Similarly, the distinction between a charged and a neutral pion only occurs in the very last phase of hadronization, and hence observables that only include charged tracks, for instance, are always IR sensitive.[m]

Two important categories of infrared safe observables that are widely used are *event shapes* and *jet algorithms*. Jet algorithms are perhaps nowhere as pedagogically described as in the 2009 ESHEP lectures by Salam [46, Chapter 5]. Event shapes in the context of hadron colliders have not yet been as widely explored, but the basic phenomenology is introduced also by Salam and collaborators in [66], with first measurements reported by CMS and ATLAS [67, 68] and a proposal to use them also for the characterization of soft-QCD ("minimum-bias") events put forth in [69].

Let us here merely emphasize that the real reason to prefer infrared safe jet algorithms over unsafe ones is not that they necessarily give very different or "better" answers in the experiment — experiments are infrared safe by definition, and the difference between infrared safe and unsafe algorithms may not even be visible when running the algorithm on experimental data — but that it is only possible to compute perturbative QCD predictions for the infrared safe ones. Any measurement performed with an infrared unsafe algorithm can only be compared to calculations that include a detailed hadronization model. This both limits the number of calculations that can be compared to and also adds an a priori unknown sensitivity to the details of the hadronization description, details which one would rather investigate and constrain separately, in the framework of more dedicated fragmentation studies.

For LHC phenomenology, the preferred IR safe algorithm for jet reconstruction is currently the *anti-kT* one [70], with size parameter R vary-

[m]This remains true in principle even if the tracks are clustered into jets, although the energy clustered in this way does provide a lower bound on Q_{UV} in the given event, since "charged + neutral > charged-only".

ing between 0.4 and 0.7, though larger sizes can be motivated in certain contexts, e.g., to look for highly energetic jets and/or the boosted decay products of high-mass objects [9, 71]. This algorithm generates circular-looking jets, so subtracting off the energy believed to be associated with the *underlying event* (UE, see section 5.2) is particularly simple.

For jet substructure, typically either the "kT" or "Cambridge/Aachen" algorithms are used, see e.g. [9, 71]. The clustering measures used in these algorithms more closely mimic the singularity structure of QCD bremsstrahlung and they are therefore particularly well suited to "unravel" a tree of QCD branchings, such as a parton shower generates. The Cambridge/Aachen algorithm may also be used to characterize the underlying event, see [72].

3. Monte Carlo Event Generators

In this section, we discuss the physics of Monte Carlo event generators and their mathematical foundations, at an introductory level. We shall attempt to convey the main ideas as clearly as possible without burying them in an avalanche of technical details. References to more detailed discussions are included where applicable. We assume the reader is already familiar with the contents of the preceding section on hard processes.

The task of a Monte Carlo event generator is to calculate everything that happens in a high-energy collision, from the hard short-distance physics to the long wavelengths of hadronization and hadron decays. Obviously, this requires some compromises to be made. General-purpose generators like HERWIG, PYTHIA, and SHERPA, start from low-order (LO or NLO) descriptions of the perturbative hard physics and then attempt to include the "most significant" corrections, such as higher-order matrix-element corrections and parton showers, resonance decays and finite-width effects, underlying event, beam remnants, hadronization, and hadron decays. Each of them had slightly different origins, which carries through to the emphasis placed on various physics aspects today:

- PYTHIA. Successor to JETSET (begun in 1978). Originated in hadronization studies. Main feature: the Lund string fragmentation model.
- HERWIG. Successor to EARWIG (begun in 1984). Originated in perturbative coherence studies. Main feature: angular-ordered parton showers.

- SHERPA. Begun in 2000. Originated in studies of the matching of hard-emission matrix elements with parton showers. Main feature: CKKW matching.

There is also a large number of more specialized generators, mainly for hard processes within and beyond the SM, including ALPGEN, CALCHEP, COMPHEP, MADGRAPH, WHIZARD, and others, and a few that offer alternative shower models, such as ARIADNE and VINCIA. The most common interface between hard-process and parton-shower generators is the Les Houches Event File (LHEF) standard, defined in [73, 74] and "spoken" by most modern generator tools.

Hard processes were the topic of section 2. In this section, we shall focus mainly on parton showers, with some brief comments on resonance decays at the end. Section 4 then concerns the matching of matrix elements and parton showers. Finally, models of hadronization and the underlying event are the topic of section 5.

Several of the discussions below rely on material from the section on Monte Carlo Event Generators in the PDG Review of Particle Physics [10] and on the more comprehensive review by the *MCnet* collaboration in [75]. The latter also contains brief descriptions of the physics implementations of each of the main general-purpose event generators on the market, together with a guide on how to use (and not use) them in various connections, and a collection of comparisons to important experimental distributions. We highly recommend readers to obtain a copy of that review, as it is the most comprehensive and up-to-date review of event generators currently available. Another useful and pedagogical review on event generators is contained in the 2006 ESHEP lectures by Torbjörn Sjöstrand [25], with a more recent update in [76].

3.1. *The Monte Carlo method*

A ubiquitous problem in fundamental physics is the following: given a source located some distance from a detector, predict the number of counts that should be observed within the solid angle spanned by the detector (or within a bin of its phase-space acceptance), as a function of the properties of the source, the intervening medium, and the efficiency of the detector. Essentially, the task is to compute integrals of the form

$$N_{\mathrm{Count}}(\Delta\Omega) = \int_{\Delta\Omega} \mathrm{d}\Omega \, \frac{\mathrm{d}\sigma}{\mathrm{d}\Omega} \, , \qquad (44)$$

with $\mathrm{d}\sigma$ a differential cross section for the process of interest.

Table 2. Relative uncertainty after n evaluations, in 1 and d dimensions, for two traditional numerical integration methods and stochastic Monte Carlo. The last column shows the number of function evaluations that are required per point, in d dimensions.

Relative uncertainty with n points	1-Dim	d-Dim	n_{eval}/point
Trapezoidal Rule	$1/n^2$	$1/n^{2/d}$	2^d
Simpson's Rule	$1/n^4$	$1/n^{4/d}$	3^d
Monte Carlo	$1/\sqrt{n}$	$1/\sqrt{n}$	1

In particle physics, phase space has three dimensions per final-state particle (minus four for overall four-momentum-conservation). Thus, for problems with more than a few outgoing particles, the dimensionality of phase space increases rapidly. At LEP, for instance, the total multiplicity of neutral + charged hadrons (before weak decays) was typically $\sim$ 30 particles, for about 86 dimensions.

If we try to generalize the standard numerical-integration methods that are highly efficient in 1D, we find that most of them give very slow convergence rates for higher-dimensional problems. For illustration, a table of convergence rates in 1 and d dimensions is given in Table 2, comparing the Trapezoidal (2-point) rule and Simpson's (3-point) rule to random-number-based Monte Carlo. In 1D, the $1/n^2$ convergence rate of the Trapezoidal rule is much faster than the stochastic $1/\sqrt{n}$ of random-number Monte Carlo, and Simpson's rule converges even faster. However, as we go to d dimensions, the convergence rate of the n-point rules all degrade with d (while the number of function evaluations required for each "point" simultaneously increases). The MC convergence rate, on the other hand, remains the simple stochastic $1/\sqrt{n}$, independent of d, and each point still only requires one function evaluation. These are some of the main reasons that MC is the preferred numerical integration technique for high-dimensional problems. In addition, the random phase-space vectors it generates can be re-used in many ways, for instance as input to iterative solutions, to compute many different observables simultaneously, and/or to hand "events" to propagation and detector-simulation codes.

Therefore, virtually all numerical cross section calculations are based on Monte Carlo techniques in one form or another, the simplest being the RAMBO algorithm [77] which can be expressed in about half a page of code and generates a flat scan over n-body phase space.[n]

[n]Strictly speaking, RAMBO is only truly uniform for massless particles. Its massive variant makes up for phase-space biases by returning weighted momentum configurations.

However, due to the infrared singularities in perturbative QCD, and due to the presence of short-lived resonances, the functions to be integrated, $|\mathcal{M}_{F+k}|^2$, can be highly non-uniform, especially for large k. This implies that we will have to be clever in the way we sample phase space if we want the integration to converge in any reasonable amount of time — simple algorithms like RAMBO quickly become inefficient for k greater than a few. To address this bottleneck, the simplest step up from RAMBO is to introduce generic (i.e., automated) importance-sampling methods, such as offered by the VEGAS algorithm [78, 79]. This is still the dominant basic technique, although most modern codes do employ several additional refinements, such as several different copies of VEGAS running in parallel (multi-channel integration), to further optimize the sampling. Alternatively, a few algorithms incorporate the singularity structure of QCD explicitly in their phase-space sampling, either by directly generating momenta distributed according to the leading-order QCD singularities, in a sort of "QCD-preweighted" analog of RAMBO, called SARGE [80], or by using all-orders Markovian parton showers to generate them (VINCIA [64, 65]).

The price of using random numbers is that we must generalize our notion of convergence. In calculus, we say that a sequence $\{A\}$ *converges* to B if an n exists for which the difference $|A_{i>n} - B| < \epsilon$ for any $\epsilon > 0$. In random-number-based techniques, we cannot completely rule out the possibility of very pathological sequences of "dice rolls" leading to large deviations from the true goal, hence we are restricted to say that $\{A\}$ converges to B if an n exists for which *the probability* for $|A_{i>n} - B| < \epsilon$, for any $\epsilon > 0$, is greater than P, for any $P \in [0, 1]$ [81]. This risk, that convergence is only given with a certain probability, is the reason why Monte Carlo techniques were named after the famous casino in Monaco.

3.2. *Theoretical basis of parton showers*

In section 2, we noted two conditions that had to be valid for fixed-order truncations of the perturbative series to be valid: firstly, the strong coupling α_s must be small for perturbation theory to be valid at all. This restricts us to the region in which all scales $Q_i \gg \Lambda_{\mathrm{QCD}}$. We shall maintain this restriction in this section, i.e., we are still considering *perturbative QCD*. Secondly, however, in order to be allowed to *truncate* the perturbative series, we had to require $\sigma_{k+1} \ll \sigma_k$, i.e., the corrections at successive orders must become successively smaller, which — due to the enhancements from soft/collinear singular (conformal) dynamics — effectively restricted us to consider only the phase-space region in which all jets are "hard and well-separated", equivalent to requiring all $Q_i/Q_j \approx 1$. In this section, we shall see how to lift this restriction, extending the applicability of perturbation theory into regions that include scale hierarchies, $Q_i \gg Q_j \gg \Lambda_{\rm QCD}$, such as occur for soft jets, jet substructure, etc.

In fact, the simultaneous restriction to all resolved scales being larger than $\Lambda_{\rm QCD}$ *and* no large hierarchies is extremely severe, if taken at face value. Since we collide and observe hadrons ($\to$ low scales) while simultaneously wishing to study short-distance physics processes ($\to$ high scales), it would appear trivial to conclude that fixed-order pQCD is not applicable to collider physics at all. So why do we still use it?

The answer lies in the fact that we actually never truly perform a fixed-order calculation in QCD. Let us repeat the factorized formula for the cross section, Eq. (28), now inserting also a function, D, to represent the fragmentation of the final-state partons into observable hadrons,

$$\frac{\mathrm{d}\sigma}{\mathrm{d}\mathcal{O}} = \sum_{i,j,f} \int_0^1 \mathrm{d}x_i \, \mathrm{d}x_j \, \mathrm{d}\Phi_f \; f_{i/h_1}(x_i, \mu_F^2) f_{j/h_2}(x_j, \mu_F^2) \times \frac{\mathrm{d}\hat{\sigma}_{ij\to f}}{\mathrm{d}\hat{\mathcal{O}}} D_f(\hat{\mathcal{O}} \to \mathcal{O}, \mu_F^2) \,, \tag{45}$$

with $\hat{\mathcal{O}}$ denoting the observable evaluated on the partonic final state, and $\mathcal{O}$ the observable evaluated on the hadronic final state, after fragmentation. Although the partonic cross section, $\mathrm{d}\hat{\sigma}_{ij\to f}$, does represent a fixed-order calculation, the parton densities, f_{i/h_1} and f_{j/h_2}, include so-called resummations of perturbative corrections *to all orders* from the initial scale of order the mass of the proton, up to the factorization scale, μ_F (see section 2.2 and/or the TASI lectures by Sterman [27]). Note that the oft-stated mantra that the PDFs are purely non-perturbative functions is therefore misleading. True, they are defined as essentially non-perturbative functions at some very low scale, $\mu_0 \sim$ a few GeV, but, if μ_F is taken large, they necessarily incorporate a significant amount of perturbative physics as well. On the "fixed-order side", all we have left to ensure in $\mathrm{d}\sigma_{ij\to f}$ is then that there are no large hierarchies remaining between μ_F and the QCD scales appearing in Φ_f. Likewise, in the final state, the fragmentation functions, D_f, include infinite-order resummations of perturbative corrections all the way *from* μ_F down to some low scale, with similar caveats concerning mantras about their non-perturbative nature as for the PDFs.

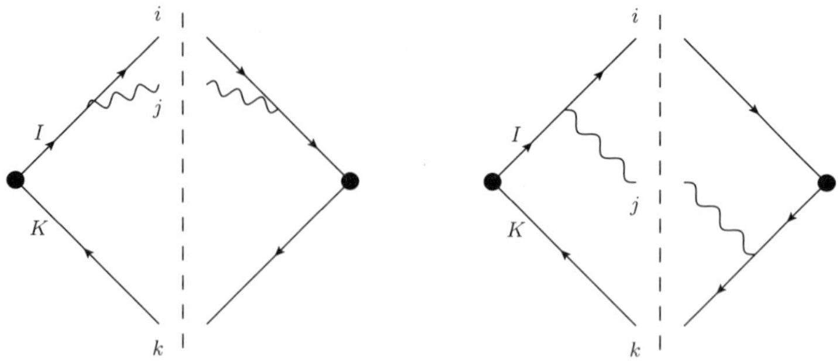

Fig. 14. Diagrams (squared) giving rise to collinear (*left*) and soft (*right*) singularities.

3.2.1. *Step One: Infinite Legs*

The infinite-order resummations that are included in objects such as the PDFs and FFs in Eq. (45) (and in their parton-shower equivalents) rely on some very simple and powerful properties of gauge field theories that were already touched on in section 2. In particular, we saw in section 2.4 that we can represent all the infrared (IR) limits of any NLO amplitude with a set of simple universal functions, based solely on knowing which partons are color-connected (i.e., have color-space index contractions) with one another.

The diagrams in Fig. 14 show the basic origin of the universal IR singularities of gauge theory amplitudes. On the left is shown a diagram (squared) in which an emission with small s_{ij} interferes with itself. In the collinear limit, $s_{ij} \to 0$, the propagator of the parent parton, I, goes on shell; the singularity of the associated propagator factor is the origin of the $1/s_{ij}$ collinear singularities. On the right is shown the interference between a diagram with emission from parton I and one with emission from parton K. The resulting term has propagator singularities when both partons I and K go on shell, which can happen simultaneously if parton j is soft. This generates the $2s_{ik}/(s_{ij}s_{jk})$ soft singularity, also called the soft eikonal factor or the dipole factor.

We now understand the fundamental origin of the IR singularities, why they are universal, and why amplitudes factorize in the soft and collinear limits — the singularities are simply generated by intermediate parton propagators going on shell, which is independent of the nature of the hard process, and hence can be factorized from it.

Thus, for each pair of (massless) color-connected partons I and K in F,

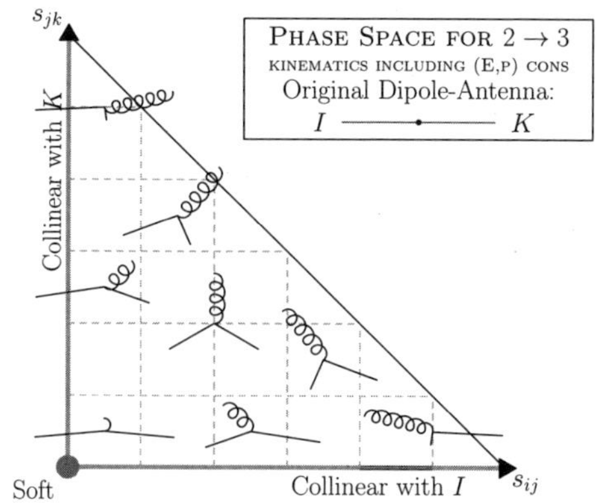

Fig. 15. Illustration of the branching phase space for $q\bar{q} \to qg\bar{q}$, with the original dipole-antenna oriented horizontally, the two parents sharing the transverse component of recoil, and the azimuthal angle ϕ (representing rotations of the emitted parton around the dipole axis) chosen such that the gluon is radiated upwards. From [64].

the squared amplitude for $F + 1$ gluon, $|\mathcal{M}_{F+1}|^2$, will include a factor

$$|\mathcal{M}_{F+1}|^2 = \underbrace{g_s^2 N_C \left(\frac{2s_{ik}}{s_{ij}s_{jk}} + \text{collinear terms} \right)}_{\text{Antenna Function}} |\mathcal{M}_F|^2 \ , \tag{46}$$

where $g_s^2 = 4\pi\alpha_s$ is the strong coupling, i and k represent partons I and K after the branching (i.e., they include possible recoil effects) and s_{ij} is the invariant between parton i and the emitted parton, j.

The branching phase space of a color dipole (i.e., a pair of partons connected by a color-index contraction) is illustrated in Fig. 15. Expressed in the branching invariants, s_{ij} and s_{jk}, the phase space has a characteristic triangular shape, imposed by the relation $s = s_{ij} + s_{jk} + s_{ik}$ (assuming massless partons). Sketchings of the post-branching parton momenta have been inserted in various places in the figure, for illustration. The soft singularity is located at the origin of the plot and the collinear regions lie along the axes.

The collinear terms for a $q\bar{q} \to qg\bar{q}$ "antenna" are unambiguous and are given in section 2.4. Since gluons are in the adjoint representation, they carry both a color and an anticolor index (one corresponding to the rows and the other to the columns of the Gell-Mann matrices), and there is therefore some ambiguity concerning how to partition collinear radiation among the two antennae they participate in. This is discussed in more detail in [64]. Differences are subleading, however, and for our purposes here we shall consider gluon antenna ends as radiating just like quark ones. The difference between quark and gluon radiation then arise mainly because gluons participate in two antennae, while quarks only participate in one. This is related to the difference between the color factors, $C_A \sim 2C_F$.

The problem that plagued the fixed-order truncations in section 2 is clearly visible in Eq. (46): if we integrate over the entire phase space including the region $s_{ij} \to 0$, $s_{jk} \to 0$, we end up with a double pole. If we instead regulate the divergence by cutting off the integration at some minimal *perturbative cutoff scale* $\mu_{\rm IR}^2$, we end up with a logarithm squared of that scale. This is a typical example of "large logarithms" being generated by the presence of scale hierarchies. Also note that the precise definition of $\mu_{\rm IR}$ is not unique. Any scale choice that properly isolates the singularities from the rest of phase space will do, with some typical choices being, for example, invariant-mass and/or transverse-momentum scales.

Before we continue, it is worth noting that Eq. (46) is often rewritten in other forms to emphasize specific aspects of it. One such rewriting is thus to reformulate the invariants s_{ij} appearing in it in terms of energies and angles,

$$s_{ij} = 2E_iE_j\,(1 - \cos\theta_{ij}) \ . \tag{47}$$

Rewritten in this way, the differentials can be partial-fractioned,

$$\frac{\mathrm{d}s_{ij}}{s_{ij}}\frac{\mathrm{d}s_{jk}}{s_{jk}} \propto \frac{\mathrm{d}E_j}{E_j}\frac{\mathrm{d}\theta_{ij}}{\theta_{ij}} + \frac{\mathrm{d}E_j}{E_j}\frac{\mathrm{d}\theta_{jk}}{\theta_{jk}} \ . \tag{48}$$

This kind of rewriting enables an intuitively appealing categorization of the singularities as related to vanishing energies and angles, explaining why they are called *soft* and *collinear*, respectively. Arguments based on this rewriting have led to important insights in QCD. For instance, within the framework of conventional parton showers, it was shown in a sequence of publications (see [82, 83] and references therein) that the destructive interference effects between two or more color-connected partons (*coherence*) can be described by using the angle of the emissions as the shower ordering variable. One should still keep in mind, however, that Lorentz non-invariant formulations come with similar caveats and warnings as do gauge non-invariant formulations of quantum field theory: while they can be practical to work with at intermediate stages of a calculation, one should be careful with any physical conclusions that rely explicitly on them.

We shall therefore here restrict ourselves to a Lorentz-invariant formalism based directly on Eq. (46), pioneered by the dipole formulation of

QCD cascades [60]. The collinear limit is then replaced by a more general *single-pole* limit in which a single parton-parton invariant vanishes (as, *for instance*, when a pair of partons become collinear), while the soft limit is replaced by one in which two (or more) invariants involving the same parton vanish simultaneously (as, for instance by that parton becoming soft in a frame defined by two or more hard partons). This avoids frame-dependent ambiguities from entering into the language, at the price of a slight reinterpretation of what is meant by collinear and soft.

In the generator landscape, *angular ordering* is used by the HERWIG [83] and HERWIG++ [84] programs, and an *angular veto* is imposed on the virtuality-ordered evolution in PYTHIA 6 [85]. Variants of the dipole approach is used by the ARIADNE [86], SHERPA [87, 88], and VINCIA [89] programs, while the $p_\perp$-ordered showers in PYTHIA 6 and 8 represent a hybrid, combining collinear splitting kernels with dipole kinematics [90]. Phase-space contours of equal value of some of these choices are illustrated in Fig. 16. During the shower evolution, each model effectively "sweeps" over phase space in the order implied by these contours. For example, a $p_\perp$-ordered dipole shower (leftmost plot in Fig. 16) will treat a hard-collinear branching as occurring "earlier" than a soft one, while a mass-ordered dipole shower (second plot) will tend to do the opposite. This affects the tower of virtual corrections generated by each shower model via the so-called Sudakov factor, discussed below.

Independently of rewritings and philosophy, the real power of Eq. (46) lies in the fact that it is *universal.* Thus, for *any* process F, we can apply Eq. (46) in order to get an approximation for $\mathrm{d}\sigma_{F+1}$. We may then, for instance, take our newly obtained expression for $F+1$ as our arbitrary process and crank Eq. (46) again, to obtain an approximation for $\mathrm{d}\sigma_{F+2}$, and so forth. What we have here is therefore a very simple recursion relation that can be used to generate approximations to leading-order cross sections with arbitrary numbers of additional legs. The quality of this approximation is governed by how many terms besides the leading one shown in Eq. (37) are included in the game. Including all possible terms, the most general form for the cross section at $F+n$ jets, restricted to the phase-space region above some infrared cutoff scale μ_{IR}, has the following algebraic structure,

$$\sigma^{(0)}_{F+n} = \alpha_s^n \left(\ln^{2n} + \ln^{2n-1} + \ln^{2n-2} + \ldots + \ln + \mathcal{F}\right) \tag{49}$$

where we use the notation $\ln^\lambda$ without an argument to denote generic functions of *transcendentality* λ (the logarithmic function to the power λ being a

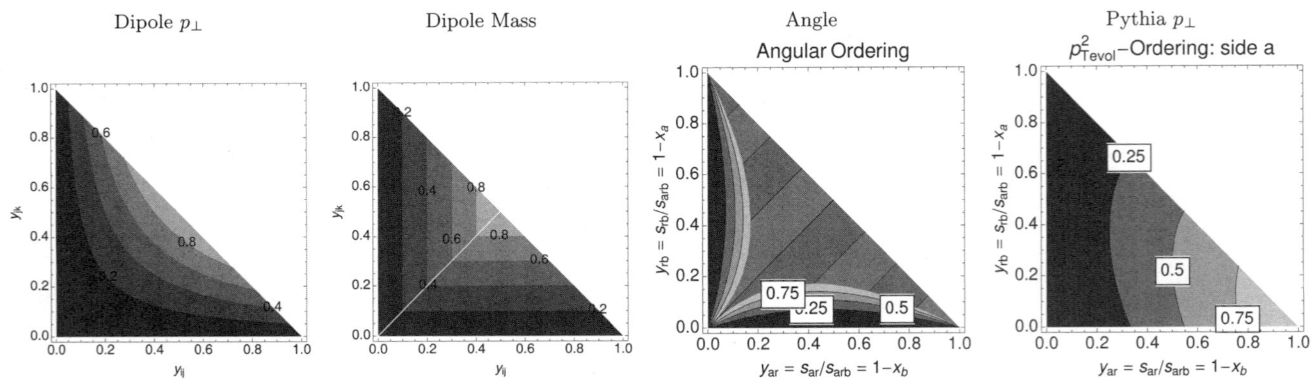

Fig. 16. A selection of parton-shower evolution variables, represented as contours over the dipole phase space. Note: the right-most variable corresponds to evolution of only one of the parents, the one with no collinear singularity along the bottom of the plot.

"typical" example of a function with transcendentality λ appearing in cross section expressions, but also dilogarithms and higher logarithmic functions[o] of transcendentality > 1 should be implicitly understood to belong to our notation $\ln^\lambda$). The last term, $\mathcal{F}$, represents a rational function of transcendentality 0. We shall also use the nomenclature *singular* and *finite* for the $\ln^\lambda$ and $\mathcal{F}$ terms, respectively, a terminology which reflects their respective behavior in the limit $\mu_{\rm IR} \to 0$.

The simplest approximation one can build on Eq. (49), dropping all but the leading $\ln^{2n}$ term in the parenthesis, is thus the *leading-transcendentality* approximation. This approximation is better known as the DLA (double logarithmic approximation), since it generates the correct coefficient for terms which have two powers of logarithms for each power of α_s, while terms of lower transcendentalities are not guaranteed to have the correct coefficients. In so-called LL (leading-logarithmic) parton shower algorithms, one generally expects to reproduce the correct coefficients for the $\ln^{2n}$ and $\ln^{2n-1}$ terms. In addition, several formally subleading improvements are normally also introduced in such algorithms (such as explicit momentum conservation, gluon polarization and other spin-correlation effects [91–93], higher-order coherence effects [82], renormalization scale choices [94], finite-width effects [95], etc), as a means to improve the agreement with some of the more subleading coefficients as well, if not in every phase-space point then at least on average. Though LL showers do not magically acquire NLL (next-to-leading-log) precision from such procedures, one therefore still expects a significantly better average performance from them than from corresponding "strict" LL analytical resummations. A side effect of this is that it is often possible to "tune" shower algorithms to give better-than-nominal agreement with experimental distributions, by adjusting the parameters controlling the treatment of subleading effects. One should remember, however, that there is a limit to how much can be accomplished in this way — at some point, agreement with one process will only come at the price of disagreement with another, and at this point further tuning would be meaningless.

Applying such an iterative process on a Born-level cross section, one obtains the description of the full perturbative series illustrated in Fig. 17. The yellow (lighter) shades used here for $k \geq 1$ indicate that the

[o]Note: due to the theorems that allow us, for instance, to rewrite dilogarithms in different ways with logarithmic and lower "spillover" terms, the coefficients at each λ are only well-defined up to reparametrization ambiguities involving the terms with transcendentality greater than λ.

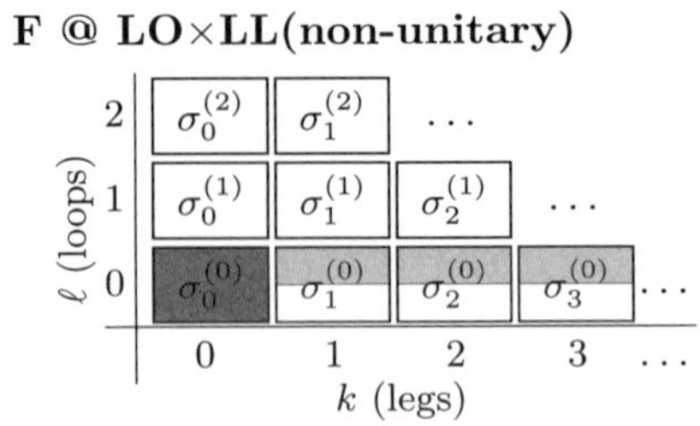

Fig. 17. Coefficients of the perturbative series covered by LO + LL approximations to higher-multiplicity tree-level matrix elements. Green (darker) shading represents the full perturbative coefficient at the respective k and ℓ. Yellow (lighter) shading represents an LL approximation to it. Half-shaded boxes indicate phase spaces in which we are prohibited from integrating over the IR singular region, as discussed in sections 2.3 and 4.

coefficient obtained is not the exact one, but rather an approximation to it that only gets its leading singularities right. However, since this is still only an approximation to infinite-order *tree-level* cross sections (we have not yet included any virtual corrections), we cannot yet integrate this approximation over all of phase space, as illustrated by the yellow boxes being only half filled in Fig. 17; otherwise, the summed total cross section would still be infinite. This particular approximation would therefore still appear to be very useless indeed — on one hand, it is only guaranteed to get the singular terms right, but on the other, it does not actually allow us to integrate over the singular region. In order to obtain a truly *all-orders* calculation, the constraint of unitarity must also be explicitly imposed, which furnishes an approximation to all-orders loop corrections as well. Let us therefore emphasize that Fig. 17 is included for pedagogical purposes only; all resummation calculations, whether analytical or parton-shower based, include virtual corrections as well and consequently yield finite total cross sections, as will now be described.

3.2.2. *Step Two: Infinite Loops*

Order-by-order unitarity, such as used in the KLN theorem, implies that the singularities caused by integration over unresolved radiation in the tree-level matrix elements must be canceled, order by order, by equal but opposite-sign singularities in the virtual corrections at the same order. That is, from Eq. (46), we immediately know that the 1-loop correction to $\mathrm{d}\sigma_F$ *must*

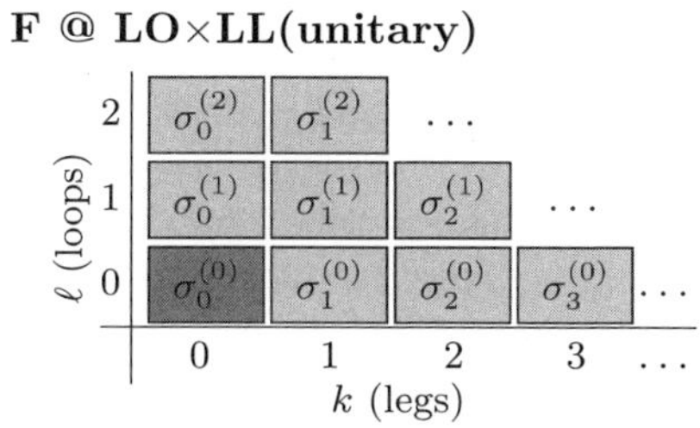

Fig. 18. Coefficients of the perturbative series covered by LO + LL calculations, imposing unitarity order by order for each $n = k + \ell$. Green (darker) shading represents the full perturbative coefficient at the respective k and ℓ. Yellow (lighter) shading represents an LL approximation to it.

contain a term,

$$2\mathrm{Re}[\mathcal{M}_F^{(0)}\mathcal{M}_F^{(1)*}] \supset -g_s^2\, N_C \left|\mathcal{M}_F^{(0)}\right|^2 \int \frac{\mathrm{d}s_{ij}\,\mathrm{d}s_{jk}}{16\pi^2 s_{ijk}} \left(\frac{2s_{ik}}{s_{ij}s_{jk}} + \text{subleading}\right) , \tag{50}$$

that cancels the divergence coming from Eq. (46) itself. Further, since this is universally true, we may apply Eq. (50) again to get an approximation to the corrections generated by Eq. (46) at the next order and so on. By adding such terms explicitly, order by order, we may now bootstrap our way around the entire perturbative series, using Eq. (46) to move horizontally and Eq. (50) to move along diagonals of constant $n = k + \ell$. Since real-virtual cancellations are now explicitly restored, we may finally extend the integrations over all of phase space, resulting in the picture shown in Fig. 18.

The picture shown in Fig. 18, not the one in Fig. 17, corresponds to what is actually done in *resummation* calculations, both of the analytic and parton-shower types.[p] Physically, there is a significant and intuitive meaning to the imposition of unitarity, as follows.

Take a jet algorithm, with some measure of jet resolution, Q, and apply it to an arbitrary sample of events, say dijets. At a very crude resolution scale, corresponding to a high value for Q, you find that everything is clustered back to a dijet configuration, and the 2-jet cross section is equal

[p]In the way these calculations are formulated in practice, they in fact rely on one additional property, called exponentiation, that allows us to move along straight vertical lines in the loops-and-legs diagrams. However, since the two different directions furnished by Eqs. (46) and (50) are already sufficient to move freely in the full 2D coefficient space, we shall use exponentiation without extensively justifying it here.

to the total inclusive cross section,

$$\sigma_{\mathrm{tot}} = \sigma_{F;\mathrm{incl}} \ . \tag{51}$$

At finer resolutions, decreasing Q, you see that some events that were previously classified as 2-jet events contain additional, lower-scale jets, that you can now resolve, and hence those events now migrate to the 3-jet bin, while the total inclusive cross section of course remains unchanged,

$$\sigma_{\mathrm{tot}} = \sigma_{F;\mathrm{incl}} = \sigma_{F;\mathrm{excl}}(Q) + \sigma_{F+1;\mathrm{incl}}(Q) \ , \tag{52}$$

where "incl" and "excl" stands for inclusive and exclusive cross sections,[q] respectively, and the Q-dependence in the two terms on the right-hand side must cancel so that the total inclusive cross section is independent of Q. Later, some 3-jet events now migrate further, to 4 and higher jets, while still more 2-jet events migrate *into* the 3-jet bin, etc. For arbitrary n and Q, we have

$$\sigma_{F+n;\mathrm{incl}}(Q) = \sigma_{F;\mathrm{incl}} - \sum_{m=0}^{n-1} \sigma_{F+m;\mathrm{excl}}(Q) \ . \tag{53}$$

This equation expresses the trivial fact that the cross section for n or more jets can be computed as the total inclusive cross section for F minus a sum over the cross sections for F + exactly m jets including all $m < n$. On the theoretical side, it is these negative terms which must be included in the calculation, for each order $n = k + \ell$, to restore unitarity. Physically, they express that, at a given scale Q, each event will be classified as having *either* 0, 1, 2, or whatever jets. Or, equivalently, for each event we gain in the 3-jet bin as Q is lowered, we must loose one event in the 2-jet one; the negative contribution to the 2-jet bin is exactly minus the integral of the positive contribution to the 3-jet one, and so on. We may perceive of this *detailed balance* as an *evolution* of the event structure with Q, for each event, which is effectively what is done in parton-shower algorithms, to which we shall return in section 3.3.

3.3. *Perturbation theory with Markov chains*

Consider again the Born-level cross section for an arbitrary hard process, F, differentially in an arbitrary infrared-safe observable $\mathcal{O}$, as obtained from

[q] *F inclusive* = F plus anything. *F exclusive* = F and only F. Thus, $\sigma_{F;\mathrm{incl}} = \sum_{k=0}^{\infty} \sigma_{F+k;\mathrm{excl}}$.

Eq. (29):

$$\left.\frac{\mathrm{d}\sigma_F^{(0)}}{\mathrm{d}\mathcal{O}}\right|_{\mathrm{Born}} = \int \mathrm{d}\Phi_F \ |\mathcal{M}_F^{(0)}|^2 \ \delta(\mathcal{O} - \mathcal{O}(\Phi_F)) \ , \tag{54}$$

where the integration runs over the full final-state on-shell phase space of F (this expression and those below would also apply to hadron collisions were we to include integrations over the parton distribution functions in the initial state), and the δ function projects out a 1-dimensional slice defined by $\mathcal{O}$ evaluated on the set of final-state momenta which we denote Φ_F.

To make the connection to parton showers, we insert an operator, $\mathcal{S}$, that acts on the Born-level final state *before* the observable is evaluated, i.e.,

$$\left.\frac{\mathrm{d}\sigma_F}{\mathrm{d}\mathcal{O}}\right|_{\mathcal{S}} = \int \mathrm{d}\Phi_F \ |\mathcal{M}_F^{(0)}|^2 \ \mathcal{S}(\Phi_F, \mathcal{O}) \ . \tag{55}$$

Formally, this operator — the evolution operator — will be responsible for generating all (real and virtual) higher-order corrections to the Born-level expression. The measurement δ function appearing explicitly in Eq. (54) is now implicit in $\mathcal{S}$.

Algorithmically, parton showers cast $\mathcal{S}$ as an iterative Markov (i.e., history-independent) chain, with an evolution parameter, Q_E, that formally represents the factorization scale of the event, below which all structure is summed over inclusively. Depending on the particular choice of shower algorithm, Q_E may be defined as a parton virtuality (virtuality-order showers), as a transverse-momentum scale ($p_\perp$-ordered showers), or as a combination of energies times angles (angular ordering). Regardless of the specific form of Q_E, the evolution parameter will go towards zero as the Markov chain develops, and the event structure will become more and more exclusively resolved. A transition from a perturbative evolution to a non-perturbative one can also be inserted, when the evolution reaches an appropriate scale, typically around 1 GeV. This scale, called the *hadronization scale*, thus represents the lowest perturbative scale that can appear in the calculations, with all perturbative corrections below it summed over inclusively.

Working out the precise form that $\mathcal{S}$ must have in order to give the correct expansions discussed in section 3.2 takes a bit of algebra, and is beyond the scope we aim to cover in these lectures. Heuristically, the procedure is as follows. We noted that the singularity structure of QCD is universal and that at least its first few terms are known to us. We also saw that we could iterate that singularity structure, using universality and unitarity, thereby bootstrapping our way around the entire perturbative series. This was illustrated by Fig. 18 in section 3.2.

Skipping intermediate steps, the form of the all-orders pure-shower Markov chain, for the evolution of an event between two Q_E scales, $Q_1 > Q_2$, is,

$$\mathcal{S}(\Phi_F, Q_1, Q_2, \mathcal{O}) = \underbrace{\Delta(\Phi_F, Q_1, Q_2)\ \delta\left(\mathcal{O} - \mathcal{O}(\Phi_F)\right)}_{F+0 \text{ exclusive above } Q_2}$$
$$\underbrace{+\sum_r \int_{Q_2}^{Q_1} \frac{\mathrm{d}\Phi^r_{F+1}}{\mathrm{d}\Phi_F}\ S_r(\Phi_{F+1})\ \Delta(\Phi_F, Q_1, Q_{F+1})\ \mathcal{S}(\Phi_{F+1}, Q_{F+1}, Q_2, \mathcal{O})}_{F+1 \text{ inclusive above } Q_2}\ , \tag{56}$$

with the so-called *Sudakov factor*,

$$\Delta(\Phi_F, Q_1, Q_2) = \exp\left[-\sum_r \int_{Q_2}^{Q_1} \frac{\mathrm{d}\Phi^r_{F+1}}{\mathrm{d}\Phi_F} S_r(\Phi_{F+1})\right]\ , \tag{57}$$

defining the probability that there is *no evolution* (i.e., no emissions) between the scales Q_1 and Q_2, according to the *radiation functions* S_r to which we shall return below. The term on the first line of Eq. (56) thus represents all events that *did not* evolve as the resolution scale was lowered from Q_1 to Q_2, while the second line contains a sum and phase-space integral over those events that *did* evolve — including the insertion of $\mathcal{S}(\Phi_{F+1})$ representing the possible further evolution of the event and completing the iterative definition of the Markov chain.

The factor $\mathrm{d}\Phi^r_{F+1}/\mathrm{d}\Phi_F$ defines the chosen phase space factorization. Our favorite is the so-called dipole-antenna factorization, whose principal virtue is that it is the simplest Lorentz invariant factorization which is simultaneously exact over all of phase space while only involving on-shell momenta. For completeness, its form is

$$\frac{\mathrm{d}\Phi^r_{F+1}}{\mathrm{d}\Phi_F} = \frac{\mathrm{d}\Phi^r_3}{\mathrm{d}\Phi_2} = \mathrm{d}s_{a1}\,\mathrm{d}s_{1b}\,\frac{\mathrm{d}\phi}{2\pi}\frac{1}{16\pi^2 s_r}\ , \tag{58}$$

which involves just one color-anticolor pair for each r, with invariant mass squared $s_r = (p_a + p_1 + p_b)^2$. Other choices, such as purely collinear ones (only exact in the collinear limit *or* involving explicitly off-shell momenta), more global ones involving all partons in the event (more complicated, in our opinion), or less global ones with a single parton playing the dominant role as emitter, are also possible, again depending on the specific algorithm considered.

The radiation functions S_r obviously play a crucial role in these equations, driving the emission probabilities. For example, if $S_r \to 0$, then $\Delta \to \exp(0) = 1$ and all events stay in the top line of Eq. (56). Thus, in regions of phase space where S_r is small, there is little or no evolution. Conversely, for $S_r \to \infty$, we have $\Delta \to 0$, implying that *all* events evolve. One possible choice for the radiation functions S_r was implicit in Eq. (46), in which we took them to include only the leading (double) singularities, with r representing color-anticolor pairs. In general, the shower may exponentiate the entire set of universal singular terms, or only a subset of them (for example, the terms leading in the number of colors N_C), which is why we here let the explicit form of S_r be unspecified. Suffice it to say that in traditional parton showers, S_r would simply be the DGLAP splitting kernels (see, e.g., [11]), while they would be so-called dipole or antenna radiation functions in the various dipole-based approaches to QCD (see, e.g., [47, 53, 60, 64, 65, 88]).

The procedure for how to technically "construct" a shower algorithm of this kind, using random numbers to generate scales distributed according to Eq. (56), is described more fully in [64], using a notation that closely parallels the one used here. The procedure is also described at a more technical level in the review [75], though using a slightly different notation. Finally, a pedagogical introduction to Monte Carlo methods in general can be found in [81].

3.4. *Decays of unstable particles*

In most BSM processes and some SM ones, an important aspect of the event simulation is how decays of short-lived particles, such as top quarks, EW and Higgs bosons, and new BSM resonances, are handled. We here briefly summarize the spectrum of possibilities, but emphasize that there is no universal standard. Users are advised to check whether the treatment of a given code is adequate for the physics study at hand.

The appearance of an unstable resonance as a physical particle at some intermediate stage of the event generation implies that its production and decay processes are treated as being factorized. This is called the *narrow width approximation* and is valid up to corrections of order Γ/m_0, with Γ the width and m_0 the pole mass of the particle. States whose widths are a substantial fraction of their mass should not be treated in this way, but rather as intrinsically virtual internal propagator lines.

For states treated as physical particles, two aspects are relevant: the mass distribution of the decaying particle itself and the distributions of its decay products. For the mass distribution, the simplest is to use a δ function at m_0. The next level up, typically used in general-purpose Monte Carlos, is to use a Breit-Wigner distribution (relativistic or non-relativistic), which formally resums higher-order virtual corrections to the mass distribution. Note, however, that this still only generates an improved picture for *moderate* fluctuations away from m_0. Similarly to above, particles that are significantly off-shell (in units of Γ) should not be treated as resonant, but rather as internal off-shell propagator lines. In most Monte Carlo codes, some further refinements are also included, for instance by letting Γ be a function of m ("running widths") and by limiting the magnitude of the allowed fluctuations away from m_0. See also [96] for an elaborate discussion of the Higgs boson lineshape.

For the distributions of the decay products, the simplest treatment is again to assign them their respective m_0 values, with a uniform (i.e., isotropic, or "flat") phase-space distribution. A more sophisticated treatment distributes the decay products according to the differential decay matrix elements, capturing at least the internal dynamics and helicity structure of the decay process, including EPR-like correlations. Further refinements include polarizations of the external states [91–93] (see also [97–99] for phenomenological studies) and assigning the decay products their own Breit-Wigner distributions, the latter of which opens the possibility to include also intrinsically off-shell decay channels, like $H \to WW^*$. Please refer to the physics manual of the code you are using and/or make simple cross checks like plotting the distribution of phase-space invariants it produces.

During subsequent showering of the decay products, most parton-shower models will preserve the total invariant mass of each resonance-decay system separately, so as not to skew the original resonance shape.

4. Matching at LO and NLO

The essential problem that leads to matrix-element/parton-shower matching can be illustrated in a very simple way. Assume we have computed the LO cross section for some process, F, and that we have added an LL shower to it, as in the left-hand pane of Fig. 19. We know that this only gives us an LL description of $F+1$. We now wish to improve this from LL to LO by

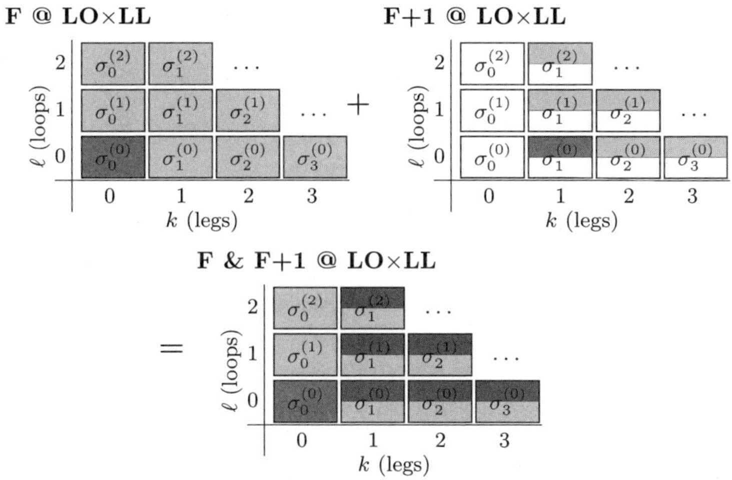

Fig. 19. The double-counting problem caused by naively adding cross sections involving matrix elements with different numbers of legs.

adding the actual LO matrix element for $F+1$. Since we also want to be able to hadronize these events, etc, we again add an LL shower off them. However, since the matrix element for $F+1$ is divergent, we must restrict it to cover only the phase-space region with at least one hard resolved jet, illustrated by the half-shaded boxes in the middle pane of Fig. 19.

Adding these two samples, however, we end up counting the LL terms of the inclusive cross section for $F+1$ twice, since we are now getting them once from the shower off F and once from the matrix element for $F+1$, illustrated by the dark shaded (red) areas of the right-hand pane of Fig. 19. This *double-counting* problem would grow worse if we attempted to add more matrix elements, with more legs. The cause is very simple. Each such calculation corresponds to an *inclusive* cross section, and hence naive addition would give

$$\sigma_{\mathrm{tot}} = \sigma_{0;\mathrm{incl}} + \sigma_{1;\mathrm{incl}} = \sigma_{0;\mathrm{excl}} + 2\,\sigma_{1;\mathrm{incl}} \,. \tag{59}$$

Recall the definition of inclusive and exclusive cross sections, Eq. (52): F *inclusive* = F plus anything. F *exclusive* = F and only F. Thus, $\sigma_{F;\mathrm{incl}} = \sum_{k=0}^{\infty} \sigma_{F+k;\mathrm{excl}}$.

Instead, we must *match* the coefficients calculated by the two parts of the full calculation — showers and matrix elements — more systematically, for each order in perturbation theory, so that the nesting of inclusive and exclusive cross sections is respected without overcounting.

Given a parton shower and a matrix-element generator, there are fundamentally three different ways in which we can consider matching the two [64]: slicing, subtraction, and unitarity. The following subsections will briefly introduce each of these.

4.1. *Slicing*

The most commonly encountered matching type is currently based on separating (slicing) phase space into two regions, one of which is supposed to be mainly described by hard matrix elements and the other of which is supposed to be described by the shower. This type of approach was first used in HERWIG [100], to include matrix-element corrections for one emission beyond the basic hard process [101, 102]. This is illustrated in Fig. 20.

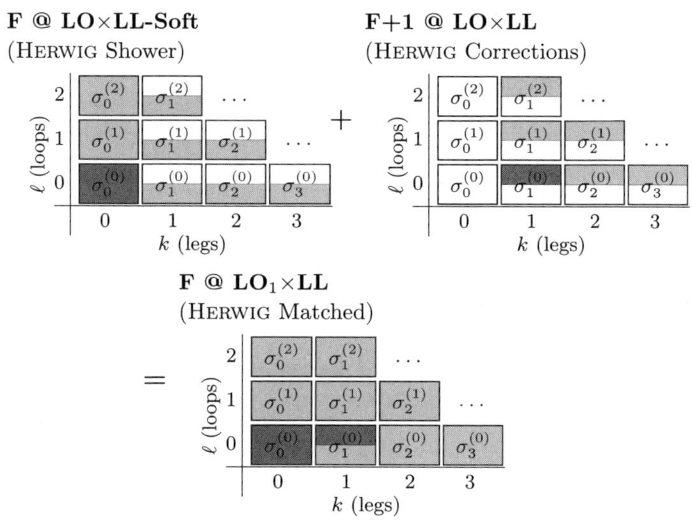

Fig. 20. HERWIG's original matching scheme [101, 102], in which the dead zone of the HERWIG shower was used as an effective "matching scale" for one emission beyond a basic hard process.

The method has since been generalized by several independent groups to include arbitrary numbers of additional legs, the most well-known of these being the CKKW [103], CKKW-L [104, 105], and MLM [106, 107] approaches.

Effectively, the shower approximation is set to zero above some scale, either due to the presence of explicit dead zones in the shower, as in HERWIG, or by vetoing any emissions above a certain *matching scale*, as in the (L)-CKKW and MLM approaches. The empty part of phase space can then be filled by separate events generated according to higher-multiplicity tree-level matrix elements (MEs). In the (L)-CKKW and MLM schemes, this process can be iterated to include arbitrary numbers of additional hard legs (the practical limit being around 3 or 4, due to computational complexity).

In order to match smoothly with the shower calculation, the higher-multiplicity matrix elements must be associated with Sudakov form factors (representing the virtual corrections that would have been generated if a shower had produced the same phase-space configuration), and their α_s factors must be chosen so that, at least at the matching scale, they become identical to the choices made on the shower side [108]. The CKKW and MLM approaches do this by constructing "fake parton-shower histories" for the higher-multiplicity matrix elements. By applying a sequential jet clustering algorithm, a tree-like branching structure can be created that at least has the same dominant structure as that of a parton shower. Given the fake shower tree, α_s factors can be chosen for each vertex with argument $\alpha_s(p_\perp)$ and Sudakov factors can be computed for each internal line in the tree. In the CKKW method, these Sudakov factors are estimated analytically, while the MLM and CKKW-L methods compute them numerically, from the actual shower evolution.

Thus, the matched result is identical to the matrix element (ME) in the region above the matching scale, modulo higher-order (Sudakov and α_s) corrections. We may sketch this as

$$\text{Matched (above matching scale)} = \overbrace{\text{Exact}}^{\text{ME}} \times \overbrace{(1+\mathcal{O}(\alpha_s))}^{\text{corrections}} , \qquad (60)$$

where the "shower-corrections" include the approximate Sudakov factors and α_s reweighting factors applied to the matrix elements in order to obtain a smooth transition to the shower-dominated region.

Below the matching scale, the small difference between the matrix elements and the shower approximation can be dropped (since their leading singularities are identical and this region by definition includes no hard jets), yielding the pure shower answer in that region,

$$\begin{aligned} \text{Matched (below matching scale)} &= \overbrace{\text{Approximate}}^{\text{shower}} + \overbrace{(\text{Exact} - \text{Approximate})}^{\text{correction}} \\ &= \text{Approximate} + \text{non-singular} \\ &\to \text{Approximate} . \end{aligned} \qquad (61)$$

This type of strategy is illustrated in Fig. 21.

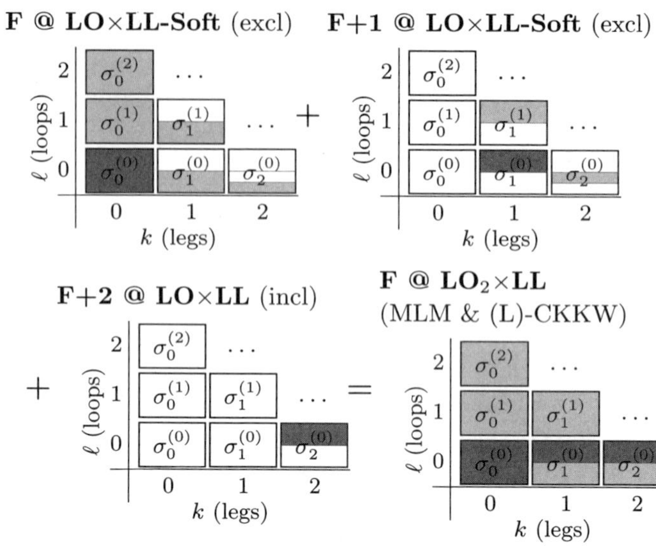

Fig. 21. Slicing, with up to two additional emissions beyond the basic process. The showers off F and $F + 1$ are set to zero above a specific "matching scale". (The number of coefficients shown was reduced a bit in these plots to make them fit in one row.)

As emphasized above, since this strategy is discontinuous across phase space, a main point here is to ensure that the behavior across the matching scale be as smooth as possible. CKKW showed [103] that it is possible to remove any dependence on the matching scale through NLL precision by careful choices of all ingredients in the matching; technical details of the implementation (affecting the $\mathcal{O}(\alpha_s)$ terms in Eq. (60)) are important, and the dependence on the unphysical matching scale may be larger than NLL unless the implementation matches the theoretical algorithm precisely [104, 105, 109]. Furthermore, since the Sudakov factors are generally computed using showers (MLM, L-CKKW) or a shower-like formalism (CKKW), while the real corrections are computed using matrix elements, care must be taken not to (re-)introduce differences that could break the detailed real-virtual balance that ensures unitarity among the singular parts [108].

It is advisable not to choose the matching scale too low. This is again essentially due to the approximate scale invariance of QCD imploring us to write the matching scale as a ratio, rather than as an absolute number. If one uses a very low matching scale, the higher-multiplicity matrix elements will already be quite singular, leading to very large LO cross sections before matching. After matching, these large cross sections are tamed by the Sudakov factors produced by the matching scheme, and hence the final cross sections may still look reasonable. But the higher-multiplicity matrix elements in general contain subleading singularity structures, beyond those accounted for by the shower, and hence the delicate line of detailed balance that ensures unitarity has most assuredly been overstepped. We therefore recommend not to take the matching scale lower than about an order of magnitude below the characteristic scale of the hard process.

One should also be aware that all strategies of this type are quite computing intensive. This is basically due to the fact that a separate phase-space generator is required for each of the n-parton correction terms, with each such sample a priori consisting of weighted events such that a separate unweighting step (often with quite low efficiency) is needed before an unweighted sample can be produced.

Finally, we note that several proposals to extend this method to NLO matrix elements were recently put forth [110, 111], and it will be interesting to see if these can be developed to a similar level as the tree-level strategies, e.g. via the Binoth Les Houches Accord [112] for interfacing one-loop programs (OLPs) with Monte Carlo codes.

4.2. *Subtraction*

Another way of matching two calculations is by subtracting one from the other and correcting by the difference, schematically

$$\text{Matched} = \overbrace{\text{Approximate}}^{\text{shower}} + \overbrace{(\text{Exact} - \text{Approximate})}^{\text{correction}} . \qquad (62)$$

This looks very much like the structure of a subtraction-based NLO fixed-order calculation, section 2.4, in which the shower approximation here plays the role of subtraction terms, and indeed this is what is used in strategies like MC@NLO [113–115], illustrated in Fig. 22. In this type of approach, however, negative-weight events will generally occur, for instance in phase-space points where the approximation is larger than the exact answer.

Negative weights are not in principle an insurmountable problem. Histograms can be filled with each event counted according to its weight, as usual. However, negative weights do affect efficiency. Imagine a worst-case scenario in which 1000 positive-weight events have been generated, along with 999 negative-weight ones (assuming each event weight has the same absolute value, the closest one can get to an unweighted sample in the presence of negative weights). The statistical precision of the MC answer would

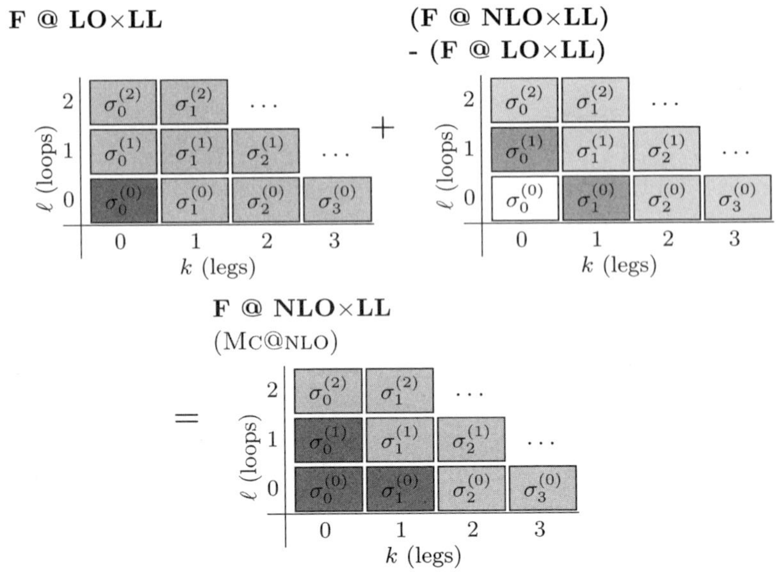

Fig. 22. MC@NLO. In the middle pane, cyan boxes denote non-singular correction terms, while the egg-colored ones denote showers off such corrections, which cannot lead to double-counting at the LL level.

be equivalent to one event, for 2000 generated, i.e., a big loss in convergence rate.

In practice, generators like MC@NLO "only" produce around 10% or less events with negative weights, so the convergence rate should not be severely affected for ordinary applications. Nevertheless, the problem of negative weights motivated the development of the so-called POWHEG approach [116], illustrated in Fig. 23, which is constructed specifically to prevent negative-weight events from occurring and simultaneously to be more independent of which parton-shower algorithm it is used with. In the POWHEG method, one effectively modifies the real-emission probability for the first emission to agree with the $F + 1$ matrix element (this is covered under unitarity, below). One is then left with a purely virtual correction, which will typically be positive, at least for processes for which the NLO cross section is larger than the LO one.

The advantage of these methods is obviously that NLO corrections to the Born level can be systematically incorporated. However, a systematic way of extending this strategy beyond the first additional emission is not

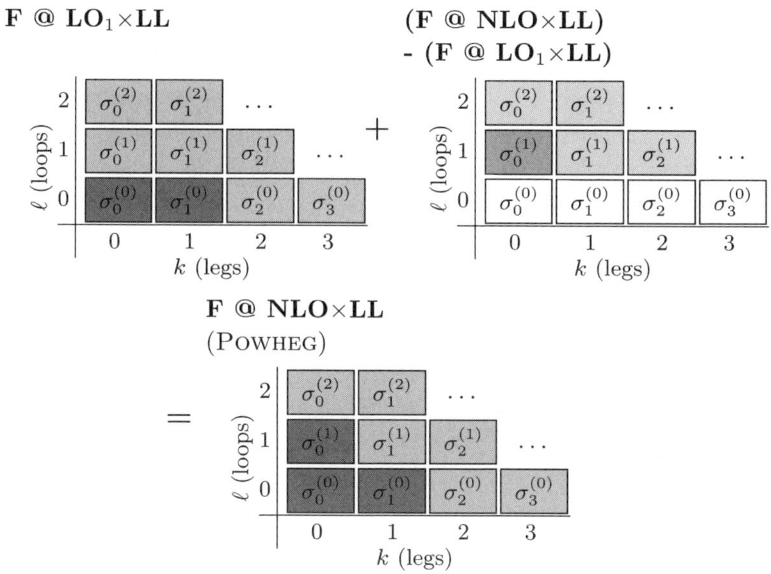

Fig. 23. POWHEG. In the middle pane, cyan boxes denote non-singular correction terms, while the egg-colored ones denote showers off such corrections, which cannot lead to double-counting at the LL level.

available, save for combining them with a slicing-based strategy for the additional legs, as in MENLOPS [117], illustrated in Fig. 24. These issues are, however, no more severe than in ordinary fixed-order NLO approaches, and hence they are not viewed as disadvantages if the point of reference is an NLO computation.

4.3. *Unitarity*

The oldest, and in my view most attractive, approach [85, 118] consists of working out the shower approximation to a given fixed order, and correcting the shower splitting functions at that order by a multiplicative factor given by the ratio of the matrix element to the shower approximation, phase-space point by phase-space point. We may sketch this as

$$\text{Matched} = \overbrace{\text{Approximate}}^{\text{shower}} \times \overbrace{\frac{\text{Exact}}{\text{Approximate}}}^{\text{correction}} . \tag{63}$$

When these correction factors are inserted back into the shower evolution, they guarantee that the shower evolution off $n-1$ partons correctly

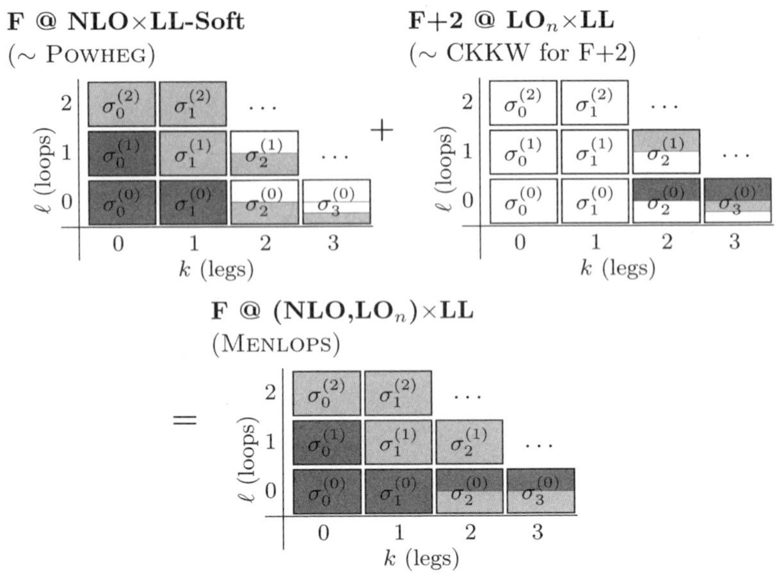

Fig. 24. MENLOPS. Note that each of the POWHEG and CKKW samples are composed of separate sub-samples, as illustrated in figs. 21 and 23.

reproduces the n-parton matrix elements, without the need to generate a separate n-parton sample. That is, the shower approximation is essentially used as a pre-weighted (stratified) all-orders phase-space generator, on which a more exact answer can subsequently be imprinted order by order in perturbation theory. Since the shower is already optimized for exactly the kind of singular structures that occur in QCD, very fast computational speeds can therefore be obtained with this method [65].

In the original approach [85, 118], used by PYTHIA [59, 119], this was only worked out for one additional emission beyond the basic hard process. In POWHEG [116, 120], it was extended to include also virtual corrections to the Born-level matrix element. Finally, in VINCIA [64, 89] it has been extended to include arbitrary numbers of emissions at tree level.

An illustration of the perturbative coefficients that can be included in each of these approaches is given in Fig. 25, as usual with green (darker shaded) boxes representing exact coefficients and yellow (light shaded) boxes representing logarithmic approximations.

Finally, two more properties unique to this method deserve mention. Firstly, since the corrections modify the actual shower evolution kernels,

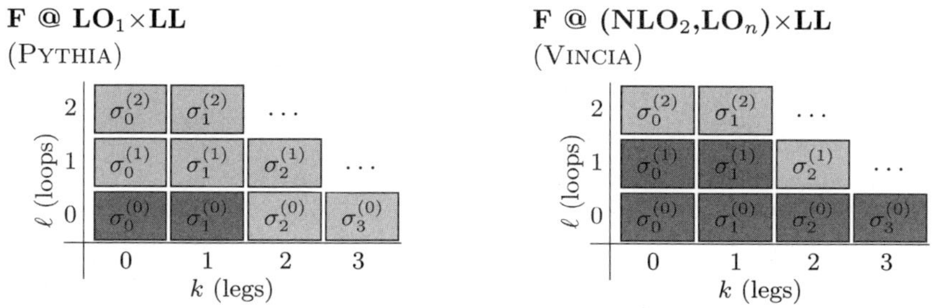

Fig. 25. PYTHIA (left) and VINCIA (right). Unitarity-based. Only one event sample is produced by each of these methods, and hence no sub-components are shown.

the corrections are automatically *resummed* in the Sudakov exponential, which should improve the logarithmic precision once $k \geq 2$ is included, and secondly, since the shower is *unitary*, an initially unweighted sample of $(n-1)$-parton configurations remains unweighted, with no need for a separate event-unweighting or event-rejection step.

5. Hadronization and Soft Hadron-Hadron Physics

We here give a very brief overview of the main aspects of soft QCD that are relevant for hadron-hadron collisions, such as hadronization, minimum-bias and soft-inclusive physics, and the so-called underlying event. This will be kept at a pedestrian level and is largely based on the reviews in [10, 75, 121].

In the context of event generators, *hadronization* denotes the process by which a set of colored partons (*after* showering) is transformed into a set of color-singlet *primary* hadrons, which may then subsequently decay further. This non-perturbative transition takes place at the *hadronization scale* $Q_{\rm had}$, which by construction is identical to the infrared cutoff of the parton shower. In the absence of a first-principles solution to the relevant dynamics, event generators use QCD-inspired phenomenological models to describe this transition.

Essentially, the problem that a hadronization model should address can be stated as follows: given a set of partons resolved at a scale of $Q_{\rm had} \sim 1$ GeV, we need a "mapping" from this set onto a set of on-shell color-singlet (i.e., confined) hadronic states. MC models do this in three steps:

(1) Map the partonic system onto a continuum of high-mass hadronic states (called "strings" or "clusters").

(2) Iteratively map strings/clusters onto discrete set of primary hadrons (string breaks / cluster splittings / cluster decays).
(3) Sequential decays into secondary hadrons (e.g., $\rho \to \pi\pi$, $\Lambda \to n\pi$, $\pi^0 \to \gamma\gamma$, ...).

The physics governing this mapping is non-perturbative and hence cannot (yet?) be solved from first principles. However, we do have some knowledge of the properties that such a solution must have. For instance, Poincaré invariance, unitarity, and causality are all concepts that apply beyond perturbation theory. In addition, lattice QCD provides us a means of making explicit quantitative studies in a genuinely non-perturbative setting (albeit only of certain questions, see section 1).

An important result in "quenched" lattice QCD[r] is that the potential of the color-dipole field between a charge and an anticharge appears to grow linearly with the separation of the charges, at distances greater than about a femtometer. This is known as "linear confinement", and it forms the starting point for the *string model of hadronization*, discussed below in section 5.1. Alternatively, a property of perturbative QCD called "preconfinement" [122] is the basis of the *cluster model of hadronization*, described in [10, 75].

In the generator landscape, PYTHIA uses string fragmentation, while HERWIG and SHERPA use cluster fragmentation. It should be emphasized that the so-called *parton level* that can be obtained by switching off hadronization in an MC generator, is not a universal concept, since each model defines the hadronization scale differently. For example, the hadronization scale can be defined by a cutoff in invariant mass, transverse momentum, or some other quantity, with different tunes using different values for the cutoff. Comparisons to distributions at this level (i.e., with hadronization switched off) may therefore be used to provide an idea of the overall impact of hadronization corrections within a given model, but should be avoided in the context of physical observables.

We use the term "soft hadron-hadron physics" to comprise all scattering processes for which a hard, perturbative scale is not required to be present.[s]

---

[r] Quenched QCD implies no "dynamical" quarks, i.e., no $g \to q\bar{q}$ splittings allowed.
[s] Note, however, that while a hard scale is not *required* to be present, it is not explicitly required to be absent either. Thus, both diffractive, minimum-bias, pile-up and underlying-event processes will have tails towards high-$p_\perp$ physics as well. For example, even $t\bar{t}$ pair production can be viewed as a tail of minimum-bias interactions, and there is a tail of diffractive processes in which hard dijets can be produced diffractively (see, e.g., [123]).

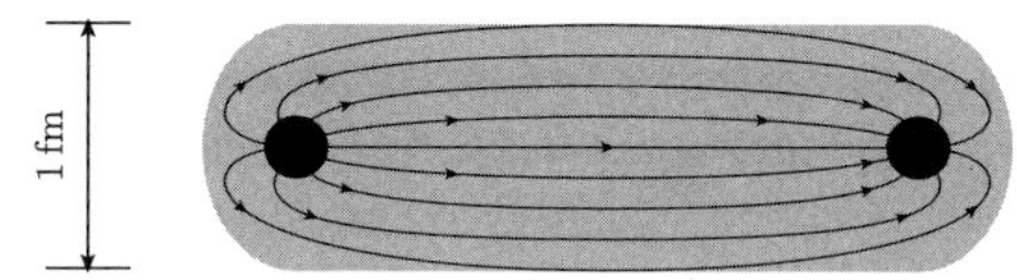

Fig. 26. Illustration of the transition between a Coulomb potential at short distances to the string-like one of Eq. (64) at large $q\bar{q}$ separations.

This includes elastic, diffractive, minimum-bias, and pile-up processes, as well as the physics contributing to the so-called underlying event. We give a brief introduction to such processes in section 5.2.

We round off with a discussion of the data constraints that enter in the tuning of Monte Carlo models in section 5.3, and give an outline of a procedure that could be followed in a realistic set-up.

5.1. *The string model of hadronization*

Starting from early concepts developed by Artru and Mennessier [124], several hadronization models based on strings were proposed in the late 1970'ies and early 80'ies. Of these, the most widely used today is the so-called Lund model, implemented in the PYTHIA code. We shall therefore concentrate on that particular model here, though many of the overall concepts would be shared by any string-inspired method. (A more extended discussion can be found in the very complete and pedagogical review of the Lund model by Andersson [125].)

Consider the production of a $q\bar{q}$ pair from vacuum, for instance in the process $e^+e^- \to \gamma^*/Z \to q\bar{q} \to$ hadrons. As the quarks move apart, linear confinement implies that a potential

$$V(r) = \kappa\, r \tag{64}$$

is asymptotically reached for large distances, r. (At short distances, there is a Coulomb term proportional to $1/r$ as well, but this is neglected in the Lund model.) This potential describes a string with tension (energy per unit length) κ. The physical picture is that of a color flux tube being stretched between the q and the $\bar{q}$, Fig. 26. From hadron mass spectroscopy the string tension κ, is known to be

$$\kappa \;\sim\; 1\,\mathrm{GeV/fm} \;\sim\; 0.2\,\mathrm{GeV}^2. \tag{65}$$

A straightforward Lorentz-invariant description of this object is provided by the massless relativistic string in 1+1 dimensions, with no transverse

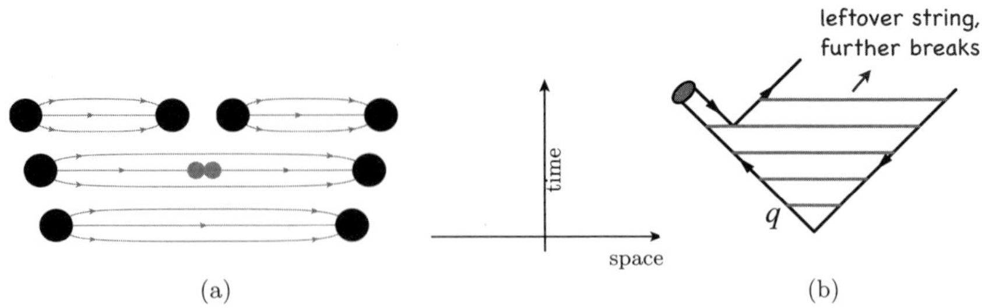

Fig. 27. (a) Illustration of string breaking by quark pair creation in the string field. (b) Illustration of the algorithmic choice to process the fragmentation from the outside-in, splitting off a single on-shell hadron in each step.

degrees of freedom. The mathematical, one-dimensional string can be thought of as parameterizing the position of the axis of a cylindrically symmetric flux tube. (Note that the expression "massless" is somewhat of a misnomer, since κ effectively corresponds to a "mass density" along the string.)

As the q and $\bar{q}$ move apart, their kinetic energy is gradually converted to potential energy, stored in the growing string spanned between them. In the "quenched" approximation, in which $g \to q\bar{q}$ splittings are not allowed, this process would continue until the endpoint quarks have lost *all* their momentum, at which point they would reverse direction and be accelerated by the now shrinking string. In the real world, quark-antiquark fluctuations inside the string field can make the transition to become real particles by absorbing energy from the string, thereby screening the original endpoint charges from each other and breaking the string into two separate color-singlet pieces, $(q\bar{q}) \to (q\bar{q}') + (q'\bar{q})$, illustrated in Fig. 27(a). This process then continues until only ordinary hadrons remain. (We will give more details on the individual string breaks below.)

More complicated multi-parton topologies including gluons are treated by representing gluons as transverse "kinks". Thus soft gluons effectively "build up" a transverse structure in the originally one-dimensional object, with infinitely soft ones absorbed into the string without leading to modifications. For strings with finite-energy kinks, the space-time evolution is then slightly more involved [125], and modifications to the fragmentation model to handle stepping across gluon corners have to be included, but the main point is that there are no separate free parameters for gluon jets. Differences with respect to quark fragmentation arise simply because quarks are only connected to a single string piece, while gluons have one on either side, increasing the energy loss per unit (invariant) time from a gluon to the string by a factor of 2 relative to quarks, which can be compared to the ratio of color Casimirs $C_A/C_F = 2.25$.

Since the string breaks are causally disconnected (as can easily be realized from space-time diagrams [125]), they do not have to be considered in any specific time-ordered sequence. In the Lund model, the string breaks are instead generated starting with the leading ("outermost") hadrons, containing the endpoint quarks, and iterating inwards towards the center of the string, alternating randomly between fragmentation off the left- and right-hand sides, respectively, Fig. 27(b). This has the advantage that a single on-shell hadron can be split off in each step, making it straightforward to ensure that only states consistent with the known spectrum of hadron resonances are produced, as will be discussed below.

The details of the individual string breaks are not known from first principles. The Lund model invokes the idea of quantum mechanical tunneling, which leads to a Gaussian suppression of the energies and masses imparted to the produced quarks,

$$\mathrm{Prob}(m_q^2, p_{\perp q}^2) \;\propto\; \exp\left(\frac{-\pi m_q^2}{\kappa}\right) \exp\left(\frac{-\pi p_{\perp q}^2}{\kappa}\right) , \tag{66}$$

where m_q is the mass of the produced quark and $p_\perp$ is the transverse momentum imparted to it by the breakup process (the antiquark obviously has the same mass and opposite $p_\perp$).

Due to the factorization of the $p_\perp$ and m dependence implied by Eq. (66), the $p_\perp$ spectrum of produced quarks in this model is independent of the quark flavor, with a universal average value of

$$\langle p_{\perp q}^2 \rangle = \sigma^2 = \kappa/\pi \sim (250\,\mathrm{MeV})^2 \ . \tag{67}$$

Bear in mind that "transverse" is here defined with respect to the string axis. Thus, the $p_\perp$ in a frame where the string is moving is modified by a Lorentz boost factor. Also bear in mind that σ^2 is here a purely non-perturbative parameter. In a Monte Carlo model with a fixed shower cutoff Q_{had}, the effective amount of "non-perturbative" $p_\perp$ may be larger than this, due to effects of additional unresolved soft-gluon radiation below Q_{had}. In principle, the magnitude of this additional component should scale with the cutoff, but in practice it is up to the user to enforce this by retuning (see section 5.3) the effective σ parameter when changing the hadronization scale. Since hadrons receive $p_\perp$ contributions from two breakups, one on either side, their average transverse momentum squared will be twice as large,

$$\langle p_{\perp h}^2 \rangle = 2\sigma^2 \ . \tag{68}$$

The mass suppression implied by Eq. (66) is less straightforward to interpret. Since quark masses are notoriously difficult to define for light quarks, the value of the strangeness suppression must effectively be extracted from experimental measurements, e.g., of the K/π ratio, with a resulting suppression of roughly $s/u \sim s/d \sim 0.2 - 0.3$. Inserting even comparatively low values for the charm quark mass in Eq. (66), however, one obtains a relative suppression of charm of the order of 10^{-11}. Heavy quarks can therefore safely be considered to be produced only in the perturbative stages and not by the soft fragmentation.

Baryon production can be incorporated in the same basic picture [126], by allowing string breaks to occur also by the production of pairs of so-called *diquarks*, loosely bound states of two quarks in an overall $\bar{3}$ representation (e.g., red + blue = antigreen). Again, the relative rate of diquark-to-quark production is not known a priori and must be extracted from experimental measurements, e.g., of the p/π ratio. More advanced scenarios for baryon production have also been proposed, in particular the so-called popcorn model [127, 128], which is normally used in addition to the diquark picture and then acts to decrease the correlations among neighboring baryon-antibaryon pairs by allowing mesons to be formed inbetween them. Within the PYTHIA framework, a fragmentation model including explicit *string junctions* [129] has so far only been applied to baryon-number-violating new-physics processes and to the description of beam remnants (and then acts to increase baryon stopping [130]).

This brings us to the next step of the algorithm: assignment of the produced quarks within hadron multiplets. The fragmenting quark (antiquark) may combine with the antiquark (quark) from a newly created breakup to produce either a vector or a pseudoscalar meson, or, if diquarks are involved, either a spin-1/2 or spin-3/2 baryon. Unfortunately, the string model is entirely unpredictive in this respect, and this is therefore the sector that contains the largest amount of free parameters. From spin counting alone, one would expect the ratio V/S of vectors to pseudoscalars to be 3, but in practice this is only approximately true for B^*/B. For lighter flavors, the difference in phase space caused by the V–S mass splittings implies a suppression of vector production. Thus, for D^*/D, the effective ratio is already reduced to about $\sim 1.0 - 2.0$, while for K^*/K and ρ/π, extracted values range from $0.3 - 1.0$. Recall, as always, that these are

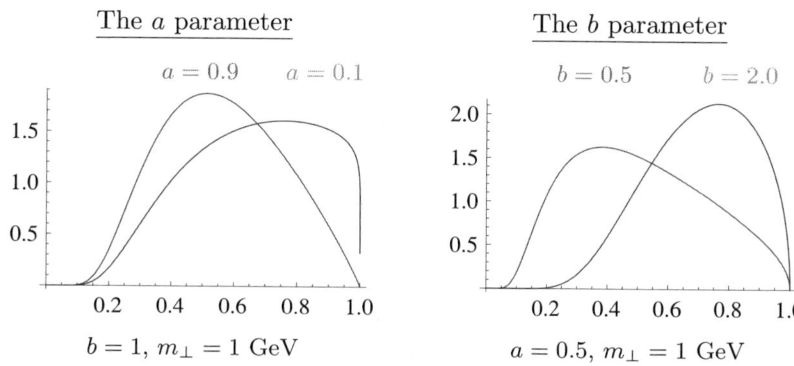

Fig. 28. The Lund symmetric fragmentation function, normalized to unity. *Left:* variation of the a parameter, from 0.1 (blue) to 0.9 (red), with fixed $b = 1\ \mathrm{GeV}^{-2}$ and $m_\perp = 1$ GeV. *Right:* variation of the b parameter, from 0.5 (red) to 2 (blue) GeV^{-2}, with fixed $a = 0.5$ and $m_\perp = 1$ GeV.

production ratios of *primary hadrons*, hence feed-down (from secondary decays of heavier hadrons) complicates the extraction of these parameters from experimental data, in particular for the lighter hadron species. The production of higher meson resonances is assumed to be low in a string framework.[t] For diquarks, separate parameters control the relative rates of spin-1 diquarks vs. spin-0 ones and, likewise, have to extracted from data, with resulting values of order $(qq)_1/(qq)_0 \sim 0.075 - 0.15$.

With $p_\perp^2$ and m^2 now fixed, the final step is to select the fraction, z, of the fragmenting endpoint quark's longitudinal momentum that is carried by the created hadron. In this respect, the string picture is substantially more predictive than for the flavor selection. Firstly, the requirement that the fragmentation be independent of the sequence in which breakups are considered (causality) imposes a "left-right symmetry" on the possible form of the fragmentation function, $f(z)$, with the solution

$$f(z) \propto \frac{1}{z}(1-z)^a \exp\left(-\frac{b\,(m_h^2 + p_{\perp h}^2)}{z}\right) , \tag{69}$$

which is known as the *Lund symmetric fragmentation function* (normalized to unit integral). The a and b parameters, illustrated in Fig. 28, are the only free parameters of the fragmentation function, though a may in principle be flavor-dependent. Note that the explicit mass dependence in $f(z)$ implies

[t]The four $L = 1$ multiplets are implemented in PYTHIA, but are disabled by default, largely because several states are poorly known and thus may result in a worse overall description when included.

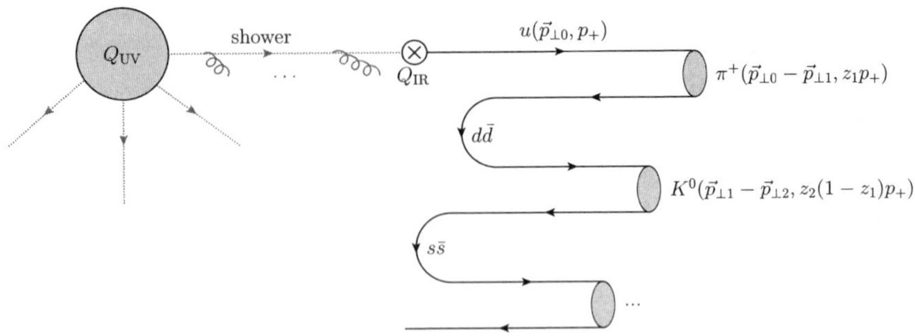

Fig. 29. The iterative selection of flavors and momenta in the Lund string-fragmentation model.

a harder fragmentation function for heavier hadrons (in the rest frame of the string).

As a by-product, the probability distribution in invariant time τ of $q'\bar{q}'$ breakup vertices, or equivalently $\Gamma = (\kappa\tau)^2$, is also obtained, with $\mathrm{d}P/\mathrm{d}\Gamma \propto \Gamma^a \exp(-b\Gamma)$ implying an area law for the color flux [131], and the average breakup time lying along a hyperbola of constant invariant time $\tau_0 \sim 10^{-23}$s [125].

The iterative selection of flavors, $p_\perp$, and z values is illustrated in Fig. 29. A parton produced in a hard process at some high scale Q_{UV} emerges from the parton shower, at the hadronization scale Q_{IR}, with 3-momentum $\vec{p} = (\vec{p}_{\perp 0}, p_+)$, where the "+" on the third component denotes "light-cone" momentum, $p_\pm = E \pm p_z$. Next, an adjacent $d\bar{d}$ pair from the vacuum is created, with relative transverse momenta $\pm p_{\perp 1}$. The fragmenting quark combines with the $\bar{d}$ from the breakup to form a π^+, which carries off a fraction z_1 of the total lightcone momentum p_+. The next hadron carries off a fraction z_2 of the remaining momentum, etc.

For massive endpoints (e.g., c and b quarks, or hypothetical hadronizing new-physics particles, generally called R-hadrons), which do not move along straight lightcone sections, the exponential suppression with string area leads to modifications of the form [132], $f(z) \to f(z)/z^{b\,m_Q^2}$, with m_Q the mass of the heavy quark. Strictly speaking, this is the only fragmentation function that is consistent with causality in the string model, though a few alternative forms are typically provided as well.

Note, however, that the term *fragmentation function* in the context of non-perturbative hadronization models is used to denote only the corrections originating from scales below the infrared cutoff scale of the parton shower. That is, the fragmentation functions introduced here are defined at an intrinsically low scale of order $Q \sim 1\,\text{GeV}$. It would therefore be inconsistent to compare them directly to those that are used in fixed-order/analytical-resummation contexts, which are typically defined at a factorization scale of order the scale of the hard process.

5.2. *Soft Hadron-Hadron processes*

5.2.1. *Elastic and Inelastic Scattering*

Elastic hadron-hadron scattering consists of all reactions of the type

$$A(p_A)B(p_B) \to A(p'_A)B(p'_B) \ , \tag{70}$$

where A and B are hadrons with momenta p_A and p_B, respectively. Specifically, the only exchanged quantity is momentum; all quantum numbers and masses remain unaltered, and no new particles are produced. Inelastic scattering covers everything else, i.e.,

$$AB \to X \neq AB \ , \tag{71}$$

where $X \neq AB$ signifies that one or more quantum numbers are changed, and/or more particles are produced. The total hadron-hadron cross section can thus be written as a sum of these two physically distinguishable components,

$$\sigma_{\text{tot}}(s) = \sigma_{\text{el}}(s) + \sigma_{\text{inel}}(s) \ , \tag{72}$$

where $s = (p_A + p_B)^2$ is the beam-beam center-of-mass energy squared.

5.2.2. *Diffractive Scattering*

If A and/or B are not elementary, the inelastic final states may be further divided into "diffractive" and "non-diffractive" topologies. This is a qualitative classification, usually based on whether the final state looks like the decay of an excitation of the beam particles (diffractive[u]), or not (non-diffractive), or upon the presence of a large rapidity gap somewhere in the final state which would separate such excitations.

Given that an event has been labeled as diffractive, either within the context of a theoretical model, or by a final-state observable, we may

[u]An example of a process that would be labeled as diffractive would be if one the protons is excited to a Δ^+ which then decays back to $p^+ + \pi^0$, without anything else happening in the event. In general, a whole tower of possible diffractive excitations are available, which in the continuum limit can be described by a mass spectrum falling roughly as $\mathrm{d}M^2/M^2$.

distinguish between three different classes of diffractive topologies, which it is possible to distinguish between physically, at least in principle. In double-diffractive (DD) events, both of the beam particles are diffractively excited and hence none of them survive the collision intact. In single-diffractive (SD) events, only one of the beam particles gets excited and the other survives intact. The last diffractive topology is central diffraction (CD), in which both of the beam particles survive intact, leaving an excited system in the central region between them. (This latter topology includes "central exclusive production" where a single particle is produced in the central region.) That is,

$$\sigma_{\mathrm{inel}}(s) = \sigma_{\mathrm{SD}}(s) + \sigma_{\mathrm{DD}}(s) + \sigma_{\mathrm{CD}}(s) + \sigma_{\mathrm{ND}}(s) , \tag{73}$$

where "ND" (non-diffractive, here understood not to include elastic scattering) contains no gaps in the event consistent with the chosen definition of diffraction. Further, each of the diffractively excited systems in the events labeled SD, DD, and CD, respectively, may in principle consist of several subsystems with gaps between them. Equation (73) may thus be defined to be exact, within a specific definition of diffraction, even in the presence of multi-gap events. Note, however, that different theoretical models almost always use different (model-dependent) definitions of diffraction, and therefore the individual components in one model are in general not directly comparable to those of another. It is therefore important that data be presented at the level of physical observables if unambiguous conclusions are to be drawn from them.

5.2.3. *Minimum Bias*

The term *minimum-bias* (MB) is an experimental term, used to define a certain class of events that are selected with the minimum possible trigger bias, to ensure they are as inclusive as possible.[v] In theoretical contexts, the term "minimum-bias" is often used with a slightly different meaning; to

[v]A typical min-bias trigger would thus be the requirement of at least one measured particle in a given rapidity region, so that all events which produce at least one observable particle would be included, which must, indeed, be considered the minimal possible bias. In principle, *everything* is a subset of minimum-bias, including both hard and soft processes. However, compared to the total minimum-bias cross section, the fraction that is made up of hard processes is only a very small tail. Since only a tiny fraction of the total minimum-bias rate can normally be stored, the minimum-bias sample would give quite poor statistics if used for hard physics studies. Instead, separate dedicated hard-process triggers are typically included in addition to the minimum-bias one, in order to ensure maximal statistics also for hard physics processes.

denote specific (classes of) inclusive soft-QCD subprocesses in a given model. Since these two usages are not exactly identical, in these lectures we have chosen to reserve the term "minimum bias" to pertain strictly to definitions of experimental measurements, and instead use the term "soft inclusive" physics as a generic descriptor for the class of processes which generally dominate the various experimental "minimum-bias" measurements in theoretical models. This parallels the terminology used in the review [75], from which most of the discussion here has been adapted. See Eq. (73) above for a compact overview of the types of physical processes that contribute to minimum-bias data samples. For a more detailed description of Monte Carlo models of this physics, see [75].

5.2.4. *Underlying Event and Multiple Parton Interactions*

In events containing a hard parton-parton interaction, the *underlying event* (UE) can be roughly conceived of as the *difference* between QCD with and without including the remnants of the original beam hadrons. Without such "beam remnants", only the hard interaction itself, and its associated parton showers and hadronization, would contribute to the observed particle production. In reality, after the partons that participate in the hard interaction have been taken out, the remnants still contain whatever is left of the incoming beam hadrons, including also a partonic substructure, which leads to the possibility of *multiple parton interactions* (MPI). Useful reviews of MPI-based MC models can be found in [75, 121]. Analytical models are mostly formulated only for double parton scattering, see e.g., [133–136].

Due to the simple fact that the remnants are not empty, an underlying event will always be there — but how much additional energy does it deposit in a given measurement region? A quantification of this can be obtained, for instance, by comparing measurements of the UE to the average activity in minimum-bias events at the same $\sqrt{s}$. Interestingly, it turns out that the underlying event is much more active, with larger fluctuations, than the average MB event. This is called the *jet pedestal effect* (hard jets sit on top of a higher-than-average "pedestal" of underlying activity), and is interpreted as follows: when two hadrons collide at non-zero impact parameter, high-$p_\perp$ interactions can only take place inside the overlapping region. Imposing a hard trigger therefore statistically biases the event sample toward more central collisions, which will also have more underlying activity.

For hard processes at the LHC at 7 TeV, the transverse energy, E_T, in the UE is about 1.5 GeV per unit ΔR area, though with large event-to-event fluctuations of order ± 1 GeV [137]. Thus, using the typical sizes for LHC jets quoted in section 2.5, and multiplying by πR^2, the E_T originating from the UE, in a cone with radius 0.4, can be estimated to be $E_{T\mathrm{UE}}(R = 0.4) \sim 0.75 \pm 0.5$ GeV, while the E_T in cones with radii 0.7 and 1.0 would be $E_{T\mathrm{UE}}(R = 0.7) \sim 2.3 \pm 1.5$ GeV and $E_{T\mathrm{UE}}(R = 1.0) \sim 4.7 \pm 3$ GeV, respectively. See [72] for a discussion on the use of jet algorithms in the characterization of the UE (including illustrations with simple analytical toy models) and these lecture notes by R. Field [138] for more discussion on the level of the UE at the Tevatron and at LHC.

5.3. *Tuning*

The main virtue of general-purpose Monte Carlo event generators is their ability to provide a complete and fully differential picture of collider final states, down to the level of individual particles. This allows them to be used as detailed — albeit approximate — theoretical references for measurements performed at accelerators like the LHC, against which models of both known and 'new' physics can be tested. As has been emphasized in these lectures, the achievable accuracy depends both on the inclusiveness of the chosen observable and on the sophistication of the simulation itself. An important driver for the latter is obviously the development of improved theoretical models, e.g., by including matching to higher-order matrix elements, more accurate resummations, or better non-perturbative models, as discussed in the previous sections; but it also depends crucially on the available constraints on the remaining free parameters of the model. Using existing data to constrain these is referred to as generator tuning.

Although Monte Carlo models may appear to have a bewildering array of independently adjustable parameters, it is worth keeping at the front of one's mind that most of these parameters only control relatively small (exclusive) details of the event generation. The majority of the (inclusive) physics is determined by only a few, very important ones, such as, e.g., the value of the strong coupling, in the perturbative domain, and the form of the fragmentation function for massless partons, in the non-perturbative one.

Armed with a good understanding of the underlying model, an expert would therefore normally take a highly factorized approach to constraining the parameters, first constraining the perturbative ones and thereafter the non-perturbative ones, each ordered in a measure of their relative significance to the overall modeling. This factorization, and carefully chosen experimental distributions corresponding to each step, allows one to concentrate on just a few parameters and distributions at a time, reducing the full parameter space to manageable-sized chunks. Still, each step will often involve more than one single parameter, and non-factorizable corrections still imply that changes made in subsequent steps can change the agreement obtained in previous ones by a non-negligible amount, requiring additional iterations from the beginning to properly tune the entire generator framework.

Recent years have seen the emergence of automated tools that attempt to reduce the amount of both computer and manpower required for this task, for instance by making full generator runs only for a limited set of parameter points, and then interpolating between these to obtain approximations to what the true generator result would have been for any intermediate parameter point, as, e.g., in the Professor tool [139, 140]. Automating the human expert input is of course more difficult. In the tools currently on the market, this question is addressed by a combination of input solicited from the generator authors (e.g., which parameters and ranges to consider, which observables constitute a complete set, etc) and the elaborate construction of non-trivial weighting functions that determine how much weight is assigned to each individual bin and to each distribution. The field is still burgeoning, however, and future sophistications are to be expected. Nevertheless, at this point the overall quality of the tunes obtained with automated methods appear to at least be competitive with the manual ones. Recent examples of tunes including uncertainty variations can be found in [108, 141, 142].

A sketch of a reasonably complete tuning procedure, without going into details about the parameters that control each of these sectors in individual Monte Carlo models, would be the following:

1) Keep in mind that inabilities of models to describe data is a vital part of the feedback cycle between theory and experiment. Also keep in mind that perturbation theory at LO$\times$LL is doing *very well* if it gets within 10% of a given IR safe measurement. An agreement of 5% should be considered the absolute sanity limit, beyond which it does not make any sense to tune further. The advent of NLO Monte Carlos may reduce these numbers slightly, but only for quantities for which one expects NLO precision to hold, see section 4. However, the sanity limit should be taken to be at least twice as large for quantities governed by non-perturbative physics. For some quantities, e.g., ones for which the underlying modeling is *known* to be poor, an order-of-magnitude agreement or worse may have to be accepted. Attempting to force Monte Carlo models to describe data far outside their domains of validity must be expected to produce similar side effects as attempting to turn a Fiat into a Ferrari merely by cranking up the engine revolutions.

2) Final-State Radiation and Hadronization: Mainly using LEP and other e^+e^- collider data. On the IR safe side, there are event shapes and jet observables, the latter including rates, resolutions, masses, shapes, and jet-jet correlations. On the IR sensitive side, special attention should be paid to the high-z tail of the fragmentation spectra, where a single hadron carries a large fraction of an entire jet's momentum, since this is the tail that is most likely to give "fake jets". Depending on the focus of the tuning, attention should also be paid to identified-particle rates and ratios, and to fragmentation in events containing heavy quarks and/or gluon jets. Usually, more weight is given to those particles that are most copiously produced, though this again depends on the focus. Finally, particle-particle correlations and baryon production are typically some of the least well constrained components of the overall modeling. The scaling properties of IR safe vs. IR sensitive contributions can be tested by comparing data at several different e^+e^- collider energies.

3) Initial-State Radiation, and so-called "Primordial[w] k_T": Here, one would in principle like to use data from DIS reactions, which are less complicated to interpret than full hadron-hadron collisions. However, due to difficulties in translating between the *ep* and *pp* environments, this is normally *not* what is done in practice. Instead, the main constraining distribution is the dilepton $p_\perp$ distribution in Drell-Yan events in hadron-hadron collisions. For any observables containing explicit jets, be aware that the underlying event can produce small horizontal shifts in jet $p_\perp$ distributions, which may in turn result in seemingly larger-than-expected vertical changes if the distributions are falling sharply. Also note that the ISR evolution is sensitive to the choice of PDFs, with caveats as discussed in section 2.2.

4) Initial-Final Connections: For example, radiation from color lines connected to the initial state and jet broadening in hadron collider environments. This is one of the most poorly controlled parts of most MC models. Keep in mind that it is *not* directly constrained by pure final-state observ-

[w]Primordial k_T: an additional soft $p_\perp$ component that is injected on top of the $p_\perp$ generated by the initial-state shower itself, see [75, Section 7.1].

ables, such as LEP fragmentation, nor by pure initial-state observables, such as the Drell-Yan $p_\perp$ spectrum, which is why we list it as a separate item here. In principle, DIS would again be a prime territory for placing constraints on this aspect at least for quark jets, but in practice more often inclusive-jet and other multi-jet processes (such as $W/Z+$jets) in hadron colliders are used. The modeling of this aspect can have important effects on specific observables, a recent example being the $t\bar{t}$ forward-backward asymmetry at the Tevatron [143].

5) Underlying Event: Good constraints on the overall level of the underlying event can be obtained by counting the summed transverse energy (more IR safe) and/or particle multiplicities and average transverse momenta (more IR sensitive) in regions *transverse* to a hard trigger jet (more IR safe) or particle (more IR sensitive), see e.g. [138]. Constraints on the *fluctuations* of the underlying event are also important, and can be obtained, e.g., by comparing to measurements of the RMS of such distributions [137]. Again, note that the UE is sensitive to the choice of PDFs [144].

6) Color (Re-)Connections and other Final-State Interactions: By Final-State Interactions, we intend a broad spectrum of possible collective effects that may be included to a greater or lesser extent in various models. These effects include: Bose-Einstein correlations (see, e.g., [145]), rescattering (see, e.g., [146]), color reconnections / string interactions (see, e.g., [147–149]), hydrodynamics (see, e.g., [150]), etc. As a rule, these effects are soft and/or non-perturbative and hence should not modify hard IR safe observables appreciably. They can, however, have *drastic* effects on IR sensitive ones, such as particle multiplicities, momentum distributions, and correlations, wherefore useful constraints are typically furnished by measurements of spectra and correlations as functions of quantities believed to serve as indicators of the strength of these phenomena (such as event multiplicity), and/or by collective-flow-type measurements. Finally, if the model includes a universal description of underlying event and soft-inclusive QCD, as many MPI-based models do, then minimum-bias data can also be used as a control sample, though one must then be careful either to address diffractive contributions properly or to include only data samples that minimize their impact.

7) Beam Remnants: Constraints on beam remnant fragmentation (see, e.g., [130]) are most easily obtained in the forward region, but, e.g., the amount of baryon transport from the remnant to a given rapidity region can also be used to probe how much the color structure of the remnant was effectively disturbed, with more baryon transport indicating a larger amount of "beam baryon blowup".

We round off by emphasizing that comparisons of specific models and tunes to data can be useful both as immediate tests of commonly used models, and to illustrate the current amount of theoretical uncertainty surrounding a particular distribution. Independently of how well the models fit the data, such comparisons also provide a set of well-defined theoretical reference curves that serve as useful guidelines for future studies. However, the conclusions that can be drawn from comparisons of individual tunes of specific models on single distributions are necessarily limited. In order to obtain more general conclusions, a strategy for a more coherent and over-arching look at both the data and the models was recently proposed in [144]. Specifically, rather than performing one global tune to all the data, as is usually done, a more systematic check on the validity of the underlying physics model can be obtained by instead performing several independent optimizations of the model parameters for a range of different phase-space windows and/or collider environments. In regions in which consistent parameter sets are obtained, with predictions that are acceptably close to the data, the underlying model can then be considered as interpolating well, i.e., it is universal. If not, a breakdown in the ability of the model ability to span different physical regimes has been identified, and can be addressed, with the nature of the deviations giving clues as to the nature of the breakdown. With the advent of automated tools making it easier to run several optimizations without much additional computing overhead, such systematic studies are now becoming feasible, with a first example given in [144].

References

[1] S. Drell and T.-M. Yan, *Phys. Rev. Lett.* **25**, 316 (1970).
[2] R. Ellis, W. Stirling and B. Webber, *QCD and collider physics* (Camb. Monogr. Part. Phys. Nucl. Phys. Cosmol., 1996).
[3] M. E. Peskin and D. V. Schroeder, *An Introduction to quantum field theory* (Addison-Wesley, 1995).
[4] T. Plehn, D. Rainwater and P. Z. Skands, *Phys. Lett.* **B645**, 217 (2007).
[5] J. Alwall, S. de Visscher and F. Maltoni, *JHEP* **0902**, p. 017 (2009).
[6] A. Papaefstathiou and B. Webber, *JHEP* **06**, p. 069 (2009).
[7] D. Krohn, L. Randall and L.-T. Wang (2011).
[8] C. K. Vermilion (2011).
[9] A. Altheimer, S. Arora, L. Asquith, G. Brooijmans, J. Butterworth *et al.*, *J. Phys. G* **G39**, p. 063001 (2012).

[10] J. Beringer *et al.*, *Phys. Rev.* **D86**, p. 010001 (2012).
[11] G. Dissertori, I. Knowles and M. Schmelling, *Quantum Chromodynamics — High energy experiments and theory* (Oxford Science Publications, 2003).
[12] S. P. Martin (1997).
[13] S. Bethke, *Eur. Phys. J.* **C64**, 689 (2009).
[14] G. Dissertori, A. Gehrmann-De Ridder, T. Gehrmann, E. Glover, G. Heinrich *et al.*, *JHEP* **0908**, p. 036 (2009).
[15] J. Alwall, M. Herquet, F. Maltoni, O. Mattelaer and T. Stelzer, *JHEP* **1106**, p. 128 (2011).
[16] A. Pukhov (2004).
[17] E. Boos *et al.*, *Nucl. Instrum. Meth.* **A534**, 250 (2004).
[18] A. Kanaki and C. G. Papadopoulos, *Comput. Phys. Commun.* **132**, 306 (2000).
[19] F. Krauss, R. Kuhn and G. Soff, *JHEP* **0202**, p. 044 (2002).
[20] M. Moretti, T. Ohl and J. Reuter (2001).
[21] W. Kilian, T. Ohl and J. Reuter, *Eur. Phys. J.* **C71**, p. 1742 (2011).
[22] A. Cafarella, C. G. Papadopoulos and M. Worek, *Comput. Phys. Commun.* **180**, 1941 (2009).
[23] M. Bähr, S. Gieseke, M. Gigg, D. Grellscheid, K. Hamilton *et al.*, *Eur. Phys. J.* **C58**, 639 (2008).
[24] T. Gleisberg and S. Hoeche, *JHEP* **0812**, p. 039 (2008).
[25] T. Sjöstrand (2006).
[26] J. C. Collins and D. E. Soper, *Nucl. Phys.* **B194**, p. 445 (1982).
[27] G. F. Sterman (1995).
[28] R. Brock *et al.*, *Rev. Mod. Phys.* **67**, 157 (1995).
[29] T. Plehn, 125 (2008).
[30] J. C. Collins, D. E. Soper and G. F. Sterman, *Nucl. Phys.* **B250**, p. 199 (1985).
[31] G. Altarelli and G. Parisi, *Nucl. Phys.* **B126**, 298 (1977).
[32] D. Mason *et al.*, *Phys. Rev. Lett.* **99**, p. 192001 (2007).
[33] A. Cooper-Sarkar (2012).
[34] S. Alekhin, K. Melnikov and F. Petriello, *Phys. Rev.* **D74**, p. 054033 (2006).
[35] E. de Oliveira, A. Martin and M. Ryskin (2012).
[36] S. Alekhin, S. Alioli, R. D. Ball, V. Bertone, J. Blumlein *et al.* (2011).
[37] A. Buckley and M. Whalley (2010), http://hepdata.cedar.ac.uk.
[38] A. Martin, W. Stirling, R. Thorne and G. Watt, *Eur. Phys. J.* **C63**, 189 (2009).
[39] J. Pumplin, D. Stump, J. Huston, H. Lai, P. M. Nadolsky *et al.*, *JHEP* **0207**, p. 012 (2002).
[40] G. Watt and R. Thorne (2012).
[41] M. Bengtsson, T. Sjöstrand and M. van Zijl, *Z. Phys.* **C32**, p. 67 (1986).
[42] S. Gieseke, *JHEP* **0501**, p. 058 (2005).
[43] Z. Bern *et al.*, p. 83 (2008).
[44] T. Kinoshita, *J. Math. Phys.* **3**, 650 (1962).
[45] T. D. Lee and M. Nauenberg, *Phys. Rev.* **133**, B1549 (1964).
[46] G. P. Salam (2010).

[47] S. Catani and M. H. Seymour, *Nucl. Phys.* **B485**, 291 (1997).
[48] S. Catani and M. H. Seymour, *Phys. Lett.* **B378**, 287 (1996).
[49] Z. Nagy, *Phys. Rev.* **D68**, p. 094002 (2003).
[50] R. Frederix, T. Gehrmann and N. Greiner, *JHEP* **09**, p. 122 (2008).
[51] D. A. Kosower, *Phys. Rev.* **D57**, 5410 (1998).
[52] D. A. Kosower, *Phys. Rev.* **D71**, p. 045016 (2005).
[53] A. Gehrmann-De Ridder, T. Gehrmann and E. W. N. Glover, *JHEP* **09**, p. 056 (2005).
[54] A. Gehrmann-De Ridder, T. Gehrmann, E. Glover and G. Heinrich, *JHEP* **0712**, p. 094 (2007).
[55] S. Weinzierl, *Phys. Rev. Lett.* **101**, p. 162001 (2008).
[56] G. Heinrich, *Int.J. Mod. Phys.* **A23**, 1457 (2008).
[57] R. Boughezal, K. Melnikov and F. Petriello, *Phys. Rev.* **D85**, p. 034025 (2012).
[58] S. Catani and M. Grazzini, *Phys. Rev. Lett.* **98**, p. 222002 (2007).
[59] T. Sjöstrand, S. Mrenna and P. Skands, *Comput. Phys. Commun.* **178**, 852 (2008).
[60] G. Gustafson and U. Pettersson, *Nucl. Phys.* **B306**, p. 746 (1988).
[61] A. Gehrmann-De Ridder and M. Ritzmann, *JHEP* **0907**, p. 041 (2009).
[62] A. Gehrmann-De Ridder, M. Ritzmann and P. Skands, *Phys. Rev.* **D85**, p. 014013 (2012).
[63] A. J. Larkoski and M. E. Peskin, *Phys. Rev.* **D81**, p. 054010 (2010).
[64] W. Giele, D. Kosower and P. Skands, *Phys. Rev.* **D84**, p. 054003 (2011).
[65] J. Lopez-Villarejo and P. Skands, *JHEP* **1111**, p. 150 (2011).
[66] A. Banfi, G. P. Salam and G. Zanderighi, *JHEP* **1006**, p. 038 (2010).
[67] V. Khachatryan *et al.* (2011).
[68] G. Aad *et al.* (2012).
[69] K. Wraight and P. Skands, *Eur. Phys. J.* **C71**, p. 1628 (2011).
[70] M. Cacciari, G. P. Salam and G. Soyez, *JHEP* **0804**, p. 063 (2008).
[71] A. Abdesselam, E. B. Kuutmann, U. Bitenc, G. Brooijmans, J. Butterworth *et al.*, *Eur. Phys. J.* **C71**, p. 1661 (2011).
[72] M. Cacciari, G. P. Salam and S. Sapeta, *JHEP* **1004**, p. 065 (2010).
[73] E. Boos *et al.* (2001).
[74] J. Alwall *et al.*, *Comput. Phys. Commun.* **176**, 300 (2007).
[75] A. Buckley, J. Butterworth, S. Gieseke, D. Grellscheid, S. Höche *et al.*, *Phys. Rept.* **504**, 145 (2011).
[76] T. Sjöstrand (2009).
[77] R. Kleiss, W. Stirling and S. Ellis, *Comput. Phys. Commun.* **40**, p. 359 (1986).
[78] G. Lepage, *J. Comput. Phys.* **27**, p. 192 (1978), Revised version.
[79] G. P. Lepage (1980), CLNS-80/447.
[80] P. D. Draggiotis, A. van Hameren and R. Kleiss, *Phys. Lett.* **B483**, 124 (2000).
[81] F. James, *Rept. Prog. Phys.* **43**, p. 1145 (1980).
[82] G. Marchesini and B. Webber, *Nucl. Phys.* **B238**, p. 1 (1984).
[83] G. Marchesini and B. R. Webber, *Nucl. Phys.* **B310**, p. 461 (1988).

[84] S. Gieseke, P. Stephens and B. Webber, *JHEP* **12**, p. 045 (2003).
[85] M. Bengtsson and T. Sjöstrand, *Nucl. Phys.* **B289**, p. 810 (1987).
[86] L. Lönnblad, *Comput. Phys. Commun.* **71**, 15 (1992).
[87] Z. Nagy and D. E. Soper, *JHEP* **10**, p. 024 (2005).
[88] S. Schumann and F. Krauss, *JHEP* **0803**, p. 038 (2008).
[89] W. T. Giele, D. A. Kosower and P. Z. Skands, *Phys. Rev.* **D78**, p. 014026 (2008).
[90] T. Sjöstrand and P. Z. Skands, *Eur. Phys. J.* **C39**, 129 (2005).
[91] J. C. Collins, *Nucl. Phys.* **B304**, p. 794 (1988).
[92] I. Knowles, *Nucl. Phys.* **B310**, p. 571 (1988).
[93] P. Richardson, *JHEP* **11**, p. 029 (2001).
[94] S. Catani, B. R. Webber and G. Marchesini, *Nucl. Phys.* **B349**, 635 (1991).
[95] M. A. Gigg and P. Richardson (2008).
[96] M. H. Seymour, *Phys. Lett.* **B354**, 409 (1995).
[97] T. Stelzer and S. Willenbrock, *Phys. Lett.* **B374**, 169 (1996).
[98] S. J. Parke and Y. Shadmi, *Phys. Lett.* **B387**, 199 (1996).
[99] J. M. Smillie and B. R. Webber, *JHEP* **0510**, p. 069 (2005).
[100] G. Corcella *et al.*, *JHEP* **01**, p. 010 (2001).
[101] M. H. Seymour, *Nucl. Phys.* **B436**, 443 (1995).
[102] M. H. Seymour, *Comp. Phys. Commun.* **90**, 95 (1995).
[103] S. Catani, F. Krauss, R. Kuhn and B. R. Webber, *JHEP* **11**, p. 063 (2001).
[104] L. Lönnblad, *JHEP* **05**, p. 046 (2002).
[105] N. Lavesson and L. Lönnblad, *JHEP* **07**, p. 054 (2005).
[106] M. L. Mangano, M. Moretti, F. Piccinini and M. Treccani, *JHEP* **01**, p. 013 (2007).
[107] S. Mrenna and P. Richardson, *JHEP* **05**, p. 040 (2004).
[108] B. Cooper, J. Katzy, M. Mangano, A. Messina, L. Mijovic *et al.* (2011).
[109] L. Lonnblad and S. Prestel, *JHEP* **1203**, p. 019 (2012).
[110] N. Lavesson and L. Lonnblad, *JHEP* **0812**, p. 070 (2008).
[111] T. Gehrmann, S. Hoeche, F. Krauss, M. Schonherr and F. Siegert (2012).
[112] T. Binoth, F. Boudjema, G. Dissertori, A. Lazopoulos, A. Denner *et al.*, *Comput. Phys. Commun.* **181**, 1612 (2010).
[113] S. Frixione and B. R. Webber, *JHEP* **06**, p. 029 (2002).
[114] S. Frixione, P. Nason and B. R. Webber, *JHEP* **08**, p. 007 (2003).
[115] S. Frixione and B. R. Webber (2008).
[116] S. Frixione, P. Nason and C. Oleari, *JHEP* **11**, p. 070 (2007).
[117] K. Hamilton and P. Nason, *JHEP* **06**, p. 039 (2010).
[118] M. Bengtsson and T. Sjöstrand, *Phys. Lett.* **B185**, p. 435 (1987).
[119] T. Sjöstrand, S. Mrenna and P. Skands, *JHEP* **05**, p. 026 (2006).
[120] S. Alioli, P. Nason, C. Oleari and E. Re, *JHEP* **06**, p. 043 (2010).
[121] P. Z. Skands (2011), In 18th European school on high-energy physics, ESHEP 2010, Raseborg, Finland, 20 Jun. - 3 Jul., 2010, arXiv:1202.1629.
[122] D. Amati and G. Veneziano, *Phys. Lett.* **B83**, p. 87 (1979).
[123] S. Navin (2010).
[124] X. Artru and G. Mennessier, *Nucl. Phys.* **B70**, 93 (1974).

[125] B. Andersson, *The Lund model* (Camb. Monogr. Part. Phys. Nucl. Phys. Cosmol., 1997).
[126] B. Andersson, G. Gustafson and T. Sjöstrand, *Nucl. Phys.* **B197**, p. 45 (1982).
[127] B. Andersson, G. Gustafson and T. Sjostrand, *Phys. Scripta* **32**, p. 574 (1985).
[128] P. Eden and G. Gustafson, *Z. Phys.* **C75**, 41 (1997).
[129] T. Sjöstrand and P. Z. Skands, *Nucl. Phys.* **B659**, p. 243 (2003).
[130] T. Sjöstrand and P. Z. Skands, *JHEP* **03**, p. 053 (2004).
[131] K. G. Wilson, *Phys. Rev.* **D10**, 2445 (1974).
[132] M. G. Bowler, *Z. Phys.* **C11**, p. 169 (1981).
[133] B. Blok, Y. Dokshitzer, L. Frankfurt and M. Strikman, *Phys. Rev.* **D83**, p. 071501 (2011).
[134] B. Blok, Y. Dokshitser, L. Frankfurt and M. Strikman, *Eur. Phys. J.* **C72**, p. 1963 (2012).
[135] J. R. Gaunt and W. J. Stirling (2012).
[136] A. V. Manohar and W. J. Waalewijn, *Phys. Rev.* **D85**, p. 114009 (2012).
[137] G. Aad *et al.*, *Phys. Rev.* **D83**, p. 112001 (2011).
[138] R. Field, *Acta Phys. Polon.* **B42**, 2631 (2011).
[139] A. Buckley, H. Hoeth, H. Lacker, H. Schulz and E. von Seggern, 267 (2009).
[140] A. Buckley, H. Hoeth, H. Lacker, H. Schulz and J. E. von Seggern, *Eur. Phys. J.* **C65**, 331 (2010).
[141] P. Z. Skands, *Phys. Rev.* **D82**, p. 074018 (2010).
[142] P. Richardson and D. Winn (2012).
[143] P. Z. Skands, B. R. Webber and J. Winter (2012).
[144] H. Schulz and P. Skands, *Eur. Phys. J.* **C71**, p. 1644 (2011).
[145] L. Lönnblad and T. Sjöstrand, *Eur. Phys. J.* **C2**, 165 (1998).
[146] R. Corke and T. Sjöstrand, *JHEP* **01**, p. 035 (2009).
[147] J. Rathsman, *Phys. Lett.* **B452**, 364 (1999).
[148] P. Z. Skands and D. Wicke, *Eur. Phys. J.* **C52**, 133 (2007).
[149] S. Gieseke, C. Rohr and A. Siodmok (2012).
[150] K. Werner, I. Karpenko and T. Pierog, *Acta Phys. Polon. Supp.* **4**, 629 (2011).

Chapter 9

Super-Tricks for Superspace

D. Bertolini, J. Thaler, and Z. Thomas

Center for Theoretical Physics, Massachusetts Institute of Technology, Cambridge, MA 02139, USA

E-mail: danbert@mit.edu, jthaler@mit.edu, ztt@mit.edu

These lectures from the TASI 2012 summer school outline the basics of supersymmetry (SUSY) in 3+1 dimensions. Starting from a ground-up development of superspace, we develop all of the tools necessary to construct SUSY lagrangians. While aimed at an introductory level, these lectures incorporate a number of "super-tricks" for SUSY aficionados, including SUSY-covariant derivatives, equations of motion in superspace, background field methods, and non-linear realizations of goldstinos.

Contents

1. Introduction

Supersymmetry (SUSY) is a well-motivated extension of the Standard Model (SM), with rich implications for collider physics and cosmology. This, however, will not be the topic of these lectures. At TASI 2012, Pierce and Papucci went into considerable depth on topics of direct phenomenological relevance, including the Minimal SUSY SM (MSSM), SUSY at colliders, and aspects of SUSY model building. Instead, the goal of these lectures is to introduce the basics of SUSY in 3+1 dimensions using the language of superspace [1, 2], an essential tool for any serious student of SUSY. In addition, these lectures will present a number of super-tricks which we have found useful in our own SUSY research.

In general, the role of symmetries is to relate the properties of different physical states. This is so familiar that we often forget how important it is. For example, an electron moving in the y-direction is distinct from an electron moving in the z-direction, and only because of rotational symmetry can we describe both modes in terms of a common electron field. Relativistic quantum field theory exhibits the full Poincaré symmetry, which relates particles traveling with different momenta and joins particles and antiparticles into a common multiplet. By packaging all of the electron creation and annihilation operators into a single quantum field $\Psi(x)$ that lives in space-time, we make the Poincaré symmetry manifest.

The amazing feature of SUSY is that it relates the properties of bosons and fermions. For example, a spin-1/2 fermionic quark field in the SM could have a spin-0 bosonic "squark" superpartner with the same gauge quantum numbers but different spin-statistics. While it is possible to describe quarks and squarks as separate quantum fields with couplings related by SUSY, it is more convenient to describe them as components of a common SUSY multiplet. Superspace introduces an auxiliary fermionic coordinate θ (and its complex conjugate $\overline{\theta}$) such that bosonic and fermionic degrees of freedom can be packaged into a single superfield $\mathbf{\Phi}(x, \theta, \overline{\theta})$, extending space-time to make SUSY manifest.

Whether or not SUSY is actually realized in nature, there are a variety of reasons one might want to relate the properties of bosons and fermions. For physics beyond the SM, SUSY can be used to regulate quantum corrections to the spin-0 Higgs potential. The mass parameter for a spin-1/2 fermion is always protected by a chiral symmetry, whereas the mass parameter for a generic spin-0 boson is quadratically sensitive to high-scale physics. When spin-1/2 and spin-0 fields are part of the same SUSY multiplet, the spin-0 boson inherits the spin-1/2 chiral mass protection. For understanding generic features of quantum field theories, SUSY theories have powerful constraints like holomorphy that make it possible to robustly predict the properties of theories even at strong coupling. Even for weakly-coupled theories, SUSY relates complicated calculations involving fermions to simpler calculations involving scalars.

Beyond the intrinsic motivations for SUSY, these lectures will emphasize how superspace simplifies the construction and manipulation of SUSY lagrangians. As we will see, the structure of $\mathcal{N} = 1$ SUSY can be largely understood from the transformation properties of the superspace coordinates θ and $\overline{\theta}$. We will briefly touch on supergravity (SUGRA) as well.

1.1. *About these lectures*

These lecture notes mimic the original four TASI 2012 lectures, but with some embellishments and corrections compared to what you can find on the online videos. The topics covered are:

- **Introducing Superspace**

- **Fermions and Sfermions**
- **SUSY Gauge Theories**
- **SUSY Breaking and Goldstinos**

For the aficionados, each lecture will feature one super-trick, namely, a technique that can be explained at the introductory level but often features prominently in advanced SUSY research. These tricks are:

- #1: SUSY-covariant derivatives. (Sec. 2.8)
- #2: Equations of motion in superspace. (Sec. 3.7)
- #3: Background field methods. (Sec. 4.6)
- #4: Non-linear goldstino multiplets. (Sec. 5.8)

Of course, superspace itself is the real super-trick, and we hope these lecture notes will help both beginning and advanced students gain familiarity and dexterity with superspace methods.

While writing these lecture notes, we relied on a number of helpful SUSY references, including Wess & Bagger [3], Martin's Supersymmetry Primer [4], Terning's Modern Supersymmetry [5], Luty's 2004 TASI lecture notes [6], and the notes of Dumitrescu and Komargodski [7]. We have also found Refs. 8–15 to be helpful SUSY resources. Unfortunately, each of these references uses a different set of conventions (and the three of us have our own differing preferences on factors of -1, i, 2, and $\sqrt{2}$), but we have tried to make these notes self-consistent.

Compared to standard treatments of SUSY, these lectures are more "ground-up". Instead of starting with the SUSY algebra, we start with the task of simply trying to put bosons and fermions into multiplets. We take the point of view that SUSY should be thought of in the context of effective lagrangians, so we make an effort to point out important dimension-5 and dimension-6 interactions that are generically present. When talking about SUSY breaking, instead of talking about specific SUSY-breaking models, we focus on the special role of the goldstino.

As a disclaimer, here are some topics that we will *not* tell you about in these lectures. We will not tell you how SUSY can solve the gauge hierarchy problem.[a] We will not tell you how SUSY naturally furnishes a dark matter candidate.[b] We will not discuss the phenomenology of the

[a]In particular, we will not tell you that SUSY ensures that the Higgs mass parameter is only logarithmically sensitive to the scale of SUSY-breaking.

[b]In particular, we will not tell you that in the MSSM a $\mathbf{Z}_2$ symmetry called R-parity is invoked to protect against proton decay, which in turn renders the lightest SUSY particle stable, and thus a dark matter candidate.

MSSM.[c] We will, however, tell you that in order for SUSY to be realized in nature, it must be spontaneously broken in a hidden sector (see Sec. 5.3). And we will pontificate in the conclusion about the status of SUSY after the 8 TeV LHC run.

1.2. *About us*

To give you some context for these lecture notes, we want to briefly say something about our SUSY background.

Jesse wrote his first paper on SUSY in 2005 [16]. Since then, he has found it enormously helpful to have a bag of tricks on hand to simplify SUSY calculations and develop SUSY intuition. The specific tricks included in these lectures were motivated by some of his recent papers.

- *Goldstini* [17, 18] — Goldstini arise when SUSY is broken independently by more than one sector. Deriving the spectrum and couplings of goldstini is made simpler using **non-linear goldstino multiplets**.
- *Flavor Mediation* [19, 20] — The (approximate) global flavor symmetries of the SM can be lifted to (spontaneously broken) gauged flavor symmetries. These flavor gauge groups can mediate SUSY breaking, and the resulting soft spectrum is most easily calculated using **background field methods**.
- *Anomaly Mediation from AdS_4 SUSY* [21] — Anomaly-mediated SUSY breaking arises from uplifting AdS_4 SUSY to flat space broken SUSY. This physics is easiest to understand in the **conformal compensator formalism of SUGRA** (which is unfortunately a super-trick that was not covered in the original TASI 2012 lectures, but is explained fairly concisely in Ref. 22). Various cross-checks are easiest to perform using **SUSY-covariant derivatives**.

Finally, the super-trick of **equations of motion in superspace** is a classic technique for constructing effective SUSY lagrangians.

Daniele and Zach were participants in the TASI 2012 summer school, and graciously agreed to help Jesse write and edit these notes. At TASI, Daniele gave a presentation about his work on visible sector SUSY breaking [23], and Zach gave a presentation about his work on AdS_4 SUSY and anomaly mediation [21].

Without further ado, let us begin our study of superspace.

[c] Among other things, we will not tell you about gauge coupling unification, the need for two Higgs doublets for anomaly cancellation, or sparticle phenomenology at the LHC.

2. Introducing Superspace

In this lecture, we will introduce the motivation and structure of superspace. After a discussion of Weyl and Grassmann notation, we show how translations in superspace can be used to determine the SUSY algebra. En route to constructing a generic SUSY lagrangian, we will introduce our first super-trick involving SUSY-covariant derivatives.

2.1. *Why superspace?*

SUSY relates the properties of bosons and fermions, but in ordinary relativistic quantum field theory, bosons and fermions are represented by very different objects. For example, a spin-0 boson is described by a complex-valued scalar field $\phi(x)$, while a spin-1/2 fermion is described by a Grassmann-valued Weyl field $\psi_\alpha(x)$ (with a Lorentz spinor index, no less). In order to make SUSY manifest, we want to somehow package bosons and fermions into a single object.

To do so, we introduce the technique of superspace, which augments the ordinary space-time coordinates with an additional Grassmann spinor θ^α (and its complex conjugate $\overline{\theta}^{\dot{\alpha}}$):

$$x^\mu \to \{x^\mu, \theta^\alpha, \overline{\theta}^{\dot{\alpha}}\}. \tag{1}$$

A field that depends on $\{x^\mu, \theta^\alpha, \overline{\theta}^{\dot{\alpha}}\}$ is called a superfield,[d] which automatically packages boson and fermion fields into multiplets. While it is possible to describe SUSY theories using ordinary space-time alone, superspace makes it simpler to identify SUSY-invariants and write SUSY lagrangians.

On one level, the Grassmann spinor θ^α simply serves as a placeholder. As an analogy, consider the case of a vector with components v_x, v_y, and v_z. This vector can be expressed either as a list of separate components or as a single vector object with the help of placeholders $\hat{x}$, $\hat{y}$, and $\hat{z}$:

$$\{v_x, v_y, v_z\} \quad \text{vs.} \quad \vec{v} = v_x\hat{x} + v_y\hat{y} + v_z\hat{z}. \tag{2}$$

The second notation allows objects with different formal properties to be summed together into a common object $\vec{v}$. Similarly, given a SUSY multiplet that contains a spin-0 boson ϕ and a spin-1/2 fermion ψ_α, we can write the SUSY multiplet either as a list of components or in a superfield:

$$\{\phi, \psi_\alpha, \ldots\} \quad \text{vs.} \quad \mathbf{\Phi} = \phi + \theta^\alpha\psi_\alpha + \ldots \tag{3}$$

[d] A true superfield must also have well-behaved properties under a SUSY transformation. This is discussed in more detail in Sec. 2.7.

Because both θ^α and ψ_α are Grassmann spinors, the two terms in $\mathbf{\Phi}$ have the same statistics and can therefore be summed together.

This analogy between vectors and superfields goes even a bit deeper. In the case of vectors, one can describe rotations either in terms of active rotations (rotation of components) or passive rotations (rotation of basis vectors). If one knows how $\hat{x}$, $\hat{y}$, and $\hat{z}$ transform under rotations, then one can easily determine how v_x, v_y, and v_z transform. In the case of SUSY, one can describe SUSY transformations in terms of manipulations of θ^α, and we will "derive" the SUSY algebra by considering the possible transformation properties of θ^α.

In terms of how θ^α transforms under SUSY, however, θ^α behaves less like a $\hat{x}$-like unit vector and more like a genuine coordinate. For example, one can consider arbitrary functions of θ^α just like one can have arbitrary functions of x_μ (though there are important differences since θ^α is a Grassmann spinor). In addition, SUSY transformations act like translations on θ^α

$$\theta^\alpha \to \theta^\alpha + \epsilon^\alpha, \tag{4}$$

just as translations on space-time coordinates act as $x^\mu \to x^\mu + \delta^\mu$. We will see explicitly how this works in Sec. 2.5.

2.2. *Invitation to two-component notation*

In these lectures we will use exclusively two-component spinor notation, also known as Weyl spinors. While it is possible to do superspace manipulations using four-component notation (as in Refs. 8, 9), Weyl spinors are far more convenient, since they are true irreducible representations of the Lorentz group. For the uninitiated, we recommend the excellent notes of Dreiner, Haber, and Martin [24]. We will give a quick motivation for why two-component notation is the "natural" notation to use, with the hope that this will give you sufficient motivation to learn this topic on your own. Throughout these notes we use the $(+,-,-,-)$ metric convention.

Consider the following way of packaging the four-vector $x_\mu = \{t, x, y, z\}$ into a two-by-two matrix

$$X = \sigma^\mu x_\mu = \begin{pmatrix} t+z & x-iy \\ x+iy & t-z \end{pmatrix}, \tag{5}$$

where

$$\sigma^\mu_{\alpha\dot{\alpha}} = \{\mathbb{1}, \sigma_1, \sigma_2, \sigma_3\} \tag{6}$$

is defined in terms of the Pauli matrices, and we will explain the reason for the dotted index in a moment. The matrix X is hermitian ($X^\dagger = X$), and its determinant is the familiar Lorentz-invariant norm[e]

$$\det X = t^2 - x^2 - y^2 - z^2 = x_\mu x^\mu. \tag{7}$$

The Lorentz group is the set of transformations that leave $x_\mu x^\mu$ invariant, or equivalently, the set of transformations that leave the determinant of X fixed while keeping $X = X^\dagger$. The most general Lorentz transformation acting on X is matrix multiplication by an arbitrary complex matrix P with determinant 1:

$$X \to P X P^\dagger. \tag{8}$$

Clearly, this transformation leaves X hermitian. Note that

$$\det P X P^\dagger = |\det P|^2 \det X, \tag{9}$$

so $\det P$ could in principle have an arbitrary phase factor, but this phase can be pulled out of the matrix P since it has no effect on the components of X. This (special linear) transformation in Eq. (8) is the group $\mathrm{SL}(2,\mathbb{C})$ which is the (double) covering group of the Lorentz group $\mathrm{SO}(3,1)$.

While Eq. (8) might seem cumbersome for describing the transformation of Lorentz four-vectors, it makes the transformations of spinors manifest. Spinors are simply objects that transform under P as

$$\psi'_\alpha = P_\alpha{}^\beta \psi_\beta, \tag{10}$$

where $\alpha = \{1, 2\}$ is called an undotted index (i.e. corresponding to the $(1/2, 0)$ representation of the Lorentz group). (Anti-)spinors with a dotted index transform in the complex conjugate representation under P^* as

$$\overline{\psi}'_{\dot\alpha} = P^{*\dot\beta}_{\dot\alpha} \overline{\psi}_{\dot\beta} = \overline{\psi}_{\dot\beta} P^{\dagger\dot\beta}{}_{\dot\alpha}, \tag{11}$$

(i.e. the $(0, 1/2)$ representation). We can build arbitrary representations of the Lorentz group by considering objects with different numbers of dotted and undotted indices. Ordinary four-vectors with a μ Lorentz index can be converted into objects with two spinor indices with the help of the σ^μ matrices from Eq. (6):

$$x_{\alpha\dot\beta} \equiv x_\mu \sigma^\mu_{\alpha\dot\beta}. \tag{12}$$

We see immediately that Eqs. (10) and (11) then imply Eq. (8), which is a nice consistency check.

[e]This unambiguously shows that $(+,-,-,-)$ is the "correct" metric signature.

The antisymmetric matrix

$$\epsilon_{\alpha\beta} = \epsilon_{\dot{\alpha}\dot{\beta}} = \begin{pmatrix} 0 & -1 \\ 1 & 0 \end{pmatrix} \tag{13}$$

is invariant under Lorentz transformations, so we can use the $\epsilon_{\alpha\beta}$ (and its inverse $\epsilon^{\alpha\beta}$) to lower (and raise) indices

$$\chi_\alpha = \epsilon_{\alpha\beta}\chi^\beta, \qquad \chi^\alpha = \epsilon^{\alpha\beta}\chi_\beta, \tag{14}$$

and analogously for dotted indices.[f] When raising and lowering indices on the σ^μ matrix, it is convenient to define

$$\overline{\sigma}^{\mu\dot{\alpha}\alpha} \equiv \epsilon^{\dot{\alpha}\dot{\beta}}\epsilon^{\alpha\beta}\sigma^\mu_{\beta\dot{\beta}} = \{\mathbb{1}, -\sigma_1, -\sigma_2, -\sigma_3\}. \tag{15}$$

Because there is a sign flip when raising and lowering indices, we have to define a convention for suppressing indices, and we use descending undotted indices and ascending dotted indices as shown below:

$$\begin{array}{cccc} \alpha & & & \dot{\alpha} \\ & \alpha & \dot{\alpha} & \end{array} \tag{16}$$

The power of two-component notation is that one can write Lorentz-invariant objects simply by matching indices (in particular avoiding the contortions necessary in Dirac notation to write down a Majorana mass). The easiest Lorentz invariants we can build are

$$\psi^\alpha\chi_\alpha \equiv \psi\chi, \qquad \overline{\psi}_{\dot{\alpha}}\overline{\chi}^{\dot{\alpha}} \equiv \overline{\psi}\overline{\chi}. \tag{17}$$

Since fermions anti-commute, we have

$$\psi\chi = \psi^\alpha\chi_\alpha = -\chi_\alpha\psi^\alpha = +\chi^\alpha\psi_\alpha = \chi\psi. \tag{18}$$

However there is an annoying minus sign when manipulating the following Lorentz-covariant combination:

$$\overline{\psi}\overline{\sigma}^\mu\chi = \overline{\psi}_{\dot{\alpha}}\overline{\sigma}^{\mu\dot{\alpha}\beta}\chi_\beta = -\chi^\alpha\sigma^\mu_{\alpha\dot{\beta}}\overline{\psi}^{\dot{\beta}} = -\chi\sigma^\mu\overline{\psi}. \tag{19}$$

It is often convenient to use the notation

$$\sigma^{\mu\nu} \equiv \frac{1}{4}(\sigma^\mu\overline{\sigma}^\nu - \sigma^\nu\overline{\sigma}^\mu), \qquad \overline{\sigma}^{\mu\nu} \equiv \frac{1}{4}(\overline{\sigma}^\mu\sigma^\nu - \overline{\sigma}^\nu\sigma^\mu). \tag{20}$$

[f] For raising and lowering indices on the ϵ tensor itself, note that $\epsilon^{\alpha\beta}\epsilon_{\beta\gamma} = \delta^\alpha{}_\gamma$, but $\epsilon^{\alpha\beta}\epsilon_{\gamma\beta} = -\delta_\gamma{}^\alpha$.

When taking the hermitian conjugate of expressions involving multiple spinors, it is conventional for all the spinors to reverse order. Combined with the hermiticity of the sigma matrices, this implies

$$(\psi\chi)^\dagger = \overline{\chi}\overline{\psi}, \tag{21}$$

$$(\psi\sigma^\mu\overline{\chi})^\dagger = \chi\sigma^\mu\overline{\psi}, \tag{22}$$

$$(\psi\sigma^\mu\overline{\sigma}^\nu\chi)^\dagger = \overline{\chi}\overline{\sigma}^\nu\sigma^\mu\overline{\psi}, \tag{23}$$

and so forth.

One can often simplify complicated expressions using the following Schouten and completeness identities:

$$A_\alpha B_\beta - A_\beta B_\alpha = \epsilon_{\alpha\beta}A^\gamma B_\gamma, \tag{24}$$

$$\sigma_{\mu\alpha\dot{\alpha}}\overline{\sigma}^{\mu\beta\dot{\beta}} = 2\delta_\alpha{}^\beta\delta_{\dot{\alpha}}{}^{\dot{\beta}}. \tag{25}$$

Together, these imply many useful Fierz identities, including

$$(\chi\psi)(\eta\lambda) = -(\chi\eta)(\lambda\psi) - (\chi\lambda)(\psi\eta), \tag{26}$$

$$(\overline{\chi}\overline{\sigma}^\mu\psi)(\eta\sigma_\mu\overline{\lambda}) = -2(\chi^\dagger\lambda^\dagger)(\psi\eta), \tag{27}$$

$$(\theta\chi)(\theta\psi) = -\frac{1}{2}(\theta\theta)(\chi\psi). \tag{28}$$

The sigma matrices satisfy identities reminiscent of the familiar relation $\{\gamma^\mu, \gamma^\nu\} = 2\eta^{\mu\nu}$ of four-component notation, namely

$$\sigma^\mu_{\alpha\dot{\alpha}}\overline{\sigma}^{\nu\dot{\alpha}\beta} + \sigma^\nu_{\alpha\dot{\alpha}}\overline{\sigma}^{\mu\dot{\alpha}\beta} = 2\eta^{\mu\nu}\delta_\alpha{}^\beta, \tag{29}$$

$$\overline{\sigma}^{\mu\dot{\alpha}\alpha}\sigma^\nu_{\alpha\dot{\beta}} + \overline{\sigma}^{\nu\dot{\alpha}\alpha}\sigma^\mu_{\alpha\dot{\beta}} = 2\eta^{\mu\nu}\delta^{\dot{\alpha}}{}_{\dot{\beta}}. \tag{30}$$

Finally, given a Dirac (four-component) spinor, one can decompose it into two Weyl spinors via

$$\Psi_D = \begin{pmatrix} \psi_\alpha \\ \overline{\psi^c}^{\dot{\alpha}} \end{pmatrix}, \tag{31}$$

and gamma matrices can be similarly decomposed

$$\gamma^\mu = \begin{pmatrix} 0 & \sigma^\mu \\ \overline{\sigma}^\mu & 0 \end{pmatrix}, \qquad \gamma_5 = \begin{pmatrix} -1 & 0 \\ 0 & 1 \end{pmatrix}. \tag{32}$$

Note that "c" in ψ^c is just a label (ψ and ψ^c are two separate quantum fields, completely unrelated if one does not invoke equations of motion), while the bar stands for complex conjugation. Both ψ and ψ^c are (left-handed) Weyl spinors.

2.3. *Grassmann coordinates*

Now that we have settled on two-component notation, we can talk about the Grassmann coordinate θ^α (and its complex conjugate $\overline{\theta}^{\dot{\alpha}}$). Here are the basics of Grassmann manipulation.

A generic Grassmann variable η (without a Lorentz index) is an anti-commuting object

$$\{\eta, \eta\} = 0 \quad \Rightarrow \quad \eta^2 = 0. \tag{33}$$

That means an arbitrary function of η can be expressed in terms of its (finite) Taylor expansion as

$$f(\eta) = a + b\eta. \tag{34}$$

There are a variety of methods to extract the components of $f(\eta)$. We will often use the notation $|_0$ to indicate setting all Grassmanns to zero

$$f(\eta)|_0 = a. \tag{35}$$

We define derivatives and integrals as acting the same way on Grassmanns:

$$\frac{\partial}{\partial \eta}(a + b\eta) = b, \qquad \int d\eta\,(a + b\eta) = b. \tag{36}$$

Note that the integral definition implies translational invariance

$$\int d\eta\, f(\eta + \epsilon) = \int d\eta f(\eta), \tag{37}$$

which has been extensively used, for example, in Ref. 18. Also, note that

$$\int d\eta\, \frac{\partial}{\partial \eta}(a + b\eta) = 0, \tag{38}$$

which will allow us to use integration by parts in superspace.

In order to package bosons and fermions into a common multiplet, we need Grassmann variables with spinor indices:

$$\theta_\alpha = \{\theta_1, \theta_2\}, \qquad \overline{\theta}_{\dot{\alpha}} = \{\overline{\theta}_1, \overline{\theta}_2\}. \tag{39}$$

These objects anti-commute with each other such that

$$\theta_1\theta_1 = \theta_2\theta_2 = 0, \qquad \theta_1\theta_2 = -\theta_2\theta_1. \tag{40}$$

We can use the ϵ matrix in Eq. (13) to form Lorentz-invariant combinations of the θs:

$$\theta^2 \equiv \theta^\alpha\theta_\alpha = \epsilon^{\alpha\beta}\theta_\beta\theta_\alpha = -2\theta_1\theta_2, \tag{41}$$

$$\overline{\theta}^2 \equiv \overline{\theta}_{\dot{\alpha}}\overline{\theta}^{\dot{\alpha}} = \epsilon^{\dot{\alpha}\dot{\beta}}\overline{\theta}_{\dot{\alpha}}\overline{\theta}_{\dot{\beta}} = 2\overline{\theta}_1\overline{\theta}_2, \tag{42}$$

$$\theta^4 \equiv \overline{\theta}^2\theta^2. \tag{43}$$

Note that the superscripts in the above relations denote powers, not components. There is a potential factor of 2 confusion that arises from the relation

$$\theta_\alpha \theta_\beta = \frac{1}{2}\epsilon_{\alpha\beta}\theta^2. \tag{44}$$

We can define derivatives and integrals just as in Eq. (36)

$$\frac{\partial}{\partial\theta^\alpha}\theta^\beta = \delta^\beta_\alpha, \qquad \frac{\partial}{\partial\overline{\theta}_{\dot\alpha}}\overline{\theta}_{\dot\beta} = \delta^{\dot\alpha}_{\dot\beta}, \qquad \int d\theta_\alpha (a + b\theta^\beta) = b\delta^\beta_\alpha. \tag{45}$$

We will often use the combinations

$$d^2\theta \equiv -\frac{1}{4}d\theta^\alpha d\theta^\beta \epsilon_{\alpha\beta}, \qquad d^2\overline{\theta} \equiv -\frac{1}{4}d\overline{\theta}_{\dot\alpha} d\overline{\theta}_{\dot\beta}\epsilon^{\dot\alpha\dot\beta}, \qquad d^4\theta \equiv d^2\theta d^2\overline{\theta}, \tag{46}$$

such that

$$\int d^2\theta\, \theta^2 = 1, \qquad \int d^2\overline{\theta}\, \overline{\theta}^2 = 1, \qquad \int d^4\theta\, \theta^4 = 1. \tag{47}$$

With all the notation out of the way, we are ready now to talk about SUSY multiplets.

2.4. *Generic SUSY multiplets*

A generic scalar supermultiplet is

$$\boldsymbol{S}(x^\mu, \theta^\alpha, \overline{\theta}^{\dot\alpha}), \tag{48}$$

which depends on the Grassmann spinor placeholders/coordinates θ^α. Throughout this paper we will use boldface letters to indicate superfields. This object is an overall Lorentz scalar, but it contains spin-0, spin-1/2, and spin-1 components. Because of the Grassmann nature of our placeholders, the Taylor expansion is exact:

$$\begin{aligned} \boldsymbol{S} = \quad & a \quad + \theta\xi \qquad + \theta^2 b \\ & + \overline{\theta}\overline{\chi} + \overline{\theta}\overline{\sigma}^\mu\theta v_\mu + \theta^2\overline{\theta}\overline{\zeta} \\ & + \overline{\theta}^2 c + \overline{\theta}^2\theta\eta \quad + \theta^4 d. \end{aligned} \tag{49}$$

We can express this more visually as

$$\boldsymbol{S} = \begin{array}{c|ccc} & \theta^0 & \theta^1 & \theta^2 \\ \hline \overline{\theta}^0 & a & \xi_\alpha & b \\ \overline{\theta}^1 & \overline{\chi}^{\dot\alpha} & v_\mu & \overline{\zeta}^{\dot\alpha} \\ \overline{\theta}^2 & c & \eta_\alpha & d \end{array} \tag{50}$$

where a, b, c, d are complex scalars, ξ, χ, ζ, η are Weyl spinors, v_μ is a complex vector, and θ^n and $\overline{\theta}^n$ stand for n powers of θ or $\overline{\theta}$. Note that $\boldsymbol{S}$ contains exactly eight (complex) fermionic and eight (complex) bosonic degrees of freedom. One might think that the superfield could include additional objects like $\theta_\alpha \theta_\beta z^{\alpha\beta}$ but one can easily show that such a term is equivalent to $\theta^2 b$ by using Eq. (44). We use the language "lower" ("higher") to refer to components that involve fewer (more) factors of θ or $\overline{\theta}$, such that a (d) is the lowest (highest) component of $\boldsymbol{S}$.

The above discussion is for a Lorentz-scalar superfield, but superfields can carry additional Lorentz structure. Indeed, there is no problem with

$$\boldsymbol{S}_\nu(x^\mu, \theta^\alpha, \overline{\theta}^{\dot{\alpha}}), \tag{51}$$

as long as the components of the superfields have extra Lorentz indices as well. When the superfield has a spinor index, this can lead to terms where the indices on θ are uncontracted due to reducible Lorentz structure. For example, the $\xi_{\alpha\beta}$ component of a superfield $\boldsymbol{S}_\beta$ could contain $\omega\epsilon_{\alpha\beta}$, leading to an expansion like

$$\boldsymbol{S}_\beta = \lambda_\beta - \omega\theta_\beta + \ldots \tag{52}$$

When discussing gauge theories in Sec. 4, we will make regular use of superfields with spinor indices.

In these notes, we will restrict ourselves to one set of Grassmann coordinates $\{\theta^\alpha, \overline{\theta}^{\dot{\alpha}}\}$, which is known as $\mathcal{N} = 1$ SUSY. Adding more sets of θ yields higher $\mathcal{N}$ SUSY, which will not be covered here.

2.5. *Translations in superspace*

In the previous subsection, we successfully packaged bosons and fermions into a common multiplet. However, as we will see in Sec. 2.7, there is a crucial difference between a package of bosons/fermions and a true superfield, which is a package of bosons/fermions that have the correct transformation properties under SUSY. This is the same distinction between a list (i.e. a collection of components) and a vector (i.e. a collection of components that transform into each other in a certain way under rotations). The real meat of SUSY is in how the components of a supermultiplet are related to each other, and we will now "derive" SUSY by considering translations in superspace.

In ordinary space-time, translational invariance

$$x^\mu \to x^\mu + \delta^\mu \tag{53}$$

is an important subset of the full Poincaré space-time symmetry. Using the Taylor expansion for infinitesimal δ^μ, this translation acts on fields as:

$$\phi(x^\mu) \to \phi(x^\mu + \delta^\mu) \tag{54}$$
$$= \phi(x^\mu) + \delta^\mu \partial_\mu \phi(x^\mu), \tag{55}$$

where we are using the notation $\partial_\mu \equiv \frac{\partial}{\partial x^\mu}$.

It is natural to define translational invariance in superspace via

$$\theta^\alpha \to \theta^\alpha + \epsilon^\alpha, \tag{56}$$

where ϵ^α is an infinitesimal Grassmann parameter. This (passive) transformation of the coordinate can be interpreted instead as an (active) transformation of the components. For example, starting with

$$\mathbf{\Phi}(\theta^\alpha) = \phi + \theta^\alpha \psi_\alpha + \ldots, \tag{57}$$

translations yield

$$\mathbf{\Phi}(\theta^\alpha + \epsilon^\alpha) = \phi + (\theta^\alpha + \epsilon^\alpha)\psi_\alpha + \ldots \tag{58}$$
$$= (\phi + \epsilon^\alpha \psi_\alpha) + \theta^\alpha \psi_\alpha + \ldots \tag{59}$$

so the components transform as $\phi \to \phi + \epsilon^\alpha \psi_\alpha$ (with ψ_α left fixed). As desired, we have related bosons to fermions! However, we know that boson and fermion kinetic terms have differing numbers of derivatives, so in order to successfully build a SUSY lagrangian, we must somehow combine θ^α translations with space-time derivatives.

The key to SUSY is that the shift of the fermionic coordinate θ^α is accompanied by a translation of the ordinary bosonic coordinate x_μ as well

$$\begin{aligned} \theta^\alpha &\to \theta^\alpha + \epsilon^\alpha, \\ \overline{\theta}^{\dot\alpha} &\to \overline{\theta}^{\dot\alpha} + \overline{\epsilon}^{\dot\alpha}, \\ x^\mu &\to x^\mu + \Delta^\mu, \end{aligned} \tag{60}$$

where

$$\Delta^\mu \equiv i\epsilon\sigma^\mu\overline{\theta} + i\overline{\epsilon}\overline{\sigma}^\mu\theta. \tag{61}$$

We could have guessed the form of Δ^μ, since this is the unique real four-vector object one can build that is linear in ϵ and has the right dimension.[g]

[g]Note that from Eq. (57), θ (and thus ϵ) must have mass dimension $[\theta] = -1/2$.

Let us now act this SUSY coordinate shift on our generic supermultiplet from Eq. (49):

$$\begin{aligned} \boldsymbol{S}(x^\mu, \theta^\alpha, \overline{\theta}^{\dot{\alpha}}) &\to \boldsymbol{S}(x^\mu + \Delta x^\mu, \theta^\alpha + \epsilon^\alpha, \overline{\theta}^{\dot{\alpha}} + \overline{\epsilon}^{\dot{\alpha}}) \\ &= \boldsymbol{S}(x^\mu, \theta^\alpha, \overline{\theta}^{\dot{\alpha}}) \\ &\quad + \left(\Delta^\mu \partial_\mu + \epsilon^\alpha \partial_\alpha + \overline{\epsilon}_{\dot{\alpha}} \overline{\partial}^{\dot{\alpha}}\right) \boldsymbol{S}(x^\mu, \theta^\alpha, \overline{\theta}^{\dot{\alpha}}), \end{aligned} \tag{62}$$

where we have used the Taylor expansion up to the first order, both for ordinary and Grassmann coordinates. Here, we are using the notation $\partial_\alpha \equiv \frac{\partial}{\partial \theta^\alpha}$ and $\overline{\partial}^{\dot{\alpha}} \equiv \frac{\partial}{\partial \overline{\theta}_{\dot{\alpha}}}$.

We see immediately that translations in superspace act in two different ways. First, the shift $\theta^\alpha \to \theta^\alpha + \epsilon^\alpha$ relates higher components of the superfield to lower components as in Eq. (58). Second, because Δ^μ contains factors of θ^α, the translation $x^\mu \to x^\mu + \Delta^\mu$ relates *derivatives* of lower components to higher components. This is crucial for relating the kinetic terms for bosons and fermions.

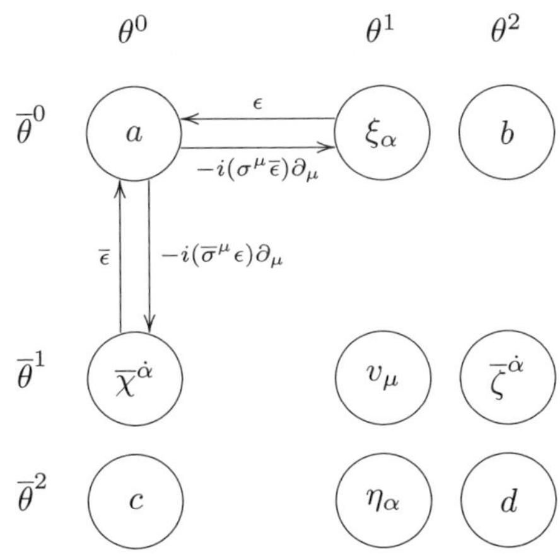

Fig. 1. Visual representation of superspace translations acting on components.

Visually, the SUSY translations act on components as in Fig. 1. Getting all the factors of two and minus signs correct in the component transformations is an exercise best done in the woodshed. The answers are

$$\begin{aligned}
\delta a &= \xi\epsilon + \overline{\chi\epsilon}, \\
\delta\xi &= 2b\epsilon - (v_\mu + i\partial_\mu a)\sigma^\mu\overline{\epsilon}, \\
\delta b &= \overline{\zeta}\overline{\epsilon} + \tfrac{i}{2}\partial_\mu \xi\sigma^\mu\overline{\epsilon}, \\
\delta\overline{\chi} &= 2c\overline{\epsilon} + (v_\mu - i\partial_\mu a)\overline{\sigma}^\mu\epsilon, \\
\delta v_\mu &= \overline{\zeta}\overline{\sigma}_\mu\epsilon - \eta\sigma_\mu\overline{\epsilon} + \tfrac{i}{2}\partial_\nu\left(\xi\sigma_\mu\overline{\sigma}^\nu\epsilon - \overline{\chi}\overline{\sigma}_\mu\sigma^\nu\overline{\epsilon}\right), \\
\delta\overline{\zeta} &= 2d\overline{\epsilon} - i\partial_\mu b\overline{\sigma}^\mu\epsilon + \tfrac{i}{2}\partial_\mu v_\nu\overline{\sigma}^\nu\sigma^\mu\overline{\epsilon}, \\
\delta c &= \epsilon\eta + \tfrac{i}{2}\partial_\mu\overline{\chi}\overline{\sigma}^\mu\epsilon, \\
\delta\eta &= 2d\epsilon - i\partial_\mu c\sigma^\mu\overline{\epsilon} - \tfrac{i}{2}\partial_\mu v_\nu\sigma^\nu\overline{\sigma}^\mu\epsilon, \\
\delta d &= \tfrac{i}{2}\partial_\mu\left(\overline{\zeta}\overline{\sigma}^\mu\epsilon + \eta\sigma^\mu\overline{\epsilon}\right),
\end{aligned} \tag{63}$$

where we have extensively used the Fierz identities of Eqs. (27) and (28) to simplify expressions like $(\overline{\theta}\overline{\sigma}^\mu\theta)(\psi\sigma_\mu\overline{\chi})$ and $(\overline{\psi}\overline{\sigma}^\mu\theta)(\overline{\theta}\overline{\sigma}^\nu\theta)$.

We see now very clearly that SUSY "rotates" bosons into fermions, and vice versa. Note that the highest (θ^4) component d transforms as a total derivative under SUSY, which will be particularly useful for building lagrangians in Sec. 2.9.

2.6. *The SUSY algebra*

Thus far, we have talked about SUSY transformations without ever mentioning the underlying SUSY algebra. Indeed, one advantage of superspace is that Eq. (60) contains the full structure of SUSY. However, it is instructive to turn the superspace translation picture into an operator picture to show that the SUSY algebra closes.

Recall that ordinary space-time translations are generated by the energy-momentum operator

$$e^{ia_\mu P^\mu} f(x^\mu) = f(x^\mu + a^\mu), \tag{64}$$

where

$$P_\mu \equiv -i\partial_\mu. \tag{65}$$

Translations are part of the full Poincaré group that includes Lorentz transformations generated by $M_{\mu\nu}$, with algebra

$$[M_{\mu\nu}, M_{\rho\tau}] = i\left(\eta_{\nu\rho}M_{\mu\tau} + \eta_{\mu\tau}M_{\nu\rho} - \eta_{\mu\rho}M_{\nu\tau} - \eta_{\nu\tau}M_{\mu\rho}\right), \tag{66}$$

$$[P_\mu, M_{\nu\rho}] = i\left(\eta_{\mu\nu}P_\rho - \eta_{\mu\rho}P_\nu\right), \tag{67}$$

$$[P_\mu, P_\nu] = 0. \tag{68}$$

Note that the explicit expression of $M_{\mu\nu}$ depends on the spin of the field it acts on. For a scalar field, for example,

$$M_{\mu\nu} = i\left(x_\mu \partial_\nu - x_\nu \partial_\mu\right). \tag{69}$$

We want to introduce new SUSY generators that implement Eq. (62), namely operators Q and $\overline{Q}$ such that

$$e^{-i\epsilon Q - i\bar{\epsilon}\overline{Q}} \boldsymbol{S}(x^\mu, \theta^\alpha, \overline{\theta}^{\dot{\alpha}}) = \boldsymbol{S}(x^\mu + \Delta^\mu, \theta^\alpha + \epsilon^\alpha, \overline{\theta}^{\dot{\alpha}} + \bar{\epsilon}^{\dot{\alpha}}). \tag{70}$$

In analogy with Eq. (64), we see immediately that

$$Q_\alpha = i\frac{\partial}{\partial\theta^\alpha} - (\sigma^\mu \overline{\theta})_\alpha \partial_\mu, \tag{71}$$

$$\overline{Q}^{\dot{\alpha}} = i\frac{\partial}{\partial\overline{\theta}_{\dot{\alpha}}} - (\bar{\sigma}^\mu \theta)^{\dot{\alpha}} \partial_\mu. \tag{72}$$

It is straightforward to show that these generators satisfy the SUSY algebra

$$\begin{aligned}
\{Q_\alpha, \overline{Q}_{\dot{\beta}}\} &= -2\sigma^\mu_{\alpha\dot{\beta}} P_\mu, \\
\{Q_\alpha, Q_\beta\} &= \{\overline{Q}_{\dot{\alpha}}, \overline{Q}_{\dot{\beta}}\} = 0, \\
[Q_\alpha, P_\mu] &= \left[\overline{Q}_{\dot{\alpha}}, P_\mu\right] = 0, \\
[M_{\mu\nu}, Q_\alpha] &= i\sigma_{\mu\nu\alpha}{}^{\beta} Q_\beta, \\
\left[M_{\mu\nu}, \overline{Q}^{\dot{\alpha}}\right] &= i\bar{\sigma}_{\mu\nu}{}^{\dot{\alpha}}{}_{\dot{\beta}} \overline{Q}^{\dot{\beta}},
\end{aligned} \tag{73}$$

thus extending the Poincaré algebra. In this way, two SUSY translations are equivalent to one ordinary space-time translation, and we sometimes refer to SUSY as being the "square root" of translations. The non-trivial commutator between SUSY and Lorentz generators just indicates that the SUSY generator is a Lorentz spinor. The SUSY algebra indeed closes, and accounting for the possibility of higher $\mathcal{N}$, one can show that SUSY is the unique extension of the Poincaré algebra. [25]

It is worth mentioning that SUSY can be present in more general space-time geometries. As we will discuss in Sec. 5.11, one can also define a SUSY algebra relevant for anti-de Sitter space (which counterintuitively is highly relevant for describing our approximately flat universe).

2.7. *What is a superfield?*

It is important to make the distinction between a true superfield and a random collection of bosons and fermions. Functions of $\{x^\mu, \theta^\alpha, \overline{\theta}^{\dot{\alpha}}\}$ are superfields, in the sense that they have well-behaved properties under a SUSY transformation:

$$\left\{x^\mu, \theta^\alpha, \overline{\theta}^{\dot{\alpha}}\right\} \to \left\{x^\mu + \Delta^\mu, \theta^\alpha + \epsilon^\alpha, \overline{\theta}^{\dot{\alpha}} + \overline{\epsilon}^{\dot{\alpha}}\right\}. \tag{74}$$

Equivalently, a (Lorentz-scalar) superfield is any object whose components transform as Eq. (63). However, if we take a superfield $\boldsymbol{S}$ and randomly manipulate its components (for example by zeroing out various components), the resulting object is generically *not* a superfield, since its (manipulated) components do not transform as Eq. (63).

Since superfields are the building blocks of SUSY lagrangians, it is important to know which objects are superfields and which are not. Eq. (62) immediately implies that any function of any number of superfields is also a superfield

$$f(\boldsymbol{S_1}(x^\mu, \theta^\alpha, \overline{\theta}^{\dot{\alpha}}), \boldsymbol{S_2}(x^\mu, \theta^\alpha, \overline{\theta}^{\dot{\alpha}}), \ldots), \tag{75}$$

since each argument of the function f has well-defined transformation properties under the shifts in Eq. (60). In particular, the sum of two superfields is also a superfield, as is the product.

Because $[P_\mu, Q_\alpha] = 0$, ordinary space-time derivatives of superfields are also superfields

$$\partial_\mu \boldsymbol{S}, \quad \Box \boldsymbol{S}, \quad \ldots \tag{76}$$

To see this, consider the SUSY transformation from Eq. (70) acting on $\partial_\mu \boldsymbol{S}$:

$$e^{-i\epsilon Q - i\overline{\epsilon}\overline{Q}} \partial_\mu \boldsymbol{S} = \partial_\mu e^{-i\epsilon Q - i\overline{\epsilon}\overline{Q}} \boldsymbol{S}. \tag{77}$$

The overall space-time derivative acts on each of the components, so Eq. (63) is preserved.

However, there are plenty of manipulations of superfields that do not result in another superfield. For example, if we multiply a superfield by a function of x^μ alone

$$f(x_\mu)\boldsymbol{S}, \tag{78}$$

the resulting object is *not* a superfield, because $f(x_\mu)$ does not have any corresponding fermionic components to rotate into. Similarly, we cannot multiply by a generic function of θ^α or $\overline{\theta}^{\dot{\alpha}}$ alone. Relevant for the next subsection is the fact that

$$\frac{\partial}{\partial \theta_\alpha} \boldsymbol{S} \tag{79}$$

is not a superfield. While $\frac{\partial}{\partial\theta_\alpha}\boldsymbol{S}$ is a collection of bosonic and fermionic components, these components do not transform as Eq. (63), which can be readily understood because $\{\frac{\partial}{\partial\theta_\alpha},\overline{Q}^{\dot\alpha}\}\neq 0$ so the equivalent manipulation in Eq. (77) is not possible.

2.8. *Super-trick #1: The SUSY-covariant derivative*

Our first super-trick is to introduce the SUSY-covariant derivative D_α. This operator is incredibly useful for SUSY manipulations, because it acts on a superfield to yield another superfield.

One way to motivate D_α is that we want an object whose lowest component is the second-lowest component of $\boldsymbol{S}$. The natural choice would be $\frac{\partial}{\partial\theta_\alpha}\boldsymbol{S}$, but we argued in Eq. (79) that this was not a superfield. We can, however, construct such an object with the help of the SUSY-covariant derivatives

$$D_\alpha = \frac{\partial}{\partial\theta_\alpha} - i(\sigma^\mu\overline{\theta})_\alpha\partial_\mu, \tag{80}$$

$$\overline{D}^{\dot\alpha} = \frac{\partial}{\partial\overline{\theta}^{\dot\alpha}} - i(\overline{\sigma}^\mu\theta)^{\dot\alpha}\partial_\mu. \tag{81}$$

It is a straightforward exercise to show that these commute with Q_α and $\overline{Q}_{\dot\beta}$. Note also that

$$\{D_\alpha,\overline{D}_{\dot\beta}\} = 2i\sigma^\mu_{\alpha\dot\beta}\partial_\mu. \tag{82}$$

Like any sensible derivative, these obey a Leibniz (product) rule

$$D_\alpha(\boldsymbol{XY}) = (D_\alpha\boldsymbol{X})\boldsymbol{Y} + \boldsymbol{X}(D_\alpha\boldsymbol{Y}). \tag{83}$$

There is one subtlety, however, because the Ds pick up a minus sign if you move them across an odd number of spinor indices:

$$D_\alpha(\boldsymbol{X}_\beta\boldsymbol{Y}) = (D_\alpha\boldsymbol{X}_\beta)\boldsymbol{Y} - \boldsymbol{X}_\beta(D_\alpha\boldsymbol{Y}). \tag{84}$$

Note that $D^3\boldsymbol{S} = \overline{D}^3\boldsymbol{S} = 0$, because $\{D_\alpha, D_\beta\} = 0$.[h]

These SUSY-covariant derivatives are useful in a variety of circumstances, justifying our designation of them as a super-trick. Let us count the ways.

---

[h]This fact is not true in AdS_4 space or SUGRA, which is part of the reason why SUGRA is so complicated.

(1) **Making new superfields.** As advertised, a SUSY-covariant derivative acting on a superfield yields another superfield. Therefore a SUSY lagrangian can depend on

$$\boldsymbol{S}, \quad D_\alpha \boldsymbol{S}, \quad D^2 \boldsymbol{S}, \quad \overline{D}_{\dot{\alpha}} \boldsymbol{S}, \quad \overline{D}^2 \boldsymbol{S}, \quad \text{etc.} \tag{85}$$

Space-time derivatives can also be written using Ds, since

$$\partial^\mu \boldsymbol{S} = -\frac{i}{4}\overline{\sigma}^{\mu\dot{\beta}\alpha}\{D_\alpha, \overline{D}_{\dot{\beta}}\}\boldsymbol{S}. \tag{86}$$

(2) **Extracting components of a superfield**. Using the notation that "$|_0$" stands for the lowest component of a supermultiplet, then

$$\boldsymbol{S}|_0 = a, \quad D_\alpha \boldsymbol{S}|_0 = \xi, \quad -\frac{1}{4}D^2\boldsymbol{S}|_0 = b, \quad \text{etc}\ldots \tag{87}$$

This gives us an alternative way to write the integrals in Eqs. (45), (46) and (47) as

$$\int d^2\theta\, \boldsymbol{S} = -\frac{1}{4}D^2\boldsymbol{S}|_0, \quad \int d^2\overline{\theta}\, \boldsymbol{S} = -\frac{1}{4}\overline{D}^2\boldsymbol{S}|_0. \tag{88}$$

We will make regular use of the equivalence

$$\int d^4\theta\, \boldsymbol{S} = \int d^2\theta \left(-\frac{1}{4}\overline{D}^2\boldsymbol{S}\right) = \frac{1}{16}D^2\overline{D}^2\boldsymbol{S}|_0. \tag{89}$$

(3) **Constructing SUSY-invariant actions**. As a special case of the last point, we can use Ds to identify the highest component of a superfield:[i]

$$\frac{1}{16}D^2\overline{D}^2\boldsymbol{S}|_0 = d. \tag{90}$$

As we showed in Eq. (63) and as we will discuss further in Sec. 2.9, the highest component of a superfield transforms as a total derivative under SUSY. Therefore, for any superfield $\boldsymbol{S}$, we can construct a SUSY-invariant action via

$$\mathcal{S} = \int d^4x\, \frac{1}{16}D^2\overline{D}^2\boldsymbol{S}|_0 + \text{h.c.} = \int d^4x\, d^4\theta\, \boldsymbol{S} + \text{h.c.} \tag{91}$$

(4) **Restricting superfields**. The generic superfield in Eq. (49) is unwieldy for most purposes since it corresponds to a reducible representation of the SUSY algebra. We usually work with irreducible representations, two of which are defined using Ds. The key multiplets

[i]The choice of $D^2\overline{D}^2\boldsymbol{S}$ rather than $\overline{D}^2D^2\boldsymbol{S}$ here (related to our convention for $d^4\theta$) is rather arbitrary. Thankfully, the only differences between the two are total derivative terms, which we will ruthlessly ignore henceforth.

are

$$\text{Chiral superfield}: \quad \overline{D}\mathbf{\Phi} = 0, \tag{92}$$

$$\text{Vector superfield}: \quad \boldsymbol{V} = \boldsymbol{V}^\dagger, \tag{93}$$

$$\text{Linear superfield}: \quad D^2\boldsymbol{L} = 0 \text{ and } \boldsymbol{L} = \boldsymbol{L}^\dagger. \tag{94}$$

Chiral multiplets are the workhorses of $\mathcal{N} = 1$ SUSY and will be the topic of Sec. 3. Vector multiplets are needed to describe SUSY gauge theories, as discussed in Sec. 4. Linear multiplets show up as conserved currents (see Ref. 7, for example), but will not appear again (with one exception) in these lectures.

(5) **Building chiral multiplets**. Given a generic superfield $\boldsymbol{S}$, one can construct a chiral superfield by acting twice with $\overline{D}$:

$$\mathbf{\Phi} = \overline{D}^2\boldsymbol{S}. \tag{95}$$

We immediately see that $\overline{D}\mathbf{\Phi} = 0$ (the defining characteristic of a chiral superfield) because $\overline{D}^3 = 0$.

(6) **Constructing the chiral projector**. Finally, a trick using Ds for true SUSY aficionados. Consider the following operator acting on a generic superfield:

$$-\frac{\overline{D}^2 D^2}{16\Box}\boldsymbol{S}. \tag{96}$$

Because $\overline{D}^3 = 0$, this is guaranteed to be a chiral multiplet (although the object is generically non-local). More interestingly, if we replace $\boldsymbol{S}$ with a chiral multiplet $\mathbf{\Phi}$ that already satisfies $\overline{D}\mathbf{\Phi} = 0$, then

$$-\frac{\overline{D}^2 D^2}{16\Box}\mathbf{\Phi} = \mathbf{\Phi}. \tag{97}$$

To see this, we make repeated use of Eq. (82). The operator $-\frac{\overline{D}^2 D^2}{16\Box}$ is known as a chiral projector, and can be useful when manipulating chiral multiplets.

We will get a chance to use each D super-trick in these notes.

2.9. *SUSY-invariant actions*

Thus far, we have defined superfields as objects living in $\{x^\mu, \theta^\alpha, \overline{\theta}^{\dot{\alpha}}\}$ superspace that transform nicely under Eq. (60). We have seen that products and sums of superfields are superfields, as are space-time and SUSY-covariant derivatives acting on superfields. We are now ready to construct SUSY-invariant actions in terms of superfields.

From Eq. (63), no component of a superfield is invariant under SUSY. For the purposes of defining a SUSY-invariant theory, though, all we need is a lagrangian that transforms as a total derivative under SUSY. Indeed, the highest component of a generic multiplet $\boldsymbol{S}$ shifts as a total derivative,

$$(-i\epsilon Q - i\overline{\epsilon}\overline{Q})\left(\frac{1}{16}D^2\overline{D}^2\boldsymbol{S}\right)\bigg|_0 = \text{total derivative}. \tag{98}$$

This is obvious from Fig. 1, which shows that the highest (θ^4) component only gets contributions from the derivatives of lower components, and it is shown explicitly in Eq. (63).

Because the lagrangian must be hermitian, we can build an arbitrary SUSY lagrangian out of the θ^4 component of a vector multiplet $\boldsymbol{V}$ satisfying $\boldsymbol{V}^\dagger = \boldsymbol{V}$,

$$\mathcal{L} = \frac{1}{16}D^2\overline{D}^2\boldsymbol{V}_{\text{comp}}|_0 = \int d^4\theta\, \boldsymbol{V}_{\text{comp}}, \tag{99}$$

where we have used the equivalence in Eq. (89). The notation $\boldsymbol{V}_{\text{comp}}$ reminds us that the vector multiplet is generically composite, meaning that it consists of products and sums of other superfields.

Certain superfield manipulations will have no effect whatsoever on the action. For example, imagine shifting $\boldsymbol{V}_{\text{comp}}$ by a chiral multiplet $\boldsymbol{\Omega}$,

$$\boldsymbol{V}_{\text{comp}} \to \boldsymbol{V}_{\text{comp}} + \boldsymbol{\Omega} + \boldsymbol{\Omega}^\dagger. \tag{100}$$

This transformation leaves $\boldsymbol{V}_{\text{comp}}$ as a vector multiplet, but because $\overline{D}^{\dot{\alpha}}\boldsymbol{\Omega} = 0$, this shift changes the action by a total derivative. Thus, we can remove any purely chiral plus anti-chiral pieces of $\boldsymbol{V}_{\text{comp}}$.[j]

While it is true that we can build an arbitrary SUSY lagrangian using vector multiplets alone, such constructions are not in general local in superspace. For example, consider a particular non-local choice for $\boldsymbol{V}_{\text{comp}}$

$$\boldsymbol{V}_{\text{comp}} = \frac{1}{4\Box}D^2\boldsymbol{\Phi}_{\text{comp}} + \text{h.c.} \tag{101}$$

where $\boldsymbol{\Phi}_{\text{comp}}$ is a (composite) chiral multiplet. Using our super-trick chiral projector from Eq. (97), the θ^4 component of the first term is

$$\frac{1}{16}D^2\overline{D}^2\frac{1}{4\Box}D^2\boldsymbol{\Phi}_{\text{comp}}|_0 = -\frac{1}{4}D^2\boldsymbol{\Phi}_{\text{comp}}|_0 = \int d^2\theta\, \boldsymbol{\Phi}_{\text{comp}}, \tag{102}$$

[j]This is only true in global SUSY. The equivalent manipulation in SUGRA (forming part of a Kähler transformation) is more complicated.

which is just the θ^2 component of the chiral multiplet $\mathbf{\Phi}_{\text{comp}}$! Indeed, both the θ^4 component of a vector multiplet and the θ^2 component of a chiral multiplet transform as total derivatives under SUSY, so both can be used to build SUSY-invariant actions. (SUSY aficionados should be mildly impressed that were able to deduce this without ever determining the components of a chiral multiplet $\mathbf{\Phi}_{\text{comp}}$.) Note that, as already mentioned, the operator $-\frac{\overline{D}^2 D^2}{16\Box}$ acts on $\mathbf{\Phi}_{\text{comp}}$ as a projector, crucially because $\mathbf{\Phi}_{\text{comp}}$ is chiral. This allows us to rewrite the θ^4 component of this non-local $\boldsymbol{V}_{\text{comp}}$ as a local term.

This saturates all the possibilities for a local action, and the most generic SUSY-invariant action one can build can be written using the lagrangian

$$\mathcal{L} = \int d^4\theta\, \boldsymbol{V}_{\text{comp}} + \left(\int d^2\theta\, \mathbf{\Phi}_{\text{comp}} + \int d^2\overline{\theta}\, \mathbf{\Phi}^\dagger_{\text{comp}} \right), \tag{103}$$

where $\boldsymbol{V}_{\text{comp}} = \boldsymbol{V}^\dagger_{\text{comp}}$ is a (composite) vector superfield and $\mathbf{\Phi}_{\text{comp}}$ is a (composite) chiral superfield satisfying $\overline{D}\mathbf{\Phi}_{\text{comp}} = 0$.

3. Fermions and Sfermions

In this lecture, we construct SUSY lagrangians for chiral multiplets. After introducing the basic properties of chiral multiplets and the free lagrangian for chiral superfields, we use effective field theory power counting to write down the most relevant interactions. We explain the super-trick of equations of motion in superspace, as well as how global symmetries work in SUSY.

3.1. *Chiral multiplets*

In order to write down concrete SUSY lagrangians, it is instructive to restrict our attention to chiral multiplets which satisfy the constraint

$$\overline{D}^{\dot{\alpha}} \mathbf{\Phi} = 0. \tag{104}$$

The complex conjugate equation

$$D^\alpha \mathbf{\Phi}^\dagger = 0 \tag{105}$$

defines an anti-chiral superfield $\mathbf{\Phi}^\dagger$. The physical states contained in a chiral multiplet are a spin-1/2 Weyl fermion and a spin-0 complex scalar (often called a sfermion). Chiral multiplets are needed to describe the quark, lepton, and Higgs sectors of the SUSY SM (and their corresponding squark, slepton, and higgsino superpartners).

One way to automatically construct a chiral multiplet is to introduce a modified space-time coordinate

$$y^\mu \equiv x^\mu + i\overline{\theta}\overline{\sigma}^\mu\theta. \tag{106}$$

A generic chiral superfield can be expressed as an arbitrary function of y^μ and θ_α (but not of $\overline{\theta}^{\dot{\alpha}}$):

$$\mathbf{\Phi}(y^\mu, \theta_\alpha). \tag{107}$$

We can check that this automatically satisfies the constraint $\overline{D}^{\dot{\alpha}}\mathbf{\Phi} = 0$, via

$$\begin{aligned}\overline{D}^{\dot{\alpha}}\mathbf{\Phi}(y^\mu,\theta) &= \left(\frac{\partial}{\partial\overline{\theta}_{\dot{\alpha}}} - i(\overline{\sigma}^\mu\theta)^{\dot{\alpha}}\partial_\mu\right)\mathbf{\Phi}(y^\mu,\theta) \\ &= \left(i(\overline{\sigma}^\mu\theta)^{\dot{\alpha}}\frac{\partial}{\partial y^\mu} - i(\overline{\sigma}^\mu\theta)^{\dot{\alpha}}\frac{\partial}{\partial y^\mu}\right)\mathbf{\Phi}(y^\mu,\theta) \\ &= 0,\end{aligned} \tag{108}$$

recalling that $\mathbf{\Phi}$ has no $\overline{\theta}$ dependence except through y^μ.

A chiral superfield has a compact expression in y-dependent components

$$\mathbf{\Phi}(y^\mu,\theta) = \phi(y) + \sqrt{2}\theta\psi(y) + \theta^2 F(y). \tag{109}$$

We will often represent such chiral superfields schematically as

$$\mathbf{\Phi} = \{\phi, \psi, F\}. \tag{110}$$

This is the smallest representation of SUSY containing a fermion (ψ) and a sfermion (ϕ). The field F is an auxiliary field which will turn out to be non-propagating; its existence is necessary to ensure that there are an equal number of bosonic and fermionic components off-shell. We can extract these components from the superfield by applying SUSY-covariant derivatives:

$$\mathbf{\Phi}|_0 = \phi, \tag{111}$$

$$D_\alpha\mathbf{\Phi}|_0 = \sqrt{2}\psi_\alpha, \tag{112}$$

$$-\frac{1}{4}D^2\mathbf{\Phi}|_0 = F. \tag{113}$$

By making the replacement $y^\mu \to x^\mu + i\overline{\theta}\overline{\sigma}^\mu\theta$ in Eq. (109) and Taylor expanding as in Eq. (54), we can express these as components with ordinary

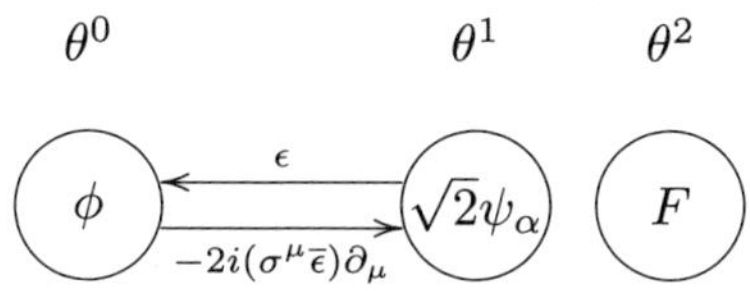

Fig. 2. Visual representation of superspace translations on a chiral multiplet.

x^μ dependence as

$$\mathbf{\Phi} = \begin{array}{c|ccc} & \theta^0 & \theta^1 & \theta^2 \\ \hline \overline{\theta}^0 & \phi & \sqrt{2}\psi_\alpha & F \\ \overline{\theta}^1 & 0 & i\partial_\mu\phi & -\frac{1}{\sqrt{2}}i\left(\overline{\sigma}^\mu\partial_\mu\psi\right)^{\dot{\alpha}} \\ \overline{\theta}^2 & 0 & 0 & -\frac{1}{4}\Box\phi \end{array} \tag{114}$$

However, it is usually more convenient to stick with the y^μ coordinates.

The SUSY transformations of a chiral multiplet follow from the transformations of y^μ and θ^α as derived from Eq. (60):

$$\theta^\alpha \to \theta^\alpha + \epsilon^\alpha, \tag{115}$$

$$y^\mu \to y^\mu + 2i\overline{\epsilon}\overline{\sigma}^\mu\theta. \tag{116}$$

Visually, this transformation acts on the components as in Fig. 2, giving rise to

$$\begin{aligned} \delta\phi &= \sqrt{2}\epsilon\psi, \\ \delta\psi &= -i\sqrt{2}(\sigma^\mu\overline{\epsilon})\partial_\mu\phi + \sqrt{2}\epsilon F, \\ \delta F &= -i\sqrt{2}\overline{\epsilon}\overline{\sigma}^\mu\partial_\mu\psi. \end{aligned} \tag{117}$$

As advertised in Sec. 2.9, the highest component (i.e. F component) of a chiral superfield transforms as a total derivative, and can therefore be used to construct a SUSY-invariant action.

Using the super-trick from Sec. 2.8, we can construct new superfields from chiral superfields by applying SUSY-covariant derivatives. Of course, $\overline{D}^{\dot{\alpha}}\mathbf{\Phi}$ vanishes by definition, but $D_\alpha\mathbf{\Phi}$ and $D^2\mathbf{\Phi}$ do not and can appear in SUSY lagrangians. We will be able to avoid a lot of messy algebra in the future by taking the anti-chiral superfield $\mathbf{\Phi}^\dagger$ and constructing a chiral superfield via

$$-\frac{1}{4}\overline{D}^2\mathbf{\Phi}^\dagger = F^\dagger(y) - i\sqrt{2}\theta\sigma^\mu\partial_\mu\overline{\psi}(y) - \theta^2\Box\phi^\dagger(y). \tag{118}$$

which in components is $-\frac{1}{4}\overline{D}^2\mathbf{\Phi}^\dagger = \{F^\dagger, -i\sigma^\mu\partial_\mu\overline{\psi}, -\Box\phi^\dagger\}$. Crucially, the components are a function of y (not y^*). We see that this is indeed a chiral superfield by recalling that $\overline{D}^3\mathbf{\Phi} = 0$. Also, it is helpful to know that products of chiral multiplets are also chiral

$$\overline{D}^{\dot{\alpha}}(\mathbf{\Phi}_1\mathbf{\Phi}_2\cdots\mathbf{\Phi}_n) = 0, \tag{119}$$

which is easy to prove because $\mathbf{\Phi}_1\mathbf{\Phi}_2\cdots\mathbf{\Phi}_n$ is a function only of y^μ and θ^α, or from the Leibniz rule of Eq. (83).

3.2. *A free SUSY lagrangian*

We are now (finally) ready to construct our first SUSY lagrangian: a free theory of a single chiral superfield. First, let us conduct some dimensional analysis. The lowest component of a chiral multiplet ϕ has mass dimension 1, while the fermionic component has mass dimension 3/2. In order for $\mathbf{\Phi}$ to have a well-defined mass dimension, we must have the following mass dimension assignments:

$$\begin{aligned}
[\mathbf{\Phi}] = [\phi] &= 1,\\
[\psi] &= 3/2,\\
[F] &= 2,\\
[\theta] &= -1/2,\\
[d\theta^\alpha] = [D_\alpha] &= 1/2.
\end{aligned} \tag{120}$$

In particular, $[d^4\theta] = 2$ and $[d^2\theta] = 1$, so in the generic lagrangian

$$\mathcal{L} = \int d^4\theta\, \boldsymbol{V}_{\text{comp}} + \left(\int d^2\theta\, \mathbf{\Phi}_{\text{comp}} + \int d^2\overline{\theta}\, \mathbf{\Phi}^\dagger_{\text{comp}}\right), \tag{121}$$

we must have $[\boldsymbol{V}_{\text{comp}}] = 2$ and $[\mathbf{\Phi}_{\text{comp}}] = 3$.

A free theory is quadratic in fields, so given a single chiral superfield $\mathbf{\Phi}$, the only possible choice for a free lagrangian is

$$\mathcal{L}_{\text{free}} = \int d^4\theta\, \mathbf{\Phi}^\dagger\mathbf{\Phi} + \left(\int d^2\theta\, \frac{1}{2}m\mathbf{\Phi}^2 + \text{h.c.}\right), \tag{122}$$

where m is a mass parameter. The only other possible quadratic term $\int d^4\theta\, \mathbf{\Phi}^2$ does not contribute to the action because it is a total derivative (see Eq. (100)).[k] The first term in Eq. (122) is called the kinetic term and the second term is called the mass term, such that $\mathcal{L}_{\text{free}} = \mathcal{L}_{\text{kinetic}} + \mathcal{L}_{\text{mass}}$.

To check that $\mathcal{L}_{\text{free}}$ corresponds with our expectations, it is instructive to find the action in components. The kinetic term is easiest to analyze by noting that

$$\mathcal{L}_{\text{kinetic}} = \int d^4\theta\, \mathbf{\Phi}^\dagger \mathbf{\Phi} = \int d^2\theta \left(-\frac{1}{4}\overline{D}^2 \mathbf{\Phi}^\dagger\right) \mathbf{\Phi}. \tag{123}$$

We already found the components of the superfield in parentheses in Eq. (118). Finding the component lagrangian is then just a matter of finding the θ^2 component of the product of two chiral multiplets. Arranging them in a grid

$$\begin{array}{r|ccc} & \theta^0 & \theta^1 & \theta^2 \\ \hline \mathbf{\Phi} & \phi & \sqrt{2}\psi & F \\ -\frac{1}{4}\overline{D}^2\mathbf{\Phi}^\dagger & F^\dagger & -i\sqrt{2}\sigma^\mu\partial_\mu\overline{\psi} & -\Box\phi^\dagger \end{array} \tag{124}$$

we need only multiply the θ^n component of the first line with the θ^{2-n} component of the second line, for $n = 0, 1, 2$.[l]

$$\mathcal{L}_{\text{kinetic}} = -\phi\Box\phi^\dagger + i\overline{\psi}\overline{\sigma}^\mu\partial_\mu\psi + F^\dagger F \tag{125}$$

$$\mathcal{L}_{\text{mass}} = \frac{1}{2}m\left(\phi F - \psi\psi + F\phi\right) + \text{h.c.} \tag{126}$$

The first line contains the usual kinetic terms for ϕ and ψ, but F is an auxiliary field since it does not have a kinetic term. The second line is not yet illuminating, but we can do the exact path integral over F by finding its equation of motion

$$F^\dagger = -m\phi. \tag{127}$$

Inserting this back into $\mathcal{L}_{\text{free}}$ yields our final answer

$$\mathcal{L}_{\text{free}} = -\phi(\Box + m^2)\phi^\dagger + i\overline{\psi}\overline{\sigma}^\mu\partial_\mu\psi - \frac{1}{2}m(\psi\psi + \overline{\psi}\overline{\psi}). \tag{128}$$

Thus, the free theory of a chiral multiplet is the theory of a free complex spin-0 scalar and a free Weyl fermion. Moreover, the mass terms for the

[k]By adding space-time derivatives, one can obtain other quadratic terms, such as $\int d^4\theta\, \mathbf{\Phi}^\dagger\Box\mathbf{\Phi}/\Lambda^2$, where Λ is some mass scale. These will always lead to terms in the component Lagrangian featuring more than two derivatives. In fact, these terms turn the auxiliary field F into a propagating field (i.e. a field with a kinetic term), suggesting that important physics was integrated out at the scale Λ. We will come back to this in Sec. 3.7.

[l]The Fierz identity of Eq. (28) tells us that $(\theta\chi)(\theta\psi) = -\frac{1}{2}\theta^2\chi\psi$, so when multiplying the two θ^1 components together, we lose the factors of $\sqrt{2}$ and pick up a sign.

scalar and fermion are related by SUSY! The $m \to 0$ limit has enhanced chiral symmetry acting on $\mathbf{\Phi}$; this protects the masses of both the fermion and the scalar at the same time. This effect helps protects radiative corrections to the Higgs boson mass in the SUSY SM.

The lagrangian in Eq. (128) contains a Majorana mass term for the single Weyl fermion. To obtain a Dirac mass term, we would need two chiral multiplets:

$$\mathcal{L}_{\text{Dirac}} = \int d^4\theta \left(\mathbf{\Phi}^\dagger \mathbf{\Phi} + \mathbf{\Phi}^{c\dagger}\mathbf{\Phi}^c\right) + \left(\int d^2\theta\, m\mathbf{\Phi}\mathbf{\Phi}^c + \text{h.c.}\right). \tag{129}$$

As before, the superscript c is just a label, and $\mathbf{\Phi}$ and $\mathbf{\Phi}^c$ are independent fields off-shell. The action in components is straightforward to find given our earlier results. We simply have two copies of the kinetic action of Eq. (125) (one for $\mathbf{\Phi}$ and one for $\mathbf{\Phi}^c$), while the mass terms are

$$\mathcal{L}_{\text{mass}} = -m^2\phi^\dagger\phi - m^2\phi^{c\dagger}\phi^c - m(\chi\chi^c + \chi^\dagger\chi^{\dagger c}) \tag{130}$$

after integrating out auxiliary fields. This theory has one massive Dirac fermion (composed of two Weyl fermions) and two complex scalars, with all fields having the same mass.

3.3. *A generic lagrangian*

Having successfully constructed a free SUSY lagrangian, we can construct more generic lagrangians using chiral multiplets. However, it is immediately clear that the number of possibilities is rather large. In the generic lagrangian in Eq. (103), $\mathbf{V}_{\text{comp}}$ and $\mathbf{\Phi}_{\text{comp}}$ can involve all sorts of functions and combinations of superfields, including ordinary derivatives and SUSY-covariant derivatives:

$$\mathbf{\Phi}, \quad D_\alpha\mathbf{\Phi}, \quad D^2\mathbf{\Phi}, \quad \partial_\mu\mathbf{\Phi}, \quad \Box\mathbf{\Phi}, \quad \partial_\mu D_\alpha\mathbf{\Phi}, \quad \text{etc.} \tag{131}$$

$$\mathbf{\Phi}^\dagger, \quad \overline{D}_{\dot\alpha}\mathbf{\Phi}^\dagger, \quad \overline{D}^2\mathbf{\Phi}^\dagger, \quad \partial_\mu\mathbf{\Phi}^\dagger, \quad \Box\mathbf{\Phi}^\dagger, \quad \partial_\mu\overline{D}_{\dot\alpha}\mathbf{\Phi}^\dagger, \quad \text{etc.} \tag{132}$$

Clearly we need some kind of organizing principle to sort out the multitude of possible terms.

Just as in non-SUSY theories, we can use effective field theory power counting to organize terms in a SUSY lagrangian by mass dimension. We will discuss relevant and marginal interactions in Sec. 3.4 and higher dimensional interactions in Sec. 3.5.

Before doing so, it is worth emphasizing that some terms will not contribute to the SUSY action. For example, we saw in Eq. (100) that a purely chiral piece of $\boldsymbol{V}_{\rm comp}$ does not contribute to the action. Similarly, we can ignore terms that only change the lagrangian by a total derivative, such as

$$\int d^4\theta\,\Box(\cdots) \to \text{total derivative.} \tag{133}$$

Less obvious is that we can ignore terms that correspond to a total SUSY-covariant derivative

$$\int d^4\theta\, D_\alpha(\cdots) \to \text{total derivative,} \tag{134}$$

which one can derive by using the fact that $D^3 = 0$. This implies that we can always do integration by parts with SUSY-covariant derivatives, which we will use to show the equivalence of different SUSY operators.

3.4. *Relevant and marginal interactions*

Relevant and marginal interactions correspond to couplings with positive or zero mass dimension. From Eq. (121), we argued that $[\boldsymbol{V}_{\rm comp}] = 2$ and $[\boldsymbol{\Phi}_{\rm comp}] = 3$. Since a linear term in $\boldsymbol{V}_{\rm comp}$ would be pure chiral, Eq. (100) tells us that it would not contribute to the action. Thus, the only choice for a renormalizable $\boldsymbol{V}_{\rm comp}$ is quadratic in chiral superfields

$$\boldsymbol{V}_{\rm comp} = \boldsymbol{\Phi}_i^\dagger \boldsymbol{\Phi}^i, \tag{135}$$

where we now have a number of chiral superfields $\boldsymbol{\Phi}^i$ labeled by i, and we sum over repeated indices. Note that we can always rotate and rescale the $\boldsymbol{\Phi}^i$ to bring $\boldsymbol{V}_{\rm comp}$ into this form. For $\boldsymbol{\Phi}_{\rm comp}$, there are three possible terms

$$\boldsymbol{\Phi}_{\rm comp} = f_i \boldsymbol{\Phi}^i + \frac{1}{2} m_{ij} \boldsymbol{\Phi}^i \boldsymbol{\Phi}^j + \frac{1}{6}\lambda_{ijk} \boldsymbol{\Phi}^i \boldsymbol{\Phi}^j \boldsymbol{\Phi}^k \equiv \boldsymbol{W}(\boldsymbol{\Phi}^i), \tag{136}$$

where the object $\boldsymbol{W}$ is called the superpotential. By Eq. (119), $\boldsymbol{W}$ itself is a (composite) chiral multiplet, and we say that $\boldsymbol{W}$ is a holomorphic function, since it only depends on chiral and not anti-chiral superfields. The mass dimensions of these couplings are $[f_i] = 2$, $[m_{ij}] = 1$, and $[\lambda_{ijk}] = 0$.

The most general renormalizable theory of chiral multiplets is often called the Wess-Zumino model:

$$\mathcal{L}_{\rm WZ} = \int d^4\theta\, \boldsymbol{\Phi}_i^\dagger \boldsymbol{\Phi}^i + \left(\int d^2\theta\, \boldsymbol{W}(\boldsymbol{\Phi}^i) + \text{h.c.} \right). \tag{137}$$

By Taylor expanding $\boldsymbol{W}$ to extract the θ^2 component, we can easily write down the component form of the lagrangian

$$\begin{aligned} \mathcal{L}_{\rm WZ} = &-\phi^i \Box \phi_i^\dagger + i\bar{\psi}_i \bar{\sigma}^\mu \partial_\mu \psi^i + F_i^\dagger F^i \\ &+ W_i F^i - \frac{1}{2} W_{ij} \psi^i \psi^j + \text{h.c.} \end{aligned} \tag{138}$$

where

$$W_i(\phi^i) \equiv \frac{\partial W}{\partial \phi^i}, \qquad W_{ij}(\phi^i) \equiv \frac{\partial^2 W}{\partial \phi^i \phi^j}, \tag{139}$$

and W_i and W_{ij} should now be thought of as scalar functions of the lowest (scalar) component of the various $\mathbf{\Phi}^i$ fields. The equations of motion for the auxiliary fields are

$$F^i = -W^{\dagger i}, \qquad F_i^\dagger = -W_i, \tag{140}$$

so the final component lagrangian is

$$\begin{aligned}\mathcal{L}_{\rm WZ} = &-\phi^i \Box \phi_i^\dagger + i\overline{\psi}_i \overline{\sigma}^\mu \partial_\mu \psi_i \\ &- W^{\dagger i} W_i - \frac{1}{2}(W_{ij}\psi^i\psi^j + W^{\dagger ij}\overline{\psi}_i\overline{\psi}_j).\end{aligned} \tag{141}$$

The $W^{\dagger i}W_i$ term corresponds to the scalar potential of the theory, which in general has linear, quadratic, cubic, and quartic interactions. The $W_{ij}\psi^i\psi^j$ term contains both mass terms for the fermions as well as Yukawa interactions between a scalar and two fermions.

We can now interpret the various terms in Eq. (136) one by one.

- f_i is a source term for F_i. It generates a constant in the scalar potential but no fermion mass terms. This type of term will be relevant for SUSY-breaking in Sec. 5.4.
- m_{ij} is a mass term for bosons and fermions. The fermion mass matrix is m_{ij} directly, but the boson mass-squared matrix is $M_{ij}^2 = m_{ik}m^k{}_j$.
- λ_{ijk} yields a quartic term for bosons and Yukawa interactions between one boson and two fermions. These two types of couplings are related in a non-trivial way by SUSY, with the quartic interaction related to the square of the Yukawa interaction.

Of course, additional terms in the scalar potential are generated when all of the f_i, m_{ij}, and λ_{ijk} are present. Apart from gauge fields (which we will discuss in the next lecture), the Wess-Zumino model accounts for all interactions present in the MSSM.

3.5. *Dimension five and six*

Continuing with our power counting, the first irrelevant operators we encounter are at dimension 5. Recalling that D^2 has mass dimension 1, the terms which show up at dimension 5 are

$$\int d^2\theta \, \frac{1}{\Lambda} \mathbf{\Phi}_1 \mathbf{\Phi}_2 \mathbf{\Phi}_3 \mathbf{\Phi}_4, \tag{142}$$

$$\int d^4\theta \, \frac{1}{\Lambda} \mathbf{\Phi}_1^\dagger \mathbf{\Phi}_2 \mathbf{\Phi}_3, \tag{143}$$

$$\int d^4\theta \, \frac{1}{\Lambda} \mathbf{\Phi}_1 D^2 \mathbf{\Phi}_2 = 4 \int d^2\theta \, \frac{1}{\Lambda} \mathbf{\Phi}_1 \Box \mathbf{\Phi}_2. \tag{144}$$

The last equality follows from the chirality of $\mathbf{\Phi}_1$ and $\mathbf{\Phi}_2$. This exhausts all the possible terms at dimension 5.[m] At low energies, $\Box$ is small, so the first two terms are often the most important deformations.

Terms like Eq. (142) are already captured by the superpotential (in the sense that Eq. (141) still holds). In order to describe the effect of terms like Eq. (143), it is helpful to introduce the Kähler potential $\boldsymbol{K}$:

$$\begin{aligned} \mathcal{L} = \int d^4\theta \, \boldsymbol{K}(\mathbf{\Phi}^i, \mathbf{\Phi}^{\dagger \bar{\jmath}}) + \int d^2\theta \, \boldsymbol{W}(\mathbf{\Phi}^i) + \text{h.c.} \\ + D_\alpha, \partial_\mu \text{ terms.} \end{aligned} \tag{145}$$

To emphasize, the Kähler potential and superpotential by themselves *do not* give the most general SUSY effective actions, but often contain the most important effects at low energies. The Kähler potential and superpotential are most useful when scalar fields can get large vacuum expectation values but space-time derivatives are small.

To derive the Lagrangian of Eq. (145) in components, the following trick is helpful. If you write a term in the Kähler potential as

$$\begin{aligned} &\int d^4\theta (\mathbf{\Phi}_1 \mathbf{\Phi}_2 \cdots)(\mathbf{\Phi}_n \mathbf{\Phi}_{n+1} \ldots)^\dagger \\ &= \int d^2\theta (\mathbf{\Phi}_1 \mathbf{\Phi}_2 \cdots) \left(-\frac{1}{4} \overline{D}^2 (\mathbf{\Phi}_n \mathbf{\Phi}_{n+1} \cdots)^\dagger \right), \end{aligned} \tag{146}$$

then we recognize $(\mathbf{\Phi}_1 \mathbf{\Phi}_2 \cdots)$ and $(\mathbf{\Phi}_n \mathbf{\Phi}_{n+1} \cdots)$ as being chiral multiplets, whose components are easy to find. Using the expression for $-\frac{1}{4}\overline{D}^2 \mathbf{\Phi}^\dagger$ from Eq. (118), it is straightforward to find the required θ^2 component. For example, consider the dimension 6 term

$$\int d^4\theta \, \frac{1}{\Lambda^2} (\mathbf{\Phi}_1 \mathbf{\Phi}_2)(\mathbf{\Phi}_3 \mathbf{\Phi}_4)^\dagger. \tag{147}$$

[m] You can convince yourself that $\int d^4\theta \, \mathbf{\Phi}_1^\dagger D^2 \mathbf{\Phi}_2 = 0$ by doing integration by parts.

Making a grid analogous to Eq. (124), we have

$$\begin{array}{r|ccc} & \theta^0 & \theta^1 & \theta^2 \\ \hline \boldsymbol{\Phi}_1\boldsymbol{\Phi}_2 & \phi_1\phi_2 & \phi_1\psi_2+\phi_2\psi_1 & \phi_1F_2+\phi_2F_1 \\ & & & -\psi_1\psi_2 \\ \hline -\frac{1}{4}\overline{D}^2(\boldsymbol{\Phi}_3\boldsymbol{\Phi}_4)^\dagger & \phi_3^\dagger F_4^\dagger+\phi_4^\dagger F_3^\dagger & \sigma\cdot\partial(\phi_3^\dagger\overline{\psi}_4 & \Box(\phi_3^\dagger\phi_4^\dagger) \\ & -\overline{\psi}_3\overline{\psi}_4 & +\phi_4^\dagger\overline{\psi}_3) & \end{array} \tag{148}$$

Among other things, this contains the four-fermion operator $\frac{1}{\Lambda^2}\psi_1\psi_2\overline{\psi}_3\overline{\psi}_4$, a type of interaction that is not present in the superpotential alone.

There are many possible terms at dimension 6. Terms like

$$\int d^2\theta \frac{1}{\Lambda^2}\boldsymbol{\Phi}_1\boldsymbol{\Phi}_2\boldsymbol{\Phi}_3\boldsymbol{\Phi}_4\boldsymbol{\Phi}_5 \tag{149}$$

are already captured by the superpotential, and terms like

$$\int d^4\theta \frac{1}{\Lambda^2}\boldsymbol{\Phi}_1\boldsymbol{\Phi}_2\boldsymbol{\Phi}_3\boldsymbol{\Phi}_4^\dagger, \qquad \int d^4\theta \frac{1}{\Lambda^2}\boldsymbol{\Phi}_1\boldsymbol{\Phi}_2\boldsymbol{\Phi}_3^\dagger\boldsymbol{\Phi}_4^\dagger, \tag{150}$$

and their hermitian conjugates are captured by the Kähler potential. Additional terms which show up at dimension 6 include

$$\int d^2\theta \frac{1}{\Lambda^2}\boldsymbol{\Phi}_1(\Box\boldsymbol{\Phi}_2)\boldsymbol{\Phi}_3, \tag{151}$$

$$\int d^4\theta \frac{1}{\Lambda^2}D^2\boldsymbol{\Phi}_1\overline{D}^2\boldsymbol{\Phi}_2^\dagger = -16\int d^4\theta \frac{1}{\Lambda^2}\boldsymbol{\Phi}_1\Box\boldsymbol{\Phi}_2^\dagger, \tag{152}$$

where the last equality follows from integration by parts and the anti-chirality of $\boldsymbol{\Phi}_2^\dagger$.

There are cases where higher dimensional terms involving ∂_μ or D_α can be very important. For example, when we talk about SUSY-breaking, the vacuum expectation value (vev) $\langle F\rangle \neq 0$. Thus, terms like $\left\langle \frac{D^2\boldsymbol{\Phi}}{\Lambda^2}\right\rangle \neq 0$, and we have to remember the possibility of order $\frac{F}{\Lambda^2}$ effects not captured by the Kähler potential $\boldsymbol{K}$ or superpotential $\boldsymbol{W}$ alone.

3.6. *The Kähler potential in components*

Though one of the goals of these notes is to emphasize the presence of higher-derivative SUSY interactions, the Kähler potential and superpotential are used so ubiquitously in the literature that one may find it useful to see the expansion in components of a Lagrangian featuring only $\boldsymbol{K}$ and $\boldsymbol{W}$. The component expansion of Eq. (145) yields:

$$\begin{aligned}
\mathcal{L} = K_{i\bar{\jmath}} &\left(\partial^\mu \phi^i \partial_\mu \phi^{\dagger\bar{\jmath}} + \frac{i}{2}\chi^i \sigma^\mu \partial_\mu \chi^{\dagger\bar{\jmath}} + \frac{i}{2}\chi^{\dagger\bar{\jmath}} \bar{\sigma}^\mu \partial_\mu \chi^i + F^i F^{\dagger\bar{\jmath}}\right) \\
&- \frac{1}{2} K_{i\bar{\jmath}k} \left(\chi^i \chi^k F^{*\bar{\jmath}} + i\chi^i \sigma^\mu \chi^{\dagger\bar{\jmath}} \partial_\mu \phi^k\right) \\
&- \frac{1}{2} K_{i\bar{\jmath}\bar{k}} \left(\chi^{\dagger\bar{\jmath}} \chi^{\dagger\bar{k}} F^i - i\chi^i \sigma^\mu \chi^{\dagger\bar{\jmath}} \partial_\mu \phi^{\dagger\bar{k}}\right) \\
&+ \frac{1}{4} K_{i\bar{\jmath}k\bar{l}} \chi^i \chi^k \chi^{\dagger\bar{\jmath}} \chi^{\dagger\bar{l}} \\
&+ W_i F^i - \frac{1}{2} W_{ij} \chi^i \chi^j + \text{h.c.} \qquad (153)
\end{aligned}$$

where the subscripts on K represent derivatives with respect to fields, e.g. $K_{i\bar{\jmath}} = \frac{\partial^2 K}{\partial \Phi^i \partial \Phi^{\dagger\bar{\jmath}}}$.

After integrating out the auxiliary fields, this expression can be rewritten in a more compact fashion by thinking of the function $K_{i\bar{\jmath}}$ as a metric $g_{i\bar{\jmath}}$ in field space: [26]

$$\begin{aligned}
\mathcal{L} = g_{i\bar{\jmath}} &\partial^\mu \phi^i \partial_\mu \phi^{\dagger\bar{\jmath}} + i g_{i\bar{\jmath}} \chi^{\dagger\bar{\jmath}} \bar{\sigma}^\mu D_\mu \chi^i \\
&+ \frac{1}{4} R_{i\bar{\jmath}k\bar{l}} \chi^i \chi^k \chi^{\dagger\bar{\jmath}} \chi^{\dagger\bar{l}} - g^{i\bar{\jmath}} W_i W^\dagger_{\bar{\jmath}} \\
&- \frac{1}{2} D_i W_j \chi^i \chi^j + \text{h.c.} \qquad (154)
\end{aligned}$$

where $g^{i\bar{\jmath}}$ is the inverse Kähler metric ($g^{i\bar{\jmath}} g_{i\bar{k}} = \delta^{\bar{\jmath}}_{\bar{k}}$, $g^{i\bar{\jmath}} g_{k\bar{\jmath}} = \delta^i_k$) and the Kähler-covariant derivative D_i, Christoffel connection Γ^k_{ij}, and curvature tensor $R_{i\bar{\jmath}k\bar{l}}$ are given by:

$$D_i V_k = V_{ik} - \Gamma^k_{ij} V_k, \qquad (155)$$

$$\Gamma^k_{ij} = g^{k\bar{l}} K_{i\bar{l}j}, \qquad (156)$$

$$R_{i\bar{\jmath}k\bar{l}} = K_{i\bar{\jmath}k\bar{l}} - g_{m\bar{n}} \Gamma^m_{ik} \Gamma^{\bar{n}}_{\bar{\jmath}\bar{l}}, \qquad (157)$$

$$D_\mu \chi^i = \partial_\mu \chi^i + \Gamma^i_{jk} \chi^k \partial_\mu \phi^k. \qquad (158)$$

In particular, the equations of motion for the auxiliary fields are

$$F^i = -g^{i\bar{\jmath}} W^\dagger_{\bar{\jmath}} + \frac{1}{2} \Gamma^i_{mn} \chi^m \chi^n. \qquad (159)$$

We will not go into any further detail regarding Kähler geometry. For the interested reader, we recommend Ref. 3 and references therein.

3.7. *Super-trick #2: Equations of motion in superspace*

Our second super-trick involves integrating out heavy mass thresholds in superspace (at tree-level). In non-SUSY theories, integrating out heavy states yields an effective theory with additional higher-dimension interactions among the light states. In SUSY theories, we could do the same manipulations with component fields, but it is more convenient to integrate out superfields at tree-level using classical superspace equations of motion.

Recall that in non-SUSY theories, the classical equations of motion are obtained by the Euler-Lagrange procedure of varying the action to find the extrema. One subtlety in superspace is that the lagrangian involves both $\int d^4\theta$ and $\int d^2\theta$ integrations, which require different constraints. To avoid having to invoke Lagrange multipliers, it is best to express everything in terms of $\int d^2\theta$ using Eq. (89):

$$\int d^4\theta \, \boldsymbol{V}_{\text{comp}} = \int d^2\theta \left(-\frac{1}{4}\overline{D}^2 \boldsymbol{V}_{\text{comp}} \right). \tag{160}$$

For simplicity, consider the Wess-Zumino model:

$$\begin{aligned} \mathcal{L}_{\text{WZ}} &= \int d^4\theta \, \boldsymbol{\Phi}_i^\dagger \boldsymbol{\Phi}^i + \int d^2\theta \, \boldsymbol{W}(\boldsymbol{\Phi}^i) + \text{h.c.} & (161) \\ &= \int d^2\theta \left(-\frac{1}{4}\overline{D}^2 \boldsymbol{\Phi}_i^\dagger \right) \boldsymbol{\Phi}^i + \int d^2\theta \, \boldsymbol{W}(\boldsymbol{\Phi}^i) + \int d^2\overline{\theta} \, \boldsymbol{W}^\dagger(\boldsymbol{\Phi}_i^\dagger). & (162) \end{aligned}$$

Written in this form, it is easy to read off the SUSY equation of motion by extremizing the action with respect to $\boldsymbol{\Phi}^i$, yielding

$$-\frac{1}{4}\overline{D}^2 \boldsymbol{\Phi}_i^\dagger + \boldsymbol{W}_i = 0, \tag{163}$$

where the first term is familiar from Eq. (118) and last term should be thought of as a full superfield. The lowest component of this expression is just the familiar $F_i^\dagger = -W_i$ from Eq. (140). The θ (θ^2) component yields the fermion (boson) equations of motion.

For a heavy supersymmetric threshold, $-\frac{1}{4}\overline{D}^2 \boldsymbol{\Phi}_i^\dagger \approx 0$ at low energies:

$$\left\langle F_i^\dagger \right\rangle = 0, \qquad \langle \sigma \cdot \partial \overline{\psi}_i \rangle \approx 0, \qquad \langle \Box \phi_i \rangle \approx 0. \tag{164}$$

Therefore, up to D^2 terms suppressed by the $\boldsymbol{\Phi}_i$ mass, we can integrate out the field $\boldsymbol{\Phi}_i$ at tree-level by simply imposing $\boldsymbol{W}_i = 0$ on superfields.

Let us try this out with three superfields $\boldsymbol{N}$, $\boldsymbol{L}$, and $\boldsymbol{H}$:

$$\boldsymbol{K} = \boldsymbol{N}^\dagger \boldsymbol{N} + \boldsymbol{L}^\dagger \boldsymbol{L} + \boldsymbol{H}^\dagger \boldsymbol{H}, \tag{165}$$

$$\boldsymbol{W} = \frac{1}{2} m \boldsymbol{N}^2 + \lambda \boldsymbol{N}\boldsymbol{L}\boldsymbol{H}. \tag{166}$$

Here, $\boldsymbol{N}$ is a heavy state we wish to integrate out to get the low energy effective theory for $\boldsymbol{L}$ and $\boldsymbol{H}$. Ignoring $-\frac{1}{4}\overline{D}^2 \boldsymbol{N}^\dagger$, the equation of motion for $\boldsymbol{N}$ is simply

$$\frac{\partial \boldsymbol{W}}{\partial \boldsymbol{N}} = m\boldsymbol{N} + \lambda \boldsymbol{L}\boldsymbol{H} = 0 \quad \Rightarrow \boldsymbol{N} = -\frac{\lambda}{m}\boldsymbol{L}\boldsymbol{H}. \tag{167}$$

Below the scale m, the leading theory is

$$\boldsymbol{K}_{\text{eff}} = \frac{\lambda^2}{m^2}(\boldsymbol{L}\boldsymbol{H})^\dagger(\boldsymbol{L}\boldsymbol{H}) + \boldsymbol{L}^\dagger \boldsymbol{L} + \boldsymbol{H}^\dagger \boldsymbol{H} + D^2 \text{ terms}, \tag{168}$$

$$\boldsymbol{W}_{\text{eff}} = -\frac{\lambda}{m}\boldsymbol{L}^2 \boldsymbol{H}^2. \tag{169}$$

In the SUSY SM, this effective superpotential can be used to describe (Majorana) neutrino masses.

Here is another example with four superfields:

$$\boldsymbol{K} = \boldsymbol{N}^\dagger \boldsymbol{N} + \boldsymbol{N}^{c\dagger} \boldsymbol{N}^c + \boldsymbol{L}^\dagger \boldsymbol{L} + \boldsymbol{H}^\dagger \boldsymbol{H}, \tag{170}$$

$$\boldsymbol{W} = m\boldsymbol{N}\boldsymbol{N}^c + \lambda \boldsymbol{N}\boldsymbol{L}\boldsymbol{H}. \tag{171}$$

Here, the superscript ^{c} is just a label to say that $\boldsymbol{N}$ and $\boldsymbol{N}^c$ are different fields with complex conjugate quantum numbers. The low energy equations of motion for $\boldsymbol{N}$ and $\boldsymbol{N}^c$ are

$$\boldsymbol{W}_N = m\boldsymbol{N}^c + \lambda \boldsymbol{L}\boldsymbol{H} = 0, \tag{172}$$

$$\boldsymbol{W}_{N^c} = m\boldsymbol{N} = 0. \tag{173}$$

Below the scale m, the effective lagrangian is

$$\boldsymbol{K}_{\text{eff}} = \frac{\lambda}{m^2}(\boldsymbol{L}\boldsymbol{H})^\dagger(\boldsymbol{L}\boldsymbol{H}) + D^2 \text{ terms}, \tag{174}$$

$$\boldsymbol{W}_{\text{eff}} = 0. \tag{175}$$

As we will discuss further in the next subsection, we could have guessed that $\boldsymbol{W}_{\text{eff}}$ would be zero because there is a U(1) symmetry under which the combination $\boldsymbol{L}\boldsymbol{H}$ has non-zero charge (and the superpotential must be a holomorphic function, so we cannot have $(\boldsymbol{L}\boldsymbol{H})(\boldsymbol{L}\boldsymbol{H})^\dagger$ in $\boldsymbol{W}$).

This trick of superspace equations of motion is often helpful for finding UV completions of low energy effective theories. Higher-dimension interactions can be generated by integrating out heavy fields, though one must always be mindful of the potential effects of $-\frac{1}{4}\overline{D}^2\boldsymbol{\Phi}_i^\dagger$ terms which we have neglected in this analysis. Of course, in order to recover all the relevant physics, we often have to integrate out fields at loop level as well, and we will see how to do that in certain cases with our third super-trick in Sec. 4.6.

3.8. *Global symmetries*

In the next lecture, we will talk about gauge theories, so it is natural to first think about how global symmetries work in SUSY lagrangians. One can have a U(1) transformation acting on the whole superfield:

$$\boldsymbol{\Phi} \to e^{iq\alpha}. \tag{176}$$

This implies every component of $\boldsymbol{\Phi}$ has charge q:

	θ^0	θ^1	θ^2
$\boldsymbol{\Phi}$	ϕ	$\sqrt{2}\psi$	F
U(1)	q	q	q

(177)

This can be used to constrain the form of the superpotential, for example. A term $\frac{1}{2}m\boldsymbol{N}^2$ in the superpotential has no U(1) symmetry, but $m\boldsymbol{N}\boldsymbol{N}^c$ respects a U(1) symmetry if $\boldsymbol{N}$ and $\boldsymbol{N}^c$ have equal and opposite charges. Similarly, one can consider chiral superfields which come in representations of non-Abelian global symmetries.

Another type of global symmetry is a $\mathrm{U}(1)_R$ symmetry, which does not commute with SUSY. Under a $\mathrm{U}(1)_R$ rotation, the superspace coordinates themselves transform, with θ ($\overline{\theta}$) having R-charge $+1$ (-1). This implies that D_α ($\overline{D}_{\dot{\alpha}}$) has R-charge -1 ($+1$), that the superpotential $\boldsymbol{W}$ has R-charge 2, and that different components of $\boldsymbol{\Phi}$ have different charges:

	θ^0	θ^1	θ^2
$\boldsymbol{\Phi}$	ϕ	$\sqrt{2}\psi$	F
$\mathrm{U}(1)_R$	r	$r-1$	$r-2$

(178)

In the context of SUGRA, R-symmetries are always broken, because the gravitino mass parameter $m_{3/2}$ has R-charge -2. But to the extent that $m_{3/2}$ is small (which may or may not be the case), $\mathrm{U}(1)_R$ symmetries can be good symmetries of SUSY lagrangians.

As an example of how symmetries can constrain a superpotential, consider a theory with (global) SU(3), SU(2), and U(1) symmetries containing superfields with quantum numbers given in Table 1. Though these lectures

Table 1. Quantum numbers of the MSSM.

	SU(3)	SU(2)	U(1)
$\boldsymbol{Q}$	**3**	**2**	$+\frac{1}{6}$
$\boldsymbol{U}^c$	$\bar{\mathbf{3}}$	**1**	$-\frac{2}{3}$
$\boldsymbol{D}^c$	$\bar{\mathbf{3}}$	**1**	$+\frac{1}{3}$
$\boldsymbol{L}$	**1**	**2**	$-\frac{1}{2}$
$\boldsymbol{E}^c$	**1**	**1**	$+1$
$\boldsymbol{H}_u$	**1**	**2**	$+\frac{1}{2}$
$\boldsymbol{H}_d$	**1**	**2**	$-\frac{1}{2}$

are not supposed to cover topics of phenomenological relevance, this example was carefully chosen (and the fields appropriately named) to have the same symmetry structure as the MSSM, albeit for only one generation.

The leading relevant and marginal interactions are given by terms in the superpotential of up to dimension 3, which could include

$$\boldsymbol{W} \supset \mu\, \boldsymbol{H}_u \boldsymbol{H}_d + \lambda^u \boldsymbol{Q}\boldsymbol{U}^c \boldsymbol{H}_u + \lambda^d \boldsymbol{Q}\boldsymbol{D}^c \boldsymbol{H}_d + \lambda^e \boldsymbol{E}^c \boldsymbol{L}\boldsymbol{H}_d. \tag{179}$$

The fermions and scalars in the $\boldsymbol{H}$ superfields (the Higgs bosons and higgsinos) receive a supersymmetric mass from the μ mass term. If $\boldsymbol{H}_u$ and $\boldsymbol{H}_d$ get vevs in their lowest component, the other three superpotential terms then yield masses for the up-type quarks, down-type quarks, and leptons, respectively. As the superpotential must be holomorphic, we need at least two Higgs doublets in order to give masses to all fermions.[n]

As written, this superpotential obeys two additional global U(1) symmetries: a baryon number symmetry under which $\boldsymbol{Q}$ and $\boldsymbol{U}^c/\boldsymbol{D}^c$ have opposite charges, and a lepton number symmetry under which $\boldsymbol{L}$ and $\boldsymbol{E}^c$ have opposite charges. However, we can easily write down terms that do not respect these symmetries:

$$\boldsymbol{W} \supset \rho \boldsymbol{L}\boldsymbol{H}_u + \lambda^{(1)}\boldsymbol{Q}\boldsymbol{D}\boldsymbol{L} + \lambda^{(2)}\boldsymbol{U}\boldsymbol{D}\boldsymbol{D} + \lambda^{(3)}\boldsymbol{L}\boldsymbol{L}\boldsymbol{E}. \tag{180}$$

With only one generation, the last two terms vanish (since the SU(2) and SU(3) indices are contracted antisymmetrically), but they are relevant in the SUSY SM with three generations. If one wanted to forbid all of these terms, one could impose baryon and lepton number symmetries explicitly.

[n]In the actual MSSM with gauge interactions, the same argument can be made on the basis of anomaly cancellation. One can have a single active Higgs doublet in the presence of higher-dimensional interactions. [27–30]

Alternatively, one could use a $U(1)_R$ symmetry to forbid the terms in Eq. (180). If one gives the Higgs doublets an R-charge of 1, and all other superfields an R-charge of $+1/2$, the problematic terms are forbidden since the resulting superpotential would not have an R-charge of 2. We do not even need a full R-symmetry to achieve the same effect, which is desirable as we generally expect continuous R-symmetries to be broken by SUSY-breaking effects (or by $m_{3/2}$ if nothing else; see Sec. 5.11). The R-symmetry contains a discrete $\mathbf{Z}_2$ subgroup called R-parity, under which the Higgs doublets have R-parity $+1$ and the other multiplets have R-parity -1. This is sufficient to forbid the terms in Eq. (180).

Of course, higher-dimension terms may be of interest or pose problems. The superpotential may contain a $\boldsymbol{L}\boldsymbol{H}_u\boldsymbol{L}\boldsymbol{H}_u$ term which gives rise to neutrino masses; we saw in Eq. (169) that this could arise from integrating out a heavy multiplet (i.e. the right-handed neutrino). With three generations, the superpotential could also contain a term like $\boldsymbol{QQQL}$, which is R-parity even but violates both baryon and lepton number.

4. SUSY Gauge Theories

In this lecture, we will derive the lagrangian for SUSY gauge theories, starting from a discussion of gauge redundancy and the vector superfield. We focus on Abelian gauge theories, with a quick discussion of the non-Abelian case in Sec. 4.5. Using the super-trick of background superfields, we will be able to derive important one-loop effects including the Konishi anomaly. We end with a brief discussion of spontaneous gauge symmetry breaking.

4.1. *Gauge redundancy and gauge invariance*

In order to describe a massless spin-1 field in non-SUSY theories, we introduce a gauge redundancy under which

$$A_\mu(x) \to A_\mu(x) + \partial_\mu \alpha(x) \tag{181}$$

describes the same physics. It is worth remembering that Eq. (181) is not a true symmetry of the theory (though we often call it a "gauge symmetry"), since it does not package physical states into multiplets with shared properties. Indeed, one can gauge fix the theory (e.g. to Coulomb gauge) to remove the redundancy at the expense of making the theory look non-Lorentz-invariant. The purpose of gauge redundancy is to allow for a manifestly local and Lorentz-invariant description of a massless spin-1 field.

The same thing will happen in SUSY lagrangians with massless spin-1 fields, except in order to make SUSY manifest, we will have to introduce *even more* redundancy. To see the reason for this, consider the action of a U(1) gauge transformation acting on a chiral superfield:

$$\boldsymbol{\Phi} \to e^{iq\alpha(x)}\boldsymbol{\Phi}. \tag{182}$$

As discussed in Sec. 2.7, because $\alpha(x)$ is a spacetime-dependent scalar and not a superfield, $\boldsymbol{\Phi}$ is no longer a superfield after this gauge transformation. In order to maintain manifest SUSY, we would like to promote the gauge parameter to a full superfield

$$\boldsymbol{\Phi} \to e^{q\boldsymbol{\Omega}}\boldsymbol{\Phi}. \tag{183}$$

To make sure that $\boldsymbol{\Phi}$ remains a chiral superfield, we need $\boldsymbol{\Omega}$ to also be a chiral superfield ($\overline{D}\boldsymbol{\Omega} = 0$), whose lowest imaginary component is the ordinary gauge parameter. A theory that is invariant under Eq. (183) will have a high degree of redundancy, but we can always gauge fix the theory to remove the extra redundant modes.

With this augmented gauge transformation, the kinetic term in Eq. (135) is clearly not invariant under U(1) transformations:

$$\boldsymbol{\Phi}^\dagger\boldsymbol{\Phi} \to \boldsymbol{\Phi}^\dagger e^{q(\boldsymbol{\Omega}^\dagger+\boldsymbol{\Omega})}\boldsymbol{\Phi}. \tag{184}$$

However, we can compensate for this by introducing a vector superfield $\boldsymbol{V} = \boldsymbol{V}^\dagger$ that transforms under a gauge transformation according to

$$\boldsymbol{V} \to \boldsymbol{V} - \frac{\boldsymbol{\Omega} + \boldsymbol{\Omega}^\dagger}{2}. \tag{185}$$

In this way, the gauge-invariant kinetic term is given by

$$\mathcal{L}_{\text{kinetic}} = \int d^4\theta\, \boldsymbol{\Phi}^\dagger e^{2q\boldsymbol{V}}\boldsymbol{\Phi}. \tag{186}$$

The vector superfield $\boldsymbol{V}$ contains a spin-1 gauge field with the desired gauge transformation property in Eq. (181). It also contains a spin-1/2 gaugino, an auxiliary D component, as well as a number of redundant components that can removed by choosing the appropriate gauge fixing for $\boldsymbol{\Omega}$.

4.2. The vector multiplet

A generic vector superfield satisfying $\boldsymbol{V} = \boldsymbol{V}^\dagger$ can be written as

$$\boldsymbol{V} = \begin{array}{c|ccc} & \theta^0 & \theta^1 & \theta^2 \\ \hline \overline{\theta}^0 & c & \chi_\alpha & N \\ \overline{\theta}^1 & \overline{\chi}^{\dot\alpha} & A_\mu & \overline{\lambda}^{\dot\alpha} - \frac{i}{2}(\overline{\sigma}^\mu \partial_\mu \chi)^{\dot\alpha} \\ \overline{\theta}^2 & N^\dagger & \lambda_\alpha - \frac{i}{2}(\sigma^\mu \partial_\mu \overline{\chi})_\alpha & \frac{1}{2}D - \frac{1}{4}\Box c \end{array} \tag{187}$$

where c is a real scalar field, χ is a spin-1/2 Weyl fermion, N is a complex scalar field, A_μ is a spin-1 gauge boson, λ is a spin-1/2 gaugino, and D is a real scalar auxiliary field.[o] Many of these components are redundant, since they can be removed by performing the transformation in Eq. (185) with

$$\boldsymbol{\Omega} = \{c, \sqrt{2}\chi_\alpha, 2N\}. \tag{188}$$

This is known as fixing to Wess-Zumino gauge, where the remaining components of $\boldsymbol{V}$ are

$$\boldsymbol{V}_{\text{WZ}} = \begin{array}{c|ccc} & \theta^0 & \theta^1 & \theta^2 \\ \hline \overline{\theta}^0 & 0 & 0 & 0 \\ \overline{\theta}^1 & 0 & A_\mu & \overline{\lambda}^{\dot\alpha} \\ \overline{\theta}^2 & 0 & \lambda_\alpha & \frac{1}{2}D \end{array} \tag{189}$$

The only remaining redundancy in $\boldsymbol{\Omega}$ that Eq. (188) does not fix is $\boldsymbol{\Omega} = i\alpha(x)$, which acts like $A_\mu \to A_\mu + \partial_\mu \alpha(x)$ because[p]

$$\boldsymbol{\Omega} = i\alpha(y) \quad \Rightarrow \quad \frac{\boldsymbol{\Omega} + \boldsymbol{\Omega}^\dagger}{2} = -\overline{\theta}\overline{\sigma}^\mu \theta \partial_\mu \alpha(x). \tag{190}$$

Thus, the physical modes in the vector multiplet are a massless spin-1 gauge field A_μ as well as the spin-1/2 gaugino λ and the auxiliary field D.

Wess-Zumino gauge explicitly breaks SUSY in the sense that Eq. (189) is not a superfield, because it has "arbitrary" zeroed out entries (see Sec. 2.7). That said, after a SUSY transformation, one can perform a compensating gauge transformation to restore the Wess-Zumino form. For determining gauge-invariant and SUSY-invariant lagrangians, we treat $\boldsymbol{V}$ as a true superfield, but when expressing $\boldsymbol{V}$ in components, we will always use Wess-Zumino gauge.

---

[o] It is unfortunate that the same symbol D is used for the SUSY-covariant derivative.
[p] This explains the need for the funny factor of 2 in Eq. (185).

If instead we wished to describe a massive spin-1 particle, then the modes in $\boldsymbol{\Omega}$ are no longer redundant and there are additional propagating modes. We will discuss this further in Sec. 4.9.

4.3. *Gauge-invariant SUSY lagrangians*

Armed with the gauge transformation properties of chiral and vector multiplets, we can write down gauge-invariant SUSY lagrangians. We have already argued for the form of the gauge-invariant kinetic term for chiral multiplets in Eq. (186). Expanding out in (Wess-Zumino gauge) components,

$$\begin{aligned}\mathcal{L}_{\text{kinetic}} &= \int d^4\theta\, \boldsymbol{\Phi}^\dagger e^{2q\boldsymbol{V}} \boldsymbol{\Phi} \\ &= \nabla_\mu \phi^\dagger \nabla^\mu \phi + i\overline{\psi}\overline{\sigma}^\mu \nabla_\mu \psi + F^\dagger F \\ &\quad - \sqrt{2}q(\phi^\dagger \psi\lambda + \phi\overline{\psi}\overline{\lambda}) + q\phi^\dagger\phi D,\end{aligned} \tag{191}$$

where $\nabla_\mu = \partial_\mu - iqA_\mu$ is the familiar gauge-covariant derivative for a field of charge q.[q] In the last line of Eq. (191), the first term corresponds to matter couplings to gauginos, and the second term will affect the scalar potential after integrating out the auxiliary field D. Interestingly, SUSY has related the gauge boson coupling to the gaugino coupling (and the size of the D-term potential).

Next, we wish to write down the SUSY version of the field strength tensor. We need an object that is simultaneously gauge invariant and a proper superfield and has the field strength $F_{\mu\nu}$ as one of its components. While $\boldsymbol{V}$ itself is a vector superfield, it is not gauge invariant.[r] To form a gauge-invariant superfield, we use the super-trick from Sec. 2.8 of creating new superfields by acting with the SUSY-covariant derivative. Under a gauge transformation and using the fact that $\boldsymbol{\Omega}^\dagger$ is anti-chiral ($D_\alpha \boldsymbol{\Omega}^\dagger = 0$)

$$D_\alpha \boldsymbol{V} \to D_\alpha \boldsymbol{V} + \frac{D_\alpha \boldsymbol{\Omega}}{2}, \tag{192}$$

so while this is not gauge invariant, we have removed much of the gauge dependence. Adding $\overline{D}$ and using Eq. (82),

$$\overline{D}_{\dot\beta} D_\alpha \boldsymbol{V} \to \overline{D}_{\dot\beta} D_\alpha \boldsymbol{V} + i(\sigma\cdot\partial)_{\alpha\dot\beta}\boldsymbol{\Omega}. \tag{193}$$

---

[q]We are working in a non-canonical normalization where the gauge coupling appears in Eq. (200).

[r]If there were no gauge symmetry, i.e. a massive gauge boson, a term like $\int d^4\theta\, m^2 \boldsymbol{V}^2$ would be allowed. See Sec. 4.9.

The resulting object is still not gauge invariant, but adding one more $\overline{D}$ will do the job:

$$-\frac{1}{4}\overline{D}^2 D_\alpha \boldsymbol{V} \to -\frac{1}{4}\overline{D}^2 D_\alpha \boldsymbol{V} + i(\sigma \cdot \partial)_{\alpha\dot{\beta}} \overline{D}^{\dot{\beta}} \Omega = -\frac{1}{4}\overline{D}^2 D_\alpha \boldsymbol{V}. \tag{194}$$

Thus, we define the gauge-invariant chiral superfield

$$\boldsymbol{W}_\alpha \equiv -\frac{1}{4}\overline{D}^2 D_\alpha \boldsymbol{V}. \tag{195}$$

This superfield carries a Lorentz index (as anticipated in Sec. 2.4), and it is manifestly chiral ($\overline{D}^{\dot{\beta}} \boldsymbol{W}_\alpha = 0$) because $\overline{D}^3 = 0$. Despite using the symbol $\boldsymbol{W}$, $\boldsymbol{W}_\alpha$ is unrelated to the superpotential from Eq. (136).

Because $\boldsymbol{W}_\alpha$ is a chiral multiplet, we can express its components using the y^μ coordinate from Eq. (106). Because it is gauge invariant, the components in any gauge are

	θ^0	θ^1	θ^2
$\boldsymbol{W}_\alpha(y^\mu, \theta)$	λ_α	$\theta_\alpha D + \frac{i}{2}(\sigma^\mu \overline{\sigma}^\nu \theta)_\alpha F_{\mu\nu}$	$i\theta^2(\sigma^\mu \partial_\mu \lambda^\dagger)_\alpha$

(196)

where $F_{\mu\nu} = \partial_\mu A_\nu - \partial_\nu A_\mu$ is the ordinary field strength.

In order to get the gauge kinetic term, we need a term quadratic in $\boldsymbol{W}_\alpha$ (so it is automatically gauge-invariant) and Lorentz-invariant. The only option is

$$\int d^2\theta \, \boldsymbol{W}^\alpha \boldsymbol{W}_\alpha, \tag{197}$$

whose component expansion is

$$2i\overline{\lambda}\overline{\sigma}^\mu \partial_\mu \lambda + D^2 - \frac{1}{2} F_{\mu\nu} F^{\mu\nu} + \frac{i}{4} \epsilon^{\mu\nu\rho\tau} F_{\mu\nu} F_{\rho\tau}. \tag{198}$$

The first term is the gaugino kinetic term, the second term shows that D is an auxiliary field, the third term is the gauge boson kinetic term, and the last term corresponds to the CP-violating Θ term. To get from Eq. (196) to Eq. (198), we have used the sigma matrix trace relation:

$$\mathrm{Tr}(\sigma^\mu \overline{\sigma}^\nu \sigma^\rho \overline{\sigma}^\tau) = 2(\eta^{\mu\nu}\eta^{\rho\tau} - \eta^{\mu\rho}\eta^{\nu\tau} + \eta^{\mu\tau}\eta^{\nu\rho} + i\epsilon^{\mu\nu\rho\tau}). \tag{199}$$

Forming a hermitian lagrangian and introducing an overall normalization, the SUSY gauge kinetic term is

$$\mathcal{L}_{\text{kinetic,gauge}} = \int d^2\theta \left(\frac{1}{4g^2} - \frac{i\Theta_{\text{CP}}}{32\pi^2} \right) \boldsymbol{W}^\alpha \boldsymbol{W}_\alpha + \text{h.c.} \tag{200}$$

whose component expansion is

$$-\frac{1}{4g^2}F_{\mu\nu}F^{\mu\nu}+\frac{1}{g^2}i\overline{\lambda}\overline{\sigma}^{\mu}\partial_{\mu}\lambda+\frac{1}{2g^2}D^2+\frac{\Theta_{\mathrm{CP}}}{64\pi^2}\epsilon^{\mu\nu\rho\tau}F_{\mu\nu}F_{\rho\tau}. \tag{201}$$

We are using the non-canonical normalization where the gauge coupling appears in the gauge boson kinetic term (and by SUSY, also in the gaugino kinetic term and auxiliary D term), not in its couplings to matter. This normalization will be of great help when discussing background superfields in Sec. 4.6. To use canonical normalization instead, simply redefine

$$\boldsymbol{V}\to g\boldsymbol{V}, \tag{202}$$

or equivalently do the same for every component of $\boldsymbol{V}$. The Θ_{CP} term in Eq. (201) corresponds to a total derivative and has no physical relevance for an Abelian gauge symmetry (but defines the vacuum in a non-Abelian theory).

Putting the pieces together, the lagrangian for a renormalizable Wess-Zumino model with a U(1) gauge symmetry is

$$\begin{aligned}\mathcal{L}_{\mathrm{WZ}}=&\int d^4\theta\sum_i\boldsymbol{\Phi}^i e^{2q_i\boldsymbol{V}}\boldsymbol{\Phi}_i^{\dagger}\\&+\int d^2\theta\,\boldsymbol{W}(\boldsymbol{\Phi}_i)+\text{h.c.}\\&+\int d^2\theta\,\frac{1}{4g^2}\boldsymbol{W}^{\alpha}\boldsymbol{W}_{\alpha}+\text{h.c.}\end{aligned} \tag{203}$$

where the field $\boldsymbol{\Phi}_i$ has charge q_i and the superpotential $\boldsymbol{W}(\boldsymbol{\Phi}_i)$ must be U(1) invariant. After integrating out the F and D auxiliary fields (and passing to a canonical basis for the gauge fields), the scalar potential of the theory is

$$V=F^{i\dagger}F^i+\frac{1}{2}D^2=W_iW^{\dagger i}+\frac{1}{2}g^2\left|\sum_i q_i\phi_i^{\dagger}\phi^i\right|^2. \tag{204}$$

4.4. *Additional gauge-invariant operators*

In the spirit of Sec. 3.5, we would like to write down additional higher-dimension (and gauge-invariant) operators that would appear in a low energy effective theory. By power counting

$$[\boldsymbol{V}]=0,\qquad[\boldsymbol{W}_{\alpha}]=3/2, \tag{205}$$

and recall that $[d^2\theta]=1$ and $[d^4\theta]=2$.

First, using Eq. (100), the following dimension-2 term is actually gauge invariant in the Abelian case

$$\mathcal{L}_{\text{FI}} = \int d^4\theta\, \kappa \boldsymbol{V}. \tag{206}$$

This is called a Fayet-Iliopoulos (FI) term, and we will encounter it again in Sec. 5.6 when we discuss SUSY breaking.

Another potential gauge-invariant term at dimension 4 is

$$\int d^4\theta\, D^\alpha \boldsymbol{W}_\alpha = 0. \tag{207}$$

There are two ways to see why this vanishes. First, this is a total SUSY-covariant derivative, so it vanishes by Eq. (134). Second, the combination $\boldsymbol{L} \equiv D^\alpha \boldsymbol{W}_\alpha$ is a linear superfield because $D^2\boldsymbol{L} = 0$ and $\boldsymbol{L}^\dagger = \boldsymbol{L}$ (see Eq. (94)). The action $\int d^4\theta\, \boldsymbol{L}$ vanishes since $\int d^4\theta\, \boldsymbol{L} = -\frac{1}{4}\int d^2\theta\, \overline{D}^2 \boldsymbol{L} = 0$.

At dimension 5, there are two types of terms, but only one of them is non-zero:[s]

$$\frac{1}{\Lambda}\int d^2\theta\, \boldsymbol{\Phi} \boldsymbol{W}^\alpha \boldsymbol{W}_\alpha, \qquad \frac{1}{\Lambda}\int d^4\theta\, D^\alpha \boldsymbol{W}_\alpha \boldsymbol{\Phi} = 0. \tag{208}$$

As with the superpotential and Kähler potential, it is convenient to introduce the gauge kinetic function $\boldsymbol{f}(\boldsymbol{\Phi}^i)$ to capture the first type of term

$$\int d^2\theta\, \boldsymbol{f}(\boldsymbol{\Phi}^i) \boldsymbol{W}^\alpha \boldsymbol{W}_\alpha. \tag{209}$$

The gauge kinetic function is a holomorphic function of chiral superfields (and therefore a chiral superfield itself), whose lowest real component is $1/4g^2$. We will address $\boldsymbol{f}(\boldsymbol{\Phi}^i)$ extensively in Secs. 4.6 and 4.7, with subtleties related to holomorphy discussed in Sec. 4.8.

Here are some representative terms at dimension 6:

$$\frac{1}{\Lambda^2}\int d^2\theta\, \boldsymbol{\Phi}_1 \boldsymbol{\Phi}_2 \boldsymbol{W}^\alpha \boldsymbol{W}_\alpha, \qquad \frac{1}{\Lambda^2}\int d^4\theta\, D^\alpha \boldsymbol{W}_\alpha \boldsymbol{\Phi}_1^\dagger \boldsymbol{\Phi}_2, \tag{210}$$

$$\frac{1}{\Lambda^2}\int d^4\theta\, \boldsymbol{W}^\alpha \sigma^\mu_{\alpha\dot{\alpha}} \partial_\mu \boldsymbol{W}^{\dagger\dot{\alpha}}, \qquad \frac{1}{\Lambda^2}\int d^4\theta\, \boldsymbol{\Phi}^\dagger \boldsymbol{W}^\alpha \boldsymbol{W}_\alpha. \tag{211}$$

The first term is already captured by the gauge kinetic function $\boldsymbol{f}(\boldsymbol{\Phi}^i)$, while the others are genuinely new terms which would appear in any realistic effective lagrangian.

[s]Because $D^\alpha \boldsymbol{W}_\alpha$ is a linear superfield, $\int d^4\theta\, D^\alpha \boldsymbol{W}_\alpha \boldsymbol{\Phi}^\dagger$ is just the complex conjugate of the second term.

4.5. *The non-abelian case*

Thus far, we have only discussed Abelian gauge theories. Non-Abelian groups involve more algebra, but the basic physics is the same. For each gauge boson A_μ^a where a labels the gauge generator T^a, we need a separate vector multiplet $\boldsymbol{V}^a$ and gauge transformation parameter $\boldsymbol{\Omega}^a$:

$$\boldsymbol{V}^a \to \boldsymbol{V}^a - \frac{\boldsymbol{\Omega}^a + \boldsymbol{\Omega}^{a\dagger}}{2} + \dots \tag{212}$$

Using the shorthand $\boldsymbol{V} \equiv \boldsymbol{V}^a T^a$ and $\boldsymbol{\Omega} = \boldsymbol{\Omega}^a T^a$, the full non-linear gauge transformation is

$$e^{2\boldsymbol{V}} \to e^{-\boldsymbol{\Omega}^\dagger} e^{2\boldsymbol{V}} e^{-\boldsymbol{\Omega}}. \tag{213}$$

The gauge covariant object

$$\boldsymbol{W}_\alpha \equiv -\frac{1}{8}\overline{D}^2 \left(e^{-2\boldsymbol{V}} D_\alpha e^{2\boldsymbol{V}}\right) \tag{214}$$

transforms in the adjoint representation of the gauge group as

$$\boldsymbol{W}_\alpha \to e^{\boldsymbol{\Omega}} \boldsymbol{W}_\alpha e^{-\boldsymbol{\Omega}}, \tag{215}$$

while chiral matter fields transform as

$$\boldsymbol{\Phi} \to e^{\boldsymbol{\Omega}^a T_\Phi^a} \boldsymbol{\Phi}, \tag{216}$$

where T_Φ^a are the generators corresponding to the appropriate representation of $\boldsymbol{\Phi}$.

The combination $\mathrm{Tr}(\boldsymbol{W}^\alpha \boldsymbol{W}_\alpha) = \frac{1}{2}\boldsymbol{W}^{\alpha a}\boldsymbol{W}_\alpha^a$ is gauge invariant[t] and the gauge kinetic lagrangian is given by

$$\mathcal{L}_{\text{gauge}} = \int d^2\theta \left(\frac{1}{4g^2} - \frac{i\Theta_{\text{CP}}}{32\pi^2}\right) \boldsymbol{W}^{\alpha a}\boldsymbol{W}_\alpha^a + \text{h.c.} \tag{217}$$

while the matter kinetic terms are

$$\mathcal{L}_{\text{kinetic}} = \int d^4\theta\, \boldsymbol{\Phi}^\dagger e^{2\boldsymbol{V}} \boldsymbol{\Phi}. \tag{218}$$

Note that the Θ_{CP} term is physical for a generic non-Abelian gauge group. Also, notice that the Fayet-Iliopoulos term from Eq. (206) is not allowed for non-Abelian gauge groups.

[t] We use the usual normalization for generators $\mathrm{Tr}(T^a T^b) = \frac{1}{2}\delta^{ab}$.

For doing phenomenological studies (such as for the SUSY SM) the typical starting point is:

$$\begin{aligned}\mathcal{L} = & \int d^4\theta\, \boldsymbol{K}(\boldsymbol{\Phi}^{\dagger\bar{i}} e^{2\boldsymbol{V}}, \boldsymbol{\Phi}^j) \\ & + \int d^2\theta\, \boldsymbol{W}(\boldsymbol{\Phi}^i) + \text{h.c.} \\ & + \int d^2\theta\, \boldsymbol{f}_{ab}(\boldsymbol{\Phi}^i) \boldsymbol{W}^{\alpha a} \boldsymbol{W}_\alpha^b + \text{h.c.} \\ & + D_\alpha, \partial_\mu \text{ terms.} \end{aligned} \tag{219}$$

where $\boldsymbol{K}$ is the Kähler potential, $\boldsymbol{W}$ the superpotential, and $\boldsymbol{f}_{ab}$ is the gauge kinetic function (now with generator indices).

4.6. *Super-tric k #3: Background superfields*

Our third super-trick is to use the background field method to calculate otherwise tedious one-loop effects directly in superspace. In non-SUSY theories, background fields are a powerful technique to calculate effective actions for scalar fields [31]. The basic idea is to treat any scalar $\phi(x)$ as if it were a constant ϕ_0 and calculate the effective action in the background ϕ_0. Up to derivative terms like $\partial_\mu \phi$, the resulting action for the ϕ_0 background can be lifted to an action for the full $\phi(x)$.

In SUSY theories, this technique becomes even more powerful since the action for a constant ϕ_0 background can be lifted to an action for a full superfield $\boldsymbol{\Phi}$. Of course, in general the lagrangian can depend on

$$\mathcal{L}(\boldsymbol{\Phi}, \partial_\mu \boldsymbol{\Phi}, D_\alpha \boldsymbol{\Phi}, \ldots), \tag{220}$$

and the background field method misses dependence involving derivatives on superfields. In practice, this is not too much of a limitation, since the Kähler potential, superpotential, and gauge kinetic functions are functions only of superfields and not their derivatives.[u]

As an example of using the background field method, consider a gauge theory with a massive "quark" $\boldsymbol{Q}\boldsymbol{Q}^c$ coupled to a chiral superfield $\boldsymbol{X}$:

$$\boldsymbol{W} = m\boldsymbol{Q}\boldsymbol{Q}^c + \lambda \boldsymbol{X}\boldsymbol{Q}\boldsymbol{Q}^c. \tag{221}$$

Assuming that $\langle \boldsymbol{X} \rangle = 0$, the quark is heavy and we can integrate it out of the spectrum. The tree-level equations of motion for $\boldsymbol{Q}$ and $\boldsymbol{Q}^c$ simply tell us that $\langle \boldsymbol{Q} \rangle = \langle \boldsymbol{Q}^c \rangle = 0$, so we need to go to loop-level to find a non-trivial

[u] An important exception shows up in the case of SUSY breaking, where $\langle D^2 \boldsymbol{\Phi} \rangle \neq 0$, so the background field method can miss certain important SUSY-breaking effects.

effective Lagrangian below the quark mass threshold m. In particular, we would like to know if any operator of the type

$$\frac{1}{\Lambda}\int d^2\theta\, \boldsymbol{X}\boldsymbol{W}^\alpha \boldsymbol{W}_\alpha \tag{222}$$

is generated after integrating out $\boldsymbol{Q}\boldsymbol{Q}^c$. Using the background field method, we can replace $\boldsymbol{X}$ with the background scalar field x_0 and ask whether $\int d^2\theta\, x_0 \boldsymbol{W}^\alpha \boldsymbol{W}_\alpha$ is generated. This in turn is equivalent to asking whether the gauge coupling for the U(1) gauge boson has any dependence on x_0.

Consider the running of the gauge coupling at one loop. Suppose $g(\Lambda)$ is the value of the gauge coupling at some high energy scale Λ above the quark mass scale m. Then at some energy $\mu < m$, the low energy gauge coupling will be

$$\frac{1}{g^2(\mu,m)} = \frac{1}{g^2(\Lambda)} + \frac{b_0}{8\pi^2}\log\frac{\Lambda}{m} + \frac{b_1}{8\pi^2}\log\frac{m}{\mu}. \tag{223}$$

Here, b_0 is the one-loop beta-function at energies $m < E < \Lambda$, which includes $\boldsymbol{Q}, \boldsymbol{Q}^c$ as degrees of freedom, while b_1 is the beta-function at energies $\mu < E < m$ where $\boldsymbol{Q}$, $\boldsymbol{Q}^c$ have been integrated out. In particular, the b coefficients are defined by

$$\frac{d\alpha}{d\log E} = \frac{b}{2\pi}\alpha^2 \tag{224}$$

at the appropriate scale E, where $\alpha = \frac{g^2}{4\pi}$.

Turning on a background value of x_0 is equivalent to shifting $m \to m + \lambda x_0$. Using the background field method, we can lift m to have dependence on the full superfield $\boldsymbol{X}$ via

$$m \to m + \lambda \boldsymbol{X}. \tag{225}$$

This in turn allows us to analytically continue the gauge coupling into superspace [32–34] by promoting the gauge coupling to a chiral superfield (i.e the gauge kinetic function):

$$\boldsymbol{f}(\boldsymbol{X}) = \frac{1}{4g^2(\mu, m+\lambda\boldsymbol{X})} = \frac{1}{4g^2(\mu,m)} - \frac{\lambda(b_0-b_1)}{32\pi^2}\frac{\boldsymbol{X}}{m} + \mathcal{O}\left(\frac{\boldsymbol{X}^2}{m^2}\right). \tag{226}$$

Indeed, the operator of Eq. (222) is generated with a coefficient proportional to the differences of beta functions.

$$\mathcal{L}_{\text{eff}} \supset \int d^2\theta \left(\frac{1}{4g^2(\mu,m)} - \frac{\lambda(b_0-b_1)}{32\pi^2}\frac{\boldsymbol{X}}{m}\right)\boldsymbol{W}^\alpha\boldsymbol{W}_\alpha + \text{h.c.} \tag{227}$$

A few comments are in order about this result:

(1) If $\langle F_X \rangle \neq 0$, then a gaugino mass is generated in Eq. (227). Indeed this is an example of soft SUSY-breaking terms generated through gauge mediation. In Refs. 32–34, the background superfield method was used extensively to recover different types of soft SUSY-breaking terms generated by integrating out some high-energy dynamics, including two-loop (and higher) effects.

(2) The background superfield method only catches leading terms in the effective Lagrangian. In particular terms involving $D_\alpha \boldsymbol{X}$ and $\partial_\mu \boldsymbol{X}$ can not be recovered with this method, as $D_\alpha m = 0$ and $\partial_\mu m = 0$.

(3) Since m is just a real parameter, one might wonder why we made the replacement $m \to m + \lambda \boldsymbol{X}$ as opposed to $m \to m + \lambda \boldsymbol{X}^\dagger$ or $m \to \sqrt{(m + \lambda \boldsymbol{X})(m + \lambda \boldsymbol{X}^\dagger)}$. Ultimately, holomorphy of the gauge kinetic function forbids any alternative replacements, though we will discuss subtleties of this argument in Sec. 4.8.

(4) The background superfield method has nothing in particular to do with SUSY and it can be used in ordinary non-SUSY theories as well. For instance, if we want to calculate the coupling of the Higgs boson to two gluons in the SM, we can apply the same procedure as above. In this case, $\mu \simeq m_h \simeq 125$ GeV and in running from the high scale Λ down to the Higgs mass, the top mass $m_t = \lambda_t v_{\text{EW}}$ is the only threshold we cross. The low energy effective lagrangian is

$$\mathcal{L}_{\text{eff}} = -\frac{1}{2g_S^2(\mu, m_t)} \text{Tr}\left(G^{\mu\nu} G_{\mu\nu}\right), \tag{228}$$

where $G_{\mu\nu}$ is the gluon field strength and

$$\frac{1}{g_S^2(\mu, m_t)} = \frac{1}{g_S^2(\Lambda)} + \frac{b_0}{8\pi^2} \log \frac{\Lambda}{m_t} + \frac{b_1}{8\pi^2} \log \frac{m_t}{\mu}. \tag{229}$$

The top mass depends on the background (physical) Higgs boson h_0 as

$$m_t \longrightarrow m_t + \frac{\lambda_t}{\sqrt{2}} h_0, \tag{230}$$

so after integrating out the top quark we generate an operator

$$\mathcal{L}_{\text{eff}} \supset \frac{\sqrt{2} g_S^2(m_h)}{48\pi^2 v_{\text{EW}}} h_0 \text{Tr}\left(G^{\mu\nu} G_{\mu\nu}\right), \tag{231}$$

where $G^{\mu\nu}$ is now canonically normalized. This operator gives the leading contribution to the Higgs-gluon-gluon vertex, with contributions coming from loops of lighter particles of mass m suppressed by powers of m/m_h. It exhibits the famous *non-decoupling* effect that the top contribution is independent of λ_t for $m_t \gg m_{h_0}$.

4.7. The Konishi anomaly

We can use the background field method to understand other loop-level effects, such as the Konishi anomaly [35, 36]. Recall that in a non-SUSY U(1) gauge theory with fermions ψ (ψ^c) with charges +1 (−1), the chiral rotation

$$\psi \to e^{+i\alpha}\psi, \qquad \psi^c \to e^{+i\alpha}\psi^c, \tag{232}$$

is anomalous. This means that if we perform the chiral rotation with a constant value of α, we must include an anomaly term to recover the same physics.

$$\mathcal{L}(\psi, \psi^c) \to \mathcal{L}(e^{+i\alpha}\psi, e^{+i\alpha}\psi^c) + \frac{\alpha}{64\pi^2}\epsilon^{\mu\nu\rho\tau}F_{\mu\nu}F_{\rho\tau}. \tag{233}$$

We can use the background field method to understand what happens if $\alpha \to \boldsymbol{\alpha}$ is promoted to a superfield. Consider the chiral rotation

$$\boldsymbol{\Phi} \to e^{\boldsymbol{\alpha}}\boldsymbol{\Phi}, \qquad \boldsymbol{\Phi}^c \to e^{\boldsymbol{\alpha}}\boldsymbol{\Phi}^c, \tag{234}$$

where $\boldsymbol{\alpha}$, $\boldsymbol{\Phi}$, and $\boldsymbol{\Phi}^c$ are all chiral multiplets. This is a valid field redefinition since it leaves the one-particle asymptotic states unchanged. However, this chiral rotation is anomalous, so to reflect the anomaly (while maintaining manifest SUSY) we must have

$$\mathcal{L}(\boldsymbol{\Phi}, \boldsymbol{\Phi}^c) \to \mathcal{L}(e^{\boldsymbol{\alpha}}\boldsymbol{\Phi}, e^{\boldsymbol{\alpha}}\boldsymbol{\Phi}^c) + \frac{1}{16\pi^2}\int d^2\theta\, \boldsymbol{\alpha}\boldsymbol{W}^\alpha\boldsymbol{W}_\alpha + \text{h.c.} \tag{235}$$

The last term, which is a generalization of the familiar chiral anomaly, is known as the Konishi anomaly. The imaginary component of $\boldsymbol{\alpha}$ corresponds to Eq. (233), while the other components give new effects required by SUSY.

To see the Konishi anomaly in action, consider the lagrangian

$$\begin{aligned}\mathcal{L} = &\int d^4\theta\, (\boldsymbol{Q}^\dagger e^{2\boldsymbol{V}}\boldsymbol{Q} + \boldsymbol{Q}^{c\dagger}e^{-2\boldsymbol{V}}\boldsymbol{Q}^c)\left(1 + \frac{\boldsymbol{X}}{\Lambda} + \frac{\boldsymbol{X}^\dagger}{\Lambda} + \cdots\right)\\ &+ \left(\int d^2\theta\, m\boldsymbol{Q}\boldsymbol{Q}^c + \int d^2\theta\, \frac{1}{4g^2}\boldsymbol{W}^\alpha\boldsymbol{W}_\alpha + \text{h.c.}\right).\end{aligned} \tag{236}$$

When we integrate out $\boldsymbol{Q}$ and $\boldsymbol{Q}^c$ at the mass scale m, we might wonder if a one loop $\boldsymbol{X}\boldsymbol{W}^\alpha\boldsymbol{W}_\alpha$ coupling is generated. Since $\boldsymbol{X}$ and $\boldsymbol{X}^\dagger$ only appear in the combination $\boldsymbol{X} + \boldsymbol{X}^\dagger$, holomorphy of the gauge kinetic function would forbid such a term. We can easily check this using the Konishi anomaly. By performing a field redefinition

$$\boldsymbol{Q} \to \boldsymbol{Q}e^{-\boldsymbol{X}/\Lambda}, \quad \boldsymbol{Q}^c \to \boldsymbol{Q}^c e^{-\boldsymbol{X}/\Lambda}, \tag{237}$$

we can remove the linear term $\boldsymbol{X} + \boldsymbol{X}^\dagger$ in the Kähler potential, but this changes the superpotential

$$m\boldsymbol{Q}\boldsymbol{Q}^c \to m e^{-2\boldsymbol{X}/\Lambda}\boldsymbol{Q}\boldsymbol{Q}^c. \tag{238}$$

In addition, because of the Konishi anomaly, we also get a term

$$\frac{1}{4g^2(\Lambda)} \to \frac{1}{4g^2(\Lambda)} - \frac{1}{16\pi^2}\frac{\boldsymbol{X}}{\Lambda}. \tag{239}$$

Integrating out $\boldsymbol{Q}, \boldsymbol{Q}^c$ using the background field method, at energies $\mu < m$ we have

$$\begin{aligned} &\frac{1}{4g^2(\mu, m e^{-2\boldsymbol{X}/\Lambda})} - \frac{1}{16\pi^2}\frac{\boldsymbol{X}}{\Lambda} \\ &\quad = \frac{1}{4g^2(\mu, m)} - \frac{b_0 - b_1}{32\pi^2}\frac{2\boldsymbol{X}}{\Lambda} + \mathcal{O}(\boldsymbol{X}^2) - \frac{1}{16\pi^2}\frac{\boldsymbol{X}}{\Lambda} \\ &\quad = \frac{1}{4g^2(\mu, m)} + \mathcal{O}(\boldsymbol{X}^2), \end{aligned} \tag{240}$$

where $b_0 - b_1 = -1$ for integrating out $\boldsymbol{Q}$ and $\boldsymbol{Q}^c$.[v] By properly including the Konishi anomaly, we get no $\boldsymbol{X}\boldsymbol{W}^\alpha\boldsymbol{W}_\alpha$ coupling as expected from holomorphy.[w]

4.8. *Holomorphy and its subtleties*

Holomorphy is a powerful tool to constrain the form of terms in the $\int d^2\theta$ part of a SUSY action. Because the $\int d^2\theta$ term only leads to a SUSY-invariant action if the integrand is a chiral superfield, this dramatically reduces the possibilities for what can appear in $\int d^2\theta$ terms.

The most celebrated applications of this principle are in superpotential non-renormalization theorems. [39, 40] The superpotential $\boldsymbol{W}$ must be a function of chiral superfields only and therefore must be holomorphic. The same restrictions also apply to couplings if one promotes them to spurions, such that a coupling λ but not its complex conjugate $\overline{\lambda}$ can appear in the superpotential. If one also charges the λ spurion under $U(1)_R$ and ordinary global symmetries, there cannot be arbitrary functions of λ in the superpotential. In fact, by ensuring that the theory has sensible limits, one can show that no new appearances of the couplings in the superpotential are

[v] The beta function contributions from a chiral multiplet include those from both the scalar and fermion components.

[w] If $m = 0$ there is no cancellation between the Konishi anomaly and the gauge coupling threshold effect, and the resulting term is related to an effect called "anomaly mediation" [37, 38].

induced by quantum corrections. That is, couplings in the superpotential do not run under perturbative renormalization group flow (though there may be non-perturbative corrections due to instantons).

Similar arguments apply to gauge theories, whose kinetic Lagrangian can also be expressed as a $d^2\theta$ integral. We can promote the gauge coupling (and $\Theta_{\rm CP}$ angle) to a spurion $\boldsymbol{S}$

$$\mathcal{L} \supset \int d^2\theta \, \boldsymbol{S} \boldsymbol{W}^\alpha \boldsymbol{W}_\alpha. \tag{241}$$

In general, one might think that $\boldsymbol{S}$ could change into an arbitrary chiral multiplet under renormalization group flow. However, the lowest imaginary component of $\boldsymbol{S}$ is $\Theta_{\rm CP}$, which cannot have any effect on physics in perturbation theory.[x] Insisting on a well-defined zero-coupling limit ($g \to 0$, $\boldsymbol{S} \to \infty$), the only allowed change under renormalization group flow is [34]

$$\boldsymbol{S} \to \boldsymbol{S} + \text{constant}, \tag{242}$$

which is how the gauge coupling runs at one loop (see Eq. (223)). Higher-loop contributions are prohibited by holomorphy plus $\Theta_{\rm CP}$, so the beta function is saturated at one loop.[y]

The subtlety of using such holomorphy arguments, though, is that some effects that would be forbidden in $\int d^2\theta$ terms can appear in $\int d^4\theta$ terms. For example, in the lagrangian

$$\int d^4\theta \, \boldsymbol{\Phi}^\dagger \boldsymbol{\Phi} + \left(\int d^2\theta \, (m\boldsymbol{\Phi}^2 + \lambda \boldsymbol{\Phi}^3) + \text{h.c.} \right), \tag{243}$$

we know by holomorphy that the (holomorphic) m and λ do not run at any loop order. However, since the Kähler potential is not holomorphic, there is no restriction on wavefunction renormalization

$$\boldsymbol{\Phi}^\dagger \boldsymbol{\Phi} \to Z(\mu) \boldsymbol{\Phi}^\dagger \boldsymbol{\Phi}, \tag{244}$$

and via a field rescaling $\boldsymbol{\Phi} \to \boldsymbol{\Phi}/\sqrt{Z(\mu)}$, the physical values of m and λ do run

$$m \to \frac{m}{Z(\mu)}, \quad \lambda \to \frac{\lambda}{Z(\mu)^{3/2}}. \tag{245}$$

Of course, holomorphy is still extremely powerful for telling us that the combination m^3/λ^2 is invariant.

[x] Non-perturbative corrections to the gauge coupling due to instantons are allowed.

[y] Note that we cannot give $\boldsymbol{V}$ any global $U(1)_R$ charge as we did for fields and couplings in the superpotential. In any interacting gauge theory, $\boldsymbol{V}$ always appears in the form $e^{\boldsymbol{V}}$ somewhere, so $\boldsymbol{V}$ cannot have R-charge. This similarly implies $\boldsymbol{S}$ cannot have R-charge, so we cannot use such arguments to rule out the constant in Eq. (242).

A similar subtlety relates to the argument that the (holomorphic) gauge coupling only runs at one-loop. Consider the effect of the field rescaling $\boldsymbol{\Phi} \to \boldsymbol{\Phi}/\sqrt{Z(\mu)}$. Via the Konishi anomaly in Eq. (235), scaling $\boldsymbol{\Phi}$ by a real number changes the effective gauge coupling, and since $Z(\mu)$ runs at all loop order, the (physical) gauge coupling must as well. The solution to this is contained in the NSVZ beta function [41–43], which differentiates the canonical gauge coupling from the holomorphic gauge coupling [44–47].

With respect to the arguments presented in Secs. 4.6 and 4.7 using background field methods, we used holomorphy to argue that the gauge kinetic function should only be a function of $\boldsymbol{X}$ and not $\boldsymbol{X}^\dagger$, which let us unambiguously continue the background value $x_0 \to \boldsymbol{X}$. This logic (which is still correct) assumed that we had to write the gauge kinetic term in the form $\int d^2\theta\, \boldsymbol{f} \boldsymbol{W}^\alpha \boldsymbol{W}_\alpha$. Consider, though, an alternative way to write the gauge kinetic term

$$\int d^2\theta\, \frac{1}{4g^2} \boldsymbol{W}^\alpha \boldsymbol{W}_\alpha + \text{h.c.} = \int d^4\theta\, \frac{1}{4g^2} \left(\boldsymbol{W}^\alpha D_\alpha \boldsymbol{V} + \boldsymbol{W}^\dagger_{\dot\alpha} \overline{D}^{\dot\alpha} \boldsymbol{V} \right), \tag{246}$$

which can be verified by using the equivalence of $\int d^4\theta$ and $\int d^2\theta \left(-\frac{1}{4}\overline{D}^2\right)$. This second expression is fully gauge-invariant (for the same reason that the second term in Eq. (208) vanished), so one might worry that we could analytically continue $1/g^2$ in an alternative way. Luckily, this argument does not hold and the right-hand side of Eq. (246) is not gauge-invariant for a generic superfield-valued $1/g^2$, so the logic in Secs. 4.6 and 4.7 is still correct, albeit for subtle reasons.

Finally, note that we can write down a gauge-invariant, non-local expression in superspace

$$\int d^4\theta\, \frac{1}{4g^2} \boldsymbol{W}^\alpha \frac{\overline{D}^2}{8\Box} \boldsymbol{W}_\alpha, \tag{247}$$

which can be shown to be equivalent to Eq. (246) using the chiral projector from Eq. (97). Here, $1/g^2$ *can* be lifted to a real vector superfield, in which case the correct analytic continuation is not $m \to m + \lambda \boldsymbol{X}$ but $m \to \sqrt{(m + \lambda \boldsymbol{X})(m + \lambda \boldsymbol{X}^\dagger)}$. Because of the logarithmic structure of the one-loop beta function, however, this way of writing the gauge kinetic term yields the same answer as Eq. (227). The reason for the equivalence is ultimately due to holomorphy, albeit now in a hidden form. See Ref. 34 for further discussion.

4.9. *Spontaneously broken gauge theories*

Before tackling SUSY-breaking theories in the next lecture, we want to briefly mention spontaneously broken gauge theories. Consider the SUSY lagrangian of a gauge theory with a chiral multiplet $\boldsymbol{Q}$ ($\boldsymbol{Q}^c$) with U(1) charge $+1$ (-1), and a neutral chiral multiplet $\boldsymbol{N}$. The superpotential for this theory is

$$\boldsymbol{W} = \lambda \boldsymbol{N}\left(\boldsymbol{Q}\boldsymbol{Q}^c - \mu^2\right). \tag{248}$$

By the $\boldsymbol{N}$ equation of motion (assuming SUSY is to remain unbroken)

$$\frac{\partial \boldsymbol{W}}{\partial \boldsymbol{N}} = \lambda(\boldsymbol{Q}\boldsymbol{Q}^c - \mu^2), \tag{249}$$

so for this to equal 0, both $\boldsymbol{Q}$ and $\boldsymbol{Q}^c$ must get vacuum expectation values which break the U(1) symmetry. (In fact, by the D-term equations of motion, $\langle \boldsymbol{Q}\rangle = \langle \boldsymbol{Q}^c\rangle$.) This will yield a spontaneously broken massive gauge theory.

To better understand the physics, consider the field redefinitions

$$\boldsymbol{Q} \equiv (\mu + \boldsymbol{R})e^{\boldsymbol{\Phi}}, \qquad \boldsymbol{Q}^c \equiv (\mu + \boldsymbol{R})e^{-\boldsymbol{\Phi}}, \tag{250}$$

where $\boldsymbol{R}$ and $\boldsymbol{\Phi}$ are chiral multiplets. The superpotential in these fields is

$$\boldsymbol{W} = 2\lambda\mu \boldsymbol{N}\boldsymbol{R} + \lambda \boldsymbol{N}\boldsymbol{R}^2, \tag{251}$$

showing that $\boldsymbol{N}$ and $\boldsymbol{R}$ get a mass of $2\lambda\mu$. Note that the field $\boldsymbol{\Phi}$ does not appear in the superpotential, since under the gauge transformation of Eq. (183), it transforms as a shift

$$\boldsymbol{\Phi} \to \boldsymbol{\Phi} + \boldsymbol{\Omega}. \tag{252}$$

Focusing only on the (as of yet massless) $\boldsymbol{\Phi}$ field, the kinetic terms for $\boldsymbol{Q}$ and $\boldsymbol{Q}^c$ become

$$\int d^4\theta\, \boldsymbol{Q}^\dagger e^{2\boldsymbol{V}}\boldsymbol{Q} + \boldsymbol{Q}^{c\dagger} e^{-2\boldsymbol{V}}\boldsymbol{Q}^c \to \int d^4\theta\, \mu^2\left(e^{2\boldsymbol{V}+\boldsymbol{\Phi}^\dagger+\boldsymbol{\Phi}} + e^{-(2\boldsymbol{V}+\boldsymbol{\Phi}^\dagger+\boldsymbol{\Phi})}\right). \tag{253}$$

We can use our gauge redundancy with $\boldsymbol{\Omega} = \boldsymbol{\Phi}$ to set all of the components of $\boldsymbol{\Phi}$ to zero, resulting in

$$\int d^4\theta\, 4\mu^2 \boldsymbol{V}^2 + \ldots \tag{254}$$

As we can no longer use the gauge redundancy to go to Wess-Zumino gauge, this is a mass term for all of the components of the vector multiplet. It contains a massive gauge boson, a Dirac fermion (formed from the two

Weyl fermions in $\boldsymbol{V}$, χ and λ), and a real scalar c. Going to canonical normalization, they all have a mass of $2g\mu$. Effectively, the now-massive spin-1 gauge boson has "eaten" the lowest imaginary component of $\boldsymbol{\Phi}$, leaving one real scalar that gets the same mass because of SUSY.

5. SUSY Breaking and Goldstinos

In this lecture, we discuss the physics of spontaneous SUSY breaking, including the standard SUSY breaking paradigm of a hidden sector coupled to a visible sector. After giving a few explicit examples of SUSY breaking theories, we highlight the important role of the goldstino and demonstrate a super-trick to calculate the goldstino couplings to matter. We conclude with a brief discussion of SUSY breaking in SUGRA.

5.1. *Spontaneous SUSY breaking*

Thus far, we have emphasized the development of SUSY from a superspace lagrangian point of view. To understand the basics of SUSY breaking, it is helpful to think in a state/operator language. If the ground state $|0\rangle$ of a theory preserves SUSY, this means that a SUSY transformation leaves the vacuum invariant

$$e^{-i\epsilon Q - i\bar{\epsilon}\overline{Q}}\,|0\rangle = |0\rangle\,, \tag{255}$$

or equivalently

$$Q^{\alpha}\,|0\rangle = 0, \quad \overline{Q}^{\dot{\alpha}}\,|0\rangle = 0 \quad \text{(SUSY vacuum)}. \tag{256}$$

If the ground state spontaneously breaks SUSY, this means that the vacuum shifts under a SUSY transformation,

$$Q^{\alpha}\,|0\rangle \neq 0 \quad \text{(SUSY-breaking vacuum)}. \tag{257}$$

We would like to find simple criteria to determine whether or not SUSY is spontaneously broken.

It is instructive to consider the vacuum energy of the theory. Using the SUSY algebra in Eq. (73), the Hamiltonian $H = P_0$ of a SUSY theory is

$$H = \frac{1}{4}(\overline{Q}_1 Q_1 + Q_1\overline{Q}_1 + \overline{Q}_2 Q_2 + Q_2\overline{Q}_2). \tag{258}$$

If SUSY is unbroken, then by Eq. (256)

$$\langle H\rangle = \langle 0|\,H\,|0\rangle = 0 \quad \text{(SUSY vacuum)}, \tag{259}$$

so the vacuum energy is zero. The converse is also true, such that a zero vacuum energy implies $Q^\alpha |0\rangle = 0$ and SUSY is unbroken. In contrast, if the vacuum energy is non-zero, then SUSY is spontaneously broken in the vacuum. In fact, because each term in Eq. (258) is an operator squared,

$$\langle H \rangle > 0 \quad \text{(SUSY-breaking vacuum)}, \tag{260}$$

so spontaneous SUSY breaking corresponds to a strictly positive vacuum energy.

Our world is clearly not supersymmetric (otherwise we would see sparticles around!), so if SUSY is realized in nature, it must be spontaneously broken. By the above logic, one might conclude that $\langle H \rangle > 0$ in our universe. Observational measures of the cosmological constant reveal that $\langle H \rangle \approx 0$, though, posing a conundrum. As we will explain in Sec. 5.11, our universe must have an underlying anti-de Sitter (AdS) SUSY, and flat space corresponds to a large breaking of that AdS_4 SUSY. To a good approximation, though, we can still use the flat space SUSY algebra for understanding the physics of SUSY and SUSY breaking.

Beyond the vacuum energy, there is one other important way to test if SUSY is broken. Because the mass-squared operator P^2 commutes with Q_α and $\overline{Q}_{\dot\beta}$, components of an irreducible SUSY multiplet must have the same mass when SUSY is unbroken. More formally, mass is a Casimir invariant of the SUSY-extended Poincaré algebra. Therefore, mass-squared splittings between states in the SUSY multiplet is evidence for (flat space) SUSY breaking.

Beyond simple mass splittings, there are other possible signatures of spontaneous SUSY breaking that can appear as terms in the low-energy effective theory. These are denoted "soft SUSY-breaking terms" because when considered by themselves, they break SUSY in a way that does not introduce quadratic divergences. We will discuss these soft terms in more detail in Sec. 5.9.

5.2. *The vacuum energy*

Now that we know that the vacuum energy offers a robust test for SUSY breaking, we can return to our lagrangian point of view. The vacuum energy in a global SUSY theory is governed by the scalar potential. As we saw in Eq. (204), for a renormalizable gauge theory

$$V = F^{i\dagger} F^i + \frac{1}{2} D^a D^a, \tag{261}$$

where (for canonically normalized gauge fields)

$$F^i = -W^{\dagger i}, \qquad\qquad D^a = -g(\phi^* T^a \phi). \tag{262}$$

If SUSY is broken, at least one F-term or D-term must have a non-zero expectation value. Conversely, if all $\langle F^i \rangle$ and $\langle D^a \rangle$ are zero, then SUSY is unbroken.

Including the effects of a Kähler potential and gauge kinetic function, the vacuum energy generalizes to

$$V = g_{i\bar{\jmath}} F^i F^{\dagger \bar{\jmath}} + \frac{1}{2}(\mathrm{Re} f_{ab}) D^a D^b, \tag{263}$$

where now

$$F^i = -g^{i\bar{\jmath}} W^{\dagger}_{\bar{\jmath}}, \qquad D^a = -(\mathrm{Re} f^{-1})^{ab}(K_i T_b \phi^i). \tag{264}$$

Here f_{ab} is the gauge kinetic function for non-canonically normalized gauge fields, $K_i = \frac{\partial K}{\partial \phi^i}$, and the Kähler metric $g^{i\bar{\jmath}}$ was introduced in Sec. 3.6. Even in this more general case, we see that to find SUSY-breaking theories, we simply need to find scenarios where $\langle F^i \rangle \neq 0$ and/or $\langle D^a \rangle \neq 0$. Adding SUSY-covariant derivatives gives additional contributions to the vacuum energy, but does not change the requirement that SUSY breaking requires a non-zero vev for at least one auxiliary field.

5.3. *The standard SUSY-breaking paradigm*

Before talking about explicit models that break SUSY, we do want to make a connection to some physics of phenomenological relevance. Clearly, if SUSY is realized in nature, it must be spontaneously broken, since we need a mass splitting between SM particles and their superpartners. Crucially, though, experimental bounds imply that most of the superpartners should be heavier than their SM counterparts (with a key exception being the top squark).

For *renormalizable tree-level* theories that spontaneously break SUSY, there is a supertrace sum rule that says that [48]

$$\mathrm{STr}(m^2) \equiv \sum_s (-1)^{2s}(2s+1)\mathrm{Tr}(m_s^2) = -2g_a \mathrm{Tr}(T^a)\langle D^a \rangle = 0, \tag{265}$$

where s represents the spin of the particle.[z] Consider the MSSM with flavor conservation (i.e. no mixing between scalars of different generations) and

[z]This last equality is obvious for a non-Abelian gauge theory with $\mathrm{Tr}(T^a) = 0$. For a U(1) gauge group, the sum of the hypercharges must vanish to avoid the gravitational anomaly.

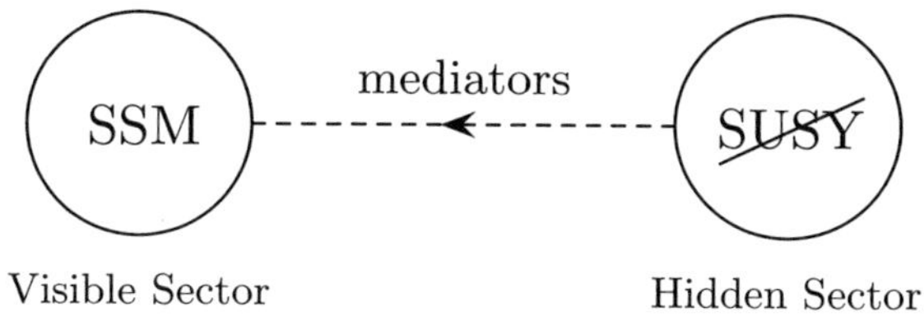

Fig. 3. Standard paradigm of a SUSY-breaking hidden sector coupled to the SUSY SM via mediators.

with no additional broken U(1) symmetries involved in SUSY breaking. For the first generation of squarks, for example, since

$$\mathrm{Tr}(\sigma^3) = 0 \quad \text{and} \quad Y_{\tilde{u}_L} + Y_{\tilde{u}_R^*} + Y_{\tilde{d}_L} + Y_{\tilde{d}_R} = 0, \tag{266}$$

Eq. (265) decouples, leading to the relation

$$m_{\tilde{u}_R}^2 + m_{\tilde{u}_L}^2 + m_{\tilde{d}_R}^2 + m_{\tilde{d}_L}^2 = 2(m_u^2 + m_d^2). \tag{267}$$

If $\mathrm{SU}(3)_C$ is to remain unbroken, this would imply light (MeV) scale superpartners, in conflict with observation. Similar arguments exist in the presence of large flavor mixings [49], even apart from the dangerous flavor-changing neutral currents they would introduce.

For these reasons, the standard SUSY-breaking paradigm is for SUSY to be broken in a "hidden sector", and the effects of SUSY breaking communicated to the SUSY SM (the "visible sector") via loop processes or higher-dimension operators. We draw this schematically as in Fig. 3. The effect of SUSY breaking on the visible sector is obviously important. But the phenomenological implications of the SUSY-breaking sector itself are typically meager (apart from the goldstino to be discussed in Sec. 5.7). For this reason, we will not focus much on specific models of SUSY breaking. In fact, the super-trick in Sec. 5.8 is aimed at abstracting the most important features of the hidden sector.

5.4. *The Polonyi model*

The absolute simplest model of SUSY breaking is the Polonyi model of one chiral multiplet $\boldsymbol{X} = \{\phi, \psi, F\}$:

$$\mathcal{L} = \int d^4\theta\, \boldsymbol{X}^\dagger \boldsymbol{X} + \int d^2\theta\, f\boldsymbol{X} + \text{h.c.} \tag{268}$$

$$= \phi^* \Box \phi + i\overline{\psi}\overline{\sigma}^\mu \partial_\mu \psi + F^\dagger F + fF + f^* F^\dagger. \tag{269}$$

After using the equation of motion $F = -f^*$, the scalar potential is simply

$$V = |f|^2, \tag{270}$$

so for any non-zero choice of f, SUSY is broken.

This should seem surprising, since apart from the vacuum energy, $\boldsymbol{X}$ looks just like a free massless SUSY multiplet. In particular, the mass of ϕ and ψ are the same. However, this model does indeed break SUSY. To convince you, consider adding a higher-dimensional term to the Kähler potential

$$\boldsymbol{K} = \boldsymbol{X}^\dagger \boldsymbol{X} - \frac{(\boldsymbol{X}^\dagger \boldsymbol{X})^2}{4\Lambda^2}, \tag{271}$$

$$\boldsymbol{W} = f\boldsymbol{X}. \tag{272}$$

The resulting scalar potential is

$$V = F^* F\left(1 - \frac{\phi^\dagger \phi}{\Lambda^2}\right) + fF + f^* F^* = \frac{|f|^2}{1 - \frac{\phi^*\phi}{\Lambda^2}}. \tag{273}$$

This potential is minimized at $\langle\phi\rangle = \langle\phi^\dagger\rangle = 0$, yielding the vacuum energy $\langle V\rangle = |f|^2$ as in Eq. (270). Expanding about the potential minimum, the scalar and fermion fields have masses

$$m_\phi^2 = \frac{|f|^2}{\Lambda^2}, \qquad m_\psi = 0. \tag{274}$$

For any finite value of Λ, we now see the mass splitting between bosons and fermions expected from SUSY breaking. Interestingly, the fermion is still massless. In fact, as we will see in Sec. 5.7, any SUSY-breaking theory has a goldstino, a massless Goldstone fermion arising from spontaneous SUSY breaking.

5.5. *Obstructions to generic F-term breaking*

The Polonyi model will be our template for SUSY-breaking models, since it contains most of the relevant physics. While we will not focus on explicit models of SUSY breaking, we do wish to convey that achieving generic F-term SUSY breaking is non-trivial.

Consider coupling a Polonyi field $\boldsymbol{X}$ to quarks $\boldsymbol{Q}$, $\boldsymbol{Q}^c$:

$$\boldsymbol{W} = f\boldsymbol{X} + \lambda \boldsymbol{X}\boldsymbol{Q}\boldsymbol{Q}^c. \tag{275}$$

You might naively think that $\langle F_X\rangle = -f^*$ and one would generate SUSY mass splittings for $\boldsymbol{Q}, \boldsymbol{Q}^c$. But instead, the vacuum shifts and SUSY stays unbroken! In particular, the set of equations

$$\begin{aligned} \boldsymbol{W}_X &= f + \lambda \boldsymbol{Q}\boldsymbol{Q}^c = 0, \\ \boldsymbol{W}_Q &= \lambda \boldsymbol{X}\boldsymbol{Q}^c = 0, \\ \boldsymbol{W}_{Q^c} &= \lambda \boldsymbol{X}\boldsymbol{Q} = 0, \end{aligned} \tag{276}$$

admits solutions with $\langle X \rangle = 0$ and $\langle QQ^c \rangle = -f/\lambda$. At those points, all F-terms vanish so the scalar potential $V = 0$ and SUSY is unbroken. With respect to the standard SUSY-breaking paradigm in Sec. 5.3, this tendency for SUSY to get restored in the presence of interactions makes it challenging to construct explicit models of SUSY breaking and mediation.

In general, if you have N chiral multiplets $\boldsymbol{X}_i$ and N equations of the type $\partial \boldsymbol{W}/\partial \boldsymbol{X}_i = 0$ to solve, then there is usually a solution and SUSY is unbroken in the true vacuum. To avoid this outcome, one can consider broken SUSY in a metastable vacuum [50]. Alternatively, one has to arrange the dynamics in such a way that the above equations cannot be simultaneously satisfied. In Ref. 51, the presence of an R-symmetry is shown to be a necessary condition for spontaneous SUSY breaking, while the spontaneous breaking of such an R-symmetry is sufficient to ensure spontaneous SUSY breaking.

5.6. *D-term breaking*

In a generic SUSY-breaking theory, non-zero F-terms will be accompanied by nonzero D-terms. For non-Abelian gauge theories, we can make the stronger statement that non-zero F-terms are required in order for the D-terms to be non-zero, such that there is no pure D-term SUSY breaking for non-Abelian gauge groups. [3]

For an Abelian gauge group, one can get D-term breaking without F-term breaking by introducing the Fayet-Iliopoulos term [52, 53] from Eq. (206)

$$\mathcal{L} \supset \int d^2\theta\, \boldsymbol{W}^\alpha \boldsymbol{W}_\alpha - \int d^4\theta\, \kappa \boldsymbol{V} \supset \frac{1}{2}D^2 - \kappa D. \tag{277}$$

After solving the D equation of motion, we find

$$D = \kappa, \qquad V = \frac{1}{2}\kappa^2. \tag{278}$$

The vacuum energy is non-vanishing, so SUSY is broken. We can see this in the spectrum by coupling this to a pair of oppositely-charged chiral superfields with a superpotential mass term

$$\boldsymbol{W} = m\boldsymbol{Q}\boldsymbol{Q}^c. \tag{279}$$

There is no way to make all three auxiliary fields (D, F_Q, and F_{Q^c}) vanish simultaneously, so SUSY is spontaneously broken. One can confirm that for $m > \sqrt{2g\kappa}$, the minimum of the potential is still at the origin and the gauge symmetry remains unbroken. The gaugino and gauge boson remain massless and there is one Dirac fermion (or two Weyl fermions) of mass m, but the scalars now have masses of[aa]

$$m^2_{Q,Q^c} = m^2 \pm 2g\kappa. \tag{280}$$

However, Fayet-Iliopoulos terms are hard to incorporate into realistic SUGRA theories. In SUGRA, Fayet-Iliopoulos terms correspond to having a gauged $\mathrm{U}(1)_R$ symmetry. [54, 55] As we will see in Sec. 5.11, demanding a $\mathrm{U}(1)_R$ symmetry makes it difficult to achieve the negligible cosmological constant observed in nature.

5.7. *The goldstino*

For any model of SUSY breaking, there is one irreducible prediction: the goldstino. Just as a spontaneously broken global symmetry always gives rise to a Goldstone boson, spontaneous SUSY breaking always gives rise to a Goldstone fermion. There are a couple of different ways to see the emergence of the goldstino.

In the operator language of Sec. 5.1, consider a vacuum that breaks SUSY, such that after performing a SUSY transformation, the vacuum state is changed

$$e^{-i\epsilon Q - i\bar{\epsilon}\overline{Q}} \left|0\right\rangle = \left|\epsilon\right\rangle . \tag{281}$$

However, because the Hamiltonian (P_0) commutes with SUSY (Q, $\overline{Q}$), the (fermionic) state $|\epsilon\rangle$ has exactly the same energy as the vacuum state. If we now imagine performing a space-time-dependent SUSY transformation

$$e^{-i\eta(x)Q - i\bar{\eta}(x)\overline{Q}} \left|0\right\rangle = \left|\eta(x)\right\rangle , \tag{282}$$

then the state $|\eta(x)\rangle$ will have a higher energy than the vacuum. But by making $\eta(x)$ closer and closer to a constant ϵ (i.e. by making $\eta(x)$ a longer and longer wavelength perturbation), we get closer and closer to the vacuum energy. This implies that there is (at least) one gapless fermionic excitation in the theory, namely the goldstino.

[aa]Since this is a renormalizable theory, the supertrace sum rule of Eq. (265) indeed holds.

A more mechanical way to see the need for a goldstino is to consider the scalar potential (with a trivial Kähler metric and no gauge interactions, for simplicity):

$$V = W^{\dagger i} W_i. \tag{283}$$

If we have SUSY breaking in a stable vacuum, then $\langle V \rangle \neq 0$ but $\frac{\partial V}{\partial \phi^i} = 0$ for all ϕ^i:

$$\frac{\partial V}{\partial \phi^j} = W^{\dagger i} W_{ij} = 0. \tag{284}$$

This can be satisfied in two ways. Either $\langle W^{\dagger i} \rangle = 0$ for all i (but then there is no SUSY breaking), or W_{ij} has a zero eigenvalue in the $W^{\dagger i}$ direction. But from Eq. (141), we see that W_{ij} is the fermion mass matrix! So there must be a massless mode

$$\eta = \frac{1}{F_{\text{tot}}} W^{\dagger i} \chi_i, \tag{285}$$

namely the goldstino. Here, $F_{\text{tot}} \equiv \langle V \rangle^{1/2}$. This logic is easily extended beyond Eq. (283) to arbitrary (perturbative) SUSY-breaking theories, and the expression with a non-trivial Kähler metric and gauge kinetic terms is

$$\eta = \frac{1}{F_{\text{tot}}} \left(F^i \chi_i + \frac{1}{\sqrt{2}} D_a \lambda^a \right), \tag{286}$$

where $F^i = -g^{i\bar{\jmath}} W^\dagger_{\bar{\jmath}}$, $D_a = -(K_i T_a \phi^i)$, and λ^a are the gauginos. Now F_{tot} includes the contribution from both F-terms and D-terms. We will encounter an even slicker way to see the goldstino in the next subsection.

Thus, the presence of a massless goldstino is guaranteed for global SUSY breaking in flat space. In Sec. 5.11, we will discuss that in SUGRA, the goldstino (from broken AdS_4 SUSY) is eaten to form the longitudinal component of the gravitino (the superpartner of the graviton).

5.8. *Supertrick #4: Non-linear goldstino multiplet*

Our last super-trick is aimed at understanding the properties and interactions of the goldstino. As inspiration for this super-trick, consider the analogy of a Higgs doublet breaking a global SU(2) symmetry. The Higgs boson gets a vev

$$\langle h \rangle = \begin{pmatrix} 0 \\ v \end{pmatrix}. \tag{287}$$

Because of this spontaneous symmetry breaking, there must be Goldstone bosons, and we can identify them by performing the broken SU(2) symmetry on the vacuum:

$$h_{\rm NL} = U \begin{pmatrix} 0 \\ v \end{pmatrix}, \qquad U \equiv e^{i\pi^a T^a / v}. \tag{288}$$

$h_{\rm NL}$ furnishes a non-linear representation of the SU(2) global symmetry, and we can recover the effective chiral Lagrangian by writing down the most generic interactions of $h_{\rm NL}$ (or U). The advantage of using this non-linear realization is that we can ignore the massive physical Higgs modes. Indeed, we can use $h_{\rm NL}$ even if SU(2)-breaking is triggered not by a perturbative Higgs doublet but by nonperturbative strong dynamics.

The same can be done for spontaneous SUSY breaking. [56–58] Consider one multiplet that gets a non-zero F component to break SUSY,

$$\langle \boldsymbol{X} \rangle = \theta^2 F. \tag{289}$$

In general, $\boldsymbol{X}$ will have a scalar component, but if SUSY is broken, we expect this state to be heavy whereas the goldstino is massless. To identify the goldstino direction, we can perform the broken SUSY transformation on the vacuum via

$$\theta^\alpha \to \theta^\alpha + \eta^\alpha, \tag{290}$$

$$y^\mu \to y^\mu + 2i\bar{\eta}\bar{\sigma}^\mu\theta, \tag{291}$$

where $\eta = \eta(x)$ is the (unnormalized) goldstino. If we treat F as a non-dynamical constant in space-time [17], the second line is irrelevant. Normalizing the goldstino via $\eta \to \eta/\sqrt{2}F$, we have the following non-linear representation of SUSY:

$$\begin{aligned} \boldsymbol{X}_{\rm NL} &= \left(\theta + \frac{\eta}{\sqrt{2}F}\right)^2 F \\ &= \frac{\eta^2}{2F} + \sqrt{2}\theta\eta + \theta^2 F. \end{aligned} \tag{292}$$

In Sec. 5.9, we will couple $\boldsymbol{X}_{\rm NL}$ to visible sector fields, treating F as a constant background field to derive soft mass terms and goldstino couplings. Even if there are multiple non-zero F-terms and D-terms in the hidden sector, the goldstino mode can be described by a single $\boldsymbol{X}_{\rm NL}$, since any non-zero F-term (or D-term) can be captured by multiplying $\boldsymbol{X}_{\rm NL}$ (or $\boldsymbol{X}^\dagger_{\rm NL}\boldsymbol{X}_{\rm NL}$) by an overall constant.

An alternative interpretation of Eq. (292) is as a constrained superfield satisfying $\boldsymbol{X}_{\rm NL}^2 = 0$ (the analog of $U^\dagger U = 1$ in the chiral Lagrangian). [58] Here, F is not a constant background and must be solved for dynamically. Neglecting terms with additional derivatives, the unique lagrangian we can write for $\boldsymbol{X}_{\rm NL}$ is

$$\mathcal{L}_{\rm NL} = \int d^4\theta\, \boldsymbol{X}_{\rm NL}^\dagger \boldsymbol{X}_{\rm NL} + \int d^2\theta\, f \boldsymbol{X}_{\rm NL} + \text{h.c.} \tag{293}$$

$$= \frac{\overline{\eta}^2}{2F^\dagger} \Box \frac{\eta^2}{2F} + i\overline{\eta}\overline{\sigma}^\mu \partial_\mu \eta + F^\dagger F + fF + F^\dagger f^*. \tag{294}$$

Solving for the F equation of motion is more complicated because of the first term, but doing so would recover terms with eight goldstinos and four derivatives that are present in the Akulov-Volkov lagrangian. [59] It is consistent to neglect those terms however, since terms that we have already neglected involving $D_\alpha \boldsymbol{X}_{\rm NL}$ and $\partial_\mu \boldsymbol{X}_{\rm NL}$ would play a role at that order. Using the equation of motion $F = -f^*$, the goldstino lagrangian is

$$\mathcal{L}_{\rm NL} = \frac{\overline{\eta}^2}{2f^\dagger} \Box \frac{\eta^2}{2f} + i\overline{\eta}\overline{\sigma}^\mu \partial_\mu \eta - |f|^2 + \ldots \tag{295}$$

This represents a massless goldstino with vacuum energy $\langle V \rangle = |f|^2$ and goldstino decay constant f. One can think of this as the low energy effective lagrangian one obtains after integrating out all of the (heavy) dynamics of the SUSY-breaking sector.

Using the non-linear goldstino multiplet to describe broken SUSY at low energies has a number of advantages. The main advantage is that one does not have to worry about the (usually irrelevant) details of how one actually breaks SUSY in the hidden sector. Whether SUSY is broken by one field or many, there is only one true goldstino, and its physical modes are completely contained in $\boldsymbol{X}_{\rm NL}$. We can also ignore the details of how the sgoldstino (scalar partner of the goldstino) is stabilized.

There is an important distinction between treating $\boldsymbol{X}_{\rm NL}$ as a constrained superfield $\boldsymbol{X}_{\rm NL}^2 = 0$ versus as an expansion around a constant $\theta^2 F$. With constrained superfields, we are guaranteed that the vacuum energy will be correlated with the goldstino decay constant (as necessitated by the SUSY algebra), whereas with fixed F, this is not the case. On the other hand, if we treat F as fixed, then we can write down direct couplings between the hidden and visible sectors without worrying about the vacuum changing. In particular, we saw in Sec. 5.5 that F-terms can shift and restore SUSY, leading to pathologies in the constrained superfield formalism. For this reason, we prefer to treat F as non-dynamical in $\boldsymbol{X}_{\rm NL}$.

5.9. *Goldstino couplings to matter*

We now use the super-trick of non-linear goldstinos to understand the leading interactions between the hidden sector and the visible sector in the standard SUSY-breaking paradigm. As we mentioned, there is very little hope to directly see states in the hidden sector, apart from their impact on soft terms for the visible sector. The one exception is the goldstino (eaten by the gravitino), which is generically light. Using our super-trick, we will show that the leading couplings of the goldstino to visible fields are universal and determined by SUSY-breaking soft terms in the visible sector. [60–62]

Let $\langle \boldsymbol{X}_{\rm NL}\rangle = \theta^2 F_{\rm hid}$ represent SUSY breaking in the hidden sector. Because $\boldsymbol{X}_{\rm NL}^2 = 0$ (and ignoring $D_\alpha \boldsymbol{X}_{\rm NL}$, $\partial_\mu \boldsymbol{X}_{\rm NL}$ terms), the possible interactions between the hidden sector and visible sector are quite restricted. Note that $[\boldsymbol{X}_{\rm NL}] = 1$ as for an ordinary chiral multiplet. Consider first the coupling of $\boldsymbol{X}_{\rm NL}$ to the gauge kinetic term (as anticipated in Sec. 4.6)

$$\mathcal{L} \supset -\int d^2\theta \, \frac{\boldsymbol{X}_{\rm NL}}{2\Lambda} W^\alpha W_\alpha + \text{h.c.} \tag{296}$$

Expanding in components yields various types of terms. The most important terms are

$$\mathcal{L} \supset -\frac{1}{2} m_\lambda \lambda\lambda + \frac{i m_\lambda}{\sqrt{2} F_{\rm hid}} \lambda \sigma^{\mu\nu} \eta F_{\mu\nu} + \text{h.c.} \tag{297}$$

where $\sigma^{\mu\nu}$ is defined in Eq. (20). The first term is a mass term for the gaugino

$$m_\lambda \equiv \frac{F_{\rm hid}}{\Lambda}, \tag{298}$$

which clearly breaks SUSY since it splits the gaugino mass from the gauge boson mass. The second term is a goldstino-gaugino-gauge boson coupling proportional to the gaugino mass m_λ. In fact, it is generically true that the goldstino couples proportional to soft masses (with deviations controlled by $D_\alpha \boldsymbol{X}_{\rm NL}$, $\partial_\mu \boldsymbol{X}_{\rm NL}$ terms).

There are additional terms in Eq. (296) which are important for consistency of the SUSY-breaking theory. If D in the visible sector gets a non-zero vacuum expectation value, then there are two additional mass terms

$$\mathcal{L} \supset \frac{D}{\sqrt{2}\Lambda} \eta\lambda + \frac{D^2}{8\Lambda F_{\rm hid}} \eta^2 + \text{h.c.} \tag{299}$$

The first term is a mixing term between the goldstino and the gaugino. This arises because the true goldstino points in the direction of Eq. (286), which includes contributions from both the visible sector and hidden sector. In general, such a mixing term will lift the mass of the goldstino, but because of the second term, the fermion mass matrix indeed has a zero eigenvalue. This highlights the usefulness of the non-linear goldstino multiplet, since without the η^2 in $\boldsymbol{X}_{\rm NL}$, the massless goldstino would only show up after adjusting the vacuum structure of the theory.

We can generalize Eq. (296) to account for all possible interactions between visible sector fields and the hidden sector surrogate $\boldsymbol{X}_{\rm NL}$. Restricting to terms that are visible sector renormalizable, the possible terms are[bb]

$$\mathcal{L} \supset -\int d^4\theta\, \frac{\widetilde{m}_i^2}{F^2} \boldsymbol{X}_{\rm NL}^\dagger \boldsymbol{X}_{\rm NL} \boldsymbol{\Phi}_i^\dagger e^{2\boldsymbol{V}} \boldsymbol{\Phi}_i - \left(\int d^2\theta\, \frac{m_\lambda}{2F_{\rm hid}} \boldsymbol{X}_{\rm NL} \boldsymbol{W}^{\alpha a} \boldsymbol{W}_\alpha^a \right.$$
$$\left. + \frac{C_i}{F_{\rm hid}} \boldsymbol{X}_{\rm NL} \boldsymbol{\Phi}_i + \frac{B_{ij}}{2F_{\rm hid}} \boldsymbol{X}_{\rm NL} \boldsymbol{\Phi}_i \boldsymbol{\Phi}_j + \frac{A_{ijk}}{6F_{\rm hid}} \boldsymbol{X}_{\rm NL} \boldsymbol{\Phi}_i \boldsymbol{\Phi}_j \boldsymbol{\Phi}_k + \text{h.c.} \right). \tag{300}$$

This yields the following soft SUSY-breaking terms in the visible sector lagrangian

$$\mathcal{L}_{\rm soft} = -\widetilde{m}_i^2 |\phi_i|^2 - \left(\frac{m_\lambda}{2} \lambda^a \lambda^a + C_i \phi_i + \frac{B_{ij}}{2} \phi_i \phi_j + \frac{A_{ijk}}{6} \phi_i \phi_j \phi_k + \text{h.c.} \right), \tag{301}$$

Each of these terms can be interpreted as follows:

- $\widetilde{m}_i^2$ gives a mass to the boson (but not the fermion) in a chiral multiplet.
- m_λ is a mass term for gauginos.
- C_i is a source term for scalars.
- B_{ij} gives a mass splitting between the scalar and the pseudoscalar in a chiral multiplet.
- A_{ijk} is a (holomorphic) three-point interaction for scalars.

As these soft terms arose by coupling the goldstino multiplet $\boldsymbol{X}_{\rm NL}$ to visible sector fields, each of these soft terms also has an associated coupling of a single goldstino

$$\mathcal{L}_\eta = \frac{1}{F_{\rm hid}} \eta \left(\widetilde{m}_i^2 \psi_i \phi_i^\dagger + C_i \psi_i + B_{ij} \psi_i \phi_j + \frac{A_{ijk}}{2} \psi_i \phi_j \phi_k \right.$$
$$\left. + \frac{i m_\lambda}{\sqrt{2}} \sigma^{\mu\nu} \lambda^a F_{\mu\nu}^a + \frac{m_\lambda}{\sqrt{2}} \lambda^a D^a \right) + \text{h.c.} \tag{302}$$

[bb]Terms like $(\boldsymbol{X}_{\rm NL} + \boldsymbol{X}_{\rm NL}^\dagger)\boldsymbol{\Phi}^\dagger\boldsymbol{\Phi}$ can be removed by a field redefinition $\boldsymbol{\Phi} \to \boldsymbol{\Phi} - \boldsymbol{\Phi}\boldsymbol{X}_{\rm NL}$. When doing that field redefinition, make sure to remember the Konishi anomaly from Sec. 4.7! This field redefinition yields an asymmetry in Eq. (302) such that there are explicit goldstino couplings to the D-term but not to the F-term.

As anticipated, these goldstino couplings are proportional to soft masses, suppressed by an overall factor of $1/F_{\text{hid}}$. Along with the SUSY interactions, the soft terms and goldstino couplings lead to the dominant phenomenology of the SUSY SM. In particular, Eq. (302) allows sparticles to decay to the corresponding particle and a goldstino (eaten by the gravitino).

If one chooses to impose a global R-symmetry on the lagrangian (as in Sec. 3.8), using $\boldsymbol{X}_{\text{NL}}$ makes it simple to determine whether the various soft terms contained in Eq. (300) respect that R-symmetry. Clearly, the soft scalar mass terms $\widetilde{m}_i^2$ respect any R-symmetry, but the other terms are less certain. As $\boldsymbol{X}_{\text{NL}}$ appears as a linear term in the superpotential (see Eq. (293)), it must have R-charge 2. This implies that the gaugino mass terms m_λ violate any R-symmetry; recall that $\boldsymbol{V}$ has R-charge 0, so $\boldsymbol{W}_\alpha$ has R-charge $+1$. The A, B, and C terms respect the R-symmetry only if the chiral superfields they multiply have net vanishing R-charge. In many models, $\boldsymbol{X}_{\text{NL}}$ multiplies the same terms that arise in an R-respecting superpotential (R-charge 2), in which case the resulting soft terms do not respect the R-symmetry.

There are more exotic ways to couple $\boldsymbol{X}_{\text{NL}}$ to matter fields using SUSY-covariant derivatives. Consider a term in the lagrangian of a non-Abelian gauge field

$$\int d^2\theta \, \text{Tr}(\boldsymbol{\Phi} \boldsymbol{W}^\alpha) \boldsymbol{W}'_\alpha, \tag{303}$$

where $\boldsymbol{\Phi}$ is a chiral multiplet in the adjoint representation and

$$\boldsymbol{W}'_\alpha = -\frac{1}{4} \overline{D}^2 D_\alpha (\boldsymbol{X}_{\text{NL}}^\dagger \boldsymbol{X}_{\text{NL}}). \tag{304}$$

Because $\langle \boldsymbol{W}'_\alpha \rangle = \theta_\alpha |F|^2$, this term induces a (SUSY-breaking) Dirac mass term between the Weyl fermion in $\boldsymbol{\Phi}$ and the gaugino in $\boldsymbol{W}^\alpha$. Such terms appear in theories of supersoft SUSY breaking [63, 64]. Note that these Dirac mass terms for gauginos (unlike the Majorana mass terms discussed above) can respect a global R-symmetry if $\boldsymbol{\Phi}$ has vanishing R-charge.

5.10. *The supercurrent*

Using the non-linear goldstino multiplet, we found that goldstino couplings were directly related to soft terms. There is a more formal way of seeing this same effect using conservation of the supercurrent.

The supercurrent is the Noether current associated with SUSY transformations [65, 66]

$$j_\alpha^\mu = (\sigma^\nu\overline{\sigma}^\mu\psi^i)_\alpha\nabla_\nu\phi_i^\dagger + i(\sigma^\mu\overline{\psi}_i)_\alpha W^{\dagger i} + \frac{1}{2\sqrt{2}}(\sigma^\nu\overline{\sigma}^\rho\sigma^\mu\overline{\lambda}^a)F_{\nu\rho}^a + \frac{i}{\sqrt{2}}g\phi^\dagger T^a\phi(\sigma^\mu\overline{\lambda}^a)_\alpha. \tag{305}$$

Note that the supercurrent has an extra α-index to match the SUSY generator Q_α. Conservation of the supercurrent implies $\partial_\mu j_\alpha^\mu = 0$.

We can isolate the goldstino contribution to the supercurrent via [61, 62]

$$j_\alpha^\mu = j_\alpha^{\mu,\text{matter}} - iF_{\text{tot}}\left(\sigma^\mu\bar{\eta}\right)_\alpha, \tag{306}$$

where $F_{\text{tot}} = \sqrt{|F_{\text{vis}}|^2 + |F_{\text{hid}}|^2}$ also includes any SUSY breaking in the visible sector. Conservation of the full supercurrent implies

$$\partial_\mu j_\alpha^\mu = 0 = \partial_\mu j_\alpha^{\mu,\text{matter}} - iF_{\text{tot}}\left(\sigma^\mu\partial_\mu\bar{\eta}\right)_\alpha. \tag{307}$$

As expected, because SUSY in the visible sector is broken, $\partial_\mu j_\alpha^{\mu,\text{matter}} \neq 0$. If we interpret Eq. (307) as an equation of motion for the goldstino, this implies that in addition to Eq. (295), the goldstino lagrangian must contain

$$\mathcal{L}_\eta \supset -\frac{1}{F_{\text{tot}}}\eta\partial_\mu j^{\mu,\text{matter}} + \text{h.c.} \tag{308}$$

This is called a Goldberger-Treiman relation from the analogous relation for couplings of Goldstone bosons of spontaneously broken global symmetries to matter currents. [67]

For a massless on-shell goldstino

$$(\partial_\mu\eta)\sigma^\mu = 0, \tag{309}$$

so after integration by parts, the second and fourth terms in Eq. (305) are irrelevant for Eq. (308), which reduces to

$$\mathcal{L}_\eta \supset \frac{1}{F_{\text{tot}}}\eta\left((\Box\psi^i)\phi_i^\dagger - \psi^i\Box\phi_i^\dagger - \frac{1}{\sqrt{2}}\sigma^{\nu\rho}(\sigma^\mu\partial_\mu\overline{\lambda}^a)F_{\nu\rho}^a + \frac{1}{\sqrt{2}}\sigma^\rho\overline{\lambda}^a\partial^\nu F_{\nu\rho}^a\right) + \text{h.c.} \tag{310}$$

Using equations of motion for the visible sector fields, we find that the three-point couplings of the goldstino are proportional to *physical* mass differences:

$$\mathcal{L} \supset \frac{m_{\phi^i}^2 - m_{\psi^i}^2}{F_{\text{tot}}}\eta\psi^i\phi_i^\dagger + \frac{B_{ij}}{F_{\text{tot}}}\eta\chi^i\psi^j\frac{im_\lambda}{\sqrt{2}F_{\text{tot}}}\eta\sigma^{\mu\nu}\lambda^a F_{\mu\nu}^a, \tag{311}$$

where, for simplicity, we have assumed unbroken gauge groups. We see that this result exactly reproduces Eq. (302) in a non-trivial way. One advantage of the supercurrent method is that it automatically accounts for effects of goldstino mixing terms like Eq. (299).

5.11. *The gravitino and AdS SUSY*

For the last topic of these lectures, we would like to go beyond global SUSY to talk a bit about supergravity (SUGRA). We have seen that in global SUSY, SUSY breaking leads to a goldstino. In supergravity, the goldstino is eaten by the gravitino to become its longitudinal components, with a mass given by

$$m_{3/2} = \frac{F_{\text{tot}}}{\sqrt{3}M_{\text{Pl}}}, \tag{312}$$

where M_{Pl} is the reduced Planck constant. This is sufficiently confusing that we want to explain (in words) what really happens, though a proof would require a real lecture on SUGRA.

In global SUSY, SUSY breaking implies $\langle V \rangle > 0$. Because our universe has $\langle V \rangle \approx 0$ though, we cannot understand SUSY breaking using the flat space SUSY algebra alone. Even before talking about any details of SUGRA, we can see a possible way out if SUSY could be realized in AdS_4 space! AdS space is a solution to Einstein's equations with negative vacuum energy

$$\Lambda_{\text{AdS}} = -3\frac{M_{\text{Pl}}^2}{\lambda_{\text{AdS}}^2}, \tag{313}$$

where λ_{AdS} is the AdS curvature. Whenever SUSY is broken, the vacuum energy increases, so if it were possible to finely balance the AdS curvature against SUSY breaking, we could have zero vacuum energy with (AdS) SUSY breaking.

Indeed, there does exist a global (a.k.a. rigid) AdS_4 SUSY algebra where λ_{AdS} is fixed, but $M_{\text{Pl}} \to \infty$. At finite M_{Pl} (namely SUGRA), space-time is dynamical so the cosmological constant depends on the vacuum structure of the theory. This happens via the scalar potential (with a trivial Kähler metric and no gauge interactions, for simplicity)

$$V \simeq |F|^2 - 3\frac{|W|^2}{M_{\text{Pl}}^2}, \tag{314}$$

where W is the superpotential itself. When SUGRA is unbroken, $\langle F \rangle = 0$ but the superpotential W contributes to the cosmological constant, yielding the AdS curvature $\lambda_{\rm AdS} = M_{\rm Pl}^2/|W|$. Even with no SUSY breaking, the gravitino (graviton partner) has a mass parameter[cc]

$$m_{3/2} = \frac{|W|}{M_{\rm Pl}^2} = \frac{1}{\lambda_{\rm AdS}}. \tag{315}$$

So $m_{3/2}$ is *not* an order parameter for SUSY breaking, rather it measures the curvature of SUSY AdS space.[dd]

Thus, we can achieve SUSY breaking with zero cosmological constant, if we delicately balance $\langle F \rangle \neq 0$ against the AdS curvature, which from Eq. (314) implies

$$F_{\rm tot} = \sqrt{3}\frac{W}{M_{\rm Pl}} \quad \text{(zero cosmological constant).} \tag{316}$$

Because SUSY is broken, there is a goldstino, but in SUGRA it is eaten to form the longitudinal components of the gravitino (with the same mass $m_{3/2}$ as Eq. (315)). At zero cosmological constant, we recover the claimed formula in Eq. (312). Crucially, the order parameter for SUSY breaking is F, and $m_{3/2}$ is related to F only by a fine-tuning.

We can therefore picture SUSY breaking in two different ways as shown in Fig. 4. We can either think about flat space SUSY being spontaneously broken to yield $V = |F_{\rm tot}|^2$, then finely adjusting $\langle W \rangle$ to return to $V \simeq 0$. Alternatively, we can think about starting with an AdS_4 SUSY algebra that would yield $V = -3M_{\rm Pl}^2/\lambda_{\rm AdS}^2$, but the vacuum spontaneously breaks SUSY to yield $V \simeq 0$.

We prefer to think in terms of uplifting AdS_4 space, because it clarifies certain SUGRA effects that would seem mysterious in flat space. Unlike in flat space, $\{Q_\alpha, Q^\beta\} \neq 0$ in AdS space. In a pseudo-flat space language, the AdS_4 SUSY algebra is given by [68–72]

$$\{Q_\alpha, \overline{Q}_{\dot\beta}\} = -2\sigma^\mu_{\alpha\dot\beta} P_\mu, \tag{317}$$

$$\{Q_\alpha, Q^\beta\} = -2i\lambda_{\rm AdS}^{-1}(\sigma^{\mu\nu})_\alpha{}^\beta M_{\mu\nu}, \tag{318}$$

where $M_{\mu\nu}$ still satisfy the algebra of the (flat space) Lorentz generators in Eqs. (66) and (67).[ee] Because we are actually in AdS space, the "transla-

[cc]Despite having a mass parameter, it only has two physical polarizations, a consequence of the AdS_4 little group.

[dd]Note that $m_{3/2}$ is an order parameter for R-symmetry breaking, since the gravitino has non-vanishing R-charge.

[ee]One should regard these indices as local Lorentz indices, and the vielbein must be used to convert to the more typical Einstein indices of AdS space.

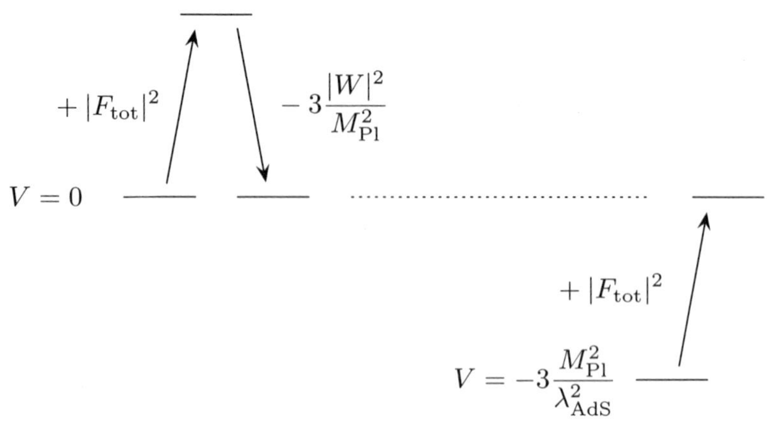

Fig. 4. Two ways to think about achieving SUSY breaking with $V \simeq 0$. The second picture makes clear the underlying AdS_4 algebra.

tion" generators P_μ are modified

$$[P_\mu, P_\nu] = -i\lambda_{\rm AdS}^{-2} M_{\mu\nu}, \tag{319}$$

$$[Q_\alpha, P_\mu] = \frac{1}{2}\lambda_{\rm AdS}^{-1}\sigma^\mu_{\alpha\dot\alpha}\overline{Q}^{\dot\alpha}. \tag{320}$$

All other commutators are the same as in flat space. Because AdS SUSY has a different algebra than flat space SUSY, there are correspondingly different properties, a few of which we will mention here:

- In global AdS SUSY, bosons and fermions in the same multiplet can be split by an amount proportional to $\lambda_{\rm AdS}^{-1}$! This happens already at tree level, giving rise to a μ/B_μ problem in theories where $m_{3/2}$ is large. [37]
- Global AdS SUSY has a boundary, and in order to maintain SUSY in AdS, certain loop effects on the boundary must be compensated by counterterms in the bulk. [73] This has an impact in SUGRA, because once SUSY is broken and AdS is uplifted to flat space, the bulk counterterms remain, giving soft masses without SUSY breaking. This contributes to an effect known as anomaly mediation. [21, 37, 38]
- When global AdS SUSY is broken, the goldstino is not massless as it is in flat space. Rather, conservation of the AdS supercurrent implies that the goldstino has a mass of $2\lambda_{\rm AdS}^{-1}$ (note the factor of 2!). [17, 71, 74] If SUSY is broken by N independent sectors, then there is a corresponding multiplicity of N "goldstini" [17]. One linear combination of these is

eaten by the gravitino of $m_{3/2}$, and the other $N-1$ pseudo-goldstinos get a mass of $2m_{3/2}$ at tree level.

Clearly, the structure of SUGRA and AdS_4 SUSY is quite rich, but beyond the level of these lectures.

6. Summary

We hope you have found these TASI lectures notes useful, regardless of your previous background in SUSY. For the complete neophyte, we have outlined the basic structure and motivation for SUSY:

- SUSY is a symmetry that relates the properties of bosons and fermions. Specially, SUSY establishes relationships between masses and couplings in the lagrangian.
- Chiral multiplets package a spin-0 complex scalar with a spin-1/2 Weyl fermion.
- Vector multiplets package a spin-1 gauge boson with a spin-1/2 gaugino fermion.
- In order to be realized in nature, SUSY must be spontaneously broken in a hidden sector, leading to mass splittings between particles and sparticles in the visible sector.

For the serious SUSY student, we have presented the following tools for constructing SUSY lagrangians:

- One can conveniently package SUSY components into a superfield $\boldsymbol{S}(x^\mu, \theta^\alpha, \overline{\theta}^{\dot{\alpha}})$ that lives in superspace.
- Superspace consists of the ordinary space-time coordinate x^μ augmented by a Grassmann coordinate θ^α (and its complex conjugate $\overline{\theta}^{\dot{\alpha}}$) with the following transformation properties

$$\theta^\alpha \to \theta^\alpha + \epsilon^\alpha, \tag{321}$$

$$\overline{\theta}^{\dot{\alpha}} \to \overline{\theta}^{\dot{\alpha}} + \overline{\epsilon}^{\dot{\alpha}}, \tag{322}$$

$$x^\mu \to x^\mu + i\epsilon\sigma^\mu\overline{\theta} + i\overline{\epsilon}\overline{\sigma}^\mu\theta. \tag{323}$$

- A generic SUSY lagrangian of chiral and vector multiplets can be written as

$$\begin{aligned}\mathcal{L} = &\int d^4\theta\, \boldsymbol{K}(\boldsymbol{\Phi}^{\dagger\bar{\imath}} e^{\boldsymbol{V}}, \boldsymbol{\Phi}^j) \\ &+ \int d^2\theta\, \boldsymbol{W}(\boldsymbol{\Phi}^i) + \text{h.c.} \\ &+ \int d^2\theta\, \boldsymbol{f}_{ab}(\boldsymbol{\Phi}^i) \boldsymbol{W}^{\alpha a} \boldsymbol{W}^b_\alpha + \text{h.c.} \\ &+ D_\alpha, \partial_\mu \text{ terms.}\end{aligned} \quad (324)$$

While the leading effects can be captured by the Kähler potential $\boldsymbol{K}$, superpotential $\boldsymbol{W}$, and gauge kinetic function $\boldsymbol{f}_{ab}$, woe unto the student who forgets the possible presence of $\partial_\mu, D_\alpha, \overline{D}_{\dot{\alpha}}$ terms in an effective SUSY lagrangian.
- SUSY breaking (in flat space) inevitably leads to a massless fermion called the goldstino. Crucially, SUSY-breaking mass splittings are always accompanied by corresponding goldstino-particle-sparticle couplings.

And for the SUSY experts, we hope that you use are able to use these super-tricks in your own research:

- SUSY-covariant derivatives, for constructing new superfields out of old superfields and simplifying the construction of SUSY lagrangians.
- Equations of motion in superspace, for finding the vacuum structure of the theory and integrating out heavy (SUSY-preserving) thresholds at tree-level.
- Background field methods, for analytically continuing static background fields to full superfields, thereby capturing important one-loop (and sometimes two-loop) effects.
- Non-linear multiplets for goldstinos, for abstracting the important features of the hidden sector and relating SUSY-breaking mass-splittings to their corresponding goldstino couplings.

To fully understand the implications of SUSY and SUSY-breaking, one needs to learn about SUGRA (unfortunately not covered in depth in these lectures). We recommend the conformal compensator formalism of SUGRA [75–77], which makes it possible to capture the leading effects of SUGRA in *global* superspace [22]. At minimum, you should be aware that if SUSY is realized in nature, then it must be SUSY in AdS_4 space, spontaneously broken to yield the (nearly) flat space SUSY-breaking vacuum we see today.

By design, these lectures have only briefly mentioned the phenomenological consequences of SUSY, since these were covered in depth in other TASI lectures. That said, it is no secret that recent analyses of 7 TeV and 8 TeV LHC data have placed stringent bounds on TeV-scale SUSY, so it is worth talking about the status of SUSY in 2012.

- First, symmetries are a powerful tool for understanding the behavior of quantum field theories, and finding genuinely new types of symmetries like SUSY is rare. Even if SUSY is not realized in nature, SUSY allows us to better understand the generic features of quantum field theory, as evidenced in these lectures by the power of SUSY background field methods.
- Second, by learning about superspace, one learns how to make symmetries manifest by introducing new "fake" coordinates. An excellent recent example of this is explained in Sundrum's notes on the AdS/CFT correspondence [78], where an extra spatial dimension "emerges" from trying to make conformal symmetries manifest. It is always valuable to stretch one's notion of space-time, and superspace in particular offers a new way to think about the relationship between bosons and fermions.
- Third, SUSY is by now the lingua franca for physics beyond the SM, and even non-SUSY extensions of the SM often share phenomenology features with SUSY. As with the language of QCD (confinement, asymptotic freedom, chiral symmetry breaking, and so on), the language of SUSY allows one to quickly explain the features of many quantum field theories.
- Finally, it is not at all clear (as of 2012) whether SUSY is really as tightly constrained as one might naively expect. The canonical flavor-blind SUSY theories with low fine-tuning of the Higgs potential are strongly disfavored, but more exotic (and some might say more generic) SUSY theories with novel flavor structures or some degree of fine-tuning are only starting to be tested. Therefore, some of the pessimism surrounding SUSY is surely misplaced, and we prefer to remain optimistic that hints of a SUSY-like theory will emerge in the 14 TeV LHC data (or elsewhere).

With that, we would like to thank all of the participants at TASI 2012 for a fun week!

Acknowledgments

We thank Martin Schmaltz for encouraging us to write these lecture notes. We thank Yoni Kahn, Grant Larsen, Matthew McCullough, and Yiming Xu for helpful comments on the manuscript. This work is supported by the U.S. Department of Energy (DOE) under cooperative research agreement DE-FG02- 05ER-41360. J.T. is supported by the DOE Early Career research program DE-FG02-11ER-41741. D.B. is partly supported by Istituto Nazionale di Fisica Nucleare (INFN) through a Bruno Rossi Fellowship.

References

[1] A. Salam and J. Strathdee, *Nucl. Phys.* **B76**, 477 (1974).
[2] S. Ferrara, J. Wess and B. Zumino, *Phys. Lett.* **B51**, p. 239 (1974).
[3] J. Wess and J. Bagger (1992).
[4] S. P. Martin (1997).
[5] J. Terning (2006).
[6] M. A. Luty, 495 (2005).
[7] T. T. Dumitrescu and Z. Komargodski, *Nucl. Phys. Proc. Suppl.* **216**, 44 (2011).
[8] S. Weinberg (2000).
[9] D. Z. Freedman and A. Van Proeyen (2012).
[10] P. Binetruy (2006).
[11] A. Signer, *J. Phys.* **G36**, p. 073002 (2009).
[12] M. Drees (1996).
[13] H. Baer and X. Tata (2006).
[14] P. Argyres (1996).
[15] M. Dine (2007).
[16] N. Arkani-Hamed, G. L. Kane, J. Thaler and L.-T. Wang, *JHEP* **0608**, p. 070 (2006).
[17] C. Cheung, Y. Nomura and J. Thaler, *JHEP* **1003**, p. 073 (2010).
[18] C. Cheung, F. D'Eramo and J. Thaler, *JHEP* **1108**, p. 115 (2011).
[19] N. Craig, M. McCullough and J. Thaler, *JHEP* **1203**, p. 049 (2012).
[20] N. Craig, M. McCullough and J. Thaler, *JHEP* **1206**, p. 046 (2012).
[21] F. D'Eramo, J. Thaler and Z. Thomas, *JHEP* **1206**, p. 151 (2012).
[22] C. Cheung, F. D'Eramo and J. Thaler, *Phys. Rev.* **D84**, p. 085012 (2011).
[23] D. Bertolini, K. Rehermann and J. Thaler, *JHEP* **1204**, p. 130 (2012).
[24] H. K. Dreiner, H. E. Haber and S. P. Martin, *Phys. Rept.* **494**, 1 (2010).
[25] R. Haag, J. T. Lopuszanski and M. Sohnius, *Nucl. Phys.* **B88**, p. 257 (1975).
[26] B. Zumino, *Phys. Lett.* **B87**, p. 203 (1979).
[27] P. W. Graham and S. Rajendran, *Phys. Rev.* **D81**, p. 033002 (2010).
[28] B. A. Dobrescu and P. J. Fox, *Eur. Phys. J.* **C70**, 263 (2010).
[29] M. Ibe, A. Rajaraman and Z. Surujon (2010).
[30] R. Davies, J. March-Russell and M. McCullough, *JHEP* **1104**, p. 108 (2011).

[31] L. Abbott, *Acta Phys. Polon.* **B13**, p. 33 (1982).
[32] G. Giudice and R. Rattazzi, *Nucl. Phys.* **B511**, 25 (1998).
[33] G. Giudice and R. Rattazzi, *Phys. Rept.* **322**, 419 (1999).
[34] N. Arkani-Hamed, G. F. Giudice, M. A. Luty and R. Rattazzi, *Phys. Rev.* **D58**, p. 115005 (1998).
[35] T. Clark, O. Piguet and K. Sibold, *Nucl. Phys.* **B159**, p. 1 (1979).
[36] K. Konishi, *Phys. Lett.* **B135**, p. 439 (1984).
[37] L. Randall and R. Sundrum, *Nucl. Phys.* **B557**, 79 (1999).
[38] G. F. Giudice, M. A. Luty, H. Murayama and R. Rattazzi, *JHEP* **9812**, p. 027 (1998).
[39] M. T. Grisaru, W. Siegel and M. Rocek, *Nucl. Phys.* **B159**, p. 429 (1979).
[40] N. Seiberg, *Phys. Lett.* **B318**, 469 (1993).
[41] V. Novikov, M. A. Shifman, A. Vainshtein and V. I. Zakharov, *Nucl. Phys.* **B229**, p. 381 (1983).
[42] V. Novikov, M. A. Shifman, A. Vainshtein and V. I. Zakharov, *Phys. Lett.* **B166**, 329 (1986).
[43] V. Novikov, M. A. Shifman, A. Vainshtein and V. I. Zakharov, *Nucl. Phys.* **B260**, 157 (1985).
[44] M. A. Shifman and A. Vainshtein, *Nucl. Phys.* **B277**, p. 456 (1986).
[45] M. A. Shifman and A. Vainshtein, *Nucl. Phys.* **B359**, 571 (1991).
[46] M. Dine and Y. Shirman, *Phys. Rev.* **D50**, 5389 (1994).
[47] N. Arkani-Hamed and H. Murayama, *JHEP* **0006**, p. 030 (2000).
[48] S. Ferrara, L. Girardello and F. Palumbo, *Phys. Rev.* **D20**, p. 403 (1979).
[49] S. Dimopoulos and H. Georgi, *Nucl. Phys.* **B193**, p. 150 (1981).
[50] K. A. Intriligator, N. Seiberg and D. Shih, *JHEP* **0707**, p. 017 (2007).
[51] A. E. Nelson and N. Seiberg, *Nucl. Phys.* **B416**, 46 (1994).
[52] P. Fayet and J. Iliopoulos, *Phys. Lett.* **B51**, 461 (1974).
[53] P. Fayet, *Nucl. Phys.* **B90**, 104 (1975).
[54] R. Barbieri, S. Ferrara, D. V. Nanopoulos and K. Stelle, *Phys. Lett.* **B113**, p. 219 (1982).
[55] S. Ferrara, L. Girardello, T. Kugo and A. Van Proeyen, *Nucl. Phys.* **B223**, p. 191 (1983).
[56] M. Rocek, *Phys. Rev. Lett.* **41**, 451 (1978).
[57] U. Lindstrom and M. Rocek, *Phys. Rev.* **D19**, 2300 (1979).
[58] Z. Komargodski and N. Seiberg, *JHEP* **0909**, p. 066 (2009).
[59] D. Volkov and V. Akulov, *Phys. Lett.* **B46**, 109 (1973).
[60] T. Clark and S. Love, *Phys. Rev.* **D54**, 5723 (1996).
[61] P. Fayet, *Phys. Lett.* **B70**, p. 461 (1977).
[62] P. Fayet, *Phys. Lett.* **B86**, p. 272 (1979).
[63] M. Dine and D. MacIntire, *Phys. Rev.* **D46**, 2594 (1992).
[64] P. J. Fox, A. E. Nelson and N. Weiner, *JHEP* **0208**, p. 035 (2002).
[65] J. Wess and B. Zumino, *Phys. Lett.* **B49**, p. 52 (1974).
[66] B. de Wit and D. Z. Freedman, *Phys. Rev.* **D12**, p. 2286 (1975).
[67] M. Goldberger and S. Treiman, *Phys. Rev.* **110**, 1478 (1958).
[68] A. Adams, H. Jockers, V. Kumar and J. M. Lapan, *JHEP* **1112**, p. 042 (2011).

[69] B. de Wit and I. Herger, *Lect. Notes Phys.* **541**, 79 (2000).
[70] B. Keck, *J. Phys.* **A8**, 1819 (1975).
[71] B. Zumino, *Nucl. Phys.* **B127**, p. 189 (1977).
[72] E. Ivanov and A. S. Sorin, *J. Phys.* **A13**, 1159 (1980).
[73] B. Gripaios, H. D. Kim, R. Rattazzi, M. Redi and C. Scrucca, *JHEP* **0902**, p. 043 (2009).
[74] I. McArthur (2013).
[75] T. Kugo and S. Uehara, *Nucl. Phys.* **B222**, p. 125 (1983).
[76] T. Kugo and S. Uehara, *Nucl. Phys.* **B226**, p. 49 (1983).
[77] S. Gates, M. T. Grisaru, M. Rocek and W. Siegel, *Front. Phys.* **58**, 1 (1983).
[78] R. Sundrum, *Phys. Rev.* **D86**, p. 085025 (2012).

Chapter 10

TASI 2012 Participants

The Students

Participant contact info and research interests

Sujeet Akula **s.akula@neu.edu**
Northeastern University
360 Huntington Ave.
111 Dana Research Center
Boston, MA 02115 USA
hep-ph/hep-th

Melinda Andrews **mgildner@sas.upenn.edu**
University of Pennsylyannia
Department of Physics and Astronomy
209 S 33rd St
Philadelphia, PA 19104 USA
Astrophysical Tests of Modified Gravity

Masha I. Baryakhtar **mbaryakh@Stanford.edu**
Varian Physics Bldg.
832 Via Pueblo Mall
Stanford, CA 94305-4060 USA
LHC Phenomenology, Inflationary Cosmology

Daniele Bertolini **danbert@MIT.edu**
Center for Theoretical Physics
Massachusetts Institute of Technology
77 Massachusets Ave
Cambridge, MA 02139 USA
Theoretical Particle Physics

Kimberly K. Boddy **kboddy@theory.caltech.edu**
California Institute of Technology
1200 E. California Blvd
MC 452-48
Pasadena, CA 91125 USA
Particle Cosmology

Joseph Bramante **josephbramante@gmail.eom**
Department of Physics and Astronomy
2505 Correa Rd
Honolulu, HI, 96826 USA
Dark Matter, Collider Phenomenology, Cosmology

Christopher Brust **cpbrust@gmail.com**
3400 N charles St.
Bloomberg 366
Baltimore, MD 21218 USA
SUSY Phenomenology and Model Building

Zachary Burell **zmburell@ua.edu**
University of Alabama
514 University Blvd
206 Gallalee Hall
Tus'caloosa, AL 35487 USA
Particle Phenomenology

Laurent Canetti **laurent.canetti@epfl.ch**
EPFL SB ITP LPPC BSP 713.3
CH-10150 Lausanne Switzerland
Leptogenesis, Dark Matter, Neutrino Physics

Mathieu Cliché **mc863@cornell.edu**
Department of Physics, LEPP
Cornell University
Ithaca, NY 14853 USA
Supersymmetry, Extra Dimensions

Jessica L. Cook **jlcook@physics.umass.edu**
Department of Physics
1126 Lederle Graduate Research Tower
University of Massachusetts
Amherst, MA 02003-9337 USA
Cosmology

Jonathan M. Cornell **jcornell@ucsc.edu**
Physics Department
University of California-Santa Cruz
Santa Cruz, CA 95064 USA
Dark Matter

Liang Dai **Dailiangpku@gmail.com**
467 Bloomberg Center
Department of Physics and Astronomy
The John Hopkins University
3400 N Charles St
Baltimore, MD 21218 USA
Cosmology, Particle Astrophysics

Valerie Domcke **Valerie.domcke@desy.de**
Deutsches Elektronen-Synchrotron (DESY)
Notkestrae 85
Hamburg DESY-TheorieGbd 2a, Raum 301
22607 Hamburg Germany
Cosmology and Particle Physics

Sean Downes **sddownes@physics.tamu.edu**
Mitchell Institute for Fundamental
Physics and Astronomy
4242 TAMU department of Physics
Texas ASLM University
College Station, TX 77843-4242 USA
Cosmology, Inflation

Gauthier Durieux **gauthier.durieux@uclouvain.be**
Center for Cosmology, Particle Physics
and Phenomenology
Universite Catholique de Louvain 2
Chemin du Cyclotron- Box L7.01.05
B-1348 Louvain-la-Nueve Belgium
Particle Physics Phenomenology

David A. Eby **Daeby@unc.edu**
Department of Physics and Astronomy
University of North Carolina-Chapel Hill
CB 3255, Phillips Hall
Chapel Hill, NC 27599-3255 USA
Model Building and High Energy Phenomenology

Catalina Gomez Sanchez **cgomez@physics.utoronto.ca**
Department of Physics
University of Toronto
60 St. George St.
Toronto, Ontario M5S 1A7 Canada

Dark Matter, Asymmetries at the LHC, Dynamical Symmetry Breaking

Garrett Goon **Garrett361@gmail.com**
Department of Physics & Astronomy
209 South 33rd Street
University of Pennsylvania
Philadelphia, PA 19104- 6396 USA

Modified Gravity, Screening Mechanisms, SSB

Jiayin Gu **jjjgu@ucdavis.edu**
Department of Physics
University of Califomia-Davis
One Shields Ave.
Davis, CA 95616 USA

High Energy Theory

Rick Gupta **guptasr@umich.edu**
Department of Physics
University of Michigan
450 Church Street
A1111 Arbor, MI 48109-1040 USA

Higgs Physics

Kristopher Healey **Healey@wayne.edu**
Wayne State University
Physics Building
666 West Hancock
Detroit, MI 48201 USA

Phenomenolgy and Heavy Flavor

Jan Heisig **jan.heisig@desy.de**
II. Institute for Theoretical Physics
University of Hamburg
Luruper Chaussee 149
22761 Hamburg Germany

Supersymmetry With Weakly Interacting LSP at the LHC and in the Early Universe

Peisi Huang **phuang7@wisc.edu**
1150 University Ave
Madison, WI 53706 USA

Supersymmetry, Jet Physics, Higgs Physics

Ran Huo **huor@uchicago.edu**
Enrico Fermi Institute
University of Chicago
5640 S. Ellis Ave.
Chicago, IL 60637 USA

Supersymmetry, Higgs Physics, Collider, Dark Matter

Ahmed Ismail **aismail@stanford.edu**
Theoretical Physics Group
MS 81 SLAC National Accelerator Laboratory
2575 Sand Hill Road
Menlo Park, CA 94025 USA

Collider Phenomenology

Pawin Ittisamai **Ittisama@msu.edu**
Department of Physics and Astronomy
Michigan State University
567 Wilson Road, Room 3208
E Lansing, M1 48824 USA

BSM Particle Theory, Strong Dynamics, Phenomenology

Austin Joyce **joyceau@SAS.Upenn.edu**
Department of Physics and Astronomy
University of Pennsylvania
209 S 33rd St, Room 2c3
Philadelphia, PA 19104 USA

Early Universe Cosmology, Field Theory on Curved Space

Durmus Karabacak **Dunnus.Karabacak@Okstate.edu**
Oklahoma State University
Physics Department
145 Physical Science Building
Stillwater, OK 74078-3072 USA

Particle Physics Phenomenology

Jack Kearney **jkmy@umich.edu**
Department of Physics
University of Michigan
450 Church St.
Ann Arbor, MI 48109-1040 USA

High-energy Phenomenology

Simon Knapen **Knapen@physics.Rutgers.edu**
Rutgers, The State University of New Jersey
Department of Physics and Astronomy
136 Frelinghuysen Rd.
Piscataway, NJ 08854 USA

Particle Theory

Grant Larsen **Larsen.ge@Berkeley.edu**
Center for Theoretical Physics
366 Le Conte Hall #7300
University of California-Berkeley
Berkeley, CA 94720-7300 USA

BSM Phenomenology, Cosmology

Matthew J. Lewandowski **mattlew@Stanford.edu**
Varian Physics Bldg.
832 Via Pueblo Mall
Stanford, CA 94305-4060 USA

Cosmology

Danang Li **Daliang.li@yale.edu**
217 Prospect Street
Department of Physics
Yale University
New Haven, CT 06511 USA

Beyond Standard Model Physics, Field Theory, Basics of Quantum Mechanics

Jiajun Liao **jliao@iastate.edu**
Department of Physics and Astronomy
Iowa State University
Ames, lA 50011 USA

Neutrino Mass and Neutrino Oscillation

Suet-Ying (Daisy) Mak **suetyinm@usc.edu**
Department of Physics and Astronomy
University of Southern California
Los Angeles, CA 90089-0484 USA
Cosmology

Elisa Mariani **elisa.kerf@gmal.com**
Center for Cosmology, Particle Physics, and Phenomenology
Universite Catholique de Louvain 2
Chemin du Cyclotron- Box L7.01.05
B-1348 Louvain-la-Nueve Belgium
Theoretical Particle Physics, Phenomenology

Gustavo Marques Tavares **gusmt@bu.edu**
Boston University
Physics Department
590 Commonwealth Ave
Boston, MA-02215 USA
Physics Beyond the Standard Model and LHC Physics

Angelo Monteux **amonteux@ucsc.edu**
ISB 329 Physics Department
University of Califomia-Santa Cruz
Santa Cruz, CA 95064 USA
Supersymmetry

Azadeh Morandinezhad Dizgah **am248@buffalo.edu**
University at Buffalo
Department of Physics
239 Fronczak Hall
Buffalo, NY, 14260-1500 USA
Cosmology

Christopher W. Murphy **cmurphy@physics.ucsd.edu**
Department of Physics 03 54
University of California-Sal Diego
9500 Gilman Drive
La Jolla, CA 92093-0354 USA
Particle Physics and Cosmology

Michael Park **q1park@gmail.com**
Department of Physics and Astronomy
Rutgers University
136 Frelinghuysen Rd.
Piscataway, NJ 08854-8019 USA

Collider Phenomenology and Quantum Field Theory

Sohyun Park **Spark@phys.ufl.edu**
University of Florida
Department of Physics
Gale Lemmerand Dr. & Museum Rd.
Gainesville, FL 32611-8440 USA

Quantum Field Theory During Inflation, Modified Gravity

Ayon Patra **ayon@okstate.edu**
145 Physical Sciences Building
Department of Physics
Oklahoma State University
Stillwater, OK 74078-3072 USA

Particle Physics Phenomenology

Erik A. Perkins **Perkins@physics.ucsb.edu**
Department of Physics
Broida Hall
University of California-Santa Barbara
Santa Barbara, CA 93106-9530 USA

High Energy Phenomenology

Jing Ren **jingren@pa.msu.edu**
Department of Physics and Astronomy
Biomedical Physical Sciences Building
567 Wilson Road, Room 3208
E Lansing, MI 48824 USA

New Physics Beyound SM, Cosmology, Inflation and Dark Energy

Rodolfo Ivan Rodriguez Montoya **rrodriguez@fis.cinvestav.mx**
AV. Instituo Politecnico
Nacional No. 2508. C01
Av. Instituo Politecnico
San Pedro Zacatenco. C.P. 07360. Mexico

Particle Physics and Cosmology

Ennio Salvioni **ennio.salvioni@cern.ch**
PH-TH, Case C01600
CERN
CH-1211 Geneva 23 Switzerland

Electroweak Symmetry Breaking, Extensions of the SM at TeV

McCullen Sandora **mesandora@ucdavis.edu**
Physics Department
UC Davis
Shields Ave.
Davis, CA 956I6 USA

Theoretical Cosmology

Marko V. Simonovic **marko.simonovic@sissa.it**
SISSA via Bonomea, 265
34136 Trieste Italy

Inflationary Cosmology

Mikhail P. Solon **mpsolon@uchicago.edu**
The University of Chicago
Department of Physics
5720 S. Ellis Ave.
Chicago, IL 60637 USA

Effective Field Theories, Dark Matter Beyond the Standard Model

Walter Tangarife Garcia **Wtang@physics.utexas.edu**
Department of Physics
University of Texas at Austin
1 University Station C1600
Austin, TX 78712-0264 USA

Supersymmetry and Cosmology

Andrea Thamm **andrea.thamm@cern.ch**
Physics Department
TH-Unit
CERN 385 Route de Meyrin
CH 1211 Geneva 23 Switzerland

BSM Phenomenology

Zachary T. Thomas **ztt@mit.edu**
Massachusetts Institute of Technology
Center for Theoretical Physics
77 Massachusetts Ave, 6-314
Cambridge, MA 02139 USA

Particle Physics and Cosmology

Mathew T. Walters **mwalters@pha.jhu.edu**
3400 N. Charles Street
Johns Hopkins University
Bloomberg 366
Baltimore, MD 21218 USA

Astroparticle, Collider Phenomenology

Daniel Wegman **Wegman.daniel@gmail.com**
Physics & Astronomy
University of Califomia-Riverside
00 University Ave.
Riverside, CA 92521 USA

Theoretical Particle Physics

Peter Winslow **Peter.winslow@gmail.com**
TRIUMF
4004 Wesbrook Mall
Vancouver, BC 2A3 Canada

Particle Phenomenology and Cosmology

Yiming Xu **ymxu@bu.edu**
590 Commonwealth Ave.
BU Physics Department
Boston, MA 02215 USA

Dark Matter Cosmology, ADS/CFT

Wei Xue **xueweiphy@gmail.com**
Ernest Rutherford Physics Building
McGill University
3600 rue University
Montreal, QC H3A 2T8 Canada

Dark Matter

David Yaylali **yaylali@hawaii.edu**
University of Hawaii
Department of Physics
2505 Correa Rd, Box 234
Honolulu, HI, 96822 USA

HEP Phenomenology, Cosmology, Quantum Gravity

Aditya Yechan Gunja **ygaditya@wayne.edu**
Physics Research Building
666 W. Hancock
Detroit, MI 48201 USA

Particle Physics

Student Presentations

Daisy Mak: *Constraints of modified gravity from galaxy clusters*

Ivan Rodriguez Montoya: *Dark matter as Bose-Einstein condensate*

David Eby: *The neutrino frontier – finite groups and mixing matrices*

Marko Simonovic: *Consistency relations in single-field inflationary models*

Zachary Burell: *Radiative symmetry breaking in the SUSY B-L model*

Azadeh Moradinezhad: *Scale-invariant primordial perturbations from alternative mechanisms to inflation*

Yiming Xu: *Model independent analysis of dark matter direct detection*

Lindy Andrews: *Testing screened modified gravity in the solar system*

Matthew Walters: *Semiconductor probes of light dark matter*

Matt Lewandowski: *The phase transition to eternal inflation*

Valerie Domcke: *Spontaneous B-L breaking to reheat the early universe*

Chris Brust: *Natural SUSY at the LHC*

Gustavo Marques Tavares: *ttbar asymmetry from a light axigluon*

Mikhail Solon: *Lorentz invariance in heavy particle effective theories and an application to dark matter*

Sujeet Ekula: *$N = 1$ 4D SUGRA Models: pheno, naturalness and the Higgs*

Jan Leisig: *Super WIMP DM: long-lived staus at the LHC*

Jing Ren: *Technicolor models in light of LHC Higgs search*

Masha Baryakhtar: *Supersymmetry in the shadow of photini*

Ran Huo: *SUSY RGE, GUT and YU*

Austin Joyce: *Non-linear realizations of conformal symmetry and effective field theory for the pseudo-conformal universe*

Pawin Ittisamai: *How to probe couplings of massive color-octet bosons*

Peisi Huang: *Natural susy phenomemology*

Andrea Thamm: *Composite Higgs models*

Ahmed Ismail: *Phenomenological MSSM, a general framework for studying minimal supersymmetry at colliders and beyond*

Jonathan Cornell: *Probing the size of the smallest dark matter halos*

Gauthier Durieux: *Observing baryon number violation at the LHC in processes involving top quarks*

Kristopher J Lealey: *Soft photon contributions to Bs $\rightarrow$ mu mu*

Garret Goon: *Geometrical and algebraic properties of galileons*

Durmus Karabacak: *Production of color sextet vector particle and its effect on single top quark production at the LHC*

Y. G. Aditya: *Searching for super-WIMPs in leptonic heavy meson decays*

Daniel Wegman: *Dark vector gauge boson model*